Readings in Training and Simulation:
A 30-Year Perspective

Edited by
Robert W. Swezey and Dee H. Andrews

Published by the
Human Factors and Ergonomics Society
P.O. Box 1369
Santa Monica, CA 90406-1369 USA
310/394-1811, Fax 310/394-2410
http://hfes.org, info@hfes.org

For Judith L. Swezey and Debra Andrews

Library of Congress Cataloging-in-Publication Data

Readings in training and simulation : a 30-year perspective / edited by Robert W. Swezey and Dee H. Andrews.

    p.   cm.

    Includes index.

    ISBN 0-945289-15-4 (alk. paper)

    1. Human engineering.   I. Swezey, Robert W.   II. Andrews, Dee H.   III. Title.

T59.7 .R43 2000

620.8'2--dc21

00-012085

Contents

Foreword

With each advance in technology, the human has become an increasingly weak link in systems. First the military, then commercial aviation, then nuclear power, and now the health care industry have all come to the same conclusion: A good human-system interface is required for easy-to-use, reliable, and safe systems. However, the application of human factors design principles is but *one* way to resolve interface issues, not the *only* way.

Training has always been a corollary to good human factors design. Building on learning theory and training models as the knowledge base, principles of training have been applied successively to the design of instructional systems, individual whole- and part-task trainers, system simulators and simulator networks, and virtual training environments.

A compelling need for the quality training of technologies' users, operators, maintainers – yes, and its builders – will always exist. This book samples 30 years of theory and practice in individual and team training drawn from the publications of the Human Factors and Ergonomics Society. It was assembled by two highly knowledgeable experts in training and simulation and provides a broad perspective on progress in individual and team training and training technology.

Harold P. Van Cott

Preface

The papers in this book were selected to emphasize important methodological and research papers on simulation and training published by the Human Factors and Ergonomics Society from the 1970s through 1999.

To select papers for inclusion in this volume, we reviewed all articles published in the Society's journal, *Human Factors,* from 1970 through 1999, and papers published in HFES Annual Meeting proceedings from 1977 through 1999. Only articles for which HFES holds copyright were considered, so although the book reflects a single organizational perspective on the world of simulation and training, that perspective emphasizes direct application of theory to applied situations, identifies and applies methodologies for use in investigating this domain, and emphasizes data-based research in drawing conclusions. Sadly, these important facets are somewhat unique in the simulation and training literatures, which often appear to contain studies that are atheoretical, lack methodological sophistication, and eschew empirical research as a basis.

Paper selection involved an iterative rating and elimination process; our objective was to include articles of foremost importance over the last 30 years while maintaining a reasonable size for the book. In the first stage, searching for articles on simulation and/or training in the target period resulted in about 170 candidates. Each was independently rated by both editors on a seven-point scale, and articles that were rated in the lower three categories by either editor were eliminated from consideration. We again independently rated the remaining papers and kept only those with a rating 6 or 7. Finally, we discussed the merits of each remaining article and agreed on the 32 that are contained in this book. It is a regrettable but unavoidable fact that more articles could not be included.

Organization of the Book

We organized the 32 articles into five parts: Transfer of Training, Training Methods, Training Devices and Simulators, Application Areas, and Training Evaluation. The Application Areas section has been further divided into three subareas: Flight Training, Maintenance Training, and Cognitive Skill Training. Several papers could have been classified in other parts, but we placed them in the technical topic area that we believed to be most appropriate to the content areas they addressed. They are arranged chronologically within parts. Papers reprinted from Annual Meeting proceedings were scanned and reformatted for this book's dimensions.

The introduction places the articles in perspective, given the major changes that have occurred in the areas of simulation and training during the past 30 years. Each part includes a brief description of its content.

We are very grateful for help provided by Lisa Dyson in the preparation of this book.

Robert W. Swezey
Leesburg, VA
Dee H. Andrews
Mesa, AZ

December 2000

Introduction

To be competitive in today's economy, corporations spend billions of dollars each year to train their employees. To be cost-effective, governments and nonprofit organizations also invest heavily in training. Throughout the world, hundreds of billions of dollars are spent annually on training, yet only a fraction of 1% of this huge investment is spent on research and development (R&D) to make training more effective. With such a tiny investment in training, it is important that high-quality, relevant research and development in this field receive the broadest possible dissemination. Our purpose in compiling this book is to highlight what we consider pertinent research and development that has come from a productive corner of the training R&D community, much of which is published by the Human Factors and Ergonomics Society (HFES).

Research on learning and its application to training have been under way for many decades, long before HFES was founded. Nevertheless, human factors/ergonomics professionals (many of whom have been HFES members) have had a significant impact on training research and the design and development of training. We have been involved with a variety of training-related professional organizations over the years, and we believe the human factors/ergonomics field brings a unique perspective to this community at large, reflected in HFES publications and meetings. We compiled these papers from the human factors/ergonomics literature on training because we believe this unique perspective can help to inform anyone involved in training on how best to design, deliver, and evaluate training.

What is this unique human factors perspective on training, and how can it add to the understanding and study of training? To help answer these questions, it might be helpful to give our definition of training:

> Training is the systematic application of scientific learning principles to produce instruction that will change behavior.

Typically, this change of behavior is desired in employees of organizations, but training is also applicable to any human endeavor in which technology is present, whether at the workplace or somewhere else.

Because of the diverse interests among human factors/ergonomics professionals (as reflected in the 21 technical interest groups within HFES), human factors training researchers are involved in the full range of training applications. One need only examine the names of the HFES technical groups to understand why human factors training researchers have such a broad outlook on training applications: Aerospace Systems, Aging, Cognitive Engineering and Decision Making, Communications, Computer Systems, Consumer Products, Educators' Professional, Environmental Design, Forensics Professional, Individual Differences in Performance, Industrial Ergonomics, Internet, Macroergonomics, Medical Systems and Rehabilitation, Safety, Surface Transportation, System Development, Test and Evaluation, Training, Virtual Environments, and Visual Performance. Given such a broad range of backgrounds and real-world interests among HFES members, one can see why training research published in HFES documents has a unique perspective when compared with other types of training research. Indeed, many of the researchers whose names appear on the papers in this book are not training psychologists or researchers but come from many diverse backgrounds. We believe that this diversity makes the impact of these papers especially important.

Because the human factors/ergonomics legacy comes from the design of systems and equipment, professionals both inside and outside the field may perceive that the interests of its professionals reside only in training people to operate and maintain such systems and equipment, which is certainly true to some extent. However, the propagation of technology into every facet of human life now makes quality training imperative in a broad range of areas that might not have traditionally been considered examples of designed systems or equipment just a short time ago. A number of the papers in this volume address training in some of these nontraditional areas.

One goal of quality human factors design is to make human interaction with systems and equipment as "user-friendly" as possible. To many, this means that an ideal system or piece of equipment, if designed with close attention to human factors principles, should require no training at all – the human-machine interface should be so well designed that its operation or maintenance should be intuitively obvious to anyone who uses it. However, as anyone interested in the training field knows, in reality, most systems and equipment of any complexity still require quality training for effective and efficient operation and maintenance. This may be because the interface design fundamentals are not yet complete or because their application is still not well understood, or a combination of both. At any rate, our experience has shown many times over that the need to provide quality training remains a major issue in today's high-technology world.

A number of the papers contained in this volume present learning algorithms that can be used to train specific skills. Learning algorithms are definitive and precise prescriptions for learning that, when followed closely, result in the desired learning. They are based mainly on learning principles derived from empirical research that is reported in the papers. Many in the learning research community have striven for years to produce such algorithms and believe that research will ultimately produce a complete set of such algorithms for any training task. However, most of the papers presented here discuss learning approaches that are neither as structured nor as data-based as learning algorithms. These approaches can best be described as heuristics for learning and/or training. As with all heuristics, they are essentially guidelines for learning. They may be based in part on empirically derived data, but they also have a strong grounding in practical training experience.

Knowing many of the authors represented in this book, we also know that many of their recommendations for successful training come from years of hard and rich experience in attempting to bring about behavioral change. We hope and believe that quality training research will continue to produce sound training algorithms, but we also believe that there will always be a place for heuristics that are derived from real-world experience.

Part 1: Transfer of Training

Transfer of training is integrally related to other training issues, such as learning, memory, retention, cognitive processing, and conditioning. These fields combine to make a large subset of the subject matter of applied psychology. In general, the term *transfer of training* concerns the way in which previous learning affects new learning or performance. The central issue involves how previous learning "transfers" to a new situation.

The effect of previous learning may function either to improve or to retard new learning. The first of these is generally referred to as *positive transfer*, the second is known as *negative transfer*. (If new learning is unaffected by prior learning, *zero transfer* is said to have occurred.) Many training programs are based on the assumption that what is learned during training will transfer to new situations and settings, most notably the operational environment. Although U.S. industries have spent an estimated $100 billion annually on training and development, 10% or less of these expenditures are thought to be applied to performance transfer in actual job situations.

There are two major historical viewpoints on transfer: *identical elements* and *transfer through principles*. The identical elements theory suggests that transfer occurs when identical elements exist in both original and transfer situations. Thus, in a new situation, a learner presumably takes advantage of what the new situation offers that is in common with the learner's earlier experiences. Alternatively, the transfer-through-principles perspective suggests that a learner need not necessarily be aware of similar elements in a situation in order for transfer to occur. Instead, previously used principles may be applied to occasion transfer. A simple example is how the principles of aerodynamics learned by the Wright brothers from flying kits were applied to airplane construction.

The problem of transfer of training is complex. It is often difficult to determine the extent to which the complex stimulus and response elements that exist in most operational situations are similar or dissimilar, until adequate description and measurement techniques are established and employed. Further, most complex environments do not simply foster either positive or negative transfer. Most environments are sufficiently complex that their components interact in ways that produce negative and/or positive transfer for each component (as well as each two-way, three-way, and *n*-way interaction).

Historically, predictions of transfer in applied settings have been based on results of basic research in learning, within the framework of stimulus-response theories. Contemporary research on information processing and memory processes involved in encoding and retrieving information during both initial task acquisition and retention offers new ways for explaining and predicting transfer effects. Literature in this area has identified several major processing factors that contribute to the prediction of transfer, including relationships between retrieval cues and encoded information, use of instructional techniques that permit the integration and abstraction of information content, organizational strategies that enhance information processing, and the automation of performance with consistent training.

This part contains five articles that address the topic of transfer of training. In the first, a classic article by Blaiwes, Puig, and Regan (1973), the *transfer effectiveness ratio* is discussed in the context of training evaluation research. In the Simon and Roscoe (1984) paper, a complex transfer experiment is reported and guidelines for conducting research of this type are offered. The Spears (1985) article presents a curve-fitting methodology for use in measuring transfer, and the Damos (1988) article applies this methodology in a transfer experiment and compares its results to the "traditional" percentage transfer measure. The final paper, by Lintern (1991), addresses the topic of *difference* measurement in transfer and presents an argument for its use.

Transfer of Training and the Measurement of Training Effectiveness

ARTHUR S. BLAIWES, JOSEPH A. PUIG, and JAMES J. REGAN, *Human Factors Laboratory, Naval Training Equipment Center, Orlando, Florida*

Transfer of training research has been conducted on actual training systems to determine: (1) the effectiveness of present training; (2) whether the training can be improved; and, (3) how the training might be improved. The present paper includes some major methodological and analytical considerations in performing this research—the experimental and descriptive models to use in investigating and expressing transfer, cost effectiveness evaluations, and aspects of the training system to be included in the study. A number of conclusions are derived from the transfer research and some popular research themes are identified. Desirable features for an applied research program for military training purposes are presented. Problems arising from the use of the transfer of training model are traced to operational constraints placed on experimental manipulation and control, and to the inadequacy of performance measurement systems. Solutions to these problems are discussed. One solution provides alternate methods to the transfer of training model for evaluating the effectiveness of a training system. Another approach recommends the employment of laboratory simulations of training or operational situations for transfer research.

INTRODUCTION

In the administration of instruction, we are subject to inefficiency and failure. We will repeat an instruction without knowing if it is an expeditious approach to our goal, whether it really leads to the goal, or even whether we would want the goal if we achieved it. The problem is compounded by the criticalness and pervasiveness of instruction in our everyday affairs. Thus, it becomes both natural and critical to confront all instructional endeavors—from those operating in social interactions and child rearing to the pedagogy of academic and vocational pursuits—with the "training effectiveness" question, *viz.*, how well does the instruction work?

The ultimate purpose of asking the training effectiveness question is to improve training.

The most definitive answers to the training effectiveness question are obtainable via the transfer of training paradigm. This procedure compares performance on operational tasks as a function of training variables. If training is long and operational performance is poor, instruction is ineffective.

There are, in addition to transfer of training, several other approaches to training evaluations, each having some shortcoming relative to the transfer method. These are covered in some detail later in this paper. Our major concern, however, is with the transfer of training methodology.

Much of what is written here applies equally well to all areas of instruction. The primary subject, however, is a special kind of training situation distinguished from others (*e.g.*, classrooms) in that devices especially designed to provide practice on the behaviors to be modified play a central role in the instructional program. One well-known, albeit relatively primitive, example of such a training device is the Link flight trainer which allows students to practice many piloting skills without leaving the ground. The military, being the largest user of training devices, employs everything from quite simple part-task procedures trainers to complex whole-task trainers of perceptual motor and decision-making skills. Training devices are used by the military in nearly every job category in which high-level skill is critical.

The type of training situation of current interest also should be distinguished from on-the-job training which also utilizes devices to provide practice on critical skills. The difference lies in the fact that training devices are not exact physical replications of all aspects of the operational jobs for which training is intended. The reasons for such departures from perfect physical fidelity are: (1) Training-effectiveness—it is contrary to good training practice to try to make training an exact replica of certain real jobs; (2) Cost-effectiveness—lower cost of training equipment allows and demands its use in place of operational components; (3) Safety—the job in all its aspects may be too dangerous to be practical in a training context; and, (4) Technological barriers—for technological and other reasons it is often not possible to duplicate the operational environment.

Thus, the scope of this paper will be limited to training on synthetic devices. Also of primary concern is that these devices be designed to simulate operational tasks. Once again, although much of what is said is also true of transfer evaluations which do not address specific operational situations, this is not the major topic.

This paper traces the evolution of some typical issues, findings, and applications that have resulted from the use of transfer of training models for training effectiveness evaluations of real-life operations. Progress has been made, but the area is still plagued with serious problems. This paper also surveys past problems, advancements made, and problems still existing in such transfer of training evaluations.

TRANSFER QUESTIONS AND CONSIDERATIONS FOR ANSWERS

At various stages in the history of training effectiveness evaluations, research results have been put to a variety of uses. All these applications, however, can be viewed as efforts to answer three major questions: (1) how effective is present training? (*i.e.*, how well do the skills acquired in training transfer to operational jobs?); (2) can the training be improved?; and, if so, (3) how might the training be improved?

To answer the first question, one needs to compare operational performance of students trained on the system with students who were not. This simple design provides valuable information on training effectiveness. There is, however, a large number of variations to this basic design for investigating transfer. Further, there is a large variety of formulas for measuring and expressing the phenomenon. The choice of design and formula must be based on what one wishes to know, as well as constraints imposed on the research by the experimental situation. Classic reviews of such issues can be found in reports by Gagne, Foster, and Crowley (1948), Murdock (1957), and Hammerton (1967).

A refinement in the measurement of transfer has been discussed recently by two investigators (Povenmire and Roscoe, 1971; Roscoe, 1971; 1972) who noted that the conventional measure of "percent transfer" does not take into account the amount of training provided prior to transfer. For example, if ground training saved 9 hours in a 45-hour pilot training curriculum, percent transfer is equal to 20 no matter how much ground training (1 hour or 100 hours) was given to save those 9 flight hours. To provide a measure which better reflects the quality of the training, they presented the "transfer effectiveness ratio" (*TER*). *TER* expresses the savings in transfer as a ratio of the difference between transfer performance (in terms of time, trials, or errors to a criterion) of control and experimental groups and a similar measure taken on the training of the experimental group. Thus, for the above example and for 9 hours training, *TER* would be $\frac{45-36}{9}$ or 1.0 (*i.e.*, one transfer hour saved for each hour on the training task); for 18 hours training, *TER* would be $\frac{45-36}{18}$ or .50 (.50 hour saved in transfer for each hour spent in training). The authors of the present paper have referred to this expression of transfer as the "substitution ratio". In past studies of flight training, this ratio has typically been .75. For example, a reduction of 3 hours of flight training is accomplished by 4 hours of flight simulator time.

The analysis, however, does not end with statements of transfer. It is more desirable to apply information on transfer toward an evaluation of the total value (or efficiency) of a training system, in terms of cost. Cost effectiveness evaluations involve integration of a complex of factors relating to the cost of training (initial system development costs and system utilization costs) and transfer performance. (Many investigators performing cost evaluations do not consider training effectiveness. This exclusion may lead to designating a trainer as cost-effective when, in fact, it does not train at all!) Along with safety, cost effectiveness is really the most critical consideration, since the least expensive training system (in the long run) is considered the best training system. A system for performing cost effectiveness evaluations is contained in Micheli, Braby, Morris, and Okraski (1972).

Two factors inevitably requiring consideration in cost effectiveness evaluations are the physical fidelity of the training device and the amount of transfer resulting from the training. As the degree of physical fidelity (*i.e.*, duplication of functional characteristics of the operational equipment) increases, the costs rise at an increasing rate. Transfer, on the other hand, does not increase much after moderate levels of physical fidelity are attained. An optimal balance between transfer and physical fidelity should be determined in order to get the most training value per dollar cost. Such are the kinds of considerations that enter into cost effectiveness evaluations.

The second and third questions are interrelated in that no one knows whether a system can be improved until at least one way of improvement is demonstrated. This latter question is a complex and problematical one. It includes issues such as: (1) modifications that should be made to the system's hardware or software; (2) the manner in which the system should be utilized (*e.g.*, team training *vs.* individual training, familiarization training *vs.* initial-skill learning *vs.* refresher training, general *vs.* specific skill training, total training *vs.* supplemental on-the-job training, *etc.*); (3) personnel qualification requirements (instructors, device operators, and students); (4) performance measures and standards (*e.g.*, how to tell when a student's training should end); and, (5) training schedules (*e.g.*, the order in which various elements of the instructional course should be presented, the duration of practice on each, *etc.*).

Early efforts at training evaluation often included the mistake of trying to assess the training effectiveness of a device, independent of consideration of certain variables outside the direct control of the training device itself. One important point that is repeatedly evident in recent research and which is reflected in the foregoing list is the importance of studying the training system as a whole, including curriculum, training personnel, trainee population, and training environment. The same training device may produce differences in training as a result of any one, or an interaction, of these outside forces. (See, *e.g.*, Meister, Sullivan, Thompson, and Finley, 1971)

RESEARCH FINDINGS AND ISSUES

A review of 30 training evaluation studies, mostly in aviation settings reported between 1939 and 1972 (Micheli and Puig, 1972), permits the following generalizations to be made, with respect to the three questions under discussion.

1. Substantial amounts of air time can be substituted by simulation time in flight training.

2. Most experimental work has been done on simple aircraft and trainers.

3. Different kinds of flight tasks transfer differentially.

4. The level of simulation and kind of trainer importantly influence transfer.

5. Careful specification of both trainer and operational tasks is necessary if transfer is to occur.

6. Motion of particular kinds affects trainee performance and transfer.

7. Adding motion and visual displays increases fidelity requirements. Coupling of these is a major issue.

8. How a device is used may influence learning and transfer to a greater degree than would trainer design.

9. Differences between training and operational equipment are necessary to exploit training technology.

10. A precise specification of operational tasks and measures of performance on operational tasks is vital to effectiveness evaluations.

Among the issues receiving high interest is that of improving a training device by modifying the device's hardware or software. One emerging aspect of this issue is the possibility of altering the transfer in such a manner that the training effects of different degrees or methods of motion simulation can be compared. Only one such study has been found to date (Ber-

geron, 1970); other studies only compared motion with no motion. (Studies along these lines are being planned by the authors utilizing a simulation of an F4 aircraft as a research vehicle.)

Another popular research theme is aimed at improving training effectiveness through changes in curricula. This research can be illustrated by an on-going evaluation of the 2F90 training device, an Operational Flight Trainer for the TA-4J aircraft (Ryan, Puig, Micheli, and Clarke, 1972). The training effectiveness of the TA-4J advanced jet trainer was evaluated by measuring transfer of training from the trainer to the operational situation. Three experimental groups were compared to each other and to a fourth group which received the normal syllabus training as a control. Of the experimental groups, one received training in flight, another group only in the trainer, and the third received only academic training on related principles of the basic instrument portion of the syllabus (that portion used in the investigation). All groups were given a checkride flight. The relative benefits of the different types of flight training were evaluated to determine the effectiveness of the trainer in training advanced Naval aviation students in a portion of their syllabus. The study demonstrated that 4.7 hours of aircraft flight time per student could be saved by substituting trainer time for aircraft time on a one-to-one basis in the basic instruments stage, a potentially significant cost saving when considering that 450 students are processed through the school annually.

The best mixture of academic training, on-the-job training, and simulator time, and the sequence in which they should be presented, involves complex experimentation. To date, very little such experimentation has been done. Many existing curricula could be improved by the evaluation of several methods by which to approach a given training objective.

A typical finding in evaluations of training systems is the lack of communication between the user of the system and the agency which developed and produced it. This communication gap seems to develop soon after the system is delivered and installed. The result usually is inefficient utilization of the device. Many applications of the device, which could be used in the training curriculum, are neglected. There are cases where training personnel are completely unaware that the trainer can be used in a particular mode of operation. In essence, many training systems are not being utilized at their maximum design capabilities.

A RESEARCH PROGRAM

In a review and analysis of some relevant psychological theories and research (Blaiwes and Regan, 1970), all research topics relating to the question "How might training be improved?" are summarized into four general categories. The research is viewed as efforts to determine: (1) which subtasks, found operationally, should be included and which excluded in the training simulation; (2) those variations in stimulus and response characteristics of the training system which should be incorporated (these two questions address the "fidelity" issue); (3) which instructional devices, materials, and methods should be introduced to improve learning and transfer; and, (4) how much generalization should be built into training devices (for example, can a single, generalized maintenance trainer serve for a class of operational equipment?). Each of these four questions has an important place in research programs on transfer of training.

Considering the research and theory reviewed with respect to military training applications, this same report discusses weaknesses in the psychology literature and identifies four desirable features for a research program which would help to remedy the observed deficiencies. They are: (1) in the interest of generality of research results, enriched and diverse experimental situations should be used

to examine the rules of transfer—in fact, experimental tasks should be treated as experimental variables; (2) longer transfer tests are required to determine the course of training effects in various stages of transfer; (3) factors influencing the difficulty of the training task should be given major experimental attention; and, (4) variations in the training objective (*e.g.*, producing immediate and system-specific capability *vs.* developing a generalized skill which would facilitate later learning of various specific skills) should be studied. Further, it is recommended that these four characteristics be incorporated in a program of research that examines learning, retention, and transfer within the same performance situation.

A related effort to define and initiate a program of research on learning, retention, and transfer for Naval training (Bernstein and Gonzales, 1971a; 1971b) started with a conference on the subject attended by training experts from universities, industry, and government. One of the results of this project was the identification of four general types of tasks of highest priority for experimental investigation in the program (based on the judged criticality of the task in Naval operations and the problems it presents in training). These tasks are: (1) procedure following; (2) decision making and problem solving; (3) vigilance; and, (4) pattern recognition.

PROBLEMS AND SOLUTIONS

One of the most common obstacles in determining the effectiveness of present training in military systems is that a no-training control group often cannot be employed to determine the amount and direction of transfer to operational situations. The safety of personnel and equipment, as well as the goals of missions, would be endangered if students were placed in certain operational jobs without certain kinds of training. In one attempt to deal with this problem, operational performance levels (on an air traffic control task) were related to the amount of training (Finley, *et al.*, 1972). This approach allowed even the lowest level of training to be safe for operational requirements. Since operational performance improved with increases in the amount of training, training on the device was judged to be effective.

Another perennial problem encountered in using the transfer of training techniques to assess training lies in the performance measures used and the criterion levels established for these measures in a given course, both of which serve to define successful student performance. Sometimes, training is considered ineffective because it does not fulfill the stated goals of a course.

Recent research (*e.g.*, Krumm and Buffardi, 1970) has shown that training may not be meeting training goals because some of the stated goals are inappropriate. This investigation revealed that student submarine-diving officers were expected to develop considerable skill at the specific function of each subordinate (to help the officer avoid giving unreasonable commands in certain circumstances). It was concluded "... that untoward emphasis (was) being placed upon developing manual skills, which have limited applicability to later job assignments" (p. 30). A recommendation was made that training goals be restated to lower the diving officers' criterion performance requirements on manual tasks of the other members of the ship's control team. This change allows more training to be devoted to tasks which are more relevant to future jobs. As a consequence of restating the goals of training, the trainer would be used more efficiently with the net effect of increased training effectiveness. Thus, answering the question "How effective is present training?" often involves restating the training goals.

Defining training goals often involves detailed task analyses of operational situations from which appropriate behavioral objectives (*i.e.*, training goals) are generated. Behavioral objectives incorporate the performance measures and criteria which allow one to determine when training goals have been met. In the statement of behavioral objectives, a major emphasis is placed on describing measures of performance in both training and operational environments which are more objective than the commonly used subjective evaluations of instructors. To obtain objective measures, much reliance is placed on computerized, automatic scoring systems in training and, where feasible, equipment instrumentation in operational settings.

The specification and collection of appropriate objective measures is an especially difficult problem in situations such as tactical decision making, where complex team interactions confound the individual measures, and where no consensus can be arrived at concerning which aspects of behavior and what consequences of action should be monitored. Thus, related to the problem of specifying appropriate performance measures and standards in statements of the goals for training is the problem of developing methods and means to acquire such performance measures, especially in operational environments.

In an effort to derive good performance measures for training evaluations, a number of different data types have been identified. For example, sometimes measurement is taken on the operations which lead up to or implement the completion of the mission, but do not describe mission completion (intermediate measures). Or, performance measures may describe functions and tasks representing mission completion (terminal measures). Intermediate measures are primarily measures of individual position performance; terminal measures reflect crew (or system) performance.

Two other kinds of performance measures may also be distinguished. In pilot training, for example, one may monitor directly the outputs a student makes (*e.g.*, eye movements), or one may monitor the movements of the aircraft in space which result, in part, from the activity of the student. Both kinds of measures have relevance to training evaluations. It is important to know such things as the number of out-of-tolerance conditions of aircraft performance, but such measures are often too gross to be influenced by some training variables. Student outputs are typically more sensitive to instructional differences and, consequently, better reflect differences in skill level among students.

Other major problems confronting researchers who perform transfer of training studies on real systems are the confounding influences stemming, for example, from: (1) specifics of operational situations which vary greatly from trial to trial, so that it is difficult to separate the influence of problem variables (*e.g.*, task difficulty) from experimental condition variables (because experimental groups uncontrollably may receive differing amounts of certain kinds of trials); (2) differences among trainee (experience level, *etc.*), team members, and equipment among experimental conditions; (3) changing crew composition; (4) differences in syllabi of instruction; and, (5) differences in school practices. Many such problems arise from the fact that operational situations usually are not amenable to experimental control and that training situations, being in existence for training, are usually available to the researcher solely on a "not-to-interfere" basis.

Summarizing, some obstacles to the evaluation of training are: (1) the danger in employing a no-training control group; (2) the difficulty in specifying appropriate performance measures and criterion levels; (3) the problem in specifying appropriate training goals and the need for task analyses; (4) the problem of recording the desired performance measures in training and (especially) operational environments; and, (5) the confounding of variables in training and transfer situations due to an inability to exercise experimental control. All of these research problems typify evaluations of military training systems. Other research settings fare somewhat better. In evaluations of academic school settings, for example, a no-

training control group frequently is readily available in the general population and, depending on the transfer task of interest, may already be performing in the operational situation. Thus, safety is seldom a factor of concern. The other problems may deter the research, more or less depending on the particular question under investigation. The most favorable conditions for the conduct of training evaluations exist in the research laboratory. Laboratory transfer of training evaluations may solve some problems but, in the process, other problems are created. The most troublesome aspects of laboratory research are associated with the employment of the types of experimental tasks and subjects which will allow generalization to real-life situations.

The possibility of circumventing the sorts of difficulties involved in performing evaluations of real-life training was considered in a prototype handbook recently developed to allow for obtaining useful evaluative information without the need for transfer of training measures (Jeantheau, 1971). It features a description of four levels of assessment, three of which allow for obtaining useful evaluative information without requiring data on the performance of personnel in operational settings. These levels are outlined in Table 1.

The first level of evaluation is qualitative. It involves examining the procedures used for training in terms of specified objectives, and examining the device design in terms of its capabilities to implement those procedures. At this level of evaluation, data are gathered from documentation review, interviews with training and operational personnel, and observation of training. Limited conclusions concerning the effectiveness of the training device can be drawn, based on the rationale that, if the training has specific training objectives, sufficient structure and control, and feedback and sequencing are governed by objective measurement, then it is more effective than if it does not have them.

The second level of evaluation, noncompara-

TABLE 1

Levels of Evaluation (from Jeantheau, 1971).

Purpose	Level 1 Qualitative	Level 2 Non-comparative Measurement	Level 3 Type A TD/Conventional Comparison	Level 3 Type B Utilization Procedure Comparison	Level 3 Type C TD Comparison	Level 4 Transfer of Training
1. Single device assessment.	Limited conclusions based on *a priori* criteria and judgment.	Demonstrate learning through measured performance improvement.	Firm conclusion about training value of single device.	Firm conclusion about relative merits of methods (Method X more effective than Method Y).	Firm conclusion about relative merits of devices (Device X more effective than Device Y).	Conclusion about transfer of training on Device X.
2. Generalization	None	Hypotheses only	Extent of conclusion about device as medium is function of experimental control and ability to characterize "conventional."	Conclusions are function of experimental control and ability to characterize methods.	Conclusions are function of experimental control and ability to characterize device.	Conclusions are function of ability to characterize device.
Key Requirements	*Access to device*	*Content valid measurement system*	*Measurement system*	*Measurement system*	*Measurement system*	*Measurement system*
	Observation of training.	Accessibility of behavior for measurement	Availability of non-TD subject group for common measurement	Ability to introduce variations in method	Comparability of training situations and methods	Ability to structure, control and measure in transfer situation

tive measurement, is the crudest form of quantitative assessment that can be made. It involves measurement of trainee performance from the beginning of training to the end of training. The gain in scores throughout training then becomes a crude expression of the effectiveness of training with the device. This level is noncomparative in the sense that performance measured in the trainer is not compared with alternate methods of training, or with training in other devices, or with performance in the operational situation.

Level three involves comparative measurement. To insure comparability among situations, control of the training curriculum is mandatory. It may require the introduction of standardized exercises, at least at the beginning and end of training, and, in some situations, for an entire training period. Within level three, there are three subtypes of evaluations which are distinguished primarily on the basis of whether comparisons are made between: a practice and no-practice group (Type A), groups receiving different treatments on the same device (Type B), or groups trained on different devices (Type C).

Transfer of training (level four) is considered to be the ultimate test of the effectiveness of training. This, of course, is the level at which behavior is observed and measured in the operational situation.

Another useful approach to the solution of many of the foregoing problems is the utilization of a generalized simulation system in which the important functional characteristics of classes of training systems can be simulated and analyzed. Such a simulator would provide the capability to assemble, in building-block fashion, various types of simulators such as procedural, part task, whole task, *etc.* Each module would be designed to provide a distinct experimental capability. Additional capability could be provided from the combined utilization of two or more modules. Conceivably, some experimentation would utilize all modules as a complex system. A facility such as this

would provide flexibility and enable the degree of simulation to be adjusted in manipulating a wide variety of training alternatives to a given objective. This feature would permit evaluation of the training effects of different kinds of individual and group practice on a single task in advance of specific trainer system development. Thus, improved training and increased cost effectiveness could be approached through research conducted on relevant problems in rigidly controlled laboratory environments, prior to system development. Although some aspects of training evaluations would still have to be conducted "in the field", this facility would replace or supplement much of the after-the-fact efforts to improve training that presently occur under the restricting conditions of real training and real operational environments.

OUTLOOK FOR THE PRESENT AND FUTURE

A number of problems connected with training and its evaluation have been identified. Some programs, which also have been identified, presently are attacking these problems, and (on a positive note) progress is being made. Programs of laboratory research supplement in-the-field investigations to provide information on the effectiveness of training and ways to improve it. Further, all sorts of research programs on training, directly or indirectly, serve to advance the methodology by which the effectiveness of training is evaluated.

All such research programs are valuable and necessary. The greatest contribution to the training field, however, will perhaps come from efforts to develop new and better ways to approach the design of training programs. One of the more promising new approaches to the advancement of training technology lies in certain cognitive notions of emerging interest which emphasize information processing activities of the performer (for a discussion on this topic, see Rigney, 1971). As such, the cognitive approach represents a viable alternative to the old "identical elements" conceptualization of transfer (Thorndike, 1904). That is, instead of analyzing transfer in terms of specific elements that are common to tasks, *per se* (as the identical elements model recommends), the cognitive approach is to look at the mental activity required to perform a given task. Thus, the identical elements model suggests that training consist of practice provided on a task which shares elements (the more the better) with the transfer task. Under the cognitive model, on the other hand, the enhancement of transfer is approached by doing whatever is necessary in training to instill, in the performer, the mental processes believed to underlie successful task performance. Such training may involve receiving instruction or practicing on a task bearing little apparent resemblance to the transfer task. Research influenced by the identical elements model of transfer typically investigates variations in the similarity of stimuli and responses between training and transfer tasks. Research topics of a cognitive inclination include investigations of cognitive models and strategies for acquiring, storing, retrieving, and utilizing information (*e.g.*, mnemonics, rehearsal, imagery) which a person can learn and later employ in performing the transfer task. Also representative of the cognitive orientation is the "learning-to-learn" approach to training, wherein people are trained to become more efficient learners.

One encouraging example of an application of the cognitive approach to training on military-like tasks is described in a recent study by Blaiwes (1973). In this exploratory study, the facility with which successive procedural tasks (*e.g.*, a mock communication controller's task) were performed was improved through providing instructions to the students on the construction and use of logical trees (a technique used for organizing the essence of procedural instructions).

Finally, while the focus of this paper has been on military training, the problems and solutions discussed are germane to all factions of the world of training—including academia and industry—and all should share and benefit mutually.

REFERENCES

Bergeron, H. P. Investigation of motion requirements in compensatory control tasks. *IEEE Transactions on Man-Machine Systems*, MMS-11, 1970, 123-125.

Bernstein, B. R. and Gonzalez, B. K. Learning, retention and transfer. Vols. I & II, Orlando, Fl: Technical Report NAVTRAVDEVCEN 68-C-0215-1, 1971a.

Bernstein, B. R. and Gonzalez, B. K. Learning, retention and transfer in military training. Orlando, Fl: Technical Report NAVTRAVDEVCEN 69-C-0253-1, 1971b.

Blaiwes, A. S. Some training factor related to procedural performance. *Journal of Applied Psychology*, 1973, 58, 214-218.

Blaiwes, A. S. and Regan, J. J. An integrated approach to the study of learning, retention, and transfer—a key issue in training device research and development. Orlando, Fl: Technical Report IH-178, 1970.

Finley, D. L., Rheinlander, F. W., Thompson, E. A., and Sullivan, D. J. Training effectiveness evaluation of Naval training devices. Part I: A study of a carrier air traffic control center training device. Orlando, Fl: Technical Report NAVTRA-EQUIPCEN 70-C-0258-1, August, 1972.

Gagne, R. M., Foster, H., and Crowley, M. E. The measurement of transfer of training. *Psychological Bulletin*, 1948, 45, 97-130.

Hammerton, M. Measures for the efficiency of simulators as training devices. *Ergonomics*, 1967, 10, 63-65.

Jeantheau, G. G. Handbook for training systems evaluation, Orlando Fl: Technical Report NAVTRADEVCEN 66-C-0113-2, 1971.

Krumm, R. L. and Buffardi, L. Training effectiveness evaluation of Naval training devices. Part I: A study of submarine diving trainer effectiveness. Orlando, Fl: Technical Report NAVTRADEVCEN 69-C-0322-1, 1970.

Meister, D., Sullivan, D. J., Thompson, E. A., and Finley, D. L. Training effectiveness evaluation of Naval training devices. Part II. A study of device 2F66A (S-2E trainer) effectiveness. Orlando, Fl: Technical Report NAVTRADEVCEN 69-C-0322-2, 1971.

Micheli, G. Analysis of the transfer of training, substitution and fidelity of simulation of training equipment. Orlando, Fl: NAVTRAEQUIPCEN

Training Analysis and Evaluation Group Report 2, 1972.

Micheli, G., Braby, R., Morris, C. L., Jr., and Okraski, H. C. Staff study on cost and training effectiveness of proposed training systems. Orlando, Fl: NAVTRAEQUIPCEN Training Analysis and Evaluation Group Report 1, 1972.

Micheli, G. and Puig, J. A. Training systems effectiveness evaluations. Orlando, FL: Unpublished Report NAVTRAEQUIPCEN, 1972.

Murdock, B. B., Jr. Transfer designs and formulas. *Psychological Bulletin*, 1957, 54, 313-326.

Povenmire, H. K. and Roscoe, S. N. An evaluation of ground-based flight trainers in routine primary flight training. *Human Factors*, 1971, 13, 109-116.

Rigney, J. W. Implications of research on internal processing operations in learning and memory for serial task training. Wash., D.C.: Office of Naval Research Technical Report No. 67, 1971.

Roscoe, S. N. Incremental transfer effectiveness. *Human Factors*, 1971, 13, 561-567.

Roscoe, S. N. A little more on incremental transfer effectiveness. *Human Factors*, 1972, 14, 363-364.

Ryan, L. E., Puig, J. A., Micheli, G. S., and Clarke, J. A. An evaluation of the training effectiveness of device 2F90, TA-4J operational flight trainer, Part I: The B stage. Orlando, Fl: Technical Report NAVTRAEQUIPCEN IH-207 August, 1972.

Thorndike, E. L. *The fundamentals of learning*. New York: Lemcke & Buechner, 1903.

Application of a Multifactor Approach to Transfer of Training Research

CHARLES W. SIMON, *Essex Corporation, Westlake Village, California, and* STANLEY N. ROSCOE,[1] *ILLIANA Aviation Sciences Limited, Las Cruces, New Mexico*

A multifactor, multicriterion transfer of training experiment involving a computer-generated horizontal tracking task was conducted to establish relationships among training and transfer scores for manual control of a maneuvering vehicle, to determine the response surfaces for training and transfer, and to demonstrate a new transfer research paradigm that makes economically feasible the simultaneous investigation of the effects of a large number of training-equipment and use variables on transfer to multiple-criterion vehicle configurations. There were 80 experimental participants, 48 of whom were trained and tested on individually unique combinations of training and transfer conditions. This study was the first to measure the training and transfer effects of as many as six training equipment and use factors in a single experiment, to examine as many as 25 training-vehicle configurations in the same experiment, to train a single individual on each of 48 training conditions, to employ multiple (3) transfer vehicle configurations, and to provide data suitable for deriving multiple-regression equations for estimating the transfer effectiveness of configurations not directly studied.

INTRODUCTION

Transfer research to evaluate pilot training simulators has been under way for more than 35 years. Williams and Flexman (1949) reported the first such experiment soon after World War II. Still, there are few simulator and training-program design principles that experts can agree on, and even these cannot be stated with sufficient precision to support confident choices of equipment and program design parameters (NATO-AGARD, 1980; Roscoe, 1980, Chapters 15-22).

There are many reasons for our limited understanding of the dependence of training ef-

fectiveness on simulator and training-program design. At best, transfer experiments involving complex simulators and real planes are logistically difficult and costly. Furthermore, classical experimental designs, while suitable for evaluating particular devices or systems, are unsuitable for research intended to identify which of a great many variables in training positively affect performance of the same task in the operational situation.

In pilot training, the number of potentially important simulator and training-program design variables is disturbingly large. They fall into various classes and often interact to a degree that defies comprehensive analysis. To study a large number of such variables in a single experiment has been unacceptably

[1] Requests for reprints should be sent to Stanley N. Roscoe, Psychology Department, New Mexico State U., Box 5095, Las Cruces, NM 88003.

expensive, if not beyond consideration. To consider only a few factors at a time inevitably produces results that are inaccurate when applied to operational situations and, because of the situation-specific nature of human behavior, cannot be generalized with confidence far from the values fixed in the experiment.

The problem of conducting transfer of training studies would be greatly simplified if one could realistically assume that transfer of simulator training to actual flight could be predicted from relative performances in the simulator. However, one can easily conceive of situations in which features that make a simulator easier to fly will not necessarily produce higher transfer, and vice versa. Until relationships are established for a wide range of such conditions, transfer experiments will be required. Demonstrating a practical way of conducting multifactor, multicriterion transfer experiments inexpensively was the main objective of this study.

APPROACH

Since the problem of obtaining good multifactor simulator design information is both economic and logistic, a "holistic" approach to experimentation is being advanced. The term *holistic* (Simon, 1979, p. 77) refers to "a philosophic point of view in the conduct of behavioral experiments that emphasizes the importance of accounting for as many potentially critical variables as possible, whether equipment, environment, subject, or temporal, controlled or uncontrolled."

The key word is "critical." Whatever their number, if critical variables are held constant in an experiment, unless the fixed values are close to those experienced operationally, findings can be grossly inaccurate when applied to the operational situation. Properly implemented, the holistic approach will yield data that are more precise, less biased, and more generalizable from laboratory to field

for far less cost than that incurred with the traditional elemental approach (Simon, 1979, Chapter II). The savings in time and resources can be considerable.

The use of multifactor designs and sequential strategies can significantly reduce the cost of doing holistic experiments (Simon, 1973, 1977, 1981). This approach was successfully implemented in a study of carrier landing performance as a function of 10 equipment/environment factors (Westra, Simon, Collyer, and Chambers, 1981). However, when many factors must be investigated, these designs may still not be sufficiently economical if extended training is involved. That is to say, even with a significant reduction in the number of data points sampled from the coordinate space, compared with the number called for by traditional factorial designs, the size of the effort may still be impractical to consider.

Transfer of training studies are especially costly because, for each data point, one or more subjects must be trained for an extended period and subsequently tested during a transfer period. In some cases, still more data must be collected for control groups. Furthermore, the nature of a transfer study requires that a different subject (or subjects) be tested at each data point, a condition that is not always necessary for performance studies in which each subject may be tested on several conditions. So what are the alternatives?

An investigator may again be tempted to resort to a series of small studies involving a few factors at a time, but this will not solve the problem. Likewise, reducing the number of factors is not a reasonable solution if all are potentially important and their relative importance is unknown. Because so little is known about the characteristics of a multifactor transfer of training performance space, or the relationships among training, transfer, and intervening variables, the present study was designed to apply a particularly economical transfer of training data collection plan and to use the resulting data to investigate the relationships between training and transfer performances.

DATA COLLECTION PLAN

A relatively simple task was employed in this exploratory study primarily to illustrate the holistic philosophy and methodology. Although the task corresponded in some degree to those of primary interest in the real world, and the simulated system involved variables believed to be important in the design of operational simulators, we would not expect the results for such a simple task to warrant generalization to complex flight operations.

Simulator

The computer-based simulator was programmed to have simplified lateral-control dynamic responses similar to an airplane. It consisted, in part, of a microcomputer, a plasma-panel matrix display, and a three-axis manual flight controller, only the lateral dimension of which was used in this experiment. It also performed automatic scoring and immediate data reduction, partial analysis, and visual display, with subsequent printout on a daisy-wheel printer.

Task

The tracking task involved the relative lateral positions of three symbols on a plasma-panel display located approximately one meter in front of the seated participant. The three symbols appeared, respectively, as an airplane-like figure viewed from the rear, a target box, and a small predictor circle (approximated by an octagon). Whenever the prediction-time variable was set at zero, the predictor symbol was superposed on the vehicle symbol and appeared as if it represented the fuselage.

Each participant was instructed to manipulate the control stick, located on the right armrest, to keep the vehicle directly on the target throughout a trial. The simulated vehicle was maneuvered by right and left deflections of the control. If the stick was deflected to the left, the vehicle responded in kind, either banking and moving left or just banking left, depending, respectively, on whether the display mode was pursuit or compensatory.

The target was driven by a forcing function generated by the summation of four sine waves, a fundamental sinusoid having a period of 41 s and its 2nd, 5th, and 13th harmonics. Trials were each 51 s long, of which the first 10 s were not scored. The true tracking error between the vehicle and the

TABLE 1

Training Phase: *Hard*, *Central*, and *Easy* Factor Values

Factors	Levels*		
	Hard (−)	Central (0)	Easy (+)
Control order (CO) in % acceleration	100	62	25
Display lag (DL) in seconds	0.30	0.15	0.00
Tracking mode (TM) in % pursuit	0	50	100
Prediction time (PT) in seconds	0.00	0.30	0.60
Control gain (CG) in unitless ratios	0.12	0.18	0.24
Number of training trials (TT)	10	20	30

* Levels designated − and + are actually coded −1 and +1, respectively, corresponding to what was assumed a priori to be the *hard* and *easy* levels. The 0 is the *central* physical position between those limits.

target was displayed unless one or the other was in saturation at the left or right edge of the screen. Because scoring was based on the displayed error, the operator could see exactly what was being scored.

Experimental Factors

The factors manipulated to form the different simulator configurations were: control order (CO), display lag (DL), tracking mode (TM), prediction time (PT), and control gain (CG). Number of training trials (TT) was a sixth experimental factor. These factors and their levels under training and transfer conditions are listed in Tables 1 and 2, respectively. It was intended that this set of factors would be rich in a variety of transfer characteristics, including high and low performances in training and both high and low positive and negative effects during transfer. This objective was not fully realized because we (and the rest of the pilot-training community) simply do not understand the relationships sufficiently well to make accurate predictions.

Control order (CO). The dynamic response of a vehicle can range from zero-order to third-order or higher. In zero-order the position of the vehicle corresponds directly to the deflection of the operator's control stick. In first-order control, stick deflection determines velocity, or rate; in second-order, acceleration; and so it goes. For this study, control orders ranged from 100% acceleration

(designated the *hard* level) to a combination of 75% velocity and 25% acceleration (the *easy* level). The *central* level combined 38% velocity and 62% acceleration.

Display lag (DL). Lags between vehicle responses and display indications can have large effects on both performance and transfer. Their negative effect on performance in a simulator is similar to that due to increasing control order, but display lags might be beneficial to transfer if learning to anticipate vehicle responses is a major consideration, as in formation flying. Typical lags can be exponential in form, as occur when noisy signals are smoothed by filtering, or they can be simple transport delays. Because transport delays are inherent in the updating of digitally generated visual displays in flight simulators, variable transport lag was selected as the second factor. The three levels, *hard* (−), *central* (0), and *easy* (+), were 0.30, 0.15, and 0.00 s, respectively.

Tracking mode (TM). Computer-based control and display manipulations can have effects on both performance during training and subsequent transfer of learning. For example, most flight tasks require compensatory control, which appears to be more than twice as inaccurate as pursuit tracking for fairly difficult courses (Roscoe, 1980). In a compensatory tracking mode only the target symbol moves to indicate tracking error relative to a fixed vehicle index, usually at the center of the display. In a pursuit tracking

TABLE 2

Transfer Phase: *Hard, Central,* and *Easy* Vehicle Configurations

	Levels*		
Transfer Vehicle Configurations	Hard (−)	Central (0)	Easy (+)
Control order (CO) in % acceleration	100	62	25
Display lag (DL) in seconds	0.30	0.15	0.00
Tracking mode (TM) in % pursuit	0	50	100
Prediction time (PT) in seconds	0.00	0.00	0.00
Control gain (CG) in unitless ratios	0.12	0.18	0.24

mode both symbols move to show the absolute positions of the vehicle and the target against fixed display coordinates. By modifying the training system to allow pursuit control, it may be possible that, because correct manual inputs are elicited earlier in training, learning will be faster and transfer effectiveness higher. For the present study, 100% pursuit was taken to be the *easy* configuration, and 100% compensatory the *hard* configuration. The *central* configuration was 50% pursuit and 50% compensatory.

Prediction time (PT). Useful predictors on flight displays can be of any order in the Taylor series (Roscoe, Corl, and Jensen, 1981). For this study, a first-order predictor, showing where the vehicle would be in a few moments with present rate maintained, was represented by a small octagon. In pursuit displays this symbol appeared as a predictor of future vehicle position; on compensatory displays it appeared as a predictor of the amount by which present error would be changed during the prediction interval. In either case, its position relative to the vehicle symbol represented imminent magnitude of error based on current vehicle velocity. A prediction interval of 0.0 s was the *hard* condition, 0.3 s was *central*, and 0.6 s was *easy*.

Control gain (CG). Control or stick gain may be described as the sensitivity of the vehicle's dynamic response to control input. For this experiment, a *central* unitless ratio of 0.18 was selected so that full deflection of the stick in a pure rate-control mode produced a vehicle velocity about 20% greater than the maximum forcing-function velocity. Thus the vehicle could always overtake the target if the operator applied full stick deflection. If this *central* value of gain is regarded as one (or 3/3), then the gain for the *hard* configuration was 2/3 as great, or 0.12, and that for the *easy* configuration was 4/3 as great, or 0.24. In other words, the gain in the *hard* condition was 66% and in the *easy* condition 133% of the *central* sensitivity.

Number of training trials (TT). To assess the incremental transfer effectiveness of increasing amounts of training (Roscoe, 1971, 1972, 1980), half of the trainees assigned to the various combinations of *hard* and *easy* conditions received 10 training trials (−), and the other half received 30 trials (+) prior to transfer. The *central* group flew 20 trials (0) with the *central* training configuration, followed by the usual 30 transfer trials with the *central* transfer configuration.

Transfer Configuration

To increase the generality of our results, we created three transfer vehicle configurations that covered one extreme to the other of task difficulty. The relationships between training and transfer levels covered extremes of *easy*-to-*easy*, *easy*-to-*hard*, *hard*-to-*easy*, and *hard*-to-*hard*. Because all factors were quantitative, the empirical data would yield descriptive equations that would make possible predictions concerning performance at any point within the limits of the experimental space. This was the type of database needed to discover training and transfer relationships across a broad multifactor space.

The compositions of the three transfer vehicles are shown in Table 2. One configuration combined the *hard* levels of four of the five simulator design (vehicle) factors to create a *hard* (−) transfer task. Another combined the corresponding *easy* vehicle factor levels to create an *easy* (+) transfer task. A *central* (0) configuration combined all *central* factor levels. Centrality was measured on the physical scale of each factor. The resulting combination produced a performance level less than halfway between those attained with the *easy* and *hard* transfer configurations, respectively.

For each transfer configuration there were eight uniquely different training vehicle configurations plus two levels of training (10 or 30 trials). Participants who transferred to the *hard* configuration after training are referred to as Group A, those who transferred to the *central* configuration as Group B, and those who transferred to the *easy* configuration as Group C. A fourth group, D, who also transferred to the *central* transfer configuration, will be described and its purpose discussed in the section on data collection.

Participants

Eighty right-handed adult male nonpilots with self-reported normal or corrected-to-normal vision participated voluntarily and without pay. Most were drawn from the Behavioral Engineering Laboratory's participant roster. These were volunteers, interested in aviation, recruited from the New Mexico State University campus and local communities. The remainder came from the Department of Psychology's participant pool and were fulfilling a course requirement by voluntarily participating in this experiment.

Data Collection

Performances at 49 coordinate points were sampled in the six-dimensional training space defined earlier. Forty-eight of these points formed a fraction of a 3×2^6 factorial design divided into three orthogonal blocks represented by Groups A, B, and C, who transferred, respectively, to the *hard, central,* and *easy* transfer vehicle configurations. From the 16 points within each block, all main effects could be estimated independently from one another and from all two-factor interactions, with the interactions confounded in strings. With data from the three blocks combined, all main effects and all two-factor interactions could be isolated from one another, although not from higher-order interactions (Connor and Young, 1961, p. 16).

The 49th point was located at the center of the fractional factorial design and was replicated eight times. Thus, eight participants, designated Group D, were trained at this *central* position and subsequently were transferred to the *central* transfer configuration

(the same as that for Group B). Centerpoint data provide an estimate of experimental error variance and, when combined with the other data, allow a test of whether or not a linear second-order model is adequate.

The coordinates for this data collection plan are given in the technical report on which this paper is based (Simon and Roscoe, 1981). A different individual was assigned to each of the 48 fractional factorial points and the eight centerpoints, making a total of 56 transfer participants who received standard instructions (again see Simon and Roscoe, 1981) and were then trained and tested in the following sequence:

Baseline trials. Each participant was given three trials with the *central* training configuration. The median of these three baseline, or "matching," scores was subsequently used to adjust training and transfer scores for individual differences in initial ability.

Masking trials. Each participant was then given two trials using his individually assigned training configuration. The purpose of this was to allow any immediate carryover effects from the baseline matching condition to the training condition to dissipate.

Training trials. Each participant was given 10, 20, or 30 trials on his specific training vehicle configuration, the number depending on his coordinate position in the six-dimensional training space.

Total instructional and "flying" time for these first three steps of the data collection ranged from approximately 25 to 45 minutes depending on the number of training trials involved.

Transfer trials. Following a 30-minute rest period away from the experimental room, each participant was brought back and tested for 30 trials on one of the three transfer vehicle configurations.

Three additional groups of eight participants each provided "control" performance levels on the three transfer vehicle configurations. Each group was given the standard instructions and was then trained on a transfer configuration without prior training on any other configuration except for the baseline trials. As with the transfer groups, baseline performance data were collected at the center of the training space followed by the two masking trials prior to the 30 training trials with the individual's transfer vehicle configuration.

MAXIMIZING AND EVALUATING DATA QUALITY

As experimental designs become more complex and capable of generating large quantities of information from relatively few data points, a necessary first step in the analysis is to examine the quality of the data. This is imperative to ensure their valid interpretation. The data in this study were examined to decide on proper analyses and to detect characteristics of the data that might distort interpretation. The more important decisions and conclusions follow:

(1) To provide more normal distributions of performance data, it was decided in advance to apply a logarithmic transformation to the RMS error scores (Draper and Hunter, 1969). Subsequent comparison of the raw RMS and transformed distributions confirmed the appropriateness of the transformation. All further discussion of performance scores will be in terms of the log RMS values.
(2) The median rather than the mean was used as the representative score for a block of trials to reduce the effect of individual outlying scores.
(3) The seven groups (A, B, C, D, and three controls) were well matched in terms of the means and variances of their baseline scores.
(4) Although the data were collected by three experimenters over a period of several weeks and at different times of day and evening, there was no evidence of bias associated with any of these potential sources.
(5) The amount of learning during training was not large, but the overall curves showed expected patterns. Individual curves, however, were characterized by large variations, both within and between configurations.
(6) The assumptions made in selecting the *hard* and *easy* levels for the factors and the transfer configurations were validated, with the single exception that the levels of prediction time did not produce performance levels close to expectations.

TRAINING AND TRANSFER: SURFACES AND EFFECTS

The data were analyzed to define the training and transfer performance surfaces as well as the individual factor effects. To describe the training and transfer surfaces, data from the 56 participants in the four experimental groups were used. Transfer effects scores were obtained by subtracting the mean performance level of the appropriate control group from each individual's transfer performance score.

Individual Initial Abilities

Prior to training, each of the 80 participants was tested for three trials on the *central* system configuration. The median of these three baseline trials was used to represent each participant's initial ability or skill level and is referred to as his matching score. The mean of the matching scores for the 80 participants was 1689.44, with a standard deviation of 82.15 and a standard error of the mean of 9.18. The smallest value was 1520, 2.06 standard deviations below the mean; the largest value was 1912, 2.71 standard deviations above the mean.

Differences among the individuals' initial abilities, as measured by the matching scores, were partialed out of all training and transfer scores. All subsequent references to training performances, transfer performances, and transfer effects are based on the adjusted values from which that portion of a score attributable to initial individual ability has been partialed out. However, a word of caution is in order: A subsequent study by Simon and Westra (1984) demonstrates the inadequacy of the covariate approach alone for reducing experimental bias due to individual subject differences.

Simon and Westra investigated the special problems of and solutions to subject-related biases, particularly as they pertain to large

TABLE 3

Analysis of Training Performance Surface: Groups A, B, C, and D Combined

Source of Variance	df	F	p	% Variance
Regression	35	79.0	0.001	94
First-order terms	8	288.5	0.001	78
2 Levels	6	367.8	0.001	75
3 Levels	2	50.7	0.001	03
Second-order terms	27	16.9	0.001	16
2 FI (2 × 2 levels)	15	19.0	0.001	10
2 FI (2 × 3 levels)	12	14.4	0.001	06
Residual	20			06
Lack of Fit	12	5.2	0.025	03
Curvature	1	95.0	0.001	03
Error	7			00

TABLE 4

Analysis of Transfer Performance Surface: Groups A, B, C, and D Combined

Source of Variance	df	F	p	% Variance
Regression	35	43.7	0.001	97
First-order terms	8	163.3	0.001	87
2 levels	6	5.7	0.025	02
3 levels	2	636.1	0.000	85
Second-order terms	27	5.7	0.025	10
2 FI (2 × 2 levels)	15	5.4	0.025	05
2 FI (2 × 3 levels)	12	5.9	0.025	05
Residual	20			03
Lack of fit	12	1.3	0.100	00
Curvature	1	29.4	0.010	03
Error	7			00

multifactor experiments. They showed that for a covariate to reduce by half the expected mean biases associated by chance with the experimental effects, the covariate (or battery of covariates) must correlate higher than 0.85 with the true validity criterion. A correlation of 0.50–higher than that found in the present study and, for that matter, in most studies— would only reduce the expected mean biases by slightly more than 13%.

Training and Transfer Surface Analyses

The median scores for the last five training trials, regardless of the total number of trials, were analyzed for the training surface characteristics. These analyses were applied to the set of 56 scores for Experimental Groups A, B, C, and D combined. Median performances on the first five transfer trials with the respective transfer vehicle configurations were analyzed for all participants, both those trained and the control groups. Summaries of the analyses for surface characteristics of the training and transfer performance data are shown in Tables 3 and 4.

Training and Transfer Surface Characteristics

When Tables 3 and 4 are compared, the following generalizations can be drawn:

(1) Both training and transfer surfaces can be approximated predominantly by a combination of first-order effects.
(2) The training surface is somewhat more complex than the transfer surface, being influenced more by two-factor interactions and showing a somewhat poorer fit.
(3) Evidence of curvature is slight in both surfaces.
(4) Experimental factors have a major influence on performance during training but numerically small effects on transfer in the presence of the large performance differences imposed by the three transfer vehicle configurations.

Interpretation of Training and Transfer Effects

The contributions of the 35 main and two-factor interaction effects during training and transfer are shown in Column 2 of Tables 5 and 6, respectively. The three-level factor, Transfer Vehicle Configuration (TVC), was divided into two orthogonal components, namely the linear and quadratic trend effects (XL and XQ, respectively) with one degree of freedom each. TVC-Linear (XL) provides a comparison of the two end levels, and TVC-Quadratic (XQ) compares the mean of the two end levels with the middle level. These interact with the other factors in the conventional manner.

Each coefficient shows the change in per-

TABLE 5

Analysis of Training Performance Effects: Groups A, B, C, and D Combined

Source of Effect*	Regression Coefficient	Standard Error	Standardized Coefficient	p
INTERCEPT	1710.9			
Control order	−123.5	12.5	−0.53	0.000
Display lag	−157.9	12.5	−0.67	0.000
Tracking mode	−36.5	12.5	−0.16	0.008
Prediction time	2.7	12.5	0.01	0.831
Control gain	−2.0	12.5	−0.01	0.872
Training trials	−8.8	12.5	−0.04	0.488
TVC, linear (XL)	−47.2	15.3	−0.16	0.006
TVC, quadratic (XQ)	16.6	10.0	0.09	0.113
CO × DL	−38.7	15.3	−0.17	0.020
CO × TM	−9.1	13.2	−0.04	0.500
CO × PT	13.1	13.2	0.06	0.335
CO × CG	−4.0	13.2	−0.02	0.764
CO × TT	6.6	13.2	0.03	0.622
CO × XL	3.5	15.3	0.01	0.820
CO × XQ	11.5	8.8	0.07	0.206
DL × TM	−29.9	13.2	−0.13	0.035
DL × PT	39.9	13.2	0.17	0.007
DL × CG	−12.6	13.2	−0.05	0.353
DL × TT	6.2	13.2	0.03	0.644
DL × XL	−3.6	15.3	−0.01	0.814
DL × XQ	−19.6	8.8	−0.12	0.038
TM × PT	19.6	13.2	0.08	0.155
TM × CG	25.2	15.3	0.11	0.115
TM × TT	−10.2	13.2	−0.04	0.452
TM × XL	−29.9	15.3	−0.10	0.065
TM × XQ	5.9	8.8	0.04	0.509
PT × CG	−19.9	13.2	−0.09	0.148
PT × TT	22.0	15.3	0.09	0.165
PT × XL	6.8	15.3	0.02	0.662
PT × XQ	−14.3	8.8	−0.09	0.119
CG × TT	−9.6	13.2	−0.04	0.477
CG × XL	−2.2	15.3	−0.01	0.887
CG × XQ	10.4	8.8	0.06	0.250
TT × XL	−20.3	15.3	−0.07	0.198
TT × XQ	−17.4	8.8	−0.11	0.063

* Sources that were included in an equation from a stepwise regression analysis with $F = 4.00$ to enter and exit are italicized.

formance per unit change in the corresponding source of variance. Thus the mean difference between the *easy* and *hard* levels of any factor can be obtained by multiplying its coefficient by two. Since performance is measured in log RMS error, a coefficient with a negative sign indicates that performance was poorer on the *hard* level of the factor associated with the coefficient. For two-factor interactions, the difference is between the means of those values in which both factors are of the same sign (+ + and − −) and those in which they have different signs (+ − and − +).

Standardized regression coefficients (SRCs) are calculated from scores normalized in units of their own standard deviations about the mean. SRCs squared approximate the proportion of total variance accounted for by each source after other sources have been

TABLE 6

Analysis of Transfer Performance Effects: Groups A, B, C, and D Combined

Source of Effect*	Regression Coefficient	Standard Error	Standardized Coefficient	p
INTERCEPT	1688.2			
Control order	−14.5	15.3	−0.04	0.355
Display lag	−32.4	15.3	−0.09	0.047
Tracking mode	31.7	15.3	0.08	0.052
Prediction time	6.5	15.3	0.02	0.678
Control gain	31.1	15.3	0.08	0.056
Training trials	0.5	15.3	0.00	0.975
TVC, linear (XL)	−408.8	18.8	−0.88	0.000
TVC, quadratic (XQ)	78.0	12.3	0.26	0.000
CO × DL	4.2	18.8	0.01	0.825
CO × TM	−20.9	16.3	−0.06	0.213
CO × PT	22.4	16.3	0.06	0.184
CO × CG	13.2	16.3	0.04	0.427
CO × TT	11.7	16.3	0.03	0.480
CO × XL	−27.6	18.8	−0.06	0.157
CO × XQ	10.4	10.8	0.04	0.350
DL × TM	−18.2	16.3	−0.05	0.277
DL × PT	27.6	16.3	0.07	0.105
DL × CG	2.2	16.3	0.01	0.895
DL × TT	35.7	16.3	0.09	0.040
DL × XL	−10.3	18.8	−0.02	0.588
DL × XQ	7.0	10.8	0.03	0.524
TM × PT	−39.1	16.3	−0.10	0.026
TM × CG	−5.3	18.8	−0.01	0.782
TM × TT	18.8	16.3	0.05	0.260
TM × XL	−58.2	18.8	−0.13	0.006
TM × XQ	25.5	10.8	0.09	0.029
PT × CG	27.4	16.3	0.07	0.107
PT × TT	−20.2	18.8	−0.05	0.296
PT × XL	22.4	18.8	0.05	0.246
PT × XQ	−14.9	10.8	−0.06	0.185
CG × TT	−46.7	16.3	−0.12	0.009
CG × XL	−20.0	18.8	−0.04	0.300
CG × XQ	24.7	10.8	0.09	0.034
TT × XL	10.0	18.8	0.02	0.599
TT × XQ	−0.7	10.8	−0.00	0.952

* Sources that were included in an equation from a stepwise regression analysis with F = 4.00 to enter and exit are italicized.

partialed out. The probability (for a two-tailed t test) that an individual coefficient of the size shown in Column 2 would occur by chance is provided in Column 5. Each source accounts for 1 degree of freedom; for these calculations, there were 20 degrees of freedom in the error term.

The italicized sources of effects in Tables 5 and 6 indicate those terms that emerge from a stepwise regression analysis in which the criterion for entry and exit is an F value of 4.00. The coefficients for these terms in the stepwise analysis will be the same as those in the full regression analysis except when one source interacts with others. In the full analysis, the reported coefficients are corrected for interaction. In the stepwise analysis, no correction is made unless both interactions in an overlapping pair are brought into the equation. Thus, the differences that may exist

TABLE 7

Stepwise Regression Analysis of Training Performance with $F = 4.00$ for Factors to Enter and Exit

Group A (Hard) Equation: $1774.67 - 115.53\ CO - 173.86\ DL + 40.83\ (CO \times TT + DL \times PT)$		
Adjusted $R^2 = 0.880$ S.E. of Coefficient $= 19.97$		$F = 37.80\ (3,12),\ p < 0.001$ S.E. of Estimate $= 78.90$
Group B (Central) Equation: $1725.73 - 146.61\ CO - 118.64\ DL$		
Adjusted $R^2 = 0.732$ S.E. of Coefficient $= 28.76$		$F = 21.50\ (2,13),\ p < 0.001$ S.E. of Estimate $= 115.04$
Group C (Easy) Equation: $1680.27 - 108.50\ CO - 181.14\ DL$		
Adjusted $R^2 = 0.715$ S.E. of Coefficient $= 33.56$		$F = 19.80\ (2,13),\ p < 0.001$ S.E. of Estimate $= 134.23$

for some terms are slight and are not important to this discussion.

In Tables 5 and 6, certain results stand out clearly:

(1) Control order and display lag show relatively large effects during training but small or marginal effects during transfer.
(2) In transfer, the three transfer vehicle configurations (Factors XL and XQ) overwhelm all other sources of variance (this was as intended, of course), with the next largest sources being interactions (Tracking Mode × TVC-Linear, Control Gain × Training Trials, and Tracking Mode × Prediction Time) rather than main effects.
(3) It would appear that fewer than 10 of the 35 isolated sources of variance in either group had a critical effect on performance, quite in line with the principle of maldistribution, a fundamental assumption in economical multifactor research. (Simon, 1973; 1977)

In Tables 7 and 8, the results from stepwise regression analyses of Groups A, B, and C, individually associated with the three vehicle configurations during transfer, are shown for both the training and transfer data. In all cases, an F of 4.00 was used as the criterion for entry to or exit from the equation. While these analyses are based on only 16 observations each, they allow an examination of the results within transfer configurations rather than between transfer configurations

and thereby eliminate comparison of interactions of other factors with TVC-linear and TVC-quadratic.

Important sources of variance. Control order and display lag have strong effects on the training performances of each group. These two factors accounted for more than 70% of the variance in each analysis. With the transfer data, the picture is different. No more than a single factor or interaction string meets the criterion for entry into the transfer equation for any group, and entries are different from group to group. Terms admitted to each of these equations account for only about 30% of the total variance. There is not sufficient information to decide why a particular factor affected a particular transfer configuration, nor to determine what caused the unexplained variability. To do so, more data would have to be obtained.

Transfer effects. The analyses of transfer effect scores yield the same coefficients as those for the transfer performance scores with three exceptions: the intercept, TVC-linear, and TVC-quadratic, the values of which become -11.1, -70.3, and $+31.7$, respectively. The reason for this is that when the average performance of the appropriate control group is removed from each transfer performance

TABLE 8

Stepwise Regression Analysis of Transfer Performance with $F = 4.00$ for Factors to Enter and Exit

Hard Vehicle Configuration (Group A) Equation: 2174.91 + 115.35 TM	
Adjusted $R^2 = 0.246$	$F = 5.90\ (1,14),\ p < 0.05$
S.E. of Coefficient = 47.47	S.E. of Estimate = 189.89
Central Vehicle Configuration (Group B) Equation: 1591.14 − 46.47 DL	
Adjusted $R^2 = 0.220$	$F = 5.24\ (1,14),\ p < 0.05$
S.E. of Coefficient = 20.30	S.E. of Estimate = 81.21
Easy Vehicle Configuration (Group C) Equation: 1357.34 + 62.80 (CO × DL + DL × TT)	
Adjusted $R^2 = 0.260$	$F = 6.33\ (1,14),\ p < 0.025$
S.E. of Coefficient = 24.95	S.E. of Estimate = 99.81

score, the differences among transfer vehicle configuration levels are diminished accordingly. However, the resulting "transfer effect" scores will be affected in the same way by the other experimental factors as will the transfer performance scores, and by exactly the same amounts.

Relations between training and transfer. To facilitate examination of factor effects across training and transfer boundaries, the 48 performance measures of Groups A, B, and C from both phases were combined. The 96 scores were subjected to a stepwise regression analysis with all the main effects and two-factor interactions also interacting with a new factor, termed experiment phase (P), with training (-1) and transfer ($+1$) as its two levels. This enabled the effects of 71 sources of variance to be isolated. There were nine main effects, 35 two-factor interactions, and 27 three-factor interactions to be considered.

Based on the criterion indicated earlier, 17 of the 71 possible terms were admitted to the equation. In combination, these accounted for 91% of the total variance of the 96 scores. These terms are listed in order of their strength of effect on performance in Table 9. Those of greatest interest are the two- and

three-factor interactions that involve both training and transfer phases. The presence of a critical interaction effect can frequently change the interpretation of a critical main effect that is part of the interaction. Of interest, are those effects that interact with Phase.

Three situations were observed for both main and interaction effects. One, an effect was not critical during training but was critical during transfer. Two, an effect was critical during training but not during transfer. In both cases, the interaction with Phase was caused by the change in magnitude of the effect, but the rank order for individual conditions remained the same. The third situation, an intrinsic interaction, is the most interesting. In this case, the magnitude of the effect stayed essentially the same, but the rank order of conditions was reversed. When this happens the implications of such interactions to training equipment design are considerable (see Simon and Roscoe, 1981, pp. 44–45).

ECONOMICAL RESEARCH

One purpose of this study was to discover relationships between training and transfer performances that might be generalizable to

TABLE 9

Results from Stepwise Regression Analysis of Combined Training and Transfer Performance Data with $F = 4.00$ to Enter and Exit (N = 96)

Source of Effect		Regression Coefficient	Proportion of Total Variance
T V Configuration (X)	Linear (L)	−227	0.376
	Quadratic (Q)	+29	
Experiment Phase × X	L	−181	0.242
	Q	+29	
Display Lag		−95	0.094
Control Order		−69	0.050
Display Lag × Phase		+63	0.041
Control Order × Phase		+55	0.031
Tracking Mode × X	L	−44	0.032
	Q	+16	
Display Lag × Prediction Time		+37	0.014
Tracking Mode × Phase		+34	0.012
Display Lag × Training Trials		+27	0.007
Control Gain × Training Time		−25	0.006
Tracking Mode × Prediction Time × Phase		−23	0.006
Display Lag × Tracking Mode		−23	0.005
Control Gain × XQ		+18	0.005

transfer of training experiments with complex pilot-training research simulators (Collyer and Chambers, 1978). Another purpose was to provide an opportunity to manipulate a database to discover more economical approaches to transfer of training experiments. Of these two goals, the latter was achieved more successfully than the former. A far simpler task was employed in the experiment than would be expected in any simulator likely to be of interest. This resulted in a less rich database than desired. Consequently, generalizing results regarding training and transfer relationships must be done cautiously and tentatively.

On the other hand, the test of a new economical approach to transfer of training research is exciting and can immediately promise better and less expensive information when applied to more complex simulation problems. After 35 years of simulation experiments using conventional transfer de-

signs, this was the first to examine more than a few equipment and training factors and multiple transfer vehicle configurations in the same experiment. The information obtained regarding both the effectiveness and the problems of this experimental plan is being employed in the planning of transfer of training experiments at the Naval Training Equipment Center in Orlando, Florida.

Holistic versus Conventional Approaches

This study illustrates the tremendous economy that can be achieved with multifactor designs that yield equivalent or better information than conventional designs. The design provided the data needed to describe the relationships among six equipment and training factors and three transfer vehicle configurations using only 48 data collection points. Eight additional data points were added to estimate the lack of fit, curvilinearity, and error. Twenty-four more measures were taken, making a total of 80, to obtain estimates of the average performance of three control groups of eight subjects each.

In the design, 49 individual training configurations were examined, and an equation was obtained that would approximate performance on any combination of training and transfer configurations within the experimental space. Although only one subject per training configuration was used in the basic design, the mean of each level of every two-level factor in the design was based on 24 measurements. The means for the three transfer vehicle configurations were based on 16 measurements each. How does this compare with more conventional approaches?

One factor at a time. Suppose we had looked at the effect of one two-level training factor on one transfer configuration at a time. Direct comparison of precision between the one-factor study and the multifactor experiment is not easy to do without knowledge of the error variance of the two studies. Still some limiting calculations can show how expensive the one-factor-at-a-time approach can be when compared with the multifactor approach. For example, instead of requiring that the mean of each level of a factor be based on 24 observations, as in our multifactor experiment, let us require only 12 observations (subjects) per experimental group in the one-factor study. Thus each factor per transfer configuration would require 24 measurements for the two experimental groups.

With six factors and three transfer configurations, we would need $24 \times 18 = 432$ measurements, plus 24 more for eight subjects in each of three control groups, or a total of 456. For this conservative number, almost six times larger than the current study, we would obtain less precise estimates of means (and mean differences) and *no* estimates of any interaction effects. Also, an often overlooked point is that because each single-factor study holds the other factors at fixed levels, the gen-

eralization of the findings to a wide range of conditions would not be appropriate or safe.

Two factors at a time. Suppose we had decided to test two factors at a time to obtain both main and two-factor interaction estimates for each of the three transfer configurations. In this plan, let us assign six subjects to each of the experimental groups. This again probably produces less precise estimates of main and interaction effects than we obtained. With six factors, there are 15 different pairs of two-factor interactions. To collect the data required to estimate all two-factor interactions for three transfer configurations, we would need 6 subjects times 4 experimental groups times 3 transfer configurations, or 72 for each pair of factors, times 15 pairs, for a total of 1080 data points. To this we add 24 more subjects, with 8 used in each of the 3 control groups. This makes a grand total of 1104 points.

With the two-factor approach, we obtain estimates of all main and two-factor interaction effects, as with the present study, but this approach requires nearly 14 times the effort of the present study. Furthermore, the two-factor approach is probably less precise in its estimates and is surely more biased because the factors not included in two-factor studies are held constant (see Simon, 1979, Section II). It can be seen that as long as the same information regarding main and two-factor interaction effects is required, a multifactor design will be cheaper and more precise and its results less biased and more generalizable than comparable few-factors-at-a-time approaches.

The Big Picture

The greatest advantage of the present design, however, is that it ties the information together in a single equation of specified precision; if we need more precision, we must collect more data. Rather than isolated segments of the total space, this design provides a description of the entire multifactor space, within which interpolations can be made with some confidence. This means that the data can be generalized anywhere within the limits of the experimental conditions included in the design, and it is easier to "hang" new data on this original frame when new factors are studied. This permits a modular database to be constructed.

INDICATIONS

As intended, the present study provides some empirical data that can be useful in planning future transfer of training experiments. However, since this is the first study of its kind, involving a simple task and with relatively little learning and transfer, one must consider the following observations as "indications," a concept of data analysis that Mosteller and Tukey (1977) bring to a high level of respectability. They write:

> The word indication is a vague concept intended to include, at one extreme, all of the classical descriptive statistics . . . but also, at another extreme, to include any hints and suggestions . . . that might prove informative to a reasonable man. . . . What indication is *not* is inference or treatment of uncertainty . . . (pp. 25, 27).

In this experiment, however, several results were observed that would be quite important in the design of transfer of training studies, provided they are generalizable:

(1) The transfer performance surface appears less complex than the training performance surface (see Tables 3 and 4).
(2) The equipment factors in the transfer regression equation (although specifically selected for their potential importance to the task under investigation) accounted for only a small proportion of the total variance.

If these patterns hold, then two important practical conclusions can be drawn:

(1) Fewer data will be required to establish a transfer performance surface, since a lower-order model is indicated. (A sequential strategy, described later in this paper, ensures the most effective and economical approach to fit the appropriate model to the data.)

(2) When prediction of transfer is critical, more rather than fewer factors should be investigated during the screening phase, including task and training factors as well as additional equipment variables not considered in this study.

Economy as a Function of Purpose

The purpose for which a study is performed can markedly affect the amount of data collection required, regardless of the number of factors being investigated. If the study is done primarily to identify the important factors, fewer data will be required than if its purpose is to say with reasonable confidence that slight differences are in fact not real differences between levels. In the latter case, the power of the test requires many more observations to allow tentative acceptance of the null hypothesis with reasonable confidence.

Similarly, fewer data are ordinarily required in experiments intended to identify critical factors than to write an equation representing a response surface. Generally, the former can be satisfied with a modified fractional factorial design; the latter requires that the data collection process continue until the model being created fits the experimental data acceptably. The present design represents a compromise between these two purposes. By adding the centerpoints, lack of fit and curvilinearity could be evaluated, yet the quadratic terms for the various factors cannot be identified individually.

Of particular interest for transfer of training experiments is the distinction between an experiment intended only to identify which factors affect transfer of training and one that must provide a measure of transfer. The coefficients of each factor (with the exception of transfer vehicle configuration) are identical for the transfer performance scores and the transfer effect scores in which the mean performance of each control group has been subtracted from the transfer performance score of the corresponding

TABLE 10

Transformations Used to Change Coded Experimental Coefficients to Similarity Coefficients for Each Factor and Condition

Coded Experimental Coefficients		Coded Similarity Coefficients			
		Fidelity (Distance)		Difficulty (Direction)	
Training Code	Transfer Code	Actual	Code	Actual	Code
-1 (hard)	-1 (hard)	0	-1	No change	0
-1	0 (central)	1	0	hard to easy	-1
-1	+1 (easy)	2	+1	hard to easy	-1
+1 (easy)	-1	2	+1	easy to hard	+1
+1	0	1	0	easy to hard	+1
+1	+1	0	-1	no change	0
0 (central)	-1	1	0	easy to hard	+1
0	0	0	-1	no change	0
0	+1	1	0	hard to easy	-1

transfer group. Thus the mean difference between levels for the individual equipment variables is the same whether or not a control group is used.

From the standpoint of economy, this means that we can predict which equipment combination will yield the highest performance on the transfer configuration without collecting independent control data. In the present study, this would have represented a savings of 30%. Roscoe and Williges (1980) have described the various ways of measuring the transfer and cost effectiveness of simulator training. However, in transfer of training studies for simulator design purposes, the raw performance scores are sufficient during the early "screening" phase, in which the purpose is simply to identify the best configuration. It is in the evaluation phase later in a program that experiments should be performed with control groups.

Predicting Transfer from Training Performance Data

If it were possible to eliminate the transfer phase of a training study, considerable savings would be elicited. This could be done if it were found that there is a close relationship between training performance scores and transfer. The observed correlation of 0.37 between the 56 training performance scores and the transfer effect scores, with 54 degrees of freedom, would be expected to occur only about five times in 1000 by chance. Nevertheless, an r of 0.37 accounts for less than 14% of the variability exhibited among the transfer effect scores and therefore has relatively little practical prediction value.

Applying Principles of Transfer to Achieve Economy

Some psychologists believe they can predict the transfer potential of simulator configurations on the basis of principles that have been proposed at one time or another regarding the necessary relationships between training and transfer (see Roscoe, 1980, Chapters 15-22). If this were in fact possible it could eliminate or at least reduce the data collection effort required to make decisions regarding the design of training simulators. However, at present, these principles are at best imprecise generalizations, largely a product of unstructured empirical observation and tenuous analogy with findings from abstract laboratory experiments on verbal and motor learning. Although the present study was not designed to isolate individual principles of transfer, some information regarding so-called transfer principles was obtained.

The training performance principle. It has been suggested that the size of transfer of training experiments might be reduced by eliminating the manipulation of factors that previously failed to show critical effects during training performance experiments. This principle may be stated as follows: *Factors that fail to influence performance during training will not critically influence performance during transfer.* Since in holistic experiments involving 10 or more factors it is not uncommon to find many noncritical effects, were this theory a valid and sufficient one, considerable savings could be made.

But there are reasons to suspect the validity of this principle. Simon (1971) has discussed how task difficulty is a hidden variable that can confound the magnitude of the effects of the manipulated variables. By way of illustration for certain situations, if a task is too easy, no differences may be observed in the ceiling performances obtained at two levels of a factor. As the task becomes more difficult, differences in performance at the two levels will begin to appear. When the task becomes too difficult, performances at the two levels will become the same again, equally bad.

In transfer, quite frequently the overall operational task will be more complex and difficult than the simulated task due to adverse environmental factors not represented in the simulation. Carryover effects, another name for intraserial transfer, can be confounded with task difficulty levels in the same way. If such conditions should in fact be present, we cannot say without a great deal of additional investigation whether effects will be larger or

smaller between training and transfer. In the present experiment, some factors had large effects during training but small transfer effects, and vice versa (Table 9).

In specific refutation of the training performance principle, there were six interaction effects that were not larger than might have been expected by chance during training but were during transfer. The coefficients in these cases increased from two to six times in magnitude. Although this study, involving a relatively simple perceptual-motor task, may not be representative of the carrier landing task, for example, or of a complex simulator's usual "fidelity" factors, it does provide an empirical indication of the danger of accepting the theory that what isn't important during training will not be important for transfer.

Lincoln (1978) performed an experiment involving four longitudinal compensatory tracking tasks that differed with regard to system dynamics. Two factors at two levels each were investigated: short-period natural frequency and damping characteristics. In discussing the results of the study (p. 88), he wrote:

> The most important result of this study concerns the differential effects of the two major dynamic variables. One of them, damping, greatly influenced task difficulty but was of little importance in determining the effects of transfer of training. The second variable, natural frequency of the system, affected performance in exactly the reverse manner. Its influence was relatively unimportant with regard to task difficulty, but it appeared to be of primary importance in determining the amount of transfer of training that occurred.

In implicit support of the holistic approach to equipment design research, Lincoln continued (p. 89):

> Unfortunately for the designer, as Muckler, Obermayer, Hanlon, and Serio (1961) have shown, the complex nature of transfer effects and the need to recognize the possibility that other variables may interact with frequency and damping, makes broad generalizations dangerous. In their second report, these investigators found that control gain settings could drastically alter the patterns of transfer that were observed. It appears, therefore, that designers

TABLE 11

Summary of Stepwise Regression Analyses of Intervening Fidelity and Difficulty Factors with an F of 4.00 to Enter and Exit

Source	Coefficients	Proportion of Variance	Standard Coefficients
Transfer Performances			
Intercept	1688.0		
Direction–TM	+157.0	0.23	0.34
Direction–CG	+157.0	0.14	0.34
Direction–CO	+106.0	0.05	0.23
		Adjusted R^2 = 0.38	
Transfer Effects			
Intercept	–11.3		
Distance–TM	+61.9	0.11	0.33
Direction–TM	+54.0	0.07	0.26
		Adjusted R^2 = 0.15	

of manual control systems are presently faced with problems of considerable complexity with only limited empirical information on which to base their design decisions. A continuing effort to untangle the interrelated effects of training and transfer would seem to be appropriate.

The similarity principle. One of the most frequently stated training and transfer generalizations is the one associated with the similarity between training and transfer conditions, commonly referred to as "fidelity" when applied to simulators. This principle may be stated as follows: *Transfer of training from simulator to aircraft is a positive function of the degree to which the simulator faithfully reflects the characteristics of the aircraft.*

The effect of this principle has been the design and development of simulators that have a high degree of physical similarity to the real airplane, a costly decision of unknown payoff. Still, it is generally recognized (but sometimes forgotten) that task similarity does not necessarily depend on a faithful physical representation of reality. What it does depend on necessarily is a faithful representation of the responses that must be learned, and stimulus conditions sufficient to elicit those responses.

The chief problem with similarity (or fidelity) in a complex system in a real-world environment is that it is not always easy to define or measure. In fact, as Simon (1979, Section VI) has noted, it is a multivariate concept that differs in form and meaning for different physical components of a complex simulation. Then too, only certain components are critical in simulating particular tasks, and as yet there are no adequate principles for deciding which are and which are not critical. The concept of similarity is further complicated by the interaction between stimulus similarity and response similarity and their differential effects on positive and negative transfer.

To make matters worse, these are not the only principles that have an effect on training and transfer, and others, such as adaptively augmented feedback (Lintern and Roscoe, 1980), may override the effects of similarity in any situation. Until these principles have been adequately dimensionalized and empirically evaluated together and in context, they will continue to offer only superficial aid in the design of complex simulators that are optimized both for pilot training and for cost of ownership, maintenance, and operation.

From the results of the present experiment, two dimensions of task similarity could be examined, *factor fidelity*, as represented by the distance between levels of the various ex-

TABLE 12

Summary of Stepwise Regression Analyses of Intervening Fidelity and Difficulty Factors with Training Performance Scores as an Independent Variable and an F of 4.00 to Enter and Exit

Source	Coefficients	Proportion of Variance	Standard Coefficients
Transfer Performances			
Intercept	874.9		
Training performance	+0.5	0.09	0.30
		Adjusted $R^2 = 0.07$	
Transfer Effects			
Intercept	−429.1		
Training performance	+0.2	0.14	0.34
Direction−TM	+57.9	0.08	0.27
Distance−TM	+51.1	0.07	0.28
		Adjusted $R^2 = 0.25$	

perimental factors, and *relative difficulty*, as represented by the different training and transfer vehicle configurations. Stated in its negative form, the fidelity principle asserts that: *The farther apart the training and transfer levels of a particular factor are, the lower the transfer effect.* The relative difficulty principle asserts that: *More positive transfer will be elicited when task difficulty shifts from hard to easy than from easy to hard.*

The transformations used to change coded experimental coefficients of each factor to "fidelity" and "difficulty" coefficients are shown in Table 10. With these "similarity" coefficients the experimental design is no longer orthogonal. Still, a stepwise regression analysis could be and was performed to see which of the six fidelity (factor-distance) and six difficulty (factor-direction) variables most influenced transfer performance. With an F of 4.00 to enter and exit the equation, only 3 of the 12 terms appeared when transfer performance scores were analyzed, and only 2 terms when transfer effect scores were the criterion.

Table 11 lists the terms that appeared, their coefficients in the equations, the incremental proportion of variance each contributed, the standard regression coefficients of each, and the adjusted R^2. When training per-

formance scores were introduced as an independent variable along with the 12 similarity variables, and a stepwise regression analysis was performed for transfer performance and transfer effect scores, the results shown in Table 12 were obtained. Several findings important for simulator design are indicated by these statistics; the positive coefficients of the similarity variables indicate that the results agreed with the principles stated earlier:

(1) For those factors that had an effect, when the factor level between training and transfer changed from hard to easy there was a greater reduction in RMS error than when it changed from easy to hard.
(2) The shorter the distance between the two levels, the greater the reduction in RMS error.
(3) The direction of change in difficulty had stronger effects than the distance of change in factor levels.
(4) The adjusted R^2 values showed that the similarity variables did not improve our prediction of transfer scores over that based on training performance scores; in fact, there was an interactive effect depending on whether training performance or similarity variables were used to predict transfer performance or transfer effect scores. None of the combinations did as well as the original equipment/training variables.

No generalizations regarding the usefulness of similarity variables as intervening factors can be drawn here. Before one can discard these approaches, however, the following limitations of this analysis must be considered: It was an adjunct effort, an afterthought, and the study was not designed to obtain this kind of information. The primary importance of this exercise lies in the fact that it illustrates what might be done were a comprehensive set of intervening factors developed and a primary investigation performed. The information obtained from such a study would help resolve the question of the contribution of intervening factors to performance and transfer predictions between simulators and test vehicles.

Sequential Strategy

In the present study, a fixed design was employed. That is, the size and form of the design, its resolution, and other characteristics were selected before the experiment began. This is not the most economical way to perform holistic experiments. Instead, only the data needed to fit the lowest-order surface should be collected initially, and the model tested to see how well it fits the empirical data. Then if the fit is poor, more data would be collected to fit the next higher-order surface. This process would continue until the fit is adequate. In the present study, the fixed design was used because of concern that the time and dollar limitations might prematurely terminate a sequential effort. Consequently, a plan was selected that would guarantee at least a Resolution V design.

What the present study suggests regarding the application of a sequential approach to transfer of training is that we should build increasingly complex models on the basis of transfer scores rather than training scores. This is indicated, at least in the current study, because the transfer surface is less complex than the training surface and therefore requires fewer data points. Then too, transfer scores are the values of ultimate interest.

Only additional effort will determine to what extent the results of the present experiment can be generalized; for our purposes, however, the approach employed was appropriate.

CONCLUSIONS

This study demonstrated an efficient and economical approach for collecting multifactor, multicriterion transfer of training data. The approach is particularly useful in the early stages of a simulator design program when many alternatives should be considered and the individual contributions of component design variables should be evaluated separately from overall simulator effectiveness. Conventional transfer of training research designs are better fitted for use at the end of a design program when a few configurations have been selected and the objective is to quantify the transfer effectiveness of each.

The application of this data collection plan provides a number of practical features not normally obtained from conventional transfer of training experiments or rational analyses by design engineers and psychologists. For example, in the early stages of a simulator design program, the actual features of the transfer criterion vehicle, and consequently the simulator requirements, may not be firm. This data collection plan provides transfer data across a broad spectrum of conditions (training and criterion) so that when the airplane features eventually become firm, relevant transfer data will be available.

Isolating the effects of potentially critical variables in the simulator provides better transfer data with which to make engineering decisions than do gross measurements of total simulator effectiveness. It enables the designer to identify negative contributions of specific components that might otherwise be hidden by positive overall results. The data collection plan, by providing multifactor data in equation form, not only allows estimates to be made of the effectiveness of simulator configurations not investigated in the study, but also provides an overview that enables trade-offs to be made more precisely between system performance and system costs.

The following results in this study have direct applications for future transfer of training efforts, provided subsequent investigations validate their generality:

(1) The transfer performance surface is approximated by a lower-order model than that required for the training performance surface and should be used as the criterion for collecting more data if a sequential data collection plan is employed to obtain maximum economy.

(2) The correlation between training and transfer performance for different design configurations was positive, but was too low for practical predictive purposes.

(3) Effects that are strong in training may not be strong in transfer, and vice versa.

(4) Intervening similarity factors combined with performance scores may increase predictability, although insufficient experimental data exist to isolate and evaluate intervening predictive factors at this time.

(5) Additional economy can be achieved by eliminating control groups in the early phases of a simulation design program when the purpose is to select those combinations of factor levels that produce the highest transfer possible rather than to measure a particular configuration's effectiveness.

ACKNOWLEDGMENTS

A number of persons made major contributions to this experiment.

Louis Corl of ILLIANA Aviation Sciences was responsible for the hardware and software to generate, display, and control the simulated horizontal steering task and for the automatic data acquisition and reduction. Jan Christopher Hull, Paul M. Simon, and Donald G. Fahrenkrog, also of ILLIANA Aviation Sciences, trained and tested the participants.

The 80 participants were volunteers from the local community of Las Cruces, New Mexico, and from the Department of Psychology, New Mexico State University.

Daniel P. Westra of the Essex Corporation helped analyze and interpret the experimental data, and Brian Nelson of the Essex Corporation helped prepare special computer software for some analyses.

Lynn Borkenhagen of ILLIANA Aviation Sciences processed the manuscript.

REFERENCES

Collyer, S. C., and Chambers, W. S. (1978). AWAVS, a research facility for defining flight trainer visual requirements. In *Proceedings of the Human Factors Society 22nd Annual Meeting.* Santa Monica, CA: Human Factors Society.

Connor, W. S., and Young, S. (1961). *Fractional factorial designs for experiments with factors at two or three levels* (Applied Mathematics Series 58). Washington, DC: U.S. Government Printing Office, National Bureau of Standards.

Draper, N. R., and Hunter, W. G. (1969). Transformations: Some examples revisited. *Technometrics, 11,* 23-40.

Lincoln, R. S. (1978). Transfer of training on manual control systems differing in short-period frequency and damping characteristics. *Human Factors, 20,* 83-89.

Lintern, G., and Roscoe, S. N. (1980). Adaptive perceptualmotor training. In S. N. Roscoe (Ed.), *Aviation psychology* (pp. 239-250). Ames, IA: Iowa State University Press.

Mosteller, F., and Tukey, J. W. (1977). *Data analysis and regression.* Reading, MA: Addison-Wesley.

Muckler, R. A., Obermayer, R. W., Hanlon, W. H., Serio, F. R., and Rockway, M. R. (1961). *Transfer of training with simulated aircraft dynamics: (II) Variations in control gain and phugoid characteristics* (WADC Technical Report 60-615). Wright-Patterson AFB, OH: Wright Air Development Center.

NATO-AGARD (North Atlantic Treaty Organization, Advisory Group for Aerospace Research and Development). (1980). *Fidelity of simulation for pilot training* (AGARD Advisory Report 159). London: Technical Editing and Reproduction, Ltd.

Roscoe, S. N. (1980). *Aviation psychology* (pp. 82-94 and 173-256). Ames, IA: Iowa State University Press.

Roscoe, S. N., Corl, L., and Jensen, R. S. (1981). Flight display dynamics revisited. *Human Factors, 23,* 341-353.

Roscoe, S. N., and Williges, B. H. (1980). Measurement of transfer of training. In S. N. Roscoe (Ed.), *Aviation psychology* (pp. 182-193). Ames, IA: Iowa State University Press.

Simon, C. W. (1971). *Considerations for the proper design and interpretation of human factors engineering experiments* (Technical Report P73-325). Culver City, CA: Hughes Aircraft.

Simon, C. W. (1973). *Economical multifactor designs for human factors engineering experiments* (Technical Report P73-326A). Culver City, CA: Hughes Aircraft. (AD A035-108).

Simon, C. W. (1977). *Design, analysis, and interpretation of screening designs for human factors engineering research* (Technical Report CWS-03-77A). Westlake Village, CA: Canyon Research Group. (AD 056-985).

Simon, C. W. (1979). *Applications of advanced experimental methodologies to AWAVS training research* (Technical Report NAVTRAEQUIPCEN 77-C-0065-1). Orlando, FL: Naval Training Equipment Center. (AD 064-332).

Simon, C. W. (1981). *Applications of advanced experimental methods to visual technology research simulator studies: Supplemental techniques* (Technical Report NAVTRAEQUIPCEN 78-C-0060-3). Orlando, FL: Naval Training Equipment Center. (AD A095-633).

Simon, C. W., and Roscoe, S. N. (1981). *Application of a multifactor approach to transfer of training research* (Technical Report NAVTRAEQUIPCEN 78-C-0060-6). Orlando, FL: Naval Training Equipment Center.

Simon, C. W., and Westra, D. P. (1984). *Handling bias from individual differences in between-subjects holistic experimental designs* (NAVTRAEQUIPCEN 81-C-0105-10). Orlando, FL: Naval Training Equipment Center.

Westra, D. P., Simon, C. W., Collyer, S. C., and Chambers, W. S. (1981). Investigation of simulator design features for carrier landing tasks. In *Proceedings of the IMAGE Generation/Display Conference II*. Williams AFB, AZ: AFHRL Operations Training Division.

Williams, A. C., Jr., and Flexman, R. E. (1949). *An evaluation of the Link SNJ operational trainer as an aid in contact flight training* (Contract N6ori-71, Task Order XVI, Technical Report 71-16-5). Port Washington, NY: Office of Naval Research, Special Devices Center.

Measurement of Learning and Transfer through Curve Fitting

WILLIAM D. SPEARS,[1] *Seville Training Systems, Pensacola, Florida*

Four constants derived by fitting equations to learning and performance data are shown to measure important aspects of learning and transfer. The constants are asymptotic level, beginning level, a rate constant, and the point of inflection of learning curves having both positive and negative acceleration. Patterns of values for the constants are stressed as bases for interpreting effects of training variables and prior preparation on learning new skills.

INTRODUCTION

The purpose of this paper is to demonstrate how four empirical constants determined by fitting curves to data can reveal important aspects of learning and transfer. The constants are A, the asymptotic level of performance; B, the beginning level of performance; K, a rate constant; and P, the point of inflection for curves that have both positively and negatively accelerated portions. These constants measure variables usually of interest, and more reliably than measures typically obtained because the constants are based on entire data patterns rather than particular individual observations within a data set. At the same time, curve fitting smooths out random irregularities in data patterns. In addition, the constants can provide measures of learning and transfer that are not available by means commonly in use, and, as illustrated, they can be used for data from individual subjects. For example, given relatively nonstringent conditions, transfer can be meaningfully quantified when no control group is available, and indicators of significant points in the progress of skill integration can be identified.

With one exception, the point of inflection in a learning (or performance) curve, the utility of constants discussed here has been recognized for some time (e.g., Deese, 1958; Woodworth and Schlosberg, 1954). However, their use has been treated only in a general sense, and no systematic examination of their utility is available. Instead, almost all curve fitting for learning data has had the validation of particular conceptions or theories of learning as a goal. The constants were not the issue. The question was whether or not a theoretically derived function fitted the data.

Different theories often led to different equations to describe data, so choices of equations to use were typically based on conceptual definitions of relationships among variables, not just on their appropriateness for describing the data at hand. The fact remains that a number of equational forms can usually describe a given set of data. From a purely empirical standpoint, the criteria for

[1] Requests for reprints should be sent to William D. Spears, Seville Training Systems, 400 Plaza Bldg., Pensacola, FL 32505.

selecting an equation are simply how well the equation fits the data and the meaningfulness of the constants.

Goodness of fit is usually assessed in terms of the mean squared error of "predicted" values (curve loci) relative to corresponding observed or actual values. This least-squares criterion was used to fit all curves discussed here. However, goodness of fit is reported as the product-moment correlation r between actual and predicted values so as to avoid the inconvenience of relating means of squared errors to the variances of the original data for interpretation. Curves were fitted and standard errors for the constants derived by the method given in Snedecor and Cochran (1980, chap. 19). To conserve space, data sets, constants for equations to fit them, significance tests, etc., are omitted unless they are needed to allow the reader to follow the discussion. These details are available from the author.

THE CONSTANTS

Asymptotic Level

The asymptotic level A refers to the limit of a dependent variable Y as the independent variable X increases indefinitely. In practice, however, A is typically closely approximated after only a few trials. It may refer to an ultimate level of achievement or to a temporary plateau. In the latter case, an "indefinitely increasing X" is meaningless except when viewed as what could be expected if factors affecting performance *did not change*. Therefore, the existence of plateaus and subsequent progress prima facie indicate that something new and of significance has happened to affect progress. Under constant external conditions, the "something" resides in the learner, an emergent level of skill integration as illustrated later, for example. In their influential text, McGeoch and Irion (1952, p. 29) emphasized that understanding when and

why plateaus occur is central to understanding learning. Hence, quantifying the level of A, and being able to specify when it occurs even in irregular data, can be important for guiding training as well as for research on the effects of training variables.

A is also of value in other respects, especially as a predictive variable. Data from Prophet (1972) help validate its predictive potentials. Subjects were 56 student pilots in fixed-wing primary flight training. On each of the first five days of their aircraft instruction, percentage of errors was recorded for each subject on the first trial of each of seven basic contact flight maneuvers. On approximately the 28th day of aircraft training the subjects completed their first checkride (35-h check). Separating the subjects into seven proficiency groups according to their 35-h checkride grades (or whether subjects washed out prior to or during the checkride), Prophet presented mean error percentage for each group on the seven basic maneuvers for each of the first five days of aircraft instruction.

Using these data, Spears (1983) fitted a hyperbolic function of the form

$$Y = \frac{AX}{X + K} \tag{1}$$

to each of 33 subsets of the data: separately for three, four, and five days' mean error percentage for each of Prophet's seven groups, plus data for the same days and four grosser groupings (the column at the left in Table 1 identifies these groups). For each curve, Y was the cumulative error percentage, X the trial day, A the asymptote, and K a rate constant peculiar to this equational form. Of the 33 rs computed for actual and predicted Y values, 27 were unity when rounded to three significant digits, and the other six were 0.999.

Table 1 shows As and their ranks for the checkride-grade groups. For the four grosser

TABLE 1

Asymptotic Level A and Rank (R) for Cumulative Error Percentages by Checkride Grade and Number of Days' Data Used to Fit Curves

Grade	3 Days		4 Days		5 Days	
Group	A	R	A	R	A	R
Washouts:						
Precheckride	6334	7	5296	7	5031	7
Checkride	1264	5	1384	6	1606	6
70-74	713	4	678	4	676	4
75-79	1356	6	1101	5	1202	5
80-84	346	1	398	2	421	2
85-89	422	3	424	3	425	3
90-94	372	2	366	1	378	1
All washouts	3811	4	3629	4	3739	4
70-79	852	3	780	3	809	3
80-89	378	2	410	2	423	2
90-94	372	1	366	1	378	1

Note: Adapted from Spears (1983).

grade groupings at the bottom of Table 1, the ranks of A correlate perfectly with the ranks of grade groups whether three, four, or five days' data were used when deriving As–even though an additional 20 or more days of aircraft trials intervened between the time the error data were acquired and the 35-h checkride. Furthermore, grades were assigned according to an evaluation scale that differed from that used for tabulating error. For the less gross grade groupings, the rank correlations were 0.93 when four or five days' data were used, and 0.79 for three days' data. Note that these correlations are based on the entire universe of grade groups, so any unreliability in the estimates is due only to unreliabilities of means in the sample.

Unfortunately, there are few published data of this sort, especially with distal performance criteria composed of measures different from those obtained during training, that are sufficient for this kind of validation. (In passing, it can be added that the major problem in preparing this paper was the lack of data in training reports sufficient even to fit curves. It is hoped that the value of curve

fitting will be presented clearly enough here that researchers will more often report detailed data whether or not they fit curves themselves.) Nevertheless, another example looks at the predictive value of A from a different perspective, in this case as a basis for quantifying transfer when no control group is available. Data for this example came from Caro, Corley, Spears, and Blaiwes (1984). Six Navy pilot trainees practiced a variety of procedural tasks in a low-fidelity cockpit procedures trainer (CPT). There were seven CPT practice sessions of approximately two hours each. Following completion of CPT training, the subjects performed 19 of the CPT tasks, plus basic flight control tasks, in the SH-3H helicopter. They were evaluated by instructors as "proficient" or "not proficient" on each trial of each task in both the CPT and SH-3H.

A control group would be needed to determine if CPT procedures training, through generalization, affected the flight control skills in the aircraft, because there were no baseline CPT data concerning these skills. However, for the 19 procedural tasks common to both CPT and SH-3H practice, a measure of transfer of CPT training to the aircraft is readily available. Letting Y represent the mean percentage of proficient ratings across the 19 tasks for successive pairs of trials, and X the mean trial per pair, a logistic (S-shaped) function

$$Y = \frac{A}{1 + ge^{-kx}} \qquad (2)$$

was fitted to the CPT acquisition data (only). Resulting curves are shown in Figure 1 separately for two subjects. Both acquisition data (CPT trial pairs 1 through 4) and transfer data (aircraft trial pairs 1 and 2) are shown. The Xs represent actual percentages of proficiency ratings for Subject 1, and the open circles the percentages for Subject 2. Corrrelations between actual and predicted CPT Y

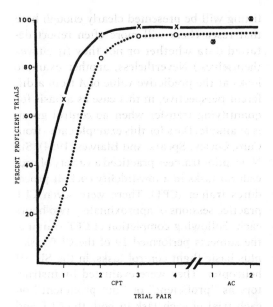

Figure 1. *Logistic curves fitted to CPT data for two subjects (data from Caro, Corley, Spears, and Blaiwes, 1984). Data points: X – Subject 1; O – Subject 2.*

values for simulator trials were essentially perfect, as is evident in the figure.

The logic of the measure of transfer is straightforward, given in this case the assumption that without prior experience in performing these tasks, the subjects could not have performed at anywhere near a proficient level in the aircraft at the outset. Then, if transfer of CPT training is 100%, aircraft proficiency measures should conform to projections of the curves for the CPT acquisition data. In other words, for 100% transfer, performance in the aircraft should progress just as though practice had continued for additional trials in the CPT. Actual measurements of transfer, then, would be based on manifest aircraft performances relative to those projected from the curves. For Subject 1, "expected" (i.e., projected from the curve) aircraft proficiency is 97%, so his transfer percentage, using the mean of 91% for his first

two aircraft trials, equals $(^{91}/_{97}) \times 100$, or 94%. For Subject 2's first two aircraft trials, manifest proficiency was also 91%, but his projected level was only 94%. Hence, transfer percentage was $(^{91}/_{94}) \times 100$, or 97%. (See Caro et al., 1984, for comparable analyses of transfer by task.)

It is not necessary to reach *A*-level performance in an original learning task to measure transfer in this way. It is necessary only to project the curves representing original performance to transfer performance. And a point to note in this measure is that projected proficiency in the transfer task would be *greater* than during original learning if asymptotic performance is not achieved in the latter. In such cases, if transfer is 100%, for example, then transfer proficiency would have to increase over that of the original learning to conform to a still-rising projection of performance. Transfer percentage should be computed accordingly.

How *A* can provide a new perspective for a controversial issue illustrates another of its uses. Nataupsky, Waag, Weyer, McFadden, and McDowell (1979) assigned 16 Air Force undergraduate pilot trainees of very little flight experience to each of two groups. Both groups practiced four flight maneuvers during each of four sorties in the Advanced Simulator for Pilot Training (ASPT). One group experienced platform motion and the other did not. Each subject was evaluated by an instructor on the last attempt on each task during each sortie. On three out of four maneuvers, mean instructor ratings of student performance across four evaluated trials were significantly higher for the motion than for the no-motion group. For two of the three significant cases, fitted curves also yielded significantly higher *A*s for motion. Fits were not close enough for the third maneuver to permit a meaningful statistical test of the higher *A* for motion. Nataupsky et al. also reported higher instructor ratings on first-trial

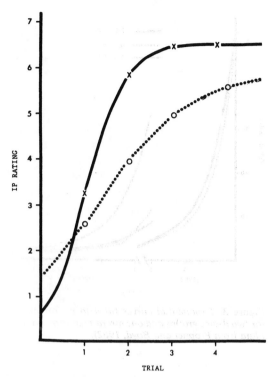

Figure 2. *Logistic curves fitted to performance on the slow flight maneuver in the ASPT (data from Nataupsky, Waag, Weyer, McFadden, and McDowell, 1979). Data point: X — Motion Group; O—No-motion Group.*

aircraft transfer performance for the motion group for each of the four maneuvers, with the motion group receiving maneuver ratings from 26% to 58% higher (mean = 46%). These aircraft differences were not significant, however.

Figure 2 illustrates the difference in simulator performance curves for slow flight, one of the three maneuvers for which mean instructor ratings for the two groups differed significantly. Correlations of values predicted from the curves and actual values were essentially perfect, as shown by the data points. A for the motion group was 6.54, whereas A for the no-motion group was 5.94, and $t(2) = 5.05, p < 0.05$. One may conclude that motion

provides cues that, when integrated into the skill, can result in higher terminal performance in the ASPT, at least for some maneuvers.

In a review of research on effects of platform motion, Waag (1981) concluded that although motion often appears to lead to superior device performance, there is no evidence that it also results in greater transfer to aircraft performance. Generally this has been true. However, in available research reports, transfer is usually equated to *early* aircraft performance, as in the study by Nataupsky et al. Woodworth and Schlosberg (1954, p. 738) cautioned against an over-dependence on early subsequent performance when assessing transfer of prior training. As they pointed out, both the subsequent rate of progress and the asymptotic level reached during the transfer task can reflect transfer. For example, Childs, Lau, and Spears (1982) presented a clearcut case in which there was negative transfer of training from a low-fidelity training device to first-day aircraft performance, but positive transfer as represented by a final (evaluation) aircraft checkride.

One reason why terminal transfer performance should be assessed is that the ultimate level of skill performance depends to a great extent on the cognitive organization of the skill, which often will not affect early performance (Spears, 1983). Childs et al. suggested that their low-fidelity training device required trainees to deliberately develop cognitive organizations of skills that would enhance ultimate performance, even though initial transfer performance suffered. A more complete explanation of the "inconsistency" is given below.

There is an additional practical reason for those interested in device transfer effects to be concerned with A as a critical aspect of transfer. Increasingly, the concern among users of simulators and other training devices is not so much to save training time as it is to achieve higher levels of operational proficiency. *A* is a direct indicator of terminal proficiency and hence is an appropriate measure to use in this context.

Beginning Level

The beginning level *B* is defined as the level of performance *before* any practice on a task in the context of interest. In other words, $B = Y_o$, the performance level for the zeroth trial X_o. Although *B*, as defined, may be viewed by some as of only theoretical interest, in fact it is of considerable applied import.

B can be estimated only by fitting a curve to data, except in situations that rarely characterize training. For example, in a simple choice experiment, a subject performs only at chance levels until some feedback is provided. In this case, first-trial proficiency estimates *B* because original success is a matter of arbitrary probabilities assigned to the various alternatives. First-trial performance in learning a complex task is rarely a satisfactory measure of *B* because it is contaminated with rate of learning *during* the first trial. (First-trial performance is also notoriously unreliable, but see later discussion.)

Data from Prophet and Boyd (1962) illustrate this point and other facets of *B*. An aircraft group practiced a set of 174 cockpit procedures only in an OV-1 aircraft. Two other groups first practiced the same procedures in training devices, one in the relatively complex, high-fidelity 2-C-9 procedures trainer and one in a low-fidelity photographic mock-up trainer. The two device groups then transitioned to the OV-1 aircraft. Each device group had five trials in its device, followed by five trials in the OV-1. The aircraft group had six trials in the aircraft. There were 10 subjects in each group, and the *Y* measures were mean error percentages on each trial by

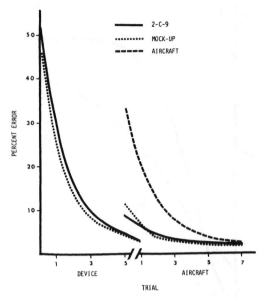

Figure 3. *Exponential curves fitted to performance for two device groups and one aircraft (control) group (data from Prophet and Boyd, 1962).*

each group in the execution of the 174 cockpit procedural tasks.

The exponential function

$$Y = h + ge^{-KX} \qquad (3)$$

was fitted to the mean error percentages for each group and period of training. The results are shown in Figure 3. Data points are omitted so as to avoid cluttering the figure, but fits were excellent ($rs \geq 0.996$) except for the 2-C-9 group's aircraft trials ($r = 0.970$). In the figure, device curves are extended to the first aircraft trial for comparison with first-trial aircraft performance, which allows one to develop a measure of transfer by comparing predicted with actual performance, as illustrated earlier. All aircraft curves are extended to the seventh "trial" to indicate the similarities among *A*s.

Considering the aircraft data first, *B*s in Figure 3 are represented by plots corresponding to Trial 5 in the devices, the "trial" preceding the first aircraft trial. The B for the aircraft group is also plotted for device Trial 5 for comparison. Transfer is indicated by the fact that (error) Bs for the 2-C-9 (8.9%) and mock-up (11.8%) groups are only 27% and 35%, respectively, as large as the B (33.3%) for the aircraft group. In this case, a similar transfer indication is obtained by comparing first-trial performances in the OV-1, where errors for both device groups are about one-third as large as those of the aircraft group. However, the B comparisons are generally to be preferred. First, each trial required from 15 to 30 minutes to complete. Unless each evaluated task was completely independent of every other—a most unlikely circumstance— learning early during a trial would result in somewhat better performance later in the trial than the subjects were actually capable of when the trial began. A second reason for preferring comparisons of Bs as a general rule is that, in contrast to these data, which show very regular first-trial performance, such performance is often quite irregular.

One reason for first-trial irregularities, especially in transfer data, is the frequent requirement to establish a cognitive-motor "interface" for the transfer conditions. (See Spears, 1983, for a detailed discussion of this point.) For example, a subject may develop a well-integrated set of, say, procedural skills in a low-fidelity device. However, the actual level of mastery may not be evident at first in an aircraft because the device-related, partially symbolic actions must be adapted to the stimulus context of the aircraft, which is somewhat interfering at first. "Interfacing" refers to such an adaptation.

Data from the Childs et al. (1982) study, referred to previously, illustrate the need for interfacing. As stated, their low-fidelity device led to inferior performance on first-day aircraft trials, even though terminal aircraft

performance favored the device group. That interfacing was involved is revealed by the fact that, of the tasks learned in the device, the device group was inferior mainly on tasks practiced the *first* aircraft day. There were seven tasks in this case, and the device group's errors were significantly greater than the aircraft group's on five of these and overall. But there were six tasks learned in the device that were practiced for the *first* time in the aircraft on the *second or later* aircraft day. Except for one task on which the aircraft group was superior, there were no significant differences on first trials of these six tasks separately, but the device group was superior overall.

These data indicate that interfacing is not only sometimes necessary to reveal transfer, but that interfacing need not be task-specific. Indeed, if it were task-specific, a question should be raised as to whether interfacing per se is the problem in early transfer performance. That is, if it occurs at all, it should facilitate—transfer to—other similar later performances, as in the data from Childs et al. Furthermore, differences among subjects in rate of and need for interfacing necessarily reduce reliability of early transfer performance.

So far, it has been shown that (1) B can be used to measure transfer; (2) in some cases, B (or first-trial performance) can be a misleading measure of transfer; and (3) A can measure aspects of transfer not manifest in B. Anticipating the later discussion of rate of learning as another measure of transfer, it can be stated that K reveals an aspect of transfer not observable in either B or A. An example from Judd (1908) makes this point vividly while providing a context for interpreting the Bs for device training in the Prophet and Boyd (1962) example as opposed to transfer Bs. One group of fifth- and sixth-grade boys was taught principles of light refraction. They and a comparable control

group then practiced hitting a target submerged 30 cm under water. The groups performed equally well. The target was then
raised so that it was only 10 cm under water,
and the subjects tried again. The pretrained
group was clearly superior the second time.
(Judd's study was repeated by Hendrickson
and Schroeder, 1941, with similar results.)
Cognitive training regarding light refraction
did not help on the first transfer effort. It did
not improve performance until the subjects
had an opportunity to interface their cognitive knowledge with a stimulus context, at
which time the effects of cognitive training
became apparent in *rate* of progress.

Except for a few studies on specific types
of cognitive pretraining, one can search in
vain for modern instances of measuring the
contribution of cognitive instruction to perceptual-motor performance via a transfer
paradigm. Paper-and-pencil tests are given
instead and then correlated with motor performances. What is typically measured by
paper-and-pencil tests does not tap the complexities of functional integration of cognitive
and motor skills, so the usual low (or zero)
correlations can be expected.

Returning to Figure 3, *device Bs* (i.e., the
zeroth trial in the *device*) were 51.0% and
45.6% errors for the 2-C-9 and mock-up
groups, respectively, and the *aircraft B* (i.e.,
the zeroth trial in the *aircraft*) was 33.3% for
the aircraft group. Device Bs for the device
groups do not differ significantly, but both
differ from that for the aircraft group ($p <$
0.05 for 2-C-9, $p < 0.01$ for mock-up). The first
question is, why were all Bs so far below
100% errors? The answer is simply that prior
cognitive (and perhaps other) instruction, as
well as generalized past experiences, gave the
subjects some head start, which means
transfer of earlier experiences. (All subjects
were qualified pilots, though inexperienced
in the OV-1.) The second question is, why was
the aircraft group's B for error percentages

lower than those of the device groups? There
are no data in this case to derive a specific
answer, but in view of the foregoing arguments, differences in required interfaces to be
established are one likely explanation. In addition, there is probably a tendency to be a
bit more careful and deliberative in one's behavior in an aircraft than in a device. Nevertheless, as illustrated in Figure 3, rate of
progress in each device was relatively rapid,
resulting in fifth-trial device performance
comparable to that of the aircraft group's
fifth trial in the aircraft.

Units for B. For many purposes it would be
desirable to express B in terms of the independent rather than dependent variable.
Such would be the case, for example, if one
wished to estimate the aircraft time "saved"
by prior device training. The task in this case
is to determine the X (trial) equivalent of B,
call it *BX* as opposed to *BY*. This can be done
easily by substituting the device groups'
transfer (i.e., aircraft) *BYs* for Y in the equation for the aircraft group and solving for X.
The equation for the aircraft group is

$$Y = 1.990 + 31.29\ e^{-0.5803X}.$$

When the 2-C-9 $BY = 8.876$ is substituted for
Y, $BX = 2.61$ trials; for the mock-up $BY =$
11.771, $BX = 2.00$ trials. So device training
could "save" aircraft usage by these amounts.
(These values can be estimated by reading
across from the respective *BYs* in Figure 3 to
the aircraft group's curve, and then down to
the *x*-axis.)

It should be pointed out that *BX* can be
negative, as can *BY* for an increasing curve.
If so, it may mean only that the function used
to fit the data was not appropriate. However,
negative transfer could result in negative Bs,
and if the function is otherwise appropriate,
negative transfer would be the interpretation.

Need for a control group. Assessing transfer
through Bs does not necessarily require a
control group, as it did by the method just

illustrated. It is necessary, however, to provide an interpretive framework for Bs. Such is easily done on a relative basis when two or more experimental groups are compared, such as the 2-C-9 and mock-up groups. Using two regimens for prior cognitive training would do just as well. Differences in BYs would reflect the relative early transfer of whatever prior training occurred. BXs could also be derived, but an interpretative framework for defining them may not be so easily established. At any rate, an adjustment on Equation 3 would provide for BXs. The trick is to alter as necessary the general constants in the equation and substitute $(X + X_o)$ for X in the exponent, where $X_o = BX$. See Mazur and Hastie (1978) for examples.

Rate Constant

Rate of learning is widely accepted as an important indicator of training efficiency. As such, it is usually measured in one way or another in both applied and theoretical studies of learning. The measures range from total time or trials to criterion to the average amount of progress during given ranges of time or trials.

Commonly used measures, although adequate for many purposes, lack the precision needed to describe an overall pattern of changes in performance during practice. The amount learned on each trial typically changes from trial to trial, and the pattern of changes itself can be of significance. Furthermore, random irregularities in data patterns often result in unreliability of usual measures, especially when average change in performance from one trial to a later trial is the concern, and rate is measured as the difference in achievement between the separate single trials.

The rate of learning constant K can overcome these problems. Being derived from the entire data pattern, it is less affected by irregularities in the data; and when used in conjunction with the equation of which it is a factor, it can identify with precision the pattern of trial-to-trial changes in amount learned. However, unlike A or B, which have the same meaning regardless of the equation used, the interpretation of K depends on the equation of which it is a part.

How K can help clarify variables affecting training is illustrated by the curves appearing in Figure 1. Subject 1's B was significantly higher than that for Subject 2. This could imply that Subject 2 gained less benefit from prior cognitive instruction. However, the curves establish that they reached similar A levels, and in comparable times. The difference must be in rate of progress. Subject 2 ($K = 1.407$) progressed more rapidly than Subject 1 ($K = 1.256$), which made up the B difference. The inference is that transfer of prior cognitive learning was probably equivalent for the two subjects; the difference between data patterns is due to more transfer being represented in B for Subject 1, but more in rate of learning for Subject 2. Thus, B and K, taken jointly, reveal not only how much transfer occurred; they can be indicative of interfacing problems as well. In such cases, Ks and differences among them can be critical measures. Suppose, for example, that K could not compensate for B differences. The conclusion would be that B reflects actual differences in transfer (ignoring A for the present), and that prior training (or general cultural experience, learning styles, etc.) does not affect rate of subsequent progress differently. When Ks make up for B differences as in the foregoing example, the inference is that one subject had less of an interfacing problem than did the other.

To see Ks' roles in identifying patterns of trial-to-trial changes, it is necessary to consider the first differential forms of their equations. The example that follows using Equations 2 and 3 also illustrates an analytic potential of fitted curves that was not apparent in the earlier discussions. Differentiating Equation 2 gives

$$\frac{dY}{dX} = KY(A - Y) \qquad (4)$$

and differentiating the *increasing* form of Equation 3 gives

$$\frac{dY}{dX} = K(A - Y) \qquad (5)$$

where dY/dX is the change in Y relative to the change in X as $dX \rightarrow 0$.

In either of these differential equations, the change in Y corresponding in a change in X (e.g., from Trial 1 to Trial 2) is some proportion of $(A - Y)$, or the difference between what "can" be learned (A) and what has already been learned (Y) up to this point. In either case, K is a proportionality constant that determines how much of what has not yet been learned will be acquired by the next $X = $ prior $X + dX$. In the case of Equation 5, K is the only multiplier in determining dY at any given time; but in Equation 4, both K and Y, the achievement up to this point, are multipliers. In other words, Equation 5 shows that its parent equation (3) simply describes an "accumulation" of learning that depends only on K and what there is left to accomplish. In contrast, Equation 4 shows that learning "snowballs"; that what has already been acquired directly enhances (multiplies) what will be accomplished during the next effort. Snowballing of learning implies intrinsic transfer, a utilization of progress-in-learning to enhance learning-in-progress.

Snowballing results in a logistic or S-shaped curve, such as Equation 2, that is so characteristic of complex human performance. (As explained in the next section, an indicator of when snowballing ceases thus becomes an important measure of progress during training and of the effects of training conditions.) Just determining whether or not a nonlogistic curve such as represented by

Equation 3 does or does not describe a set of data as well as Equation 2 can provide insights into the nature of the learning or transfer that is occurring. For example, a moment's reflection regarding problems of interfacing prior cognitive learning with motor performance as discussed earlier reveals that a beginning level B, originally depressed because the cognitive knowledge cannot be brought fully to bear immediately, will be followed by snowballing if the developing interface "catalyzes" the transfer, which, as suggested earlier, an interface would be expected to do.

Equations 4 and 5 reveal that K is positively correlated with the trial-to-trial changes in Y. This is not always the case. For example, inspection of the hyperbolic Equation 3 shows that its K is negatively correlated which changes in Y. Hence, one must interpret large and small Ks according to what they imply regarding fast versus slow progress.

Inflection Point

The inflection point P of a curve is (normally) the X value at which positive (or negative) acceleration of a curve becomes negative (or positive). In increasing sigmoid curves, the early portions are concave upward, indicating that rate of progress is greater from one trial to another (snowballing). In the later portions, the curves are concave downward, which means that, though progress still occurs, the rate of progress is less and less trial by trial. P is the point at which concavity reverses. Intuitively, it might be considered the point at which snowballing ceases, and perhaps the honing of skills begins. An example in a later section relates the cessation of snowballing (i.e., P) to a significant stage in the process of skill integration.

P is meaningful only for equations that can have changes from positive to negative acceleration, or vice versa. Mathematically, *P* is the *X* value, call it *PX*, at which the *second* differential of the equation for a curve equals zero. But the *Y* equivalent, *PY*, of *PX* can be readily determined by entering *PX* into the fitted equation; and *PY* thus measures the level of achievement at the time snowballing (or whatever) ceases.

An example from Smith, Waters, and Edwards (1975) illustrates how effects of prior cognitive training can be represented by *P*. They gave one group of 15 student pilots intensive cognitive pretraining to aid in recognizing four flight segments. A control group of the same size had only normal instruction (not defined). Figure 4 shows the logistic Equation 2 fitted to the data for each group on 14 subsequent aircraft rides. Data points are omitted, but the *r* for fit was 0.998 in each case. Table 2 presents the derived constants. *A*s are of no present interest since both groups closely approximated maximum possible performance. Also, *K*s are almost

identical, as are *PY*s. However, *BY*s and *PX*s differed radically, with *p* substantially less than 0.001 in each instance. Briefly, the cognitive pretraining not only gave the group receiving it a head start on initial transfer to the performance context, the snowballing was completed sooner and for a comparable level of achievement at the time.

In the present example, there was no advantage to cognitive pretraining insofar as *PY* was concerned. However, it should be apparent that an advantage in both *PX* and *PY* would mean that prior learning results not only in more rapid, say, skill integration, but also in more skill components being integrated when *P* occurs. Also, when *PX*s are the same but *PY*s differ, the conclusion would be that there is a difference in the number of skill components integrated when snowballing ceases, which may also be reflected in *A* or *B* levels, or both. Hence, all four constants, *A*, *B*, *K*, and *P*, taken together, become bases for insights into what goes on during

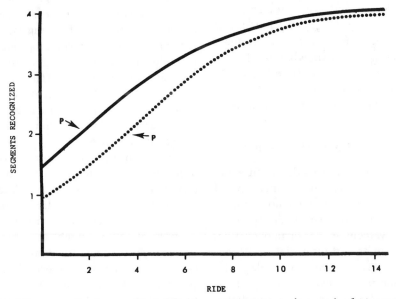

Figure 4. *Logistic curves fitted to number of flight segments recognized on each of 14 successive aircraft rides by a group receiving cognitive pretraining and a control group (data from Smith, Waters, and Edwards, 1975).*

TABLE 2

Constants for Curves in Figure 4

Group	A	BY	K	PX	PY
Pretraining	4.10	1.43	0.341	1.84	2.05
Normal training	4.07	0.87	0.354	3.69	2.04

learning and transfer; and given appropriate experimental controls, for interpreting the effects of any number of types of variables on learning and transfer.

It is possible to get a negative PX, and in many cases it can be meaningful (but rarely can a negative PY). Such would occur when a logistic equation is fitted to a data pattern that has only negative acceleration. Given adequate a priori grounds, a negative PX would indicate that snowballing ceased during learning prior to that represented by the data pattern. Such might happen through earlier cognitive instruction that needed no interfacing with equipment. In some comparisons, a negative PX would definitely be interpretable. For example, if control group data have a sigmoid pattern, but not a device group upon transfer, the inference is that device training resulted in pretransfer achievement beyond the cessation of snowballing. Hence, sigmoid curves should be fitted for both groups in the transfer setting, and PX differences should take the negative value into account.

TRACKING SKILL INTEGRATION

A number of data patterns do not fit equations of a form to provide useful measures of A, B, K, and P. These data have no apparent regularity. Even though there may be an overall improvement in performance, progress may be minor and intermixed with apparent day-to-day or trial-to-trial degradations in performance. Occasionally, perhaps often, the irregularities are due to unreliable measures or to seriously inadequate control

of factors other than learning that affect them; but as will be shown, the irregularities may be related to learning processes of considerable interest.

The problem may be the inappropriateness of measures, at least for the stage of training at which they are used. This could be the case when an overall task proficiency measure, such as catching the restraining cable during a carrier landing, for example, is used to assess carrier landing performance by novices. (During precarrier preparation, touch-and-go landings are practiced on a field runway that is marked off as a carrier—Field Carrier Landing Practice, or FCLP; a Landing Signal Officer, or LSO, evaluates the touchdowns, gives wave-offs, etc., according to the judged adequacy of the approach and touchdown.) Catching a wire, or an LSO's belief that the cable would be engaged, can be a rather chaotic indicator of performance, as is evident in the performance curves provided by Brictson (1978). In preparing for the pilot study described below, similar irregularities were found in LSOs' evaluations of more than 100 Navy pilots during FCLP trials. Yet, as will be evident, measures focusing on skill components rather than terminal proficiency can be highly systematic in such a situation.

During the pilot study (Isley and Spears, 1982), and prior to FCLP attempts, seven Student Naval Aviators practiced simulated carrier and field landings in the Visual Technology Research Simulator (VTRS) at the Naval Training Equipment Center in Orlando, Florida. Each pilot had a different combination of conditions: simulated carrier or simulated field-carrier landing strip; day or night visual scene; wide or narrow field of view; circling or straight-in approach. The subjects flew from 32 to 48 landing attempts each in the VTRS. Computerized measures of accuracy of touchdown were as chaotic as those referred to above. One subject never achieved a successful touchdown in 32 attempts. Yet, the LSO monitoring his performance at an off-board station saw steady improvement. When asked the basis for his judgments, the LSO remarked "he's beginning to put it together," and then explained briefly that, although no successful landings were made, he could discern that the subject was making progress on the various component skills and was beginning to integrate them.

The other six subjects achieved various numbers of touchdowns, but none of several automated measures at touchdown revealed significant progress, even for 48 trials. However, there were also automated measures of percentages of time in tolerance for three variables during various segments of the landing approach: line-up (LU) with the centerline of the carrier or field; angle of attack (AOA); and glideslope (GS). For any given subject, plots of percentages of time in tolerance for these three variables revealed substantial progress. It was even clearly apparent that the subject who had not achieved any touchdowns *was* "putting it together." *But*, it was necessary to identify characteristics in performance patterns that could reveal progress. Specifically, for the first 30 trials or so, when a curve rose for, say, LU, it dropped for AOA or GS, and frequently both. This was true not only across trials, but across segments within a trial. When computed, intercorrelations of percentages for the three measures tended to be predominantly negative until Trial 20 or so, when they approached zero. Then, later in practice, curves for two, and then all three, began to rise together, and the *r*s were generally positive. The only variation in this pattern was due to the circling versus straight-in approach. For the latter, the subjects began a trial close to the desired approach path, and they tended to start high on one measure but low on the other two. Immediately, however, the favored skill component degraded while one or both of the

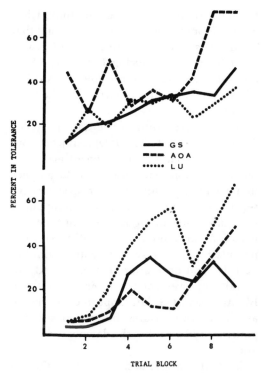

Figure 5. *Percentages of time in tolerance for line-up (LU), angle of attack (AOA), and glideslope (GS) during a segment of an approach to a carrier landing in a simulator: straight-in approach (top); circling approach (bottom) (data from Isley and Spears, 1982).*

others improved. Subjects who flew a circling approach generally began low on all three measures; but then one measure increased radically, only to fall when the pattern just described began. The general patterns, one straight-in (top) and one circling (bottom) approach, are shown in Figure 5, plotted by blocks of five trials. These plots illustrate the patterns found in detailed analyses for separate approach segments, individual trials, and successive segments within a trial. For all subjects, mean percentages for all three variables tended to drop below some early level, resulting in no *apparent* overall improvement even by Trial 20 or later. It should be added that in a companion study, Isley, Spears, Prophet, and Corley (1982) found that when LSOs were asked to assess LU, AOA, and GS separately during FCLP, the LSOs' measures showed similar patterns. (The LSOs were not told the purpose of the measures.)

The interpretation of the data patterns seems obvious. At the outset, subjects concentrated on one skill component at a time but shifted from one component to another from trial to trial and even during successive segments of a single trial. The component given attention had relatively high performance. At some point, two or all three components began to receive attention simultaneously, and only after this happened did overall improvement begin. As a result, the LU, AOA, and GS data appeared as chaotic as touchdown measures, but the consistent r patterns indicated that the data were systematic.

Shelnutt, Spears, and Prophet (1983) further demonstrated the regularities in these data. Combined LU, AOA, and GS means from the lower half of Figure 5 were fitted with Equation 2. In fitting the curve, it was necessary to begin with Trial Block 2, because the lack of improvement from Block 1 to 2 is not mathematically consistent with what happened later. It was also necessary to consider the likelihood of a plateau, and the procedure used illustrates how plateaus can be identified in a pattern of data. Beginning with all data points after the first block, successive fits were made, dropping the latest remaining trial block each time until a very close fit was obtained. The solid curve shown in Figure 6 resulted. Such a procedure can result in a valid identification of a plateau only if a very good fit is eventually obtained for more than three successive data points (because three degrees of freedom are lost in determining A, B, and K), *and* the fit is very

poor if additional data are included. Inspection of Figure 6 reveals that these conditions are met for Blocks 2 through 6. For the final fit, the r between actual and predicted Ys was 0.9997.

The two constants of most present interest are A, the level of the plateau, and PX, the point at which positive acceleration ceases. As is evident in Figure 6, a plateau, approximated by Trial Block 4, extended through Block 6, a total of 10 trials beyond Block 4. Comparison with the lower plot in Figure 5 suggests that a plateau occurred because the subject had not included AOA in the skill integration, and so he relaxed attention to GS and especially LU to focus more on AOA. By Block 7, performance on all three was similar, and then progress on all three followed for the next block of five trials.

However, performance at Block 7 was significantly *below* the level of the plateau. Such might be expected, given (1) a fairly low level of integration at the plateau ($A = 32\%$ mean time in tolerance); and (2) a need to make some sacrifices to focus on a component on which performance had lagged. Nevertheless, overall progress suddenly spurted, suggesting that through the preceding struggles the subject had "put it together" to some extent. Furthermore, the patterns for Blocks 7 to 9 in Figures 5 and 6 suggest that another plateau was imminent. (A curve fitted to these three data points yielded $A = 46\%$.) Considering the low level of the first plateau, further struggling to coordinate LU, AOA, and GS could be expected (see Block 9 in Figure 5). Hence, a second and probably one or two more brief plateaus would be likely.

$PX = 3.3$ trial blocks was a precursor of the first plateau, as it should be, as negative acceleration of the curve began at this point. However, Figure 5 relates PX to what was going on: the last joint rise in all three components until after the plateau. There was

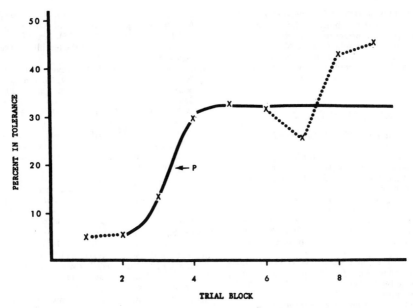

Figure 6. *Logistic curve (solid line) fitted to means of LU, AOA, and GS percentages in tolerance as shown in bottom plot in Figure 5 (data from Shellnutt, Spears, and Prophet, 1983). Xs are data points.*

further improvement after *PX* in LU, but at the expense of AOA and to some extent GS.

CONCLUDING COMMENTS

The foregoing discussion of curve fitting was restricted to pilot performance and to only three equational forms, and only practice trial numbers or equivalents were used for the independent (*X*) variables. However, there are a number of other equations that are useful for describing a wide variety of learning and performance data (see Lewis, 1960; Restle and Greeno, 1970), and *X* can represent any one of a number of independent variables, as it often has in experiments on learning. Furthermore, curve fitting has value to the applied researcher in areas other than learning. Experimental reports on perception, choice behavior, and decision making typically use curve fitting not only to describe what has occurred but analytically

to identify facets of the behavior and influences of experimental variables as well (see Baird and Noma, 1978; Restle and Greeno, 1970).

As has been true of most uses of curve fitting in studies of learning, the equations used in other areas of psychology typically reflect the theoretical positions of the researchers, and thus may appear of limited value to an eclectic applied researcher. But, as with learning equations, the mathematical formulations in the other areas need not involve theory. The point is that the equations describe actual behavior. And the applied researcher's only concerns in choosing an equation to use with a set of data are, how good is the fit? Are the constants empirically meaningful?

ACKNOWLEDGMENTS

Portions of this research were supported by the Naval Training Equipment Center, Contract Numbers N61339-

78-C-0113 and N61339-80-D-0009. The views expressed in
this paper are those of the author and should not be interpreted as necessarily representing the official views or
policies, either express or implied, of the Naval Training
Equipment Center or the United States Government. The
author thanks Wallace W. Prophet for his careful review
of an early draft of the manuscript and for his helpful
comments.

REFERENCES

Baird, J. C., and Noma, E. (1978). *Fundamentals of scaling
and psychophysics.* New York: Wiley.

Brictson, C. A. (1978, April). *A-7 training effectiveness
through performance analysis* (Tech. Report NAVTRAE-
QUIPCEN 75-C-0105-1). Orlando, FL: Naval Training
Equipment Center.

Caro, P. W., Corley, W. E., Spears, W. D., and Blaiwes,
A. S. (1984, June). *Training effectiveness evaluation and
utilization demonstration of a low cost cockpit procedures trainer* (Tech. Report NAVTRAEQUIPCEN 78-C-
0113-3). Orlando, FL: Naval Training Equipment
Center.

Childs, J. M., Lau, J. R., and Spears, W. D. (1982, June).
An empirical assessment of multiengine flight training
(Tech. Report TR 82-04). Pensacola, FL: Seville Re-
search Corporation.

Deese, J. (1958). *The psychology of learning.* New York:
McGraw-Hill.

Hendrickson, G., and Schroeder, W. H. (1941). Transfer of
training in learning to hit a submerged target. *Journal
of Educational Psychology, 32,* 205-213.

Isley, R. N., and Spears, W. D. (1982, April). *Phase I pilot
study: VTRS transfer of training experiment* (Tech. Report TR 82-03). Pensacola, FL: Seville Research Corporation.

Isley, R. N., Spears, W. D., Prophet, W. W., and Corley,
W. E. (1982, April). *VTRS transfer of training experiment: Phase I—Experimental design* (Tech. Report TR
82-02). Pensacola, FL: Seville Research Corporation.

Judd, C. H. (1908). The relation of special training to general intelligence. *Educational Review, 36,* 28-42.

Lewis, D. (1960). *Quantitative methods in psychology.* New
York: McGraw-Hill.

McGeoch, J. A., and Irion, A. L. (1952). *The psychology of
human learning.* New York: Longmans, Green.

Mazur, J. E., and Hastie, R. (1978). Learning as accumulation: A reexamination of the learning curve. *Psychological Bulletin, 85,* 1256-1274.

Nataupsky, M., Waag, W. L., Weyer, D. C., McFadden,
R. W. and McDowell E. (1979, November). *Platform
motion contributions to simulator training effectiveness:
Study III—Interaction of motion with field-of-view.*
(AFHRL-TR-79-25). Brooks AFB, TX: Air Force Sys-
tems Command.

Prophet, W. W. (1972, April). *Performance measurement in
helicopter training and operations* (Professional Paper
10-72). Alexandria, VA: Human Resources Research
Organization.

Prophet, W. W., and Boyd, H. A., Jr. (1962, June). *Relative
effectiveness of Device 2-C-9 and a photographic cockpit
mock-up device in teaching ground cockpit procedures
for the AO-1 aircraft* (informal report). Fort Rucker, AL:
U.S. Army Aviation Human Research Unit.

Restle, F., and Greeno, J. G. (1970). *Introduction to mathematical psychology.* Reading, MA: Addison-Wesley.

Shelnutt, J. B., Spears, W. D., and Prophet, W. W. (1983,
June). *VTXTS training effectiveness measures.* (Tech. Report TR 83-18). Pensacola, FL: Seville Research Corporation.

Smith, B. A., Waters, B. K., and Edwards, B. J. (1975, December). *Cognitive pretraining of the T-37 overhead
traffic pattern* (AFHRL-TR-75-72). Brooks AFB, TX: Air
Force Systems Command.

Snedecor, G. W., and Cochran, W. G. (1980). *Statistical
methods* (7th ed.). Ames, IA: Iowa State University
Press.

Spears, W. D. (1983, November). *Processes of skill performance: A foundation for the design and use of training
equipment* (Tech. Report NAVTRAEQUIPCEN 78-C-
0113-4). Orlando, FL: Naval Training Equipment
Center.

Waag, W. L. (1981, January). *Training effectiveness of visual and motion simulation* (AFHRL-TR-79-72). Brooks
AFB, TX: Air Force Systems Command.

Woodworth, R. S., and Schlosberg, H. (1954). *Experimental
psychology* (rev. ed.). New York: Holt.

Determining Transfer of Training Using Curve Fitting

Diane L. Damos, Department of Human Factors, University of Southern California, Los Angeles, California

The purpose of this paper is to demonstrate the measurement of learning and transfer using a curve-fitting technique discussed in a 1985 *Human Factors* article by Spears. The data were collected during an experiment that determined if rotation skills could become automated with practice and if the skills could transfer between stimuli. The dependent variables of interest were the slope and intercept of the regression equation relating correct reaction time and degrees of rotation. Curve fitting was accomplished using a common statistical package, BMDP, and an IBM-XT. The curve-fitting technique showed large initial transfer of training on several variables that did not affect the asymptotic level of performance. In contrast standard transfer of training calculations indicated small positive transfer.

INTRODUCTION

In 1985, Spears described a new method of determining transfer of training by using curve-fitting techniques. Spears maintained that more information about transfer can be obtained from curve-fitting techniques than from traditional methods because each of the parameters describing the curve may reflect a different aspect of transfer. Specifically, Spears proposed that four parameters – the rate of learning, the beginning level (the performance level before any practice), the asymptote, and the inflection point – could provide the investigator with independent information about the effect of prior practice on the performance of the transfer task. Additionally, Spears demonstrated that meaningful. quantitative measures of transfer could be obtained without a control group.

Despite the apparent advantages of the curve-fitting technique, few human factors practitioners have adopted this method. Many explanations can be offered for the lack of enthusiasm for curve fitting. One explanation is that the technique appears complicated, requiring a sophisticated statistical package and knowledge of advanced mathematical techniques.

A second concerns the "mapping" of the results between the traditional methods and the curve fitting technique. The beginning level, the rate of learning, and the asymptotic level appear to have some relation to percent transfer although the nature of the relation and the redundancy between the three measures is not clear. The inflection point appears to have no direct relation to any traditional measure of transfer.

This paper has two purposes. First, it demonstrates that curve fitting is a relatively easy technique that does not require expensive statistical packages and large computer systems. Second, it compares measures of transfer obtained using curve fitting with the traditional percent transfer measure calculated on the same data.

The data to be analyzed are from an experiment examining the transfer of mental rotation skills. This experiment was designed to determine if mental rotation skills became automatic with practice and if rotation skills developed using one stimulus transferred to a different stimulus. Cooper and Shepard's (1973) paradigm was used to present stimuli and analyze data. Unlike most of the earlier research on mental rotation, the dependent measures of

Reprinted from Proceedings of the Human Factors and Ergonomics Society 32nd Annual Meeting (pp. 1276–1279).

interest in this experiment were the slope and intercept of the best-fitting linear equation relating the correct reaction time to the absolute degrees of rotation. The slope reflects the speed of the mental rotation of the stimulus while the intercept reflects stimulus encoding, comparison of the encoded figure with a figure held in memory, and response selection and execution. If mental rotation becomes automatic with practice, the slope should decrease to 0.0 ms/degree of rotation. The intercept should also decrease because of increased familiarity with the response device. Transfer of mental rotation skills may be evident in either the slope or the intercept.

A brief description of the experimental procedure is given below. Interested readers should contact the author for more details.

METHOD

Subjects

Thirty right-handed males between 18 and 35 completed the experiment. All subjects were native English speakers and none had any flight training. Subjects were paid $5.50/hour for participating and could win performance bonuses as described below.

Apparatus

A Micro PDP 11 computer generated all stimuli, recorded and processed the subject's responses, and timed all trials. The stimuli were displayed on a Tektronix 4125 color graphics terminal. Subjects responded by pressing keys on two identical 4×4 matrix-type keypads. The subject sat approximately 110 cm from the display screen.

Task

The task required the subject to distinguish between a shape and its mirror image as quickly as possible. If the standard shape was presented, the subject pressed a key under his right index finger. If the mirror image was presented, he pressed a key under his left index finger. Stimuli could be presented at one of six orientations: 0 degrees (upright), 60, 120, 180, 240, and 300 degrees of clockwise rotation. Three different stimuli were used: capital F,

capital G, and a 24-point abstract shape (Stimulus # 29, Vanderplas and Garvin, 1959). The stimulus presentation was unpaced.

Design

Subjects were randomly assigned to one of three groups. Group 1, an experimental group, received training using the letter G and transferred to the 24-point figure. Group 2, also an experimental group, trained using the letter F and transferred to the 24-point figure. Group 3, a control group, practiced only with the 24-point figure.

Procedure

All testing began on a Monday. Subjects in Groups 1 and 2 completed one session on each of ten consecutive week days. Subjects in Group 3 completed one session on each of five consecutive week days. Thirty-six trials were administered in each session. Each trial consisted of five presentations of the standard shape and five presentations of the mirror image at each of the six orientations. The order of presentation of the stimuli were random within each trial. All subjects saw the same order of stimuli.

The percentage of correct responses and the average correct reaction time (RT) was presented as feedback to the subject at the end of each trial. The subject received a $0.25 bonus for each trial in which he exceeded his own previous best performance. All instructions were taped and immediately preceded the relevant condition.

RESULTS

Preliminary Data Analysis

As noted earlier, the dependent variables of interest were the slope and the intercept of the function relating correct RT to the degrees of absolute rotation (degrees from upright without regard to direction). Thus, slopes and intercepts had to be calculated first, then described using the curve-fitting technique. Each experimental session was divided in two blocks, early versus late, with 18 trials per block. Best fitting linear equations were calculated for the standard and mirror responses on a subject-by-subject basis for each block using the PC version of BMDP1R (Dixon, 1985). For illustrative purposes, the results of the curve-fitting technique will be compared to the traditional technique only for the standard response equations.

Curve-fitting Technique

No theory suggests the type of learning curve (exponential, logarithmic, etc.) resulting from the transfer of rotational skills. Spears (1985) notes that the usefulness of an equation depends on how well the equation fits the data and the meaningfulness of the parameters. The first problem then was to find a family of curves that could be meaningfully interpreted and fit the data well. The search was limited to exponential, logarithmic, and hyperbolic functions because these are the functions most often used to describe psychomotor and cognitive learning curves. The curves to be fit were obtained by averaging the values of beta and alpha across subjects within groups for each block of both the training and transfer phase of the experiment. Figure 1 shows the resulting

curves for the averaged betas during the transfer phase of the experiment. Figure 2 shows the corresponding alpha values.

The curves were then compared to graphs of families of exponential, logarithmic, and hyperbolic curves (Lewis, 1960). Visual inspection of the data indicated that an exponential equation of the form

$$dv = c \; exp \; (gx) + h \qquad (1)$$

Where:
   c, g, and h are parameters to be fit
   x is the block number
   dv is the dependent variable (alpha or beta)

might be appropriate for fitting the curves.

Determining the goodness of fit for a nonlinear equation is not straightforward (Draper and Smith, 1966). From a practical viewpoint, however, the correlational approach described by Spears offers an easy-to-calculate method of determining if the fit of an equation is sufficient. Spears (personal communication) suggests that if the product-moment correlation

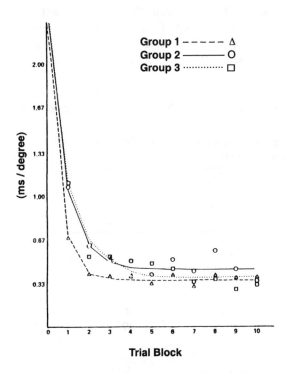

Figure 1. Betas as a function of trial block for the abstract figure.

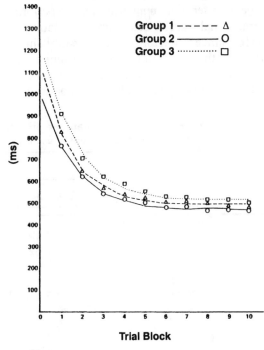

Figure 2. Alphas as a function of trial block for the abstract figure.

between the data and the predicted values is at least 0.95, the equation provides a sufficient fit to the data.

The averaged beta and alpha values were submitted to the PC version of BMDP3R. The correlations between the observed and predicted beta values for the Groups 1, 2, and 3 were .94, .97, and .95, respectively. All three groups had correlations of .99 between the predicted and observed alpha values. Despite the one relatively low correlation of .94, the exponential equations were used for determining transfer. Estimates of the beginning level, asymptotic level, and learning rate for the transfer phase of the experiment are shown in Table 1. The beginning level represents the y-intercept of the exponential function (the level of performance before any practice) and is the sum of the *c* and *h* parameters in Equation 1. The asymptotic level is represented by the *h* parameter; the learning rate, by the *g* parameter.

Traditional Transfer of Training Calculations

The second purpose of this experiment was to compare estimates of transfer obtained from curve fitting to estimates obtained using traditional transfer-of-training measures. To perform a traditional transfer of training calculation, a performance criterion must be established. Because this experiment was designed for curve fitting analysis, an arbitrary criterion had to be chosen post hoc to allow a comparison of

curve-fitting and traditional measures of transfer of training. The author decided to select a value that most of the subjects reached. A beta value less than 0.4 ms/degree was unrealistic because the estimated asymptotes for beta (see Table 1) were all greater than or equal to 0.4 ms/degree. The smallest possible slope that allowed at least 50% of the subjects in each group to reach criterion was 0.7 ms/degree. Nine subjects reached criterion in Group 1; eight, in Group 2; and seven, in Group 3.

Transfer was calculated using the equation

$$\text{Percentage of transfer} = \frac{C - E}{C + E} \qquad (2)$$

where:

C is the average time to criterion for the control group

E is the average time to criterion for the experimental group (Murdock, 1957)

The percentage transfer was 4.2% for Group 1 and 11.0% for Group 2 (Equation 2 provides a conservative estimate of transfer).

Using the same logic described above, 530 ms was selected as the criterion for the alpha values. This allowed all of the subjects in Group 1 and nine of the subjects in Groups 2 and 3 to reach criterion. The percentage transfer was 13.0% for Group 1 and 3.7% for Group 2.

To examine transfer using the curve fitting technique, the beginning level and the asymp-

TABLE 1: The Learning Rate, Beginning Level, and Asymptotic Level of the Alpha and Beta Curves for Groups 1, 2, and 3 During Transfer

	Parameter		
	Learning	Beginning	Asymptotic
Group 1			
Beta	−1.19	2.0	.5
Alpha	−.62	538	470
Group 2			
Beta	−1.81	2.0	.4
Alpha	−.70	663	490
Group 3			
Beta	−1.02	1.9	.4
Alpha	−.62	736	510

totic value are expressed as percentages of the corresponding values of the control group. The beginning alpha value for Group 1 was 73% that of Group 3. The corresponding value for Group 2 was 90%. The beginning beta for both Groups 1 and 2 was 105% of that of Group 3. The asymptotic alpha values for Groups 1 and 2 were 92% and 96% that of Group 3, respectively. The corresponding beta values were 125% and 100%. To compare the learning rates in a manner consistent with the above calculations (large numbers represent good rather than poor performance), the Group 3 learning rates were divided by the rates of the two experimental groups. The learning rate for the alpha values of Group 1 was 100% that of Group 3. The corresponding value for Group 2 was 89%. The learning rates for the beta values of Groups 1 and 2 were 86% and 56% that of Group 3, respectively.

DISCUSSION

The two methods of calculating transfer of training, the traditional method and the curve-fitting technique, provide very different information on the nature of transfer. The traditional method indicates small, positive transfer for both alpha and beta as a result of prior training with alphabet letters. The investigator, however, obtains no insight concerning the source of the transfer.

In contrast, the curve-fitting technique provides a detailed look at both the amount of transfer that has occured and the source of this transfer. The data from this experiment show high positive transfer for the beginning alpha values, not for the betas. The most plausible explanation for this result is that response selection and execution were aided by previous training although the rotational skills were not. The results concerning the learning rates indicate, however, that some aspect of the rotational skills did transfer; the learning rates for beta were much faster for Groups 1 and 2 than for Group 3. It appears then that subjects in Groups 1 and 2 had to learn to rotate abstract figures, but that they learned this skill more quickly than subjects in the control group. Finally, the curve-fitting technique indicates that the transfer from alphabet letters to the abstract figure was transitory; estimates of asymptotic performance reveal few differences between the groups.

ACKNOWLEDGEMENT

This research was supported under contract No. N00014-86-K-0119 from the Naval Medical Research and Development Command. Dr. Guy Banta was the contract monitor.

REFERENCES

Cooper, L., and Shepard, R. (1973). Chronometric studies of the rotation of mental images. In W. Chase (Ed.), *Visual information processing*. New York: Academic Press.

Dixon, W. (1985). *BMDP statistical software*. Berkeley, CA: University of California Press.

Draper, N. and Smith, H. (1966). *Applied regression analysis*. New York: Wiley.

Lewis, D. (1960). *Quantitative methods in psychology*. New York: McGraw-Hill.

Murdock, B. (1957). Transfer designs and formulas. *Psychological Bulletin, 54*, 313–324.

Spears, W. (1985). Measurement of learning and transfer through curve fitting. *Human Factors, 27*, 251–266.

Vanderplas, J., and Garvin, E. (1959). The association value of random shapes. *Journal of Experimental Psychology, 57*, 147–154.

An Informational Perspective on Skill Transfer in Human-Machine Systems

GAVAN LINTERN,[1] *University of Illinois, Savoy, Illinois*

Differentiation of perceptual invariants is proposed as a theoretical approach to explain skill transfer for control at the human-machine interface. I propose that sensitivity to perceptual invariants is enhanced during learning and that this sensitivity forms the basis for transfer of skill from one task to another. The hypothesis implies that detection and discrimination of critical features, patterns, and dimensions of difference are important for learning and for transfer. This account goes beyond other similarity conceptions of transfer. To the extent that those conceptions are specific, they cannot account for effects in which performance is better following training on tasks that are less rather than more similar to the criterion task. In essence, this is a theory about the central role of low-dimensional informational patterns for control of behavior within a high-dimensional environment, and about the adjustment of an actor's sensitivity to changes in those low-dimensional patterns.

INTRODUCTION

An actor will often find a new task easier as a result of prior experience with a different task. This effect, which is referred to as *skill transfer*, suggests that specific skills or capabilities acquired by practice with one task can be employed in the performance of another task. A task may be understood as a problem to be solved or a goal to be achieved. In the research to be reviewed the task will be accomplished via the actor's manipulation of a mechanical or electronic device. The manner in which the task is to be accomplished will be determined by the dynamics of the controlled device and the relationships at the interfaces among actor, device, and environment. Thus the characteristics of the controlled device, the means of activating the system, and knowledge about the nature of the task may all affect transfer.

If two tasks are highly related, there will be high transfer; if they are essentially unrelated, there will be no transfer. In some circumstances transfer can be negative—that is, performance on the transfer task will be poorer than if there had been no pretraining at all. This occurs when the tasks are related but differ radically in some critical aspect. In other circumstances prior experience with a special training task can result in better performance than would equivalent prior training with the transfer task itself. Thus transfer can be enhanced by the use in training of carefully planned distortions of the criterion task.

The goal of this paper is to develop an ac-

[1] Requests for reprints should be sent to Gavan Lintern, University of Illinois, Institute of Aviation, Q5, Aviation Research Lab, #1 Airport Road, Savoy, IL 61874.

count that will form the basis for understanding skill transfer at the human-machine interface. That goal is pursued primarily by a review and analysis of research from the task domain of manual control (i.e., tracking or vehicular control), but observations are offered on how the insights derived from this analysis could be applied to a wider range of tasks. Established approaches to skill transfer are first reviewed. These do not provide a satisfactory account of existing data or useful guidance for researchers and training specialists. A contrasting informational perspective is developed around the concept of a perceptual invariant as introduced by J. J. Gibson (1979). This perspective is then applied to a set of transfer data from the manual control literature. Ideas drawn from the perceptual differentiation theory of E. J. Gibson (1969) are used to develop a description of a learning process that is consistent with the informational perspective.

SIMILARITY AND TRANSFER

It is self-evident that skill transfer is based on some type of similarity between training experience and operational experience. That idea has been prevalent at least since Thorndike (1903) formulated his law of identical elements. The basic notion is that transfer between two tasks will occur to the extent that they share common components, even though they may differ in many other respects. In applied training environments some notion such as similarity to the operational task or quantity of fidelity is often invoked to justify trainer design features. This approach has, for example, dominated discussion of motion and visual systems for flight simulators (Comstock, 1984; Cyrus, 1978; Needham, Edwards, and Prather, 1980; Spooner, Chambers, and Stevenson, 1980).

A contemporary cognitive-based justification of the high-fidelity approach may be found in the instance memory theory of skilled behavior put forth by Logan (1988a, 1988b). In that theory each experience with an event establishes an episodic memory trace. There is stochastic variability for recall speed of individual traces, so that as the number of stored traces for a specific behavioral activity increases, performance becomes more skilled because there is an increase in the likelihood of quickly retrieving an appropriate trace. One strong implication of instance memory theory is that training situations should be as similar as possible to those encountered in the field (Logan, 1988a, p. 590). The emphasis would be on fidelity in both training programs and training devices.

Nevertheless, this account does not acknowledge that some dimensions of fidelity do not contribute to transfer effectiveness. Few would argue, for example, that transfer is enhanced if the color of an aircraft and its companion simulator are identical. A cognitive theory might specify that the attention-demanding aspects of the task are the ones that need to be faithfully represented in the training environment (G. D. Logan, personal communication, June 1990). In applied training circles, appeals to psychological fidelity (Goldstein, 1986) or functional equivalence (Baudhuin, 1987) imply recognition of the fact that some identities are irrelevant to transfer. Although these notions may constitute some progress in relation to a naive view of similarity, they do not offer a specific statement about the similarities that are important.

From another perspective, similarity and fidelity theories fail because some data show better transfer if the training task differs on specific dimensions from the criterion or transfer task. For example, the addition of visual augmented feedback can speed acquisition of landing skill in an aircraft simulator (Lintern, 1980), and transfer to crosswind landings is actually better if no crosswind is used in training (Lintern, Roscoe, and Sivier,

1990). Wightman and Sistrunk (1987), in an investigation of part-task training for teaching carrier landings, have shown that a backward-chaining procedure can enhance transfer to the whole task. In these experiments deliberate departures from similarity actually enhanced transfer to the criterion task.

More generally, the exploration of special instructional strategies is based on the assumption that transfer will not be degraded and may be enhanced by planned departures from similarity (Lintern, 1989; Wightman and Lintern, 1985). Thus similarity, as it is normally viewed, is not a sufficient element of a conceptual approach to skill transfer. The challenge remains to formalize the notion of similarity in some way that will account for skill transfer effects and that will permit useful exploitation of that notion in applied training programs. The success of this endeavor depends on distinguishing identities that are critical to transfer from those that are irrelevant.

One approach to distinguishing dimensions of similarity can be found in the response surface proposed by Osgood (1949). Although that surface was modeled entirely on the basis of verbal learning research, Holding (1976) reviewed it for its applicability to action skills. Holding's model indicates that maximum stimulus and response similarity should yield maximum positive transfer but that variation in stimuli will not always degrade transfer. However, the distinction between stimulus and response is not easily sustained in manual control or in many other action skills in which responses generate important perceptual information.

Gick and Holyoak (1987) have developed a theory of cognitive transfer for verbal skills in which rules constitute the elements that are transferred. Those rules are generally explicit, which cannot account for manual control, but there are implicit rules as well. Presumably the abstract rules central to

Schmidt's (1975) motor skill theory are also implicit. In that theory the rules are incorporated into generalized motor programs that guide behavior. Thus similarity for transfer could be embodied in common motor programs or implicit rules. Motor program or rule-based views might explain data that show that some identities are irrelevant to transfer (and also explain those data that show better transfer from less similar tasks) in terms of how well the rules or programs are established by the training conditions (Schmidt and Young, 1987). However, the type of task-related information that leads to development of a rule or motor program is not specified. Learning principles that could account for the transfer effects to be found in the domain of manual control also remain unspecified. It is the nature of that information and of the learning principles that is the focus of this paper.

INFORMATIONAL INVARIANTS

The transfer perspective developed here draws on the ecological theory of perception as developed by J. J. Gibson (1979). I argue that manual control skills are supported by information derived from critical relationships in the task environment. Such relationships have specific values that remain invariant while the actor is performing correctly but which have different values when he or she is performing poorly. With aircraft landings, for example, one or more relationships will be invariant if the pilot remains on the designated approach glide slope. Those relationships will change, however, if the pilot deviates from the designated glide slope, and the pilot's ability to recognize that deviation and to correct for it depends on his or her sensitivity to changes in the relationships that specify correct or accurate control.

Gibson (1979) characterizes such relationships as *invariants*. Within a changing sea of information, the individual character of an event will be perceived because of some specifiable property or relational value. That property or relation will remain constant across events that are perceived as similar but will differ between events that are perceived as different. Thus an invariant is a property of an event that remains unchanged as other properties change: that which specifies the persistent character of the event (Gibson, 1979). Following Stoffregen and Riccio (1988), it may be viewed as a lawful relationship between patterns of stimulation and properties of the task.

A complete description of any natural state or event would require reference to enormous, even infinite, detail. One useful strategy for a scientific endeavor is to reduce the infinite detail in the domain of inquiry to a succinct and meaningful description—that is, to offer a low-dimensional description of nature (Feynman, 1967). Any characterization by way of invariants is a low-dimensional description of a high-dimensional event. Because invariants distinguish differences and link identities (Cutting, 1986), they may provide an appropriate low-dimensional basis for a theory of transfer.

Reference to an intriguing demonstration by Johansson (1973) will illustrate the power of invariants. His observers viewed a set of lights that were located on the main joints of an otherwise invisible human's limbs. While the human was stationary the lights appeared as a random assemblage, but the nature of the activity and the fact that it was generated by a human could be perceived if that human engaged in activities such as walking or hopping. At the cessation of activity the lights again appeared as a random assemblage. The manner in which the lights changed relative to one another in time and space specified the activity and the humanness of the actor. Within the experimental literature this effect is referred to as *biological motion*.

Invariants for Perception

Not all invariants are functional. It is often possible to identify relationships that meet the formal requirements but which have no effect on behavior, perhaps because the changes that correspond to changes in an event are below threshold or because a different invariant is selected by the observer (Cutting, 1986). One important challenge for perceptual research is to identify functional invariants (those that support perceptual judgments).

Cutting (1986) has argued that some perceptual judgments can be supported by cross ratios formed from the spatial separation of elements within objects. For example, a ratio formed from the distances among elements on a rotating surface will be invariant if the surface is rigid but not otherwise. This cross ratio might be used for judgments of rigidity, and Cutting's data verify that hypothesis. Invariants for judgment of "time to contact"— that is, the time that will elapse before an approaching object will reach an observer (Lee, 1976; Lee, Young, Reddish, Lough, and Clayton, 1983)—and those that support judgments pertaining to the ballistic trajectory of a ball (Todd, 1981) have also been specified analytically and shown to affect perceptual judgments.

Even the seemingly complex phenomenon of biological motion might succumb to analysis by way of invariants. Cutting, Profitt, and Kozlowski (1978) have argued that the perception of biological motion is based on information about centers of moment. All rigid objects that move in contact with the ground have at least one center of rotation about which the motions of other elements on the rigid object move in arcs. For complex, multisegmented objects there will be multiple centers of moment. During motion, rotations around the centers of moment are symmetric and periodic. Centers of moment for the human body can be calculated on the basis of sizes of body parts. Cutting et al. (1978) showed how centers of moment can account for many puzzling biological motion effects, such as the ability to detect walking motion merely from ankle movements and the ability to distinguish male from female walkers. Centers of moment appear to offer a useful basis for a low-dimensional description of complex, high-dimensional events.

THE CONTROL OF ACTION

For J. J. Gibson's ecological psychology, invariants provide the information for action; in other words, the control of action is informed by perceptual invariants (Gibson, 1958). The term *visual kinesthesis* refers to the fact that moving observers coordinate their behavior with visual information from the environment. That information will be specific to the layout or the structural properties of the environment and also to relative motions or kinematic properties generated by an observer's own actions. The emphasis in this paper is on manual control—that is, the operation of dynamic, interactive control systems such as vehicles, vehicle simulators, and laboratory tracking systems. As a subcategory of human action it should be amenable to the type of analysis recommended by Gibson.

Invariants for Manual Control

For invariants to be of use in a theory of human action, it must be possible to isolate them analytically and to test their effects on behavior. An invariant to judge the speed of self-motion might be derived from the rate of optic flow. Where speed and distance to visible surfaces remain constant, the relative rate at which visible elements in a scene pass by a moving observer will generally be constant. The perceived invariance of flow can, however, be disrupted by inhomogeneity of element distribution. Denton (1980) evaluated

the influence of this latter factor in a driving simulator with a visual display. Pattern distortions in which elements became more compressed throughout a trial caused subjects to reduce their simulated speed despite instructions to keep it constant. This experiment indicates that an invariant relative rate of visual flow can be used to maintain velocity and that distortions of the invariant will have predictable effects.

Mertens (1981) has examined the information available for aircraft landings. He identified the form ratio (the ratio of projected runway length to projected breath of the distant end of the runway) as an invariant that might support landings. Distortions of this ratio within a simulator were shown to bias the glide slope judgments of experienced pilots in the expected direction. Although we are as yet a considerable distance from accounting for the full range of human behavior by analysis of the relevant invariants, the studies of Denton (1980) and of Mertens (1981) indicate that perceptual invariants can be isolated analytically and that their effects on behavior can be demonstrated.

An Invariant for Transfer

A contrast between experiments by Gordon (1959) and Briggs and Rockway (1966), in which transfer was tested from pursuit to compensatory and from compensatory to pursuit displays, suggests that an invariant for the transfer of manual control skills may be found within a regular or predictable forcing function. A pursuit display is one in which a cursor, under the control of the actor, is used to track a target that is disturbed by a forcing function. A compensatory display has only one moving cursor, and the actor's task is to maintain it at a fixed target position while it is being disturbed by a forcing function. In these two experiments subjects were trained with either a pursuit or a compensatory task, and in transfer they either continued with their training task or switched to the alternative version of the task.

Gordon (1959) used a simple cloverleaf pattern for the tracking course. Tracking in the pursuit mode was generally better than tracking in the compensatory mode (Figure 1). In transfer from pursuit to compensatory tracking, performance initially regressed to a level obtained from subjects who had first trained with and then continued with compensatory tracking, but it improved quickly toward a level obtained from subjects who were now tracking the pursuit task. This result suggests that pursuit training is better for compensatory tracking than is an equivalent amount of compensatory training.

Briggs and Rockway (1966) used a more complex pattern for their tracking course (the sum of 0.092-Hz, 0.153-Hz, and 0.247-Hz sine waves). Although there were clear differences between pursuit and compensatory performances in both training and transfer, there were no large differential effects on transfer to either display as a function of the training manipulation (Figure 2). In particular, those transferring to pursuit from compensatory tracking did not perform better than did those who tracked with the compensatory system throughout. Apparently Gordon's subjects could better learn skills required for

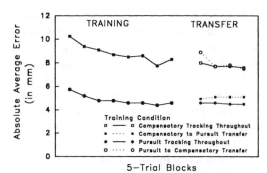

Figure 2. *Transfer between pursuit and compensatory displays with a complex course. Half of the subjects from each training group transferred to the alternative condition. (Adapted from Briggs and Rockway, 1966.)*

compensatory tracking from practice on a pursuit task than from practice on a compensatory task, but Briggs and Rockway's could not.

Given the differences between experimental apparatus and procedures in these two experiments, any statement about the differences in results must remain speculative. Nevertheless, within the terms of the theory outlined here, these contrasting results may be understood by considering the difference between pursuit and compensatory tracking and the differences in course complexity of the different tracking systems. Compensatory displays almost always present a more difficult task than do pursuit displays, in large part because effects of the disturbance or course cannot be distinguished visually from control effects. A compensatory display will increase the difficulty an actor has in visually differentiating the effects on system behavior of control movements and a disturbance. Thus an actor is unlikely to acquire sensitivity to repeating regularities in the forcing function as quickly with a compensatory display as with a pursuit display. If, however, sensitivity to these regularities could be acquired during pursuit tracking, subsequent performance on the compensatory version of the task should be enhanced.

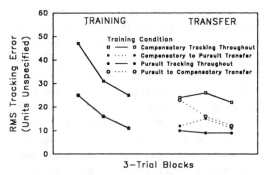

Figure 1. *Transfer between pursuit and compensatory displays with a simple course. Half of the subjects from each training group transferred to the alternative condition. (Adapted from Gordon, 1959.)*

Considering the relatively small amounts of practice allowed in each of these experiments, practice with the pursuit version of the task could have established awareness of the regularities in the relatively simple cloverleaf pattern used by Gordon but not of the more complex sum-of-sines pattern used by Briggs and Rockway. This contrast between the two experiments implicates the pattern of the course as an important factor in transfer. In terms of invariants it is a low-dimensional relationship within the pattern that is critical.

The view of an invariant as a low-dimensional description of a high-dimensional event is important. Any discussion of an invariant as a high-dimensional description of a high-dimensional event would render the concept superfluous and would constitute a high-fidelity theory of transfer. The episodic memory trace of instance theory (Logan, 1988a), being a copy of an event, is a high-dimensional representation of that event. Instance theory is not consistent with the view that low-dimensional invariants support transfer. One relatively straightforward interpretation of Gordon's data, which is consistent with Logan's (1988a) theory, is that subjects had learned all details of the pattern; that is, they had learned to exploit a high-dimensional description of the critical information. In contrast, perceptual invariants such as cross ratios (Cutting, 1986), form ratios (Mertens, 1981), and centers of moment (Cutting et al., 1978) are of a low-dimensional form. To be consistent with the concept of invariants, it must be possible to develop a low-dimensional description of a forcing function which specifies the relationships essential for transfer.

Schmidt and Young (1987) argued that it is the timing relationships in a patterned response that remain invariant. Thus a low-dimensional description of a pattern to be followed in a tracking task is more likely to be based in the temporal relationships between adjustments in control actions than in the amplitudes or total duration of those actions. The results of two pursuit rotor experiments (Lordahl and Archer, 1958; Namikas and Archer, 1960) bear on this issue. Subjects were transferred from high or low speeds to moderate speeds of rotation or from high or low amplitudes to moderate amplitudes of rotation. A penalty was incurred in transfer to a new speed of rotation but not to a new amplitude of rotation. Because speed and amplitude of rotation represent frequency and amplitude of the forcing function, these results suggest that the functional invariant is related to frequency but not to amplitude. These data suggest that a low-dimensional description of the similarity underlying transfer might be based in the timing relationships of the course, and that a fully detailed description is not required for transfer.

Environmental Information

One implication of the hypothesis outlined for this paper is that natural environments contain structural information to guide behavior. In a test of this notion, Lintern et al. (1990) used a light aircraft simulator to teach flight-naive subjects to land. Training was conducted either with a pictorial representation of a normal airport scene or with a symbolic display that guided subjects through the required maneuver but offered no realistic representation of any natural or cultural features. The two groups performed equally well in training, but those trained on the symbolic display performed poorly in transfer (Figure 3). Although the specific nature of the information that guided behavior was not isolated in this experiment, low-dimensional relationships such as form ratio (Mertens, 1981), projected aimpoint-to-horizon distance (i.e., the distance between the runway aimpoint and the horizon projected onto a

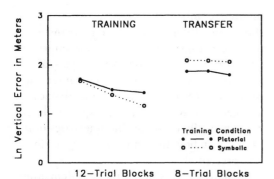

Figure 3. *Transfer of aircraft landing skills from a symbolic or pictorial display. In transfer, subjects were required to execute an approach, roundout, flare, and touchdown to a display identical to that used for the pictorial training group. (Adapted from Lintern et al., 1990.)*

plane at some arbitrary distance in front of the pilot and perpendicular to the line of sight; Langewiesche, 1944), or horizon-aimpoint angle (Lintern and Liu, in press) to be found in a pictorial scene might perform that function.

LEARNING AS DIFFERENTIATION

In the discussion so far I have advanced claims about the nature of what is learned during acquisition of an action skill but have referred only tangentially to the learning process. For example, I argued that Gordon's (1959) subjects learned invariant relationships within the forcing function more effectively in the pursuit tracking mode because the pattern of the course was more perceptible. It will now be useful to consider the nature of learning in more depth. A view of learning consistent with the informational perspective is that of perceptual differentiation, a process by which information becomes more discriminable (E. J. Gibson, 1969). Being a process whereby what was once perceived as the same is now perceived as different, it is analogous to cellular biology, where *differentiation* refers to the process by which

formerly identical cells acquire unique characteristics.

Perceptual thresholds are central to any theory of perception based on invariants because without reference to thresholds, there is no reference to the capabilities of the actor (Cutting, 1986). An actor must be able to perceive changes in perceptual information that is to be held invariant during execution of a task, and the concept of threshold is employed to account for the fact that the perceptual resolution of such changes is finite. The perceptual differentiation hypothesis implies that an actor's difference thresholds for recognition of changes or patterns in the workspace can be lowered through experience. There is a progressive development of the actor's sensitivity to constancies within tasks and differences between them. Learning moves through a course of progressively finer discriminations of important information, and regularities become informative via processes leading to differentiation along dimensions of variation. Previously vague impressions become increasingly specific as they relate to task requirements. From this perspective skilled behavior must exploit perceptual information, and skill develops via a process of becoming sensitive to it.

In her discussion of differentiation, E. J. Gibson (1969) emphasized objects and events in the environment external to the observer. At the human-machine interface there is a wider range of concerns. Here learning will involve the increasing differentiation of previously confusable information, which may reside in any element of the task environment, even in response production. The difference between skilled and unskilled behavior can therefore be at least partially characterized as differential sensitivity to task-related information.

Gibson's (1969) review suggests that certain instructional techniques speed differentiation, particularly those that draw attention to the distinguishing perceptual invariants. Techniques that contrast different values of distinguishing properties, abstract them, or accentuate them should be useful. In some cases it may help to offer advice about what to look for (Biederman and Shiffrar, 1987). Nevertheless, even highly skilled actors are often unaware of the information they use to support their own activities, and it is unlikely that much of the information for skilled activity is sufficiently explicit for that type of instruction. Implicit information might be learned more readily via special instructional techniques that enhance or clarify it. Additionally, anything that conceals or diverts attention from critical information will impede the differentiation process.

DIFFERENTIATION FOR MANUAL CONTROL

Established theories of transfer cannot account for those effects that show enhanced transfer from less similar training. Asymmetric transfer of this type has already been encountered in the study by Gordon (1959). As is consistent with the differentiation view, the pursuit display appears to have clarified the pattern of the course to be followed. That clearer presentation enhanced learning of critical invariants, which then resulted in superior control. The experiments reviewed in the following sections demonstrate similar asymmetric effects.

Augmented Feedback

A number of manual control experiments have examined the use of supplementary visual or auditory information during training. The supplementary information is used to provide on-line feedback to subjects during their performance of the task. The data from these experiments do not always favor the use of augmented feedback. Several experiments have shown that it can speed acquisition of the task and enhance transfer to nonaug-

mented conditions, whereas others have shown no differential effect in transfer to nonaugmented conditions, or sometimes even a decrement. From an analysis of approximately 30 tracking studies, Lintern and Roscoe (1980) argued that the effectiveness of augmented feedback training appears to depend critically on some characteristics of the task and on how the supplementary feedback was presented.

In many early studies the supplementary feedback was presented when the subjects were tracking correctly. The data tended to show a strong performance advantage during training but no transfer advantage. Occasionally the training advantage reversed in transfer so that the loss of supplementary cues appeared to set subjects back further than if they had practiced with a nonaugmented condition. Lintern and Roscoe (1980) argued that subjects who became dependent on augmenting information would not perform well in its absence. In terms of the differentiation hypothesis, augmenting information may distract attention from some task-related invariants that offer a significant learning challenge.

Later studies presented the supplementary feedback when subjects were tracking incorrectly, and it has usually been this manipulation that has shown transfer benefits from training with augmented feedback. The contrast between augmented feedback conditions that show positive differential transfer and those that do not suggests some insights about differentiation of perceptual invariants in manual skill. In particular, it seems that an off-target augmentation would not distract attention from critical invariants that specify correct performance or that would support accurate tracking. In contrast, on-target information is likely to distract attention from at least some of the critical invariants. Even when enhanced transfer is found following training with on-target augmentation, the enhancement is usually not as strong as that obtained with off-target augmentation (e.g., Williams and Briggs, 1962).

Two studies, one by Kinkade (1963) and the other by Williams and Briggs (1962), are particularly informative for the differentiation hypothesis. Kinkade (1963) examined the effects of visual noise (random, high-frequency, two-dimensional movements of the cursor) and augmented feedback (supplied by auditory clicks when the subjects were tracking correctly) on the acquisition of skill with a compensatory tracking task. A relatively simple forcing function (the sum of 0.1-Hz and 0.2-Hz sine waves) was used. When present, the noise was superimposed on the forcing function. Its maximum amplitude was approximately one-third the maximum amplitude of the forcing function. The visual noise conditions of training (either some or none) were continued without change in transfer. For no visual noise, augmented feedback training helped performance in training and in transfer to no augmentation (Figure 4). Thus it appears that augmented feedback can serve to clarify invariants of the forcing function except when the regularities in that forcing function are obscured by a confounding influence such as noise.

Williams and Briggs (1962) used the same apparatus (and forcing function) and no-noise condition used by Kinkade. In addition to the control and on-target augmented feedback conditions tested by Kinkade, they examined off-target augmented feedback. Training with augmented feedback produced a transfer benefit to nonaugmented conditions, but the benefit from the off-target schedule was greater than from the on-target schedule (Figure 5). The implication of this experiment is that although augmented feedback may enhance differentiation of course-related invariants by encouraging subjects to reproduce the correct tracking pattern, there is also some potential for it to mask or distract attention from those invariants. In this case those subjects trained in the on-target condition may have come to depend at least partly on the augmenting information to inform them when they were tracking correctly.

More generally, it seems important for skill acquisition that subjects attend to the natural task invariants when they are tracking correctly in order to promote sensitivity to the range of invariant information that supports correct performance. Thus augmented feedback provided when the subject is tracking correctly may divert attention to artificial

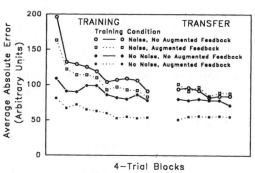

Figure 4. *Effects of augmented feedback on transfer of a continuous tracking task. Visual noise was included as a nontransfer variable. (Adapted from Kinkade, 1963.)*

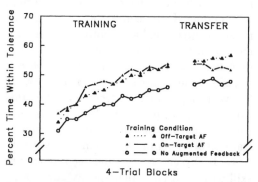

Figure 5. *Effects of augmented feedback on transfer of a continuous tracking task. All subjects transferred to a nonaugmented condition. (Adapted from Williams and Briggs, 1962.)*

information. In contrast, off-target augmented feedback can serve to guide the subject back to correct performance and thereby increase opportunities for attending to the information that specifies correct performance, but it should not divert attention from the information that specifies accurate control.

Nevertheless, on-target augmentation is not always detrimental to transfer (Kinkade, 1963; Williams and Briggs, 1962). Lintern (1980) postulated that the tendency for on-target augmenting information to mask or distract attention from those invariants that specify the actual state of the controlled cursor (or vehicle) will be accentuated to the extent that the normal information of this type is obscure or less compelling. Thus the advantage of off-target augmenting feedback over on-target (and continuous) augmenting feedback would be enhanced when the task-related invariants are obscure because the tendency for the supplementary information to attract attention would be stronger.

In a test of this notion, subjects were taught to land a light aircraft simulator. The difficulties with landing experienced by a beginning student are primarily attributable to the obscurity of the information that can be used to guide the aircraft along the correct approach path (Langewiesche, 1944). Training was conducted with continuous, off-target, or no augmented feedback. Both types of augmentation helped training performance, but only the advantage accruing from the off-target schedule carried over to the transfer test with no augmented feedback. Thus this experiment supports the hypothesis that some forms of augmentation will impede learning of perceptual invariants that are obscure. It also emphasizes the need to acquire skill with invariants that specify current device state in relation to its desired state. One such invariant is the horizon-aimpoint angle, which has been shown to affect glide slope

performance of experienced pilots (Lintern and Liu, in press).

Clarification of System Response

Two other flight training experiments are also relevant to the argument developed here. In the first, Lintern, Thomley-Yates, Nelson, and Roscoe (1987) taught military pilots a visually supported bombing skill in a flight simulator with either a relatively detailed pictorial scene or a grid pattern. Those trained on the pictorial scene performed better on subsequent transfer to the grid pattern than did those trained with the grid pattern itself. In the second experiment Lintern et al. (1990) used a light aircraft simulator to teach crosswind landings to flight-naive subjects. Some subjects were trained with no crosswind and others were trained with 5 knots of crosswind. On transfer to a 5-knot crosswind condition, those subjects trained without crosswind performed better than did those subjects trained with it.

From the informational perspective it can be argued that the better visual information provided by Lintern et al. (1987) in the more detailed scene clarified the relationships between control actions and system response. In the experiment of Lintern et al. (1990), the presence of crosswind during training would confound perception of the relationships between control actions and system response, thereby slowing learning of critical invariants related to system control. Thus data from both experiments suggest that clarification of system response will lead to more effective learning of kinematic invariants associated with dynamic control.

DISCUSSION

The transfer observed following training on a task that differs in some specific respects from the criterion task is often better than transfer observed following equivalent training on the criterion task itself. The informational account offered here is, to my knowledge, the first synthesis of such transfer effects. From an identical elements perspective (Thorndike, 1903), an instance memory perspective (Logan, 1988a), or a response versus stimulus similarity perspective (Holding, 1976), those transfer effects must be viewed as atypical phenomena. In contrast, the appeal to informational invariants forms a basis for a detailed characterization of the task and, when combined with the differentiation hypothesis, one that provides a theoretical account within which these puzzling transfer data become explicable.

The development of a coherent account of skill transfer has proved to be a struggle throughout the past century of psychological and educational research. Commencing with the notion of formal discipline (Woodrow, 1927), a variety of views have emerged, none of which has been able to provide an account that offers a clear research agenda or comprehensive training principles. Major training research programs are sometimes planned under the explicit assumption that contemporary theory should be ignored (e.g., Donchin, 1989). At other times research programs are built around such limited tasks that their relevance to common human activity is questionable (e.g., Logan, 1988b). Reviews of transfer research tend to emphasize behavioral outcomes and also to lament obvious deficiencies but offer little in the way of solid theoretical integration (e.g., Baldwin and Ford, 1988; Briggs, 1969; Lintern and Gopher, 1978; Wightman and Lintern, 1985). The account put forth here constitutes an attempt to resolve these problems. Although the emphasis has been on transfer in manual control, J. J. Gibson's ecological theory is posed as an account of normal human activity. Thus the informational perspective should be relevant to issues of skill transfer with all types of human-machine systems that are of concern in the field of human factors.

Invariants as a Basis for Transfer

The central claim put forth in this paper is that transfer can occur only when critical similarities are maintained across the training and transfer tasks. I have argued that informational invariants constitute properties that define critical similarities and that they are essential components of all tasks that can be learned. If critical invariants (specifically, those that pose a meaningful learning challenge) remain unchanged, transfer will be high even when many other features of the environment, context, or task are changed. The learning aspect of this theory was described in terms of differentiation, a process whereby perceptual thresholds are modified. If an operator's perceptual sensitivity to critical invariants can be improved, that enhanced sensitivity will serve to facilitate transfer. A basic assumption of this discussion is that initial contact with any unnatural control environment will almost invariably require the development of sensitivity to new invariants.

According to E. J. Gibson (1969), procedures that accentuate invariants to be learned will enhance differentiation. Thus clarification or enhancement of invariants that offer a significant learning challenge— such as those associated with a pattern within a forcing function, complex control-display relationships, or relationships between informational properties in the environment and the desired state of the controlled system—will speed skill acquisition. The concealment or distortion during training of those critical invariants will actually impede learning. On the other hand, emphasis on invariants that are already well learned, easily learned, or nonfunctional will not affect transfer.

Transfer Theory

There are points of contact with other theories, but one implication of the informational account outlined here is that other attempts to account for transfer of skill through an appeal to some conception of similarity are either incomplete or inadequate. This account shares with Logan's (1988b) instance memory theory a concern with information from an actor's environment but differs in its emphasis on low-dimensional properties of that environment. It also has something of the character of the response surfaces developed by Osgood (1949) and Holding (1976) but offers a different conceptualization of the task features that must be considered.

In that it could be viewed as an account of the information that is internalized as a motor program or a rule, this informational theory might be seen to complement internal process theories, such as the rule-based approach of Gick and Holyoak (1987) or the motor-program approach of Schmidt and Young (1987). That view is not, however, consistent with the ecological program of J. J. Gibson (1979). From the ecological perspective, an appeal to rules or motor programs adds nothing in the way of explanatory power and constitutes little more than an alternative description of observable properties in behavior or in the environment. Such an appeal does not offer a more parsimonious, more general, or more fundamental description, nor does it offer an explanation (Gibson, 1973, 1976).

Nevertheless, some form of change or reorganization internal to the actor is accomplished during learning. A clear understanding of that change or reorganization would, in all likelihood, contribute considerably to our understanding of skill transfer. There is, however, no consensus among behavioral scientists about how to characterize that change or even about the scientific strategy that is most likely to reveal it. The most common strategy—to postulate a hypothetical construct that appears to account for important data trends—raises a serious difficulty for a theory of transfer. Although implications for transfer can be drawn from a hypothetical construct, failure to find the anticipated effects can be accommodated by adjustment of the theoretical formulation. As is evident within the general field of cognitive psychology, endless variations on hypothetical constructs can be forwarded to account for diverse effects.

In the current context the ecological critique amounts to a claim that an internal process, if rigorously specified, will be defined in terms of objective informational properties to be found in the actor-task environment. In that case the postulated internal process adds nothing useful to the informational account. A less rigorous specification of internal process will rely heavily on hypothetical constructs that cannot be evaluated directly and that provide nothing in the way of compelling predictions. Circularity becomes a significant problem in that a transfer construct (e.g., functional equivalence, motor program) can be specified only in terms of behavioral data that the construct is presumed to explain.

In contrast, information is derived from real properties in the task environment. Those properties can be measured objectively and can be distorted, enhanced, or removed. An appropriate task analysis should reveal the relevant informational invariants and provide specific predictions relating to the effects of adjusting that information. J. J. Gibson's ecological program places a heavy burden on the scientist to identify the informational properties that affect behavior. For transfer theory, specification of informational invariants that support transfer is a central requirement. From the ecological perspective, success in that regard will provide considerably more insight into the complexities of skill transfer than will any amount of theorizing about internal process or structure.

Transfer Research

There has been considerable confusion about transfer because experiments that should be able to demonstrate it often do not. There is generally only a vague notion of what can be transferred and what might promote that transfer. The informational perspective suggests a strategy that should correct this unsatisfactory situation. In terms of this perspective it should be possible to specify by analysis the information that is expected to support transfer. It should then be possible to demonstrate predictable performance effects of distorting or concealing that information. It should also be possible to demonstrate predictable effects on transfer of specialized pretraining with that information in isolation or of pretraining on the whole task with that information distorted or concealed.

Specifically, for a task such as landing a light aircraft, it should be possible to identify structural relationships in the environment which specify whether or not the pilot is maintaining correct lineup and glide slope. The horizon-aimpoint angle is one such invariant property that is used for glide slope control (Lintern and Liu, in press). It should also be possible to identify dynamic relationships within the flight control system which specify whether or not the pilot has stable control over the aircraft. Speed of system response to control inputs is likely to incorporate one invariant relationship critical for transfer (Lintern and Garrison, in press). Beginning flight students may already be sensitive to some important invariants, but if the task poses a significant learning challenge, at least some invariants will not be perceived

well, and sensitivity to them will have to be enhanced through instruction and practice. It is those invariants that must be identified and assessed in terms of their effect on transfer. More generally, the challenge for transfer research is to identify informational invariants by analysis, to demonstrate their effect on behavior, and to assess their effects on transfer.

Issues for Applied Training

It should be apparent from consideration of the data reviewed in this paper that the development of a training system requires a detailed analysis of the tasks to be learned. As a first step, invariant relationships that support transfer must be identified and faithfully reproduced in a training system. In the design of a flight simulator, for example, faithful representation of the horizon-aimpoint angle may be critical (e.g., Lintern and Liu, in press). It is not uncommon, however—especially when inexpensive computers are used for the generation of a visual display—to model a relatively small world. That results in a distorted horizon-aimpoint angle that does not remain invariant for a constant angle of a landing approach and which, by the perspective offered here, is likely to compromise transfer effectiveness.

Considering that we know little about invariants that support behavior, nothing presented in the development of this theory should be taken as implying that identification of relevant informational invariants is straightforward. The few data that have emerged from research stimulated by J. J. Gibson's ecological program indicate that invariants will be based in abstract relationships (Lintern and Liu, in press; Mark, 1987; Warren and Whang, 1987). The results of the biological motion research suggest that invariants will be discovered in unexpected forms—in particular, ones that are not well anticipated by the bulk of the traditional research on perceptual issues. Identification of critical perceptual invariants is likely to pose a continuing challenge for designers of instructional programs and devices.

Once the relevant invariants have been identified and represented appropriately in the training system, a second step is to implement instructional strategies that will speed their learning. There is at least some indication in the research discussed here of the types of instructional principles that could be useful. From the review of E. J. Gibson (1969), clarification and accentuation seem to be important principles. It will, however, require careful evaluation to verify that these types of strategies can be extended to the diversity of complex tasks found in operational human-machine systems and to tune them for maximum effectiveness in specific circumstances. Thus considerable work remains before the implications of this theory can be readily transferred to operational training. Nevertheless, the promise is that a systematic development of the concepts presented here can, in the long term, have a substantial effect on training effectiveness.

ACKNOWLEDGMENTS

This work was supported by the Basic Research Office of the Army Research Institute under contract MDA 903-86-C-0169. Michael Drillings is the technical monitor. John M. Flach, Joseph D. Hagman, Arthur F. Kramer, William M. Mace, Neville Moray, Gary E. Riccio, Thomas A. Stoffregen, and Christopher D. Wickens reviewed earlier drafts of this paper.

REFERENCES

Baldwin, T. T., and Ford, J. K. (1988). Transfer of training: A review and directions for future research. *Personnel Psychology, 41*, 63–105.

Baudhuin, E. S. (1987). The design of industrial and flight simulators. In S. M. Cormier and J. D. Hagman (Eds.), *Transfer of learning* (pp. 217–237). San Diego, CA: Academic.

Biederman, I., and Shiffrar, M. M. (1987). Sexing day-old chicks: A case study and expert systems analysis of a difficult perceptual-learning task. *Journal of Experimental Psychology: Learning, Memory, and Cognition, 13*, 640–645.

Briggs, G. E. (1969). Transfer of training. In E. A. Bilodeau (Ed.), *Principles of skill acquisition* (pp. 205–234). New York: Academic.

Briggs, G. E., and Rockway, M. R. (1966). Learning and performance as a function of the percentage of pursuit component in a tracking display. *Journal of Experimental Psychology, 71*, 165–169.

Comstock, J. R., Jr. (1984). The effects of simulator and aircraft motion on eye scan behavior. In *Proceedings of the Human Factors Society 28th Annual Meeting* (pp. 128–132). Santa Monica, CA: Human Factors Society.

Cutting, J. E. (1986). *Perception with an eye for motion.* Cambridge: MIT Press.

Cutting, J. E., Proffitt, D. R., and Kozlowski, L. T. (1978). A biomechanical invariant for gait perception. *Journal of Experimental Psychology, 4*, 357–372.

Cyrus, M. L. (1978). *On the role of motion systems in flight simulators for flying training* (AFHRL-TR-78). Williams Air Force Base, AZ: Air Force Human Resources Laboratory.

Denton, G. G. (1980). The influence of visual pattern on perceived speed. *Perception, 9*, 393–402.

Donchin, E. (1989). The learning strategies project: Introductory remarks. *Acta Psychologica, 71*, 1–15.

Feynman, R. S. (1967). *The character of physical law.* Cambridge: MIT Press.

Gibson, E. J. (1969). *Principles of perceptual learning and development.* Englewood Cliffs, NJ: Prentice-Hall.

Gibson, J. J. (1958). Visually controlled locomotion and visual orientation in animals. *British Journal of Psychology, 49*, 182–194.

Gibson, J. J. (1973). Direct visual perception: A reply to Gyr. *Psychological Bulletin, 79*, 396–397.

Gibson, J. J. (1976). The myth of passive perception: A reply to Richards. *Philosophy and Phenomenological Research, 37*, 234–238.

Gibson, J. J. (1979). *The ecological approach to visual perception.* Boston: Houghton Mifflin.

Gick, M. L., and Holyoak, K. J. (1987). The cognitive basis of knowledge transfer. In S. M. Cormier and J. D. Hagman (Eds.), *Transfer of Learning* (pp. 9–46). San Diego, CA: Academic.

Goldstein, I. L. (1986). *Training in organizations: Needs assessment, development, and evaluation* (2nd ed.). Monterey, CA: Brooks/Cole.

Gordon, N. B. (1959). Learning a motor task under varied display conditions. *Journal of Experimental Psychology, 57*, 65–73.

Holding, D. H. (1976). An approximate transfer surface. *Journal of Motor Behavior, 8*, 1–9.

Johansson, G. (1973). Visual perception of biological motion and a model for its analysis. *Perception and Psychophysics, 14*, 201–211.

Kinkade, R. G. (1963). *A differential influence of augmented feedback on learning and on performance* (U.S. Air Force AMRL Technical Documentary Report 63-12).

Langewiesche, W. (1944). *Stick and rudder: An explanation of the art of flying.* New York: McGraw-Hill.

Lee, D. N. (1976). A theory of visual control of braking based on information about time-to-collision. *Perception, 5*, 437–459.

Lee, D. N., Young, D. S., Reddish, P. E., Lough, S., and Clayton, T. M. H. (1983). Visual timing in hitting an accelerating ball. *Quarterly Journal of Experimental Psychology, 35A*, 333–346.

Lintern, G. (1980). Transfer of landing skill after training with supplementary visual cues. *Human Factors, 22*, 81–88.

Lintern, G. (1989). The learning strategies program: Concluding remarks. *Acta Psychologica, 71*, 301–309.

Lintern, G., and Garrison, W. (in press). Transfer effects of scene content and crosswind in landing. *International Journal of Aviation Psychology.*

Lintern, G., and Gopher, D. (1978). Adaptive training of perceptual-motor skills. Issues, results, and future directions. *International Journal of Man-Machine Studies, 10,* 521–551.

Lintern, G., and Liu, Y.-T. (in press). Explicit and implicit horizons for simulated landing approaches. *Human Factors.*

Lintern, G., and Roscoe, S. N. (1980). Visual cue augmentation in contact flight simulation. In S. N. Roscoe (Ed.), *Aviation psychology* (pp. 227–238). Ames: Iowa State University Press.

Lintern, G., Roscoe, S. N., and Sivier, J. (1990). Display principles, control dynamics, and environmental factors in pilot performance and transfer of training. *Human Factors, 32,* 299–317.

Lintern, G., Thomley-Yates, K. E., Nelson, B. E., and Roscoe, S. N. (1987). Content, variety, and augmentation of simulated visual scenes for teaching air-to-ground attack. *Human Factors, 29,* 45–59.

Logan, G. D. (1988a). Automaticity, resources, and memory: Theoretical controversies and practical implications. *Human Factors, 30,* 583–598.

Logan, G. D. (1988b). Toward an instance theory of automatization. *Psychological Review, 95,* 492–527.

Lordahl, D. S., and Archer, J. E. (1958). Transfer effects on a rotary pursuit task as a function of first task difficulty. *Journal of Experimental Psychology, 56,* 421–426.

Mark, L. S. (1987). Eyeheight-scaled information about affordances: A study of sitting and stair climbing. *Journal of Experimental Psychology: Human Perception and Performance, 13,* 361–370.

Mertens, H. W. (1981). Perception of runway image shape and approach angle magnitude by pilots in simulated night landing approaches. *Aviation, Space, and Environmental Medicine, 52,* 373–386.

Namikas, G., and Archer, J. E. (1960). Motor skill transfer as a function of intertask interval and pre-transfer task difficulty. *Journal of Experimental Psychology, 59,* 109–112.

Needham, R. C., Edwards, B. J., and Prather, D. C. (1980). Flight simulation in air-combat training. *Defense Management Journal, 16*(4), 18–23.

Osgood, C. E. (1949). The similarity paradox in human learning: A resolution. *Psychological Review, 56,* 132–143.

Schmidt, R. A. (1975). A schema theory of discrete motor skill learning. *Psychological Review, 82,* 225–260.

Schmidt, R. A., and Young, D. E. (1987). Transfer of movement control in motor skill learning. In S. M. Cormier and J. D. Hagman (Eds.), *Transfer of learning* (pp. 47–79). San Diego, CA: Academic.

Spooner, A. M., Chambers, W. S., and Stevenson, B. S. (1980). Visual simulation in Navy training. *Defense Management Journal, 16*(4), 26–32.

Stoffregen, T. A., and Riccio, G. E. (1988). An ecological theory of orientation and the vestibular system. *Psychological Review, 95,* 3–14.

Thorndike, E. L. (1903). *Educational psychology.* New York: Lemcke and Buechner.

Todd, J. T. (1981). Visual information about moving objects. *Journal of Experimental Psychology: Human Perception and Performance, 7,* 795–810.

Warren, W. H., and Whang, S. (1987). Visual guidance of walking through apertures: Body-scaled information for affordances. *Journal of Experimental Psychology: Human Perceptions and Performance, 13,* 371–383.

Wightman, D. C., and Lintern, G. (1985). Part-task training for tracking and manual control. *Human Factors, 27,* 267–283.

Wightman, D. C., and Sistrunk, F. (1987). Part-task training strategies in simulated carrier landing final-approach training. *Human Factors, 29,* 245–254.

Williams, A. C., and Briggs, G. E. (1962). On-target versus off-target information and the acquisition of tracking skill. *Journal of Experimental Psychology, 64,* 519–525.

Woodrow, H. (1927). The effect of the type of training upon transference. *Journal of Educational Psychology, 18,* 159–172.

Part 2: Training Methods

Those who develop and conduct training must consider not only how to achieve quality training outcomes but also how well the trained skills will endure. Some trainers view retention as any duration that is longer than the intertrial interval used at the time of training. Although variables certainly exist which produce differences in rates of initial learning, many appear to have no residual influence over longer retention intervals. For example, learning is often enhanced when content information is high in "meaningfulness," but once the material is learned, meaningfulness has little influence on retention.

The list of variables known to influence the retention of trained skills is relatively short. They include degree of original learning, characteristics of the learning task, and the instructional methods and strategies used during training. The most important determinant of retention, however, appears to be the extent of original learning. In general, the greater the degree of learning, the slower the rate of forgetting. This relationship is so strong that it has prompted some researchers to argue that any variable that leads to high initial levels of learning will facilitate skill retention. Systematic relationships between degree of original learning and retention have occurred in a wide variety of contextual domains, including motor, procedural, and verbal learning tasks.

The organization and complexity of content material also appears to have a powerful influence on learning. The efficiency of acquisition, retrieval, and retention appears to depend on how the content material is organized. Content domains organized according to a conceptually logical basis appear to result in greater acquisition and retention and to be more generalizable than material that is not so organized.

Research has also identified numerous variables that constitute strategies for skill and knowledge acquisition and retention, including spaced reviews, massed/distributed practice, part/whole learning, and feedback. Techniques that help learners to build mental models they can use to generate retrieval cues, recognize externally provided cues, and/or generate or reconstruct information have also been shown to facilitate learning and retention. Three general strategies for enhancing long-term retention include reminding learners of currently possessed knowledge that is related to the material to be learned, ensuring that training makes repeated use of the information presented, and providing for and encouraging elaboration of the material during training.

Factors shown to negatively influence learning include proactive or retroactive interference by competing material that has been previously or subsequently learned, and the events encountered by individuals during the time between training and the retention test. Information acquired during this interval may impair retention, whereas rehearsal or re-exposure may facilitate retention. Additional factors influencing skill and knowledge retention include the length of the retention interval, the methods used to assess retention, and individual difference variables among trainees.

Part 2 contains seven articles on training methods. The first, by Cannon-Bowers, Tannenbaum, Salas, and Converse (1991), offers a *linkage* framework as an attempt to move away from faddism in training and toward a logical basis for integrating theory and practice. In the second paper, Swezey, Perez, and Allen (1991) describe how several training strategies and the use of motion contribute to enhancing acquisition and transfer in a maintenance training context. The Whaley and Fisk (1993) article presents the results of a series of studies that address part- versus whole-task training effects on memory-dependent tasks. The Volpe, Cannon-Bowers, Salas, and Spector (1996) paper concerns the topic of cross-training in teams. In the fifth article, Schmidt and Wulf (1997) address concurrent visual feedback and provide implications of such feedback for training and simulation. The Najjar (1998) article offers a number of principles derived from the literature of psychology, instructional design, and other areas for designing user interfaces. In the final article, Salas, Prince, Bowers, Stout, Oser, and Cannon-Bowers (1999) briefly review the topic of crew resource management (CRM) training and discuss a general methodology, developed by the U.S. Navy, for design and delivery of training to address this problem area.

Toward an Integration of Training Theory and Technique

JANIS A. CANNON-BOWERS,[1] *Naval Training Systems Center, Orlando, Florida,* SCOTT I. TANNENBAUM, *State University of New York, Albany, New York,* EDUARDO SALAS, *Naval Training Systems Center, Orlando, Florida,* and SHAROLYN A. CONVERSE, *North Carolina State University, Raleigh, North Carolina*

Reviewers of the training literature have generally concluded that training theory and practice are not well integrated, and that research findings are not often translated into useful training methods. In an effort to bridge the gap between training theory and practice, an organizing framework for conceptualizing training research is presented. The purpose of the framework is to highlight the linkages between training-related theory and technique in the areas of training analysis, design, and evaluation. The linkages are described in detail, and illustrated via consideration of research into mental models. We hope that the framework will lead to future research programs that enhance the transition of training research from theory into practice, and integrate more fully these two perspectives.

INTRODUCTION

Reviews of the training literature over the past 20 years have painted an increasingly optimistic picture of the field. In the early 1970s John Campbell charged that "by and large, the training and development literature is voluminous, nonempirical, nontheoretical, poorly written, and dull" (1971, p. 565). Goldstein, in 1978, asserted further that the field of training "appears to be dominated by a fads approach" (Goldstein, 1978, p. 134) that emphasized the development of tech-

niques. Examination of the table of contents of the *Training and Development Handbook* (Craig, 1976) bolsters Goldstein's 1978 claim; the vast majority of chapters presented training methods or techniques, with little (if any) grounding in theory, and even less empirical evidence of their validity. In 1984, Wexley was slightly more optimistic, but maintained that a number of crucial areas of training were still in need of systematic study, including an understanding of the impact of organizational factors in training, the relationship between task analysis and program design, the implications of individual differences for instructional strategies, and the factors that facilitate transfer of training to the workplace (Wexley, 1984). By 1988, Latham had concluded that the training literature *had* become more theoretical, but that train-

[1] Requests for reprints should be sent to Janis A. Cannon-Bowers, Code 262, Naval Training Systems Center, 12350 Research Parkway, Orlando, FL 32826-3224.

The views expressed herein are those of the authors and do not reflect the official position of the organizations with which they are affiliated.

ing practitioners still largely ignored (i.e., did not apply) results from the research literature. In the most recent review of the field, Tannenbaum and Yukl (in press) noted improvements in integration and conceptual development, but they also suggested that a great deal more is possible and needed.

At present, then, it appears that the challenge for those in the training community is still to provide better avenues of transition from theory into practice. According to Redding (1990), this is difficult because the two communities (i.e., researchers and practitioners) exist separately and are not well coordinated (or, perhaps, are unwilling to coordinate). This is unfortunate because we believe that the field of training cannot advance fully until the two perspectives come together. Training practitioners are more likely to succeed given a solid theoretical and empirical foundation upon which to design training. This foundation should allow practitioners to be better able to determine why or how training was effective or ineffective and, in turn, to be better able to make necessary adjustments or predict effectiveness in other settings. In this respect, we agree with Kurt Lewin that there is nothing more practical than a good theory (see Marrow, 1969).

With respect to researchers, we believe that meaningful research is best accomplished when it considers fully the practical implications associated with it. Furthermore, research that focuses directly on practical concerns—specific training methods, implementation and delivery issues, organization-specific factors, and the like—is as critical to the development of the training field as is more traditional, laboratory-based research. By researching *training in practice*, crucial lessons can be folded back into the field as a means to refine, confirm, or expand theoretical developments. Moreover, given the current (and probably, the future) fiscal constraints prevalent in both the public and private sectors, applications-oriented research may be the only kind that will be supported.

Overall, we maintain that everyone associated with the field of training—researchers and practitioners alike—can benefit from a stronger linkage between training-related theory and technique. In an effort to foster such linkages, we offer a framework that is designed to make explicit the linkages between training-related theory and technique. Following this, we use the framework to trace the progression of research in a particular area—mental model theory. This example was selected to demonstrate how the framework can be used in considering the linkages between theory and technique and, we hope, in stimulating similar thinking in subsequent research.

A FRAMEWORK FOR LINKING TRAINING-RELATED THEORY AND TECHNIQUE

In an effort to bring various areas of training theory and practice together, the following sections describe an organizing framework for conceptualizing training research (it should be noted that throughout this paper we use the term "theory" loosely, referring to conceptual developments as well as fully defined theories). The proposed framework is designed to make explicit the relationships among various areas of training theory and between training theory and practice. It incorporates three related questions pertinent to training research: (1) What should be trained (i.e., what is the nature of the knowledge, skills, and abilities that must be trained)? (2) How should training be designed? (3) Is training effective, and if so, why? Each of these questions has a number of conceptual developments and applied techniques associated with it. Some of these are depicted in Figure 1.

RESEARCH QUESTIONS	TRAINING-RELATED THEORY	TRAINING-RELATED TECHNIQUE
What Should Be Trained?	**CELL 1** Nature of Expertise (Rasmussen, 1979) Mental Models & Knowledge Structures (Rouse & Morris, 1986) Teamwork Skills (Prince et al., in press) Expert/Novice Differences (Chase & Simon, 1973; Glaser, 1989) Taxonomies of Task Requirements (Fleishman & Quaintance, 1984) Meta-Cognition (Bereiter & Scardamalia, 1985) Information Processing Theory (Schneider & Shiffrin, 1977)	**CELL 2** Cognitive Task Analysis (Redding, 1989) Team Task Analysis (Levine & Baker, 1991) Job Analysis Methods (Levine, 1983) Needs Analysis Methods (Goldstein, Braverman & Goldstein, 1991) Protocol Analysis (Ericsson & Simon, 1984) Critical Incidents (Flanagan, 1954) Future-Oriented Task Analysis (Schneider & Konz, 1989)
How Should Training Be Designed?	**CELL 3** Information Processing Theory (Schneider & Shiffrin, 1977; Fisk & Eggemeier, 1988) Skill Acquisition (Anderson, 1985; Ackerman, 1987) Social Learning Theory (Bandura, 1986) Team Performance (Salas et al., in press) Learning Principles (Gagne 1970; Kyllonen & Alluisi, 1987) Taxonomy of Learning Skills (Kyllonen & Shute, 1989) Fidelity (Hays & Singer, 1989) Meta-Cognition (Bereiter & Scardamalia, 1985) Expectancy Theory (Vroom, 1964) Action Control (Kuhl, 1985) Mental Models (Kieras, 1988)	**CELL 4** Behavior Modeling (Goldstein & Sorcher, 1974) Simulators/Training Devices (Cream et al., 1978) Networked Training (Alluisi, 1991) Part-Task Training (Wightman & Lintern, 1985) Feedback (Ilgen et al., 1979) Relapse Prevention (Marx, 1982) Self-Management (Manz & Sims, 1989) Computer Based Instruction (Crawford & Crawford, 1978) Games & Simulations (Thornton & Cleveland, 1990) Action Learning (Revans, 1982)
Is Training Effective and Why?	**CELL 5** Individual Differences (Noe & Schmitt, 1986; Cronbach & Snow, 1977) Hierarchy of Evaluation (Kirkpatrick, 1976; Tannenbaum et al., 1991) Transfer of Training (Baldwin & Ford, 1988) Multi-Component Approach (Cannon-Bowers et al., 1989) Attitude/Behavior Relations (Ajzen & Fishbein, 1980) Training Climate (Rouillier & Goldstein, 1991) Self-Fulfilling Prophecy (Eden, 1990) Evaluability Assessment (Rutman, 1980) Content Validity (Ebel 1977; Guion, 1978) Attribution Theory (Kelley, 1972)	**CELL 6** Scale Development (Anastasi, 1988) Utility Analysis (Cascio, 1991) Quasi-Experimental Methods (Cook, Campbell & Peracchio, 1991) Work Sample Tests (Asher & Sciarrino, 1974) Team Performance Measures (Coovert & McNelis, in press) Critical Incidents (Morgan et al., 1986) Content Validation Strategies (Lawshe, 1975) Mental Model Measurement (Moore & Gordon, 1988; Schvaneveldt, 1990) Program Evaluation (Rossi & Freeman, 1982) Walk Through Performance Testing (Hedge & Lipscomb, 1987)

Figure 1. *Framework for linking training-related theory and technique.*

The concepts and techniques listed in each of the six cells in Figure 1 are intended to provide examples of research topics that characterize that cell, but they by no means represent exhaustive lists. The arrows or *linkages* between categories indicate that research can cross the boundaries from theory to technique and between theoretical areas; it is our contention that this type of research holds the greatest potential for advancing the field of training. Some of the entries have already been applied in the training context, whereas others are prime for application. Overall, the framework suggests that research can be conducted in both training theory and training techniques, so that (1) theoretical findings can be translated into specific training techniques and (2) the study of techniques can help to confirm/refine/expand related theory. The following sections explain in more detail the linkages depicted in Figure 1.

EXPLANATION OF THE LINKAGES

Cell 1 ↔ Cell 2. Understanding the nature of knowledge, skills, abilities (KSAs), and expertise associated with a task will help to determine methods appropriate to establish specific task requirements and training needs (1 → 2). In turn, research into specific task analysis techniques can help to confirm and/or refine theories about the nature of knowledge, skills, abilities, and expertise associated with a particular task (2 → 1).

Cell 1 ↔ Cell 3. Understanding the nature of knowledge, skills, abilities, and expertise associated with a task will help generate hypotheses regarding the nature of learning and skill acquisition for the task and the type of training that will be most effective (1 → 3). In turn, research that establishes the nature of learning and skill acquisition on a particular task can help to confirm and/or refine theories about the nature of KSAs and expertise for that task (3 → 1).

Cell 3 ↔ Cell 4. Understanding the nature of learning and performance for a task will help to determine techniques/methods appropriate to train it (3 → 4). In turn, research regarding particular training methods can help confirm and/or refine theories about learning and skill acquisition for a task (4 → 3).

Cell 3 ↔ Cell 5. Understanding the nature of learning and skill acquisition for a task will help generate hypotheses regarding why training is effective or ineffective, and help to establish the parameters of effectiveness (i.e., determine the conditions under which training is likely to be effective) (3 → 5). In turn, research that addresses the effectiveness of training can help to confirm/refine theories about learning and skill acquisition (5 → 3).

Cell 5 ↔ Cell 6. Understanding why training may be effective (i.e., the parameters of and causes underlying effectiveness) will help establish techniques/methods to assess the effectiveness of training (5 → 6). In turn, research on training effectiveness assessment techniques can help to confirm and/or refine theories about why training is effective and under what conditions it is likely to be effective (6 → 5).

Cell 1 ↔ Cell 5. Understanding the nature of KSAs and expertise for a task will help generate hypotheses about why training is effective and under what conditions it will be effective (1 → 5). In turn, research that addresses the causes and parameters of training effectiveness can help to confirm and/or refine theories about the nature of KSAs and expertise for a task (5 → 1).

Cells 2, 4, and 6. It should be noted that moving from Cell 2 to Cell 4 to Cell 6 is a sequence consistent with a systems approach to training design (see Goldstein, 1980). In principle, we support the sequence of first determining training needs, then designing training, and finally evaluating training, even though we did not include arrows connecting these cells. In fact, from a practitioner's perspective, this is exactly the sequence of steps we would recommend. However, the current state of training research does not provide the practitioner with a variety of *theoretically based, empirically tested* methods for training analysis, design, and evaluation. Ideally, the field will advance to the point where research can better support the 2 → 4 → 6 sequence advocated by the traditional systems approach to training.

In summary, it is our contention that the field of training can best be advanced by designing problem-oriented research programs that span the categories shown in Figure 1. The purpose of the framework, then, is simply to explicate linkages between training-related theory and technique so that transition and exchange between these two perspectives may be enhanced.

ILLUSTRATION OF
FRAMEWORK RELATIONSHIPS

The following sections should help illustrate the linkages between cells depicted in Figure 1 and demonstrate that conducting research that crosses boundaries between cells can advance knowledge about training. We focus on research in one area—mental models—so that a more detailed description of the relationship between training-related theory and technique can be presented. We do not present a comprehensive review of the literature in this area, but rather, we highlight work that illustrates most clearly the linkages explicated in the framework.

Mental Models and Complex Performance (Cell 1)

Researchers in several disciplines (e.g., cognitive psychology, human factors, cognitive science, educational psychology) have hypothesized that humans interact effectively with their environment by organizing knowledge into meaningful patterns that are stored in memory. The term *mental model* is often

used to refer to these knowledge structures or schemata (see Rouse and Morris, 1986). As mental models contain information about both stimuli and the consequences of responses to the stimuli, mental models are believed to guide the comprehension of, and responses to, information in the environment.

Mental models are considered to be dynamic cognitive representations that perform a number of important functions in complex task performance. According to Rouse and Morris (1986), mental models perform several functions: they help people describe, explain, and predict events in the environment. In terms of description and explanation, mental models serve a heuristic function: they speed the rate of comprehension by allowing situations, objects, relationships, and environments to be classified in terms of their most salient or important features (Palmer, 1975). Further, the dynamic nature of mental models allows people to create causal event chains via mental manipulation of model parameters, providing a basis to predict future events and assess the consequences of an action before it is taken (Johnson-Laird, 1983).

Research into the nature of mental models and their relationship to task performance has led to several important conclusions regarding how training content can be specified (Cell 1 → 2 linkage), how training can be best designed (Cell 1 → 3 linkage), and how training can be best evaluated (Cell 1 → 5 linkage). In addition, research concerning specific analysis, training, and evaluation techniques (Cells 2, 4, and 6) have helped to clarify notions about related underlying constructs (Cells 1, 3, and 5). The following sections highlight several areas of research that illustrate these linkages.

Mental Models and Cognitive Task Analysis (Cell 1 ↔ 2)

Research and theorizing about mental models over the past few years has contributed to the development of new methods for conducting task analysis. A family of techniques, generally referred to as *cognitive task analysis techniques*, seeks to specify the content and structure of knowledge required for successful task performance (see Redding, 1989, 1990). Cognitive task analysis techniques differ from more traditional task analysis techniques on several bases: (1) they are concerned with the organization of requisite knowledge and the relationships among important concepts, (2) they emphasize skill development and progression of knowledge structures from novice to expert, and (3) they segment jobs/tasks according to skill-based components rather than behavioral or temporal components (Redding, 1989).

An interesting theoretical question that has affected cognitive task analysis development relates to the progression of knowledge structures as people move from being novices to being experts on a task. Briefly, researchers have hypothesized that "expert" models are not simply more elaborate or accurate than "novice" models. Instead, evidence suggests that expert mental models are fundamentally different than novice models. Experts appear to organize knowledge around "deep" underlying principles, whereas novices organize knowledge around "shallow" surface features (Chi, Feltovich, and Glaser, 1981). Early learning involves the acquisition of declarative knowledge (i.e., knowledge about facts, relationships), with mastery of verbal terms and elemental rules (Glaser, 1990; Rasmussen, 1986). As learning advances, changes in the organization of the mental model occur as knowledge becomes more procedural in nature (i.e., focusing on how the task is performed) (Anderson, 1987). In addition, Larkin (1983) maintains that as people become more expert, their mental models move from being representational to being abstract. Others maintain that expert mental models are pattern-oriented (Dreyfus and Dreyfus, 1979),

more highly integrated, and stored in larger chunks.

The implication of these theoretical contentions is that experts may have limited access to their own knowledge structures—that is, they may have difficulty explaining why or how they are performing a task (Cooke and McDonald, 1987; Rouse and Morris, 1986). In fact, several researchers have found that experts have difficulty verbalizing their mental models (e.g., Broadbent, Fitzgerald, and Broadbent, 1986; Van Heusden, 1980; Whitfield and Jackson, 1982). This calls into question techniques that require experts to verbalize their strategy or thinking process during or after a task (Redding, 1989; Rouse and Morris, 1986). Furthermore, Sanderson (1989) found that subjects' ability to verbalize their internal models varied as a function of task and training parameters. As a result of these findings, "indirect" cognitive task analysis methods have been developed to document expert models (see Redding, 1989; Rouse and Morris, 1986). These methods include, for example, psychological scaling techniques (Cooke and McDonald, 1987; Zubritzky and Coury, 1987), cognitive simulation (Roth and Woods, 1990), and question probe techniques (Moore and Gordon, 1988).

Findings described above illustrate how theorizing about, and researching the nature of, expertise and skilled performance has contributed to development of techniques for specifying what must be trained (a Cell 1 → 2 linkage in the framework). Likewise, research that examined how verbal reports can be elicited and scored has contributed to the understanding of how verbal knowledge and task performance are related (illustrating a Cell 2 → 1 linkage). Sanderson's (1989) work demonstrated, for example, that the pattern of verbal responses (i.e., the extent to which verbalizations supported a particular mental model of the system) was more informative in assessing the conditions under which a verbal report/performance association could be expected than was a simple tally of correct verbal responses.

Mental Models and Training (Cell 1 ↔ 3 and Cell 3 ↔ 4)

Research into training system design in the context of mental models has addressed a number of questions, including how should information be presented so that an accurate model is developed? Several researchers maintain in this regard that training should present an explicit *conceptual model* of the material to be trained. According to Mayer (1989), a conceptual model contains words or diagrams that highlight the "major objects and actions in a system as well as the causal relations among them" (p. 43). It is similar to what Ausubel (1960) called an *advance organizer* (i.e., specially designed introductory material that provides trainees with the "ideational scaffolding" needed to incorporate and retain targeted material). According to Mayer (1989), conceptual models improve learning for three reasons. First, they help to direct and focus a trainee's attention on the important components and relationships in a system. Second, they help the trainee to organize incoming information (thereby increasing "internal connectedness"). Third, they help the trainee to integrate incoming information with existing relevant knowledge (thereby increasing "external connectedness").

Research that has investigated the utility of presenting conceptual models to trainees has been generally supportive. For example, Bayman and Mayer (1984) found that trainees who were not presented with a conceptual model of a four-function calculator employed incorrect strategies during the posttest. Other research has also supported this result, generally finding that unguided practice on a system can lead to development of inaccurate and/or incomplete mental models (Frederik-

sen and White, 1989). On the other hand, some research has shown that teaching only the theories or principles of a system does not necessarily lead to enhanced performance either (e.g., Kieras and Bovair, 1984). An important question for training design then becomes, under what conditions does the presentation of a conceptual model improve task performance?

Recently, Kieras (1988) attempted to answer this question in the domain of device operation. Based on several studies in this area (e.g., Gentner and Gentner, 1983; Kieras and Bovair, 1984), Kieras maintains that presenting a conceptual model is useful when the mental model allows the user to infer the exact procedures for operating the device, or when it is necessary to generalize the model to situations that were not trained explicitly. However, Kieras (1988) contends that presenting a conceptual model will not be useful when procedures are easily learned by rote mastery, when the device is so simple that the trainee does not need to make inferences about how the device works, when the conceptual model is too difficult or complicated for the trainee to acquire, or when the model does not support inferences that the trainee needs to make regarding device functioning. (This work illustrates the Cell 1 → 3 linkage in the framework.) In addition, Kieras's (1988) contentions regarding the utility of presenting explicit models in training has conceptual implications for the way in which a task may be analyzed (Cell 3 → 1). Specifically, he maintains that determining whether a conceptual model will be useful in training must rest on a careful analysis of the nature of the *inferences* a trainee will be required to draw about the system during task performance.

In another experiment, Eylon and Reif (1984) were interested in the effect of different formats for presenting information in the conceptual model. They hypothesized that presenting hierarchically organized material in training would be superior to presenting material organized in a linear fashion. This hypothesis regarding how to design training was based on conceptual developments that suggested that expert knowledge structures are hierarchically organized (Cell 1 → 3). Eylon and Reif's results indicated that subjects who were presented with hierarchically organized material performed significantly better on complex problems, whereas no differences existed between groups on simple problems. Their study also illustrates how research into training strategies can confirm or refine theories about the nature of knowledge, skills, and expertise underlying task performance (Cell 3 → 1). In this case, the finding that hierarchically organized material enhanced task performance is consistent with the notion that experts may develop hierarchically organized mental models.

To date, several training systems have been developed based on mental model theory and related concepts (see Halff, Hollan, and Hutchins, 1986; Redding and Lierman, 1990), illustrating the Cell 3 → 4 linkage in the framework. For example, Hutchins, McCandless, Woodworth, and Dutton (1984) developed a system (called **MOBOARD**) that is consistent with notions that effective learning will occur when new knowledge is presented in a manner that is compatible with existing knowledge structures (Ausubel, 1960; Mayer, 1979).

Other attempts to incorporate principles of mental model theory into specific training techniques have hypothesized that training systems that simulate the system to be trained will be effective because they are consistent with the dynamic nature of mental models (e.g., Greeno, 1989). White (1984) provided data that supported use of this type of computer simulation in training. In addition, a recent study by Augustine and Coovert (in press) demonstrated that computer anima-

tion of a complex system was a more effective means to train subjects how to use a mainframe computer than was a less dynamic system.

Mental Models in Training Evaluation (Cell 3 ↔ 5, Cell 5 ↔ 6)

Research into mental models and training system design also has implications for training evaluation. The most direct implication of mental model research for evaluation is that a viable criterion for training may be the existence of an accurate, fully articulated mental model at the conclusion of training (Cell 3 → 5). For example, Moore and Gordon (1988) used a conceptual graph technique (via question probes) to diagnose trainee comprehension. By comparing trainees' conceptual graphs to those of experts, these researchers were able to determine the incorrect inferences made by trainees in solving problems. Gill, Gordon, Moore, and Barbera (1988) successfully employed a similar technique to compare two different instructional techniques, as did K. Kraiger (personal communication, June 1991). In addition, these and other studies (e.g., Goldsmith and Johnson, 1990) provide an example of how conceptual notions about evaluating training can provide data regarding the development of training evaluation techniques (Cell 5 → 6 linkage).

Mental Models in Training Analysis and Evaluation (Cell 1 ↔ 5)

Investigating the nature of knowledge, skills, abilities, and expertise from the mental model perspective has crucial implications for determining the parameters and underlying causes of training effectiveness (Cell 1 ↔ 5 linkage). The work described above has contributed to the generation of hypotheses about how, and under what conditions, mental models may change over the course of training (Cell 1 → 5). Particularly important in this area is research that tracks changes in knowledge structures as learning progresses (an example of this approach can be found in Naveh-Benjamin, McKeachie, Lin, and Tucker, 1986). In addition, research that addresses the conditions under which particular mental models will develop can refine theories about the nature of expertise (Cell 5 → 1). An example of this is provided by Sanderson (1989). As noted earlier, she developed a method of scoring *patterns* of verbal responses to determine which of several mental models subjects were using to solve problems. This allowed her to assess the parameters of task practice that led to the development of specific models, contributing to a fundamental understanding of how mental models may develop.

CONCLUSIONS

As a field, training lies at the crossroads of science and application. While academics and researchers conduct scientific studies of learning and skill acquisition, U.S. industry and government spend upwards of $200 billion per year to train the workforce. Yet reviews of the field continue to call for greater integration of research and practice. In response to these concerns, we presented a framework for thinking about training research that makes explicit the potential linkages between training-related theory and technique. Clearly, there are many theoretical and conceptual developments with implications for training practice, a few of which were noted in the framework. Likewise, research on various training-related techniques can provide valuable insights into underlying constructs and theory. However, these connections are sometimes overlooked or dismissed as irrelevant. This is unfortunate because research that links theory and technique holds great potential for advancing both the science and the application of training.

The possibilities are there, as evidenced by some of the mental model research we described. Theory and technique appear to be supporting each other in that research domain; other research areas could have been selected to illustrate the linkages as well. Overall, however, the current body of training research does not provide the practitioner with a full complement of theoretically based, empirically tested methods for training analysis, design, and evaluation. Further advancements are possible and desirable. We hope that by highlighting some potential linkages between training-related theory and technique, we will encourage additional research that spans cells in the framework and strengthens the connection between science and application.

REFERENCES

Ackerman, P. L. (1987). Individual differences in skill learning: An integration of psychometric and information processing perspectives. *Psychological Bulletin, 102,* 3–37.

Anderson, J. R. (1985). *Cognitive psychology and its implications.* New York: Freeman.

Anderson, J. R. (1987). Methodologies for studying human knowledge. *Behavioral and Brain Science, 10,* 476–477.

Ajzen, I., and Fishbein, M. (1980). *Understanding attitudes and predicting social behavior.* Englewood Cliffs, NJ: Prentice-Hall.

Alluisi, E. A. (1991). The development of technology for collective training: SIMNET, a case history. *Human Factors, 33,* 343–362.

Anastasi, A. (1988). *Psychological testing* (6th ed.). New York: MacMillan.

Asher, J. J., and Sciarrino, J. A. (1974). Realistic work sample tests: A review. *Personnel Psychology, 27,* 519–534.

Augustine, M. A., and Coovert, M. D. (in press). Simulation and information order as influences in the development of mental models. *Association for Computing Machinery SIGCHI Bulletin.*

Ausubel, D. P. (1960). The use of advance organizers in learning and retention of meaningful material. *Journal of Educational Psychology, 51,* 267–272.

Baldwin, T. T., and Ford, J. K. (1988). Transfer of training: A review and directions for future research. *Personnel Psychology, 41,* 63–101.

Bandura, A. (1986). *Social foundations of thought and action.* Englewood Cliffs, NJ: Prentice-Hall.

Bayman, P., and Mayer, R. E. (1984). Instructional manipulation of users' mental models for electronic calculators. *Instructional Journal of Man-Machine Studies, 20,* 189–199.

Bereiter, C., and Scardamalia, M. (1985). Cognitive coping strategies and the problem of "inert" knowledge. In S. Chipman, J. W. Segal, and R. Glaser (Eds.), *Thinking and learning skills: Current research and open questions* (Vol. 2, pp. 65–80). Hillsdale, NJ: Erlbaum.

Broadbent, D. E., Fitzgerald, P., and Broadbent, M. H. P. (1986). Implicit and explicit knowledge in the control of complex systems. *British Journal of Psychology, 77*, 33–50.

Campbell, J. P. (1971). Personnel training and development. *Annual Review of Psychology, 22*, 565–602.

Cannon-Bowers, J. A., Prince, C., Salas, E., Owens, J. M., Morgan, B. B., and Gonos, G. H. (1989). Determining aircrew coordination effectiveness. In *Proceedings of the Interservice/Industry Training Systems Conference* (Vol. 1, pp. 128–135). Washington, DC: American Defense Preparedness Association.

Cascio, W. F. (1991). *Costing human resources: The financial impact of behavior in organizations* (3rd ed.). Boston, MA: Kent.

Chase, W. G., and Simon, H. A. (1973). The mind's eye in chess. In W. G. Chase (Ed.), *Visual information processing* (pp. 215–281). New York: Academic.

Chi, M. T. H., Feltovich, P. J., and Glaser, R. (1981). Categorization and representation of physics problems by experts and novices. *Cognitive Science, 5*, 121–152.

Cook, T. D., Campbell, D., and Peracchio, L. (1991). Quasi-experiments. In M. D. Dunnette and L. M. Hough (Eds.), *Handbook of industrial/organizational psychology* (2nd ed., Vol. 2, pp. 491–576). Palo Alto, CA: Consulting Psychologists.

Cooke, N. M., and McDonald, J. E. (1987). The application of psychological scaling techniques to knowledge elicitation for knowledge-based systems. *International Journal of Man-Machine Studies, 26*, 533–550.

Coovert, M. D., and McNelis, K. (in press). Team decision making and performance: A review and proposed approach employing Petri nets. In R. W. Swezey and E. Salas (Eds.), *Teams: Their training and performance.* Norwood, NJ: Ablex.

Craig, R. L. (Ed.). (1976). *Training and development handbook: A guide to human resource development* (2nd ed.). New York: McGraw-Hill.

Crawford, A. M., and Crawford, K. S. (1978). Simulation of operational equipment with a computer-based instructional system: A low cost training technology. *Human Factors, 20*, 215–224.

Cream, B. W., Eggemeier, F. T., and Klein, G. A. (1978). A strategy for the development of training devices. *Human Factors, 20*, 145–158.

Cronbach, L. J., and Snow, R. E. (1977). *Aptitudes and instructional methods: A handbook for research on interactions.* New York: Irvington.

Dreyfus, H. L., and Dreyfus, S. E. (1979). *The psychic boom: Flying beyond the thought barrier* (Tech. Report 79-3). Berkeley, CA: University of California, Operations Research Center.

Ebel, R. L. (1977). Prediction? Validation? Construct validity? *Personnel Psychology, 30*, 55–63.

Eden, D. (1990). *Pygmalion in management.* Lexington, MA: Lexington.

Ericsson, K. A., and Simon, H. A. (1984). *Protocol analysis: Verbal reports as data.* Cambridge, MA: MIT Press.

Eylon, B. S., and Reif, F. (1984). Effects of knowledge organization on task performance. *Cognition and Instruction, 1*, 5–44.

Fisk, A. D., and Eggemeier, F. T. (1988). Application of automatic/controlled processing theory to training tactical command and control skills: 1. Background and analytic methodology. In *Proceedings of the Human*

Factors Society 32nd Annual Meeting (pp. 1227–1231). Santa Monica, CA: Human Factors Society.

Flanagan, J. C. (1954). The critical incident technique. *Psychological Bulletin, 51*, 327–358.

Fleishman, E. A., and Quaintance, M. K. (1984). *Taxonomies of human performance.* Orlando, FL: Academic.

Fredericksen, J., and White, B. (1989). An approach to training based upon principled task decomposition. *Acta Psychologica, 71*, 89–146.

Gagné, R. M. (1970). *The conditions for learning.* New York: Holt, Rinehart, and Winston.

Gentner, D., and Gentner, D. R. (1983). Flowing waters or teeming crowds: Mental models of electricity. In D. Gentner and A. L. Stevens (Eds.), *Mental models* (pp. 99–129). Hillside, NJ: Erlbaum.

Gill, R., Gordon, S., Moore, J., and Barbera, C. (1988). The role of conceptual structures in problem solving. In *Proceedings of the 1988 Annual Meeting of the American Society of Engineering Education* (pp. 583–590). Washington, DC: ASEE.

Glaser, R. (1989). Expertise and learning: How do we think about instructional processes now that we have discovered knowledge structures? In D. Klahr and K. Kotovsky (Eds.), *Complex information processing* (pp. 269–282). Hillsdale, NJ: Erlbaum.

Glaser, R. (1990). The reemergence of learning theory within instructional research. *American Psychologist, 45*, 29–39.

Goldsmith, T. E., and Johnson, P. J. (1990). A structural assessment of classroom learning. In R. W. Schvaneveldt (Ed.), *Pathfinder associative network: Studies in knowledge organization.* Norwood, NJ: Ablex.

Goldstein, A. P., and Sorcher, M. (1974). *Changing supervisor behavior.* New York: Pergamon.

Goldstein, I. L. (1978). The pursuit of validity in the evaluation of training programs. *Human Factors, 20*, 131–144.

Goldstein, I. L. (1980). Training in work organizations. *Annual Review of Psychology, 31*, 229–272.

Goldstein, I. L., Braverman, E. P., and Goldstein, H. W. (1991). The use of needs assessment in training systems design. In K. Wexley (Ed.), *Developing human resources* (Vol. 5, pp. 35–75). Washington, DC: BNA Books.

Greeno, J. (1989). Situations, mental models, and generative knowledge. In D. Klahr and K. Kotovsky (Eds.), *Complex information processing* (pp. 285–318). Hillsdale, NJ: Erlbaum.

Guion, R. M. (1978). Content validity in moderation. *Personnel Psychology, 31*, 205–214.

Halff, H. M., Hollan, J. D., and Hutchins, E. L. (1986). Cognitive science and military training. *American Psychologist, 41*, 1131–1139.

Hays, R. T., and Singer, M. J. (1989). *Simulation fidelity in training system design.* New York: Springer-Verlag.

Hutchins, E., McCandless, T., Woodworth, G., and Dutton, B. (1984). *Maneuvering board training system: Analysis and redesign* (Tech. Report NPRDC-TR-84-19). San Diego, CA: Navy Personnel Research and Development Center.

Hedge, J. W., and Lipscomb, M. S. (Eds.). (1987). *Walk through performance testing: An innovative approach to work sample testing.* (Tech. Report AFHRL-TP-87-8). Brooks AFB, TX: Training Systems Division, Air Force Human Resources Laboratory.

Ilgen, D. R., Fisher, C. D., and Taylor, M. S. (1979). Consequences of individual feedback on behavior in organizations. *Journal of Applied Psychology, 64*, 349–371.

Johnson-Laird, P. (1983). *Mental models*. Cambridge, MA: Harvard University Press.

Kelley, H. H. (1972). Attribution in social interaction. In E. E. Jones, D. E. Kanhouse, H. H. Kelley, S. Valins, and B. Weiner (Eds.), *Attribution: Perceiving the causes of behavior*. Morristown, NJ: General Learning.

Kieras, D. E. (1988). What mental model should be taught: Choosing instructional content for complex engineering systems. In J. Psotka, L. D. Massey, and S. A. Mutter (Eds.), *Intelligent tutoring systems: Lessons learned* (pp. 85–111). Hillsdale, NJ: Erlbaum.

Kieras, D. E., and Bovair, S. (1984). The role of a mental model in learning to control a device. *Cognitive Science, 8*, 255–273.

Kirkpatrick, D. L. (1976). Evaluation of training. In R. L. Craig (Ed.), *Training and development handbook: A guide to human resources development*. New York: McGraw-Hill.

Kuhl, J. (1985). Volitional mediators of cognition-behavior consistency: Self-regulatory processes and action versus state orientation. In J. Kuhl and J. Beckman (Eds.), *Action control: From cognition to behavior* (pp. 11–39). Berlin: Springer-Verlag.

Kyllonen, P. C., and Alluisi, E. A. (1987). Learning and forgetting facts and skills. In G. Salvendy (Ed.), *Handbook of human factors* (pp. 124–153). New York: Wiley.

Kyllonen, P. C., and Shute, V. J. (1989). A taxonomy of learning skills. In P. L. Ackerman, R. J. Sternberg, and R. Glaser (Eds.), *Learning and individual differences: Advances in theory and research* (pp. 117–163). New York: Freeman.

Larkin, J. H. (1983). The role of problem representation in physics. In D. Gentner and A. L. Stevens (Eds.), *Mental models* (pp. 75–98). Hillsdale, NJ: Erlbaum.

Latham, G. P. (1988). Human resource training and development. *Annual Review of Psychology, 39*, 545–582.

Lawshe, C. H. (1975). A quantitative approach to content validity. *Personnel Psychology, 28*, 563–575.

Levine, E. L. (1983). *Everything you ever wanted to know about job analysis*. Tampa, FL: Mariner.

Levine, E. L., and Baker, C. V. (1991, April). *Team task analysis: A procedural guide and test of the methodology*. Paper presented at the annual meeting of the Society of Industrial and Organizational Psychology, St. Louis, MO.

Manz, C., and Sims, H. (1989). *Superleadership: Leading others to lead themselves*. New York: Simon and Schuster.

Marrow, A. J. (1969). *The practical theorist: The life and work of Kurt Lewin*. New York: Basic Books.

Marx, R. D. (1982). Relapse prevention for managerial training: A model for maintenance of behavior change. *Academy of Management Review, 7*, 433–441.

Mayer, R. E. (1979). Twenty years of research on advance organizers: Assimilation theory is still the best predictor of results. *Instructional Science, 8*, 133–167.

Mayer, R. E. (1989). Models for understanding. *Review of Educational Research, 59*, 43–64.

Moore, J. L., and Gordon, S. C. (1988). Conceptual graphs as instructional tools. In *Proceedings of the Human Factors Society 32nd Annual Meeting* (pp. 1289–1293). Santa Monica, CA: Human Factors Society.

Morgan, B. B., Glickman, A. S., Woodard, E. A., Blaiwes, A. S., and Salas, E. (1986). *Measurement of team behaviors in a navy environment*. (Tech. Report TR-86-014). Orlando, FL: Naval Training Systems Center.

Naveh-Benjamin, M., McKeachie, W. J., Lin, Y., and

Tucker, D. G. (1986). Inferring students' cognitive structures and their development using the "ordered tree technique." *Journal of Educational Psychology, 78*, 130–140.

Noe, R. A., and Schmitt, N. (1986). The influence of trainee attitudes on training effectiveness: Test of a model. *Personnel Psychology, 39*, 497–524.

Palmer, S. E. (1975). The effects of contextual scenes on the identification of objects. *Memory and Cognition, 3*, 519–526.

Prince, C., Chidester, T. R., Cannon-Bowers, J. A., and Bowers, C. (in press). Aircrew coordination: Achieving teamwork in the cockpit. In R. Sweezey and E. Salas (Eds.), *Teams: Their training and performance*. New York: Ablex.

Rasmussen, J. (1979). *On the structure of knowledge—A morphology of mental models in a man-machine system context* (Tech. Report Riso-M-2192). Roskilde, Denmark: Riso National Laboratory.

Rasmussen, J. (1986). *Information processing and human-machine interaction: An approach to cognitive engineering*. New York: North-Holland.

Redding, R. E. (1989). Perspectives on cognitive task-analysis: The state of the state of the art. In *Proceedings of the Human Factors Society 33rd Annual Meeting* (pp. 1348–1352). Santa Monica, CA: Human Factors Society.

Redding, R. E. (1990). Taking cognitive task analysis into the field: Bridging the gap from research to application. In *Proceedings of the Human Factors Society 34th Annual Meeting* (pp. 1304–1308). Santa Monica, CA: Human Factors Society.

Redding, R. E., and Lierman, B. (1990). Development of a part-task, CBI trainer based upon a cognitive task analysis. In *Proceedings of the Human Factors Society 34th Annual Meeting* (pp. 1337–1341). Santa Monica, CA: Human Factors Society.

Revans, R. W. (1982). *The origin and growth of action learning*. Hunt, England: Chatwell-Bratt, Bickley.

Rossi, P. H., and Freeman, H. E. (1982). *Evaluation: A systematic approach* (2nd ed.). Beverly Hills, CA: Sage.

Roth, E. M., and Woods, D. D. (1990). Analyzing the cognitive demands of problem-solving environments: An approach to cognitive task analysis. In *Proceedings of the Human Factors Society 34th Annual Meeting* (pp. 1314–1317). Santa Monica, CA: Human Factors Society.

Rouillier, J. Z., and Goldstein, I. L. (1991, April). *Determinants of the climate for transfer of training*. Paper presented at the annual meeting of the Society of Industrial and Organizational Psychology, St. Louis, MO.

Rouse, W. B., and Morris, N. M. (1986). On looking into the black box: Prospects and limits in the search for mental models. *Psychological Bulletin, 100*, 349–363.

Rutman, L. (1980). *Planning useful evaluations*. Beverly Hills, CA: Sage.

Salas, E., Dickinson, T. L., Converse, S. A., and Tannenbaum, S. I. (in press). Toward an understanding of team performance and training. In R. W. Swezey and E. Salas (Eds.), *Teams: Their training and performance*. Norwood, NJ: Ablex.

Sanderson, P. M. (1989). Verbalizable knowledge and skilled task performance: Association, dissociation, and mental models. *Journal of Experimental Psychology, 15*, 729–747.

Schneider, B., and Konz, A. (1989). Strategic job analysis. *Human Resource Management, 28*, 51–63.

Schneider, W., and Shiffrin, R. M. (1977). Controlled and automatic human information processing: I. Detection, search, attention. *Psychological Review, 84,* 1–66.

Schvaneveldt, R. W. (Ed.). (1990). *Pathfinder associative networks: Studies in knowledge organization.* Norwood, NJ: Ablex.

Tannenbaum, S. I., Mathieu, J. E., Salas, E., and Cannon-Bowers, J. A. (1991, April). *An examination of the factors that influence training effectiveness: A model and research agenda.* Paper presented at the Annual Meeting of the Society for Industrial and Organizational Psychology, St. Louis, MO.

Tannenbaum, S. I., and Yukl, G. (in press). Training and development in work organizations. *Annual Review of Psychology.*

Thornton, G. C., III, and Cleveland, J. N. (1990). Developing managerial talent through simulation. *American Psychologist, 45,* 190–199.

Van Heusden, A. R. (1980). Human prediction of third-order autoregressive time series. *IEEE Transactions on Systems, Man and Cybernetics, SMC-10,* 38–43.

Vroom, V. H. (1964). *Work and motivation.* New York: Wiley.

Wexley, K. N. (1984). Personnel training. *Annual Review of Psychology, 35,* 519–551.

White, B. (1984). Designing computer games to help physics students understand Newton's laws of motion. *Cognition and Instruction, 1,* 69–108.

Whitfield, D., and Jackson, A. (1982). The air traffic controller's "picture" as an example of a mental model. In G. Johannsen and J. E. Rijnsdorp (Eds.), *Analysis, design and evaluation of man-machine systems* (pp. 45–52). London: Pergamon.

Wightman, D. C., and Lintern, G. (1985). Part-task training for tracking and manual control. *Human Factors, 27,* 267–283.

Zubritzky, M. C., and Coury, B. G. (1987). Multidimensional scaling as a method for probing the conceptual structure of state categories: An individual differences analysis. In *Proceedings of the Human Factors Society 31st Annual Meeting* (pp. 107–111). Santa Monica, CA: Human Factors Society.

Effects of Instructional Strategy and Motion Presentation Conditions on the Acquisition and Transfer of Electromechanical Troubleshooting Skill

ROBERT W. SWEZEY,[1] *InterScience America, Sterling, Virginia,* RAY S. PEREZ, *U.S. Army Research Institute, Alexandria, Virginia, and* JOHN A. ALLEN, *George Mason University, Fairfax, Virginia*

Three instructional strategy conditions and the presence or absence of visually presented motion during instruction were manipulated, and their effects on the acquisition and transfer of electromechanical troubleshooting performance were investigated. In this study use of visually presented motion during training (as opposed to static display presentation) was not found to enhance either maintenance performance or transfer on troubleshooting tasks. However, results indicated that subjects trained via a procedure-based training strategy performed more accurately, but slower, than did subjects whose training consisted of conceptual information concerning system structure and function for a reference performance task. When a transfer task was used as the criterion measure, results indicated that training that included conceptual information concerning a system's structure and/or function improved performance. Results thus suggested that some level of generic structure and functional knowledge is required for cross-domain transfer on cognitive troubleshooting tasks, and that this information should include both general procedures for troubleshooting and declarative information on the structure and function of the systems of interest.

INTRODUCTION

Despite a rich history of research, a variety of important issues remain of concern to those involved in training technical skills. This is particularly true for programs designed to teach maintenance performance, which requires complex activities such as those involved in troubleshooting electromechanical systems. The development of appropriate training strategies and media presentations to best facilitate acquisition of such skills is high on the list of concerns in this area. A related issue is the extent to which skills acquired in one situation will transfer to another. Indeed, the fact that technicians are frequently required to solve one-time

---

[1] Requests for reprints should be sent to Robert W. Swezey, Clarks Ridge Rd., Rt. 3, Box 142, Leesburg, VA 22075.

problems on unfamiliar equipment in unique situations underscores skill transfer as a practical concern.

In the military many complex weapons systems incorporate sophisticated technologies that tend to evolve over time. Modifications occur after a system has been used in the field and the training for maintenance and operation has been completed. Further, personnel attrition and turnover make it difficult to ensure that a cadre of experienced, well-trained technicians is constantly at hand. In short, on strictly practical grounds, there is ample justification for concern about how to best train maintenance personnel.

The research described in this article is the second in a series of studies designed to address some of these issues. It was conducted as part of an ongoing program of research aimed at studying the ways in which training system variables may influence the effectiveness of electrical and mechanical troubleshooting performance. In a previous article Swezey, Perez, and Allen (1988) reported on a series of studies that manipulated such parameters as degree of hands-on practice of maintenance skills, effects of various instructional delivery systems, and the contribution of technical documentation and job aids to complex maintenance performance. In a companion article Allen, Hays, and Buffardi (1986) addressed the unique and independent effects of physical and functional fidelity of simulation on troubleshooting performance and transfer. A comprehensive review of the fidelity literature has been provided by Alessi (1988), and Schlechter (1986) has reviewed the general effects of computer-based instruction on complex task performance.

Training Strategies: Skill Acquisition and Transfer

As Baldwin and Ford (1988) have noted, a large proportion of the empirical research on transfer has focused on ways of improving training programs by incorporating general learning principles. Four of these—the concept of identical elements, the teaching of general principles, stimulus variability, and the effects of practice—have received the most attention. Although the Baldwin and Ford review emphasized research in the organizational training literature, similar principles have been used in research focusing on technical training, such as that found in the maintenance skills area. Of particular interest are studies aimed at determining whether a strategy of teaching task-specific procedures—as opposed to one emphasizing general principles/knowledge (theory) procedures—will produce superior acquisition, retention, and transfer.

Research addressing the issue of appropriate training strategies for facilitating skill acquisition and transfer has been extensive. Morris and Rouse (1985), for example, have reviewed studies focusing on the effects of various instructional strategies on transfer. A general finding was that training programs that emphasized the learning of procedures, as compared with those that emphasized theory, were often more efficient (i.e., required less training time) but that the resulting skills frequently did not generalize to other situations or equipment. Conversely, training regimens emphasizing theoretical knowledge, though generally less efficient in terms of time, frequently have resulted in greater transfer to similar tasks. In view of the fact that many findings in this area are equivocal, Morris and Rouse noted that it is difficult to select a single approach as the most effective. Indeed, the suggestion has been offered that many of the studies may have suffered from methodological and/or theoretical flaws. For example, Morris and Rouse have noted that experimental treatments often lack consistent operational definitions, and thus many training programs mix procedural and conceptual elements of a task in unknown ways.

A suggestion was also made to the effect that integrated training approaches, in which procedural and theoretical knowledge is manipulated in carefully defined ways, should be investigated.

Although practical advice concerning how such information might best be integrated, organized, structured, and sequenced is sparse, contemporary cognitive science and recent models of the transfer process offer some suggestions on the topic (e.g., see Gray and Orasanu, 1987). Kieras (1987), for example, has emphasized the importance of requiring trainees to acquire a deeper understanding of the operation of the equipment with which they work—that is, a knowledge of functional and structural properties. He has suggested that understanding both "how" and "what" machine components work may serve to facilitate the development of a mental model of the task at hand. This, in turn, may facilitate transfer. Within his theory of skill acquisition, Anderson (1981, 1983, 1987; Singley and Anderson, 1989) has proposed a two-stage theory of problem-solving skills in which *declarative* knowledge (i.e., concepts, facts, principles) is first acquired, followed by *procedural* knowledge (i.e., the "hows" of a task). In this view declarative knowledge gives rise to procedural knowledge. The contention is that declarative knowledge is flexible and therefore accessible to a large number of problem-solving methods.

In the present study we have attempted to identify several elements that might enhance transfer from one troubleshooting situation to another. These include knowledge concerning such things as troubleshooting strategies and processes, principles of electronic circuits and thermodynamics, functional knowledge of the system, and so on. In maintenance situations, the training required to enable complex task performance may include at least four components: theoretical knowledge of the functional interrelationships among equipment elements as well as the causal aspects of their operation; structural knowledge of the component parts of equipment; procedural knowledge of the specific steps involved in operating and/or maintaining the equipment; and the provision for hands-on practice of the required tasks. Typically maintenance training curricula, as practiced by Army schools and elsewhere, tend to emphasize the structural and procedural portions of the knowledge base, with much less emphasis on conceptual and functional knowledge.

A primary focus of the present study, therefore, was to investigate effects of conceptual versus procedural training strategies on troubleshooting performance within a specific context. *Procedural* training was defined as providing step-by-step instructions on precisely how to complete a prescribed troubleshooting task, whereas *generic system structure and function* (GSSF) training was defined as providing instruction on conceptual and structural aspects of engine operation. A third strategy, *integration* (I), was also defined to provide instruction that combined training material from both training approaches.

A substantial literature base (see Rouse, 1982) suggests that training via an integrated strategy framework should result in superior maintenance performance during transfer (i.e., untrained) tasks because it provides both a conceptual understanding of the subject domain and a pragmatic prescription for performance. Similarly, a considerable amount of real-world maintenance experience suggests that a strictly procedural training strategy often produces superior troubleshooting performance in specific applied situations (see Pieper, Swezey, and Valverde, 1970). An important purpose of the present study, therefore, was to address several aspects of these issues experimentally.

The Use of Motion in Instruction

Although the choice of an effective training strategy is an essential issue in maintenance situations, media selection decisions are also important. This is increasingly true given that the numbers and kinds of media presentations available to the instructional designer continue to increase steadily. To date, however, research has not provided decision makers with practical, valid, and dependable guidelines for making such choices, at least in terms of instructional effectiveness (Higgins and Reiser, 1985; Levie, 1975). One potentially significant media attribute that has received only limited attention is motion (Llaneras, Swezey, and Allen, 1989), and of the work that has been done, equivocal findings have been widespread. For example, some researchers have found that dynamic presentations enhance maintenance training (e.g., Spangenberg, 1973), whereas others have found that they do not (e.g., Laner, 1954). It nevertheless seems possible that important components of the skill of troubleshooting include changes in the functional relationships among elements over time (e.g., fuel traveling from one component of the fuel system to another). One way to present such relationships is to use motion in the training materials. In view of this, a secondary aim of the present study was to focus on the effects of dynamic (i.e., videotape, motion-based) versus static (i.e., 35 mm slides) media presentations on maintenance task performance. Interactions between instructional strategy conditions and the presence or absence of motion in training presentations were of particular interest.

Army Maintenance Training and Evaluation Simulation System (AMTESS)

The research conducted in this study was performed on the Army's computer-based AMTESS training devices. The overall purpose of the AMTESS program is to accumulate data on ways in which training system variables may influence the effectiveness of electronic and mechanical maintenance performance. Such information can then be used to develop recommendations and, ultimately, specifications and guidelines for the design and use of simulation-based maintenance training systems.

Briefly, the AMTESS training devices were designed to meet an existing need for general-purpose maintenance training equipment. They offer the capability to provide training in a wide variety of maintenance skill areas that are pertinent and generalizable to Army jobs. They include both two-dimensional cathode ray tube displays that present training material and three-dimensional simulators for hands-on practice and testing. Major components include an instructor station with a central computer, a separate student station, and modular three-dimensional interchangeable components. Previous research activity using the devices as a test bed for research on training device features was reported in Swezey et al. (1988).

METHOD

Subjects

Subjects in this study consisted of 120 college undergraduates (49 males and 71 females) ranging in age from 17 to 30 years, with a mean age of 19. All subjects were drawn from introductory psychology courses at George Mason University and received $5.00 per hour plus course credit for their participation.

Design

The research design employed in this study consisted of a 3 × 2 factorial design with repeated measures. Subjects were randomly assigned to one of three instructional strategy conditions identified as *procedural* (P) training, *generic system structure and function* (GSSF) training, or the *integration* (I) of these two conditions. On the second dimension subjects were assigned to one of two motion conditions in which the training program was presented either via videotape, which showed motion, or via a static presentation using randomly accessed 35 mm slides with a synchronized audiotape. Data were collected at the end of training and again following a one-week retention interval.

Task

The task selected for use in the development of the training conditions involved insertion of one of two predetermined malfunctions into the engine simulator, either of which caused the engine to fail to start. Components in the AMTESS diesel engine simulator which could prevent the engine from starting included the starter motor and fuel pump, both of which are terminal points in their respective engine subsystems. The troubleshooting task selected for use in this study, therefore, was defined such that a malfunction could be programmed into either the starter motor or the fuel pump, both of which prevented the engine from starting.

The task required that each subject first try to start the engine simulator. When it failed to start, the subject's task was to determine whether the malfunction occurred in the starter or in the fuel subsystem, an activity termed *between-stage troubleshooting*. Once it was determined which subsystem contained the malfunction, the task required that each subject troubleshoot specific components of that subsystem until the malfunctioning component was identified. This was accomplished by performing a series of branched observations throughout the subsystem, in which components were checked until the final malfunctioning component was identi-

fied. This process was termed *within-stage troubleshooting*.

Training Strategy Conditions

Procedural (P) condition. In the P condition, subjects received step-by-step training on how to troubleshoot relevant components of the AMTESS diesel engine. The training provided specific, directive instruction on the precise tasks required to successfully troubleshoot the fuel and electrical systems on the engine simulator. Toward this end an approach based on instructional systems development was taken (see Branson et al., 1975), in which the following items were developed: task list, task analysis, training objectives, performance objectives, and criterion-based test (of the performance objectives). This approach was used to develop two forms of the P condition (i.e., dynamic and static versions). In the dynamic condition training was delivered via videotape (and thus was capable of showing motion), whereas in the static condition identical instruction was delivered via a static (i.e., slide/tape) presentation. Visuals presented in the P conditions emphasized actual photography (presented either via VHS videotape in the dynamic condition or via randomly accessible 35 mm slides in the static condition) of the operating diesel engine simulator, and employed both highlighting and zooming to emphasize equipment operating procedures.

Generic system structure and function (GSSF) condition. The GSSF condition provided (again, in both static and dynamic versions) instructional information on what the relevant engine system components were and how they operated in their system context. To this end, using specially prepared graphic diagrams and schematics, information concerning the structure (e.g., what the fuel pump is) and function (how it operates) was presented to subjects. In the GSSF motion-based condition, the schematic material identical to that in the static condition was presented via videotaped dynamic graphics that showed, for instance, the flow of fuel through the system. Photographic displays of the actual system were specifically avoided, as was direct procedural information on the operation of the specific reference system of interest.

Integrated (I) condition. The third experimental training condition (I) provided instructional information that included components extracted from both of the other programs (P and GSSF). Thus both system-specific procedural information, which incorporated direct equipment photographs, and component structural/functional information via graphics and line drawings, were provided in the I condition. Again, both static and dynamic versions were developed. The information presented in the I condition consisted of direct excerpts from the P and GSSF material, which were integrated by presenting P information followed by GSSF information for each specific system component (e.g., the battery) for fuel system components first, followed by electrical system components. (Note that the identical order of presentation of system information was maintained for all instructional conditions: equipment start-up information, followed by between-stage troubleshooting information, followed in turn by within-stage troubleshooting information for the fuel or electrical subsystems.)

All subjects in all conditions were provided with workbook-based problem questions that addressed the specific information provided in their condition. All instructional conditions—both static and dynamic—required approximately 45 min (±5 min) to complete, including completion of the workbook exercises. (The I condition subjects received selected material from the P and GSSF conditions, which, when combined and redundancies eliminated, also required approximately 45 min to complete.)

Performance Measures

Performance measures employed in this study consisted of four classes of measures, as follows: (1) a hands-on, criterion-based reference task administered on the AMTESS diesel simulator; (2) a similar hands-on transfer task administered on a separate (and considerably different) AMTESS simulator; (3) a second, abstract, transfer task, which assessed strategic troubleshooting performance administered via a series of abstract, computer-based problems; and (4) a multiple-choice conceptual knowledge task, which assessed electrical and fuel system structure and function. All performance measures were administered both immediately following completion of the training program and again after a one-week retention interval.

Diesel engine reference task. The diesel engine performance test consisted of a criterion-referenced, hands-on task administered on the AMTESS simulator. The task required performance of a relatively complex troubleshooting activity in which the subject was required, in the following order, to (1) perform a predetermined procedure for starting the engine; (2) establish in which of two engine subsystems (fuel system or starter system) a previously inserted malfunction existed which prevented the engine from starting (between-stage troubleshooting); (3) troubleshoot the engine subsystem in which the malfunction was presumed to exist (within-stage troubleshooting); and (4) identify the malfunctioning component. Malfunctions were inserted into alternative subsystems in a counterbalanced fashion over the two sessions in which each subject participated.

Performance data were obtained on two classes of measures: time and errors. These measures and their scoring schemes were based on those used in previous studies (e.g., Swezey et al., 1988). Time measures were computed on the number of seconds required to complete each portion of the task, and error measures quantified the number of errors made while troubleshooting, as well as the number of checks made during the troubleshooting process. In addition, the sequence of all responses was recorded.

Hands-on transfer task. This task used a separate AMTESS engine simulator as a test device for measuring transfer of training to an independent (but comparable) task. The transfer device was a relatively low-fidelity, non-diesel-powered engine simulator composed of a starter subsystem, a generator, and a voltmeter. Specifically, the existing device consisted of an instrument panel, positive and negative leads, grounds, starter motor, batteries, and starter solenoid. The instrument panel included a starter button, battery switch, starter switch, and battery/generator gauge. The device was slightly modified to represent an appropriate transfer task. Because a fuel subsystem was not represented in the extant device, a mock-up of such a system was built using a plywood foundation with simulated engine parts. The resultant fuel subsystem included a fuel pump, fuel filter, carburetor, throttle control (on the instrument panel), connections, and fuel lines.

The modified training device, constructed as indicated earlier, provided a vehicle for measuring transfer of training to a comparable engine troubleshooting task. The resulting task was similar in concept to the diesel engine reference task but required performance on a separate, nondiesel engine device.

Scoring for this transfer task was identical to that for the diesel engine reference task. Time in seconds was recorded for each stage, and errors and checks were recorded and scored using the same criteria as in the reference task.

Abstract transfer task. A second, more abstract transfer task was also developed. This task employed a computer-based, diagnostic troubleshooting task that consisted of a series of hypothetical troubleshooting problems. Each problem was composed of a configuration of components, outputs, and interconnections, each of which was represented by a different symbol (circles, squares, connecting lines, arrows). In each problem one component was not working. Subjects were required to check components on the screen until they believed that they had isolated the malfunctioning component. If a working component was checked, then the computer emitted a beep, indicating that the component was working properly. When the subject encountered the actual malfunctioning component (the correct response), the screen disappeared and the next problem was presented.

This task provided a vehicle for examining subjects' troubleshooting strategies in an abstract, generic context. It was programmed to record detailed accounts of every response made by a subject. For the present purposes, measures of efficiency were collected. These measures included time to problem solution and number of component checks made during the problem-solving process.

Conceptual knowledge task. A written knowledge test was developed to assess subjects' acquisition and retention of the knowledge base provided in the instruction. The test was organized into three sections, reflecting the order in which instruction was presented. These sections addressed overall engine functions and the fuel subsystem or starter subsystem. Subjects were required to identify components on diagrams of each subsystem and then to answer multiple-choice questions about the functions of each subsystem.

Additional Measures

The problem sets used in the criterion tasks were validated in previous studies in this series (see Swezey, Perez, and Allen, 1988, for a discussion of this issue) and were counterbalanced in terms of order of presentation across the one-week retention interval. In addition to the measures described earlier, demographic data were also collected from all subjects, including age, gender, academic year status, academic major, and estimates of familiarity with diesel engine maintenance. If subjects were found to be familiar with diesel engine maintenance issues, they were not included in the study.

RESULTS

The primary analyses conducted in this study consisted of a series of 3 × 2 MANOVA computations (three instructional strategy conditions × static vs. dynamic presentation conditions) and supporting post hoc multiple comparison tests. Separate MANOVAs were computed for two testing sessions (end of training and one-week follow-up).

Reference Task Performance

At the end of training significant differences were identified among the instructional strategy conditions on reference task performance for time required to complete the task, $F(2,114) = 59.90$, $p < 0.001$; total number of errors committed, $F(2,114) = 13.07$, $p < 0.001$; total number of checks performed $F(2,114) = 41.59$, $p < 0.001$; and on the number of correct checks performed, $F(2,114) = 15.41$, $p < 0.001$. Post hoc Scheffé test results, which provided direct two-way comparisons among the instructional strategy conditions, showed that the direction of the significant differences cited earlier favored the GSSF and/or I conditions over the P condition on time ($p < 0.05$), errors ($p < 0.05$), and total checks ($p < 0.05$). On the number of *correct* checks variable, however, Scheffé test results favored the P condition ($p < 0.05$). Thus on reference task performance at the end of training, subjects in the P group took longer to perform the task than did I of GSSF group subjects. Further, they made more errors and

performed more total checks than did subjects in either of the other two groups. However, subjects in the P group also made significantly more *correct* checks than did subjects in either of the other groups. No differences occurred among GSSF and I group performance on any major reference task dependent variable at the end of training ($p > 0.05$).

Almost identical results occurred for subjects in the various instructional strategy groups upon retesting (with a different malfunction inserted) after a one-week retention interval on reference task performance. No meaningful differences occurred on reference task performance ($p > 0.05$) among subjects in the static and dynamic presentation conditions either at end of training or at the one-week follow-up testing. Similarly, no major Instruction × Presentation interactions were observed ($p > 0.05$) for either testing session. Means and standard deviations for experimental groups on the reference task are shown in Table 1.

Hands-On Transfer Task Performance

Performance on the hands-on transfer task showed significant differences among the instructional conditions at the end of training for the following variables: time, $F(2,114) = 6.31$, $p < 0.003$; total errors, $F(2,114) = 4.76$, $p < 0.01$; and correct checks, $F(2,114) = 6.19$, $p < 0.003$. No differences were observed among the instructional strategy groups on the total number of checks made ($p > 0.05$). As was the case with the reference task, no differences occurred between the static and dynamic presentation groups, nor were there any significant Presentation × Instructional Strategy interactions ($p > 0.05$) on any of the primary dependent variables. Scheffé (and Tukey) post hoc test results indicated that where significant instructional strategy condition main effects occurred (reported earlier), all favored the I condition over the P

TABLE 1

Means and Standard Deviations on Reference Task Performance Measures for Experimental Groups

			Performance Measures			
			Time	Errors	Checks	Correct Checks
Instructional Strategy Conditions						
P	EOT	M	527	17	19	9
		SD	202	10	9	5
	1 wk	M	431	16	17	8
		SD	244	9	10	4
GSSF	EOT	M	227	13	8	4
		SD	187	5	6	3
	1 wk	M	117	14	7	4
		SD	75	5	5	3
I	EOT	M	136	9	6	5
		SD	109	4	5	4
	1 wk	M	101	11	6	4
		SD	65	6	5	3
Motion Conditions						
Static	EOT	M	320	14	12	5
		SD	252	8	10	4
	1 wk	M	220	14	11	5
		SD	210	7	9	4
Dynamic	EOT	M	273	12	10	6
		SD	223	7	7	5
	1 wk	M	213	13	10	5
		SD	220	7	9	4

P = procedural; GSSF = generic system structure and function; I = integration. EOT = end of training data collection session. 1 wk = one week follow-up data collection session.

condition ($p < 0.05$). In the case of the total errors dependent variable, a significant difference was also demonstrated which favored the I condition over the GSSF condition ($p < 0.05$).

Performance on the transfer task after the one-week interval showed significant differences among instructional strategy conditions for the total time variable, $F(2,114) = 6.15$, $p < 0.003$, and total errors variable, $F(2,114) = 3.15, p < 0.05$, but not for the numbers of total and correct checks performed. No Presentation Mode or Instructional Strategy × Presentation Mode interactions were significant ($p > 0.05$). Post hoc tests indicated that the two significant strategy condition

main effects that did occur favored the I and GSSF groups over the P group for the total time measure ($p < 0.05$) and the I group over the P group ($p < 0.05$) for the total errors measure. Table 2 shows means and standard deviations for the experimental groups on the hands-on transfer task.

Thus on the transfer task, groups whose training included information on generic system structure and function (I and GSSF groups) exhibited superior performance at the end of training compared with the P condition, whereas this effect, though still evident, was less pronounced after a one-week retention interval. The only exception to this was on total checks made during transfer task

TABLE 2

Means and Standard Deviations on Hands-On Transfer Task Performance Measures for Experimental Groups

			Performance Measures			
			Time	*Errors*	*Checks*	*Correct Checks*
Instructional Strategy Conditions						
P	EOT	M	394	6	10	5
		SD	260	6	6	3
	1 wk	M	277	8	11	5
		SD	250	9	8	4
GSSF	EOT	M	308	7	11	6
		SD	236	6	7	3
	1 wk	M	154	7	11	6
		SD	122	8	8	3
I	EOT	M	216	4	10	8
		SD	159	4	5	4
	1 wk	M	159	4	10	7
		SD	126	5	7	4
Motion Conditions						
Static	EOT	M	301	5	9	6
		SD	239	5	5	3
	1 wk	M	216	8	11	6
		SD	197	9	9	4
Dynamic	EOT	M	310	6	11	6
		SD	228	6	7	3
	1 wk	M	177	5	10	6
		SD	169	7	7	4

P = procedural; GSSF = generic system structure and function; I = integration. EOT = end of training data collection session. 1 wk = one week follow-up data collection session.

performance; in this case no differences were exhibited among instructional strategy condition groups either at the end of training or after the one-week interval.

Abstract Transfer Task Performance

For the abstract transfer task, data were collected on the time required to reach the correct problem solution and on the total number of components that were checked during the solution process only. No data were available on so-called correct checks, nor were data available on hands-on errors for this task. Results indicated that few significant main effects and no significant interactions occurred for performance on this task

($p > 0.05$), indicating that the task was not sufficiently sensitive to differentiate among the instructional strategy and presentation conditions either at the end of training or following the one-week delay. Table 3 shows means and standard deviations for the experimental groups on the abstract transfer task.

Conceptual Knowledge Task Performance

Results of the administration of the conceptual knowledge test indicated that significant main effects occurred across instructional strategy conditions for all measured knowledge components. No significant differences ($p > .05$) occurred between static versus dynamic presentation modes, or for Strategy ×

TABLE 3

Means and Standard Deviations on Abstract Transfer Task Performance Measures for Experimental Groups

			Performance Measures	
			Solution Time	Component Checks
Instructional Strategy Conditions				
P	EOT	M	394	10
		SD	260	6
	1 wk	M	277	11
		SD	250	8
GSSF	EOT	M	308	11
		SD	236	7
	1 wk	M	154	11
		SD	122	8
I	EOT	M	216	10
		SD	159	5
	1 wk	M	159	10
		SD	126	7
Motion Conditions				
Static	EOT	M	301	9
		SD	239	5
	1 wk	M	216	11
		SD	197	9
Dynamic	EOT	M	310	11
		SD	228	7
	1 wk	M	177	10
		SD	169	7

P = procedural; GSSF = generic system structure and function; I = integration. EOT = end of training data collection session. 1 wk = one week follow-up data collection session.

Presentation Mode interactions for tests administered either at the end of training or after the one-week follow-up period. Four scores were computed on this test: knowledge of overall engine function, fuel system knowledge, starter system knowledge, and total test score.

Across instructional strategy conditions, the following significant results occurred at the end-of-training test administration: overall engine knowledge, $F(2,114) = 8.73$; $p < 0.001$; fuel system knowledge, $F(2,114) = 16.08$; $p < 0.001$; starter system knowledge, $F(2,114) = 11.71$; $p < 0.001$; and total test items correct $F(2,114) = 8.73$; $p < 0.001$. In all cases post hoc analyses showed that these differences favored the I condition over the P and/or GSSF conditions ($p < 0.05$). Results

identical to those obtained at the end-of-training conceptual knowledge test administration also occurred for test readministration following the one-week retention period. Means and standard deviations for the experimental groups on the conceptual test are shown in Table 4.

Thus results demonstrated that for overall system knowledge, information on both procedural and generic system structure and function components (as were included in the I condition) was necessary; however, exclusive emphasis on either P or GSSF information was insufficient to present adequate overall system knowledge.

Intercorrelations among Variables

A great deal of information was collected on intercorrelations among the various dependent measures used in this study and among the demographic data collected from subjects and the various dependent variable measures. In general, moderate to low intercorrelations ($r = -0.36$ to $+0.33$) were observed among the classes of dependent variables (i.e., reference task measures vs. transfer task measures), indicating that although commonalities existed, the tasks were in fact unique. Similar levels of correlations occurred among abstract transfer task variables and the reference and transfer task measures ($r = -0.33$ to $+0.32$; tables of correlations among dependent variables are available from the first author).

With respect to the existence of relationships among subject demographic variables and the major dependent measures used in the study, correlations were computed in order to identify preexisting relationships that might serve to contaminate and thereby reduce the internal validity of the performance measures used as dependent variables. An arbitrary criterion (0.70 in either direction) was established in order to differentiate meaningful from unmeaningful variable intercorrela-

TABLE 4

Means and Standard Deviations on Conceptual Knowledge Task Performance
Measures for Experimental Groups

			Number Correct			
			Total	*Fuel*	*Starter*	*General*
			Instructional Strategy Conditions			
P	EOT	*M*	27.70	12.32	9.03	6.35
		SD	7.67	3.2	3.2	2.3
	1 wk	*M*	29.30	12.53	10.27	6.5
		SD	6.75	3.11	2.83	1.95
GSSF	EOT	*M*	30.83	13.08	10.33	7.43
		SD	6.08	2.53	2.54	2.19
	1 wk	*M*	30.13	12.53	10.18	7.43
		SD	6.82	3.32	2.75	1.93
I	EOT	*M*	35.73	15.57	11.85	8.3
		SD	5.10	2.11	1.96	1.67
	1 wk	*M*	35.9.	15.73	11.95	8.32
		SD	5.26	1.78	2.26	1.8
			Motion Conditions			
Static	EOT	*M*	31.42	13.53	10.63	7.33
		SD	6.79	2.64	2.36	1.98
	1 wk	*M*	31.98	13.54	11.25	7.37
		SD	6.33	2.74	2.53	1.62
Dynamic	EOT	*M*	31.42	13.83	10.18	7.38
		SD	7.52	2.67	2.96	2.13
	1 wk	*M*	31.60	13.63	10.45	7.47
		SD	7.54	2.78	2.74	2.17

P = procedural; GSSF = generic system structure and function; I = integration. EOT = end of training
data collection session. 1 wk = one week follow-up data collection session.

tions. This criterion was chosen (as opposed to default significance levels) because such values merely identify relationships that are significantly different from a criterion of no relationship (i.e., zero) and are themselves not meaningful in this situation. With the criterion set at 0.70, no meaningful relationships were noted among individual difference variables and dependent measures used in this study. (Considerable additional data are available on these issues in Swezey, Llaneras, Allen, and Perez, 1989.)

DISCUSSION

Motion Conditions

For the vast majority of the analyses computed, no performance differences were observed on any of the dependent variables measured in this study among subjects who received static versus dynamic presentations of the training material, regardless of the instructional strategy condition employed. The few static versus dynamic comparisons that did show significant differences across groups (2 on the abstract transfer task and 1 on the total errors measure in the hands-on reference performance task, $p < 0.05$, of 26 such comparisons reported) favored the dynamic over the static condition. These comparisons may be attributable to Type I errors, as the level of significance was marginal in all three cases.

In general, therefore, the data collected in this study did not support the requirement for use of motion in troubleshooting and maintenance training. Llaneras, Swezey, and Allen (1989) have indicated that use of motion in training is usually considered appropriate when one or more of the following characteristics applies:

(1) it is a defining characteristic of a concept to be taught
(2) the activity involves natural human movements
(3) The activity requires simultaneous motion in different directions
(4) the activity is unfamiliar to the learner
(5) the activity is not readily described accurately with words
(6) the material contains figure-ground relationships that are critical and potentially complex
(7) the learner population consists of low-aptitude, uneducated, or retarded persons, or other cultural/language barriers exist.

Although in the present situation Characteristics 6 and 7 probably did not apply, certainly Characteristics 1 through 5 were applicable at some level to the reference maintenance task trained in this study. Llaneras, Swezey, and Allen (1989) suggested that actual data-based studies that address the motion issue in maintenance tasks are extremely rare, and in the few studies that do exist, equivocal findings have occurred. Herein lies an area that is fertile for research. The data obtained in this study, however, do not support the requirement for a motion base in training cognitive troubleshooting tasks.

Instructional Strategy Conditions

As regards the effects of instructional strategy conditions on either reference or transfer task performance, the data obtained in this study suggest several possible conclusions. Of interest here were the effects of training strategy (P, GSSF, or I), not merely on performance of the reference task trained but also on performance of an independent transfer task in the same general domain. In the present case specific diesel engine troubleshooting training was applied both to actual troubleshooting of the AMTESS diesel engine simulator and to a similar task on a separate nondiesel engine simulator.

First, concerning the hands-on reference task on the AMTESS diesel simulator, results of this study suggested that the procedurally trained (P) group performed most accurately, making more correct checks than either the GSSF or I groups. The P group also took longer, however, making more total checks and committing more errors in the process than did subjects trained via the GSSF or I strategies. Thus for accuracy on a reference task, P training appears superior to conceptually based (GSSF or I) training; however, if speed is the criterion, it does not. In most maintenance situations it is typically argued that accuracy is a more important criterion than speed.

When one considers which type of training strategy best transfers to a comparable (but untrained) task, it appears that some conceptual knowledge of the system's structure and/or function is required. In the present study sufficient structural/functional knowledge was apparently included in the I condition, considering that increased emphasis on these topics in the exclusive GSSF condition produced no performance increment (over I-trained subjects) on the transfer task, even though both GSSF- and I-trained subjects performed better than the P-trained subjects on virtually all performance criteria.

Thus in situations in which generic troubleshooting skills are required (as opposed to single, system-specific skills), some level of structural/functional training appears necessary. This appears to be the case regardless of the specific criterion (speed, accuracy, etc.) employed and confirms earlier research results (e.g., Rouse, 1982) as well as intuition on this matter. In fact, the moderate zero-order

Pearson correlation coefficients between conceptual knowledge test scores and hands-on performance measures for reference and transfer tasks both for the end-of-training and one-week follow-up data collection session provided a moderate level of direct prediction between conceptual knowledge and task performance (see Swezey, Llaneras, Allen, and Perez, 1989, for further details).

In the present study differences among groups on the abstract transfer task were not apparent, either from an instructional strategy or from a presentation mode perspective. This result may have been caused by a major dissimilarity across instructional domains from that being measured by the hands-on transfer task, where differences among groups did exist. It also may be the case that none of the instructional conditions used in this study sufficiently addressed abstract aspects of troubleshooting performance to be detected by this measure.

Finally, subjects in the I condition generally performed best, as expected, on the conceptual knowledge task, which contained both system-specific information (included in the P and I instructional conditions) and generic structure/function information (included in the GSSF and I conditions). This measure was essentially a domain mastery test for the I condition training.

Results of the study thus suggested that some level of generic structure/function knowledge is required for cross-domain transfer on cognitive troubleshooting tasks and, further, that this knowledge should include both general procedures for troubleshooting and declarative information on the structure and function of the systems of interest.

The fact that subjects in the I and GSSF conditions showed strong performance on the transfer task can be interpreted as support for the notion (of Anderson and associates) that knowledge of declarative information (principles, theory, conceptual information) is an important ingredient in facilitating transfer. Further, the fact that subjects in the P condition did not transfer well confirms this. However, because subjects in the P condition showed superior performance on the reference task, some justification is also provided for the traditional wisdom of training task-specific content for direct reference tasks (Pieper et al., 1970). Perhaps a good balance is a mixture of the two (as in the I condition), as suggested by the work of Rouse and associates. Of note, however, is that this kind of training may require considerably more effort on the part of instructional designers with respect to such issues as the levels of content taught and the sequencing of the instruction. Although not reported here, additional studies in this series which have investigated those issues (i.e., Llaneras, Swezey, Allen, and Perez, 1989; Llaneras, Swezey, and Perez, 1989) have, however, found that sequencing effects may be less important than the instructional strategy and delivery conditions for the types of maintenance skills discussed here.

ACKNOWLEDGMENTS

The research reported in this study was supported under Contract No. MDA-903-85-C-0444 from the U.S. Army Research Institute to Science Applications International Corporation, where the first author was working at the time this research was conducted. The views presented in this paper are those of the authors and do not necessarily represent the opinions of the U.S. government. The authors wish to acknowledge the important contributions made to this work by Carol A. Tolbert, Mark Barnes, Robert E. Llaneras, O. K. Park, and R. B. Pearlstein.

Because of space limitations, only a few of the many statistical analyses computed during this project were reported. Interested readers are invited to consult the study report (Swezey, Llaneras, Allen, and Perez, 1989) on this project for further details.

REFERENCES

Alessi, S. M. (1988). Fidelity in the design of instructional simulations. *Journal of Computer-Based Instruction, 15*(2), 40–47.

Allen, J. A., Hays, R. T., and Buffardi, L. C. (1986). Maintenance training simulator fidelity and individual differences in transfer of training. *Human Factors, 28*, 497–509.

Anderson, J. R. (1981). Knowledge compilation: Mechanisms for the automatization of cognitive skills. In J. R. Anderson (Ed.), *Cognitive skills and their acquisition.* Hillsdale, NJ: Erlbaum.

Anderson, J. R. (1983). *The architecture of cognition.* Cambridge, MA: Harvard University Press.

Anderson, J. R. (1987). Compilation of weak-method problem solutions. *Psychological Review, 94,* 192–210.

Baldwin, T. T., and Ford, J. K. (1988). Transfer of training: A review and directions for future research. *Personal Psychology, 41,* 63–105.

Branson, R. K., Rayner, G. T., Coxx, J. L., Furman, J. P., King, F. J., and Hannum, W. J. (1975). *Interservice procedures for instructional systems development: Executive summary and model* (ADAO19486). Fort Benning, GA: U.S. Army Combat Arms Training Board.

Gray, W. D., and Orasanu, J. M. (1987). Transfer of cognitive skills. In S. M. Cormier and J. D. Hagman (Eds.), *Transfer of learning: Contemporary research and applications* (pp. 183–215). San Diego, CA: Academic.

Higgins, N., and Reiser, R. (1985). Selecting media for instruction: An exploratory study. *Journal of Instructional Design, 8*(2), 6–10.

Kieras, D. (1987). *The role of cognitive simulation models in the development of advanced training and testing systems* (TR-87/ONR-23). Ann Arbor: University of Michigan.

Laner, S. (1954). The impact of visual aid displays showing a manipulative task. *Quarterly Review of Experimental Psychology, 6*(3), 95–106.

Levie, W. H. (1975). How to understand instructional media. *Viewpoints, 51*(5), 25–42.

Llaneras, R. E., Swezey, R. W., and Allen, J. A. (1989). Motion as an instructional feature in maintenance training. In *Proceedings of the Human Factors Society 33rd Annual Meeting* (pp. 1305–1309). Santa Monica, CA: Human Factors Society.

Llaneras, R. E., Swezey, R. W., Allen, J. A., and Perez, R. S. (1989). *Effects of procedural and conceptual information strategy sequencing on acquisition and transfer of troubleshooting performance* (Tech. Report SAIC-89-03-178). McLean, VA: Science Applications International Corp.

Llaneras, R. E., Swezey, R. W., and Perez, R. S. (1989). *Effects of content sequencing on acquisition and transfer of troubleshooting performance* (Tech. Report SAIC-89-04-178). McLean, VA: Science Applications International Corp.

Morris, N. M., and Rouse, W. B. (1985). Review and evaluation of empirical research in troubleshooting. *Human Factors, 27,* 503–530.

Pieper, W. J., Swezey, R. W., and Valverde, H. H. (1970). *Learner-centered instruction (LCI): Volume VII. Evaluation of the LCI approach* (Report 16, AFHRL-TR-70-1). Wright-Patterson Air Force Base, OH: Air Force Human Resources Laboratory.

Rouse, W. B. (1982). A mixed-fidelity approach to technical training. *Journal of Educational Technology Systems, 11*(2), 103–115.

Schlecter, T. M. (1986). *An examination of the research evidence for computer-based instruction in military training* (Tech. Report 722). Fort Knox, KY: U.S. Army Research Institute for the Behavioral and Social Sciences.

Singley, M. K., and Anderson, J. R. (1989). *The transfer of cognitive skill.* Cambridge, MA: Harvard University Press.

Spangenberg, R. (1973). The motion variable in procedural learning. *AV Communications Review, 21,* 419–436.

Swezey, R. W., Llaneras, R. E., Allen, J. A., and Perez, R. S. (1989). *Effects of instructional strategy and presentation mode conditions on acquisition and transfer of troubleshooting performance* (Tech. Report SAIC-89-01-178). McLean, VA: Science Applications International Corp.

Swezey, R. W., Perez, R. S., and Allen, J. A. (1988). Effects of instructional delivery system and training parameter manipulations on electromechanical maintenance performance. *Human Factors, 30,* 751–762.

Effects of Part-Task Training on Memory Set Unitization and Retention of Memory-Dependent Skilled Search

CHRISTOPHER J. WHALEY *and* ARTHUR D. FISK,[1] *Georgia Institute of Technology, Atlanta, Georgia*

Two experiments were conducted to examine the effects of part-task training on the acquisition and retention of a memory-dependent skill. Participants received extensive practice on a semantic category, memory/visual search task in one of three training conditions. To assess the effects of part-task training on memory element unitization, subjects trained on one third, one half, or all of the memory set elements during any given training session. Transfer tests requiring whole-task performance provided one index of training effectiveness. The results suggest that consistent memory sets can be unitized even if part-task training is used. Indeed, part-task training was as effective as whole-task training when immediate transfer was assessed. Part-task training produced retention performance equivalent to whole-task training when retention performance was determined by both target and distractor learning. Retention performance was superior for part-task training compared with whole-task training when performance was based on only target learning.

INTRODUCTION

The goal of the present research was two-fold. First, we were interested in extending our understanding of consistent component training (Eggemeier, Fisk, Robbins, and Lawless, 1988; Fisk and Eggemeier, 1988) of high-performance skills (Schneider, 1985) in situations .requiring memory-dependent skills (Logan, 1988). As one examines real-world skills such as those involving perceptual and judgment decision-making skills (e.g., air traffic control, battle management), a need for part-task training of memory-dependent skills becomes apparent. Second, we wanted to extend our documentation and understanding of the long-term retention of automatic processing. In particular, we were interested in the potential influence of part-task training on skill retention in memory-dependent search-detection tasks.

A substantial literature has emerged over the last 15 years that points to automatic component processes as fundamental building blocks of high-performance skills (for a review, see Fisk and Rogers, 1992; Logan, 1985; Schneider, 1985; Shiffrin, 1988).

[1] Requests for reprints should be sent to either author, School of Psychology, Georgia Tech, Atlanta, GA 30332-0170.

Although skill and automaticity are not synonymous terms, the basic premise of that body of research is that most complex skills are tasks that are dominated by automatic component processes. Furthermore, that research has demonstrated that only consistent task components result in automatic component processes. Research examining automatic/controlled processing and skill acquisition is yielding a growing list of principles of human information processing by which to guide developers of part-task training programs (e.g., Fisk, Lee, and Rogers, 1991; Schneider, 1985).

Consistent and Varied Mapping

A fundamental principle derived from research on the development of high-performance skills is that consistent task components should be isolated as candidates for part-task training (for real-world examples, see Eggemeier, Fisk, Robbins, Lawless, and Spaeth, 1988; Fisk and Eboch, 1989; Fisk and Gallini, 1989; Regian and Schneider, 1990). In their seminal series investigating controlled search and automatic detection, Schneider and Shiffrin (1977; Shiffrin and Schneider, 1977) demonstrated differences in performance that were modulated according to whether training was consistent or variable. This has been referred to as *consistently* or *variably mapped training*.

More precisely, in a consistently mapped (CM) situation the individual always deals with (i.e., attends to, responds to, or uses information from) a stimulus, or class of stimuli, in the same manner. CM training conditions result in dramatic performance improvements (see Schneider and Shiffrin, 1977; Shiffrin and Schneider, 1977, for details) and the eventual development of performance characteristics indicative of automatic processing. Variably mapped (VM) training situations are those in which practice is inconsistent; that is, the response or

degree of attention devoted to the stimulus changes from one stimulus exposure to another. Pure VM training conditions result in little performance improvement.

Mechanisms of Automatism

Many theories of skill development that posit performance and processing requirements changing as a function of practice are based on the modal view of a strength representation of knowledge (e.g., Anderson, 1982, 1983; Dumais, 1979; LaBerge and Samuels, 1974; MacKay, 1982; Schneider, 1985; Schneider and Detweiler, 1987; Shiffrin and Czerwinski, 1988; but see Logan, 1988, for a nonstrength theory). All of these theories propose that some increase and/or decrease in "strength" is responsible for the improvement in performance observed in tasks in which substantive learning occurs (CM tasks, within our paradigm).

The concept of strength varies among models but is generally related to the importance or significance of a stimulus, set of stimuli, rule, or connection (e.g., between nodes). In the present series of experiments we examined the development of skilled search/detection performance. Many investigations have provided evidence that search performance is determined by the strength of target stimuli relative to the strength of the distractor stimuli (e.g., Dumais, 1979; Prinz, 1979; Rogers, 1992; but see Fisher and Tanner, 1992, for an alternative view).

In addition to the development of attention-attracting strength, associative learning is a necessary component of performance improvement in search/detection tasks (see Schneider and Detweiler, 1987, for a review). Our view of associative learning is not new, and it has been precisely specified by other investigators (e.g., see McClelland, Rumelhart, and Hinton, 1986; Schneider and Fisk, 1984; Shiffrin and Schneider, 1977, for a general review). We assume memory to be a large collection of interconnected nodes. Associative learning is reflected in the modification of the activation patterns among these nodes. Stimulus information that is concurrently activated in short-term storage will become associated if the coactivation consistently occurs across numerous learning trials. Once a set of information nodes becomes associated, or *unitized*, a single representation can be extracted to represent the set. Such learning can facilitate search performance by allowing the comparison of a single, unitized memory set, rather than by forcing an item-by-item search through memory (Shiffrin, 1988).

Unitization and Part-Task Training

Skilled performance for memory-dependent tasks results partially from the ability to unitize situation-dependent categories of information. Unfortunately, little is known about the influence of part-task training on the unitization process. Hence, it is crucial to understand whether or not breaking the to-be-unitized set into manageable segments (task simplification) affects skill development. If training a subset of to-be-unitized elements compromises the unitization process, then the ability to train for automatization of memory-dependent skills using part-task training would be constrained. Hence, the present series of experiments compares skill development of a memory-dependent task under different part-task training conditions with whole-task training.

Retention of Search-Detection Skill

Another objective of the present research is to examine retention of skilled search-detection performance. Many real-world skills are often used infrequently. But when required, less-than-superior performance can lead to devastating consequences, hence the importance of understanding retention characteristics of given classes of skills (or components of skills). Also, understanding the

characteristics of performance retention within a given learning domain has been shown to be valuable for understanding the structure of learning within that domain (e.g., see Bahrick, 1979, 1984; Fisk and Hodge, 1992; Healy, Fendrich, and Proctor, 1990; Kolers, 1976; Rabbitt, Cumming, and Vyas, 1979; Salasoo, Shiffrin, and Feustel, 1985). Similarly, retention tests seem to be sensitive measures of degree or strength of learning (e.g., Bahrick, 1984). Measures of skill retention after some period of inactivity may be a better determinant of training program success than would a measure of performance during or just after training (Schendel and Hagman, 1991).

Overview of Experiment

The two experiments that follow were designed to address the issue of the effectiveness of part-task training for hybrid memory/visual search tasks. In addition, retention of trained performance as a function of memory-set segmentation was examined. Participants in the experiments searched for words from specified semantic categories. A variation of the pure-part method with simplification (Wightman and Lintern, 1985) was applied to two of the three training conditions. For each experiment subjects were required to simultaneously search for exemplars from two to six categories (depending on the experimental condition) through a succession of rapidly presented displays. Subjects progressed to a faster display speed as their accuracy levels surpassed a preset criterion. Thus subjects were trained at the limit of their perceptual abilities.

For expository purposes each experiment is divided into two parts: training and transfer and retention. The purpose of Experiment 1A was to test whether varying the memory load (i.e., breaking the to-be-unitized set into component parts) during training affected transfer performance. Subjects were trained with varying memory loads and then transferred to the whole task. Both target and distractor learning could facilitate transfer performance in this experiment. Experiment 2A replicated the first experiment, except that only target learning could support transfer performance. Hence, with results from both experiments, an assessment can be made of the effects of part-task training in memory-dependent tasks on target and on distractor learning. Experiments 1B and 2B were conducted to examine retention of trained performance 30 days following training. Experiment 1B examined retention as a function of part-task training for both target and distractor learning. Experiment 2B examined retention of only target learning.

The experimental task was explicitly designed as a conceptual analog of aspects of battle management tasks. However, the results can generalize to other tasks requiring rapid, skillful search for categories of events with responses contingent on event detection. The results should generalize to tasks in which memory load is high but the potential for memory restructuring through training is possible and in which retention of trained performance is a necessity.

EXPERIMENT 1

In Experiment 1A a comparison of part-task training (simplification and pure-part training) and whole-task training was conducted using a training/transfer design. The experiment was designed to allow subjects to benefit from both target and distractor learning. The retention phase (Experiment 1B) addressed the stability of performance after a period of inactivity and whether the performance of the part-trained groups remained equivalent to that of the whole-trained group.

Experiment 1A Method

Subjects. Seven undergraduate females and eleven undergraduate males were paid for their participation in the experiment. All subjects had at least 20/40 far and near vision. Participants were administered three subscales (Vocabulary, Digit-Symbol Substitution, and Digit Span) of the Wechsler Adult Intelligence Scale-Revised (WAIS-R; Wechsler, 1981). The averaged WAIS-R scaled scores were representative of those of the average population: Vocabulary = 13.00 (range: 10 to 17), Digit Span = 12.17 (range: 7 to 18), and Digit-Symbol Substitution = 11.72 (range: 7 to 16).

Apparatus. Individual Epson Equity I+ personal computers were programmed with Psychological Software Tools's Microcomputer Experimental Language (Schneider, 1988) to present the appropriate stimuli, collect responses, and control timing of the display presentations. Standard Epson monochrome monitors (Model MBM 2095-E) connected to Epson multimode graphics adapters were used to display the stimuli. Subjects were tested at individual subject stations with ambient pink noise at approximately 55 dB(A) to mask outside noise.

Stimuli. Target and distractor stimuli were chosen from the taxonomic category norms compiled by Battig and Montague (1969). Six categories were used for the target sets, and eight categories were used for the distractor sets (note that the stimulus items were either targets or distractors; that is, CM). Target set items consisted of the following semantically unrelated categories (Collen, Wickens, and Daniele, 1975): countries, earth formations, fruits, human body parts, occupations, and reading material. Distractor set items consisted of the semantically unrelated categories of articles of clothing, furniture, human dwellings, musical instruments, relatives, type of vehicles, units of time, and weapons. Each category contained six exemplars, and all words appeared in capital letters.

Adaptive multiple frame task. A "multiple frame" procedure was used to present successive "frames" of stimuli (exemplars from the target and distractor categories). Each frame consisted of 3 exemplars, and eight frames were presented, for a total of 24 exemplars per trial. On a positive trial only one of the 24 exemplars was a target exemplar, with the remaining 23 exemplars drawn from the distractor categories. On a negative trial none of the words was an exemplar from the target categories.

In order to train each subject to his or her performance limit, an adaptive, multiple-frame procedure was used. The procedure was based on tasks used previously in the search literature (e.g., Schneider and Shiffrin 1977; Sperling, Budiansky, Spivak, and Johnson, 1971), except that frame time (time from the onset of one word display to the onset of the next display of three words) changed as a function of the subject's performance after each block (i.e., group) of 30 trials. Adaptive frame times were set in response to the accuracy performance of each individual subject.

For each training condition the initial frame time was set to 940 ms because of the high memory load in the six category conditions. Throughout training each subject's performance (accuracy level) determined the frame speed for the following block of trials. If the subject reached an accuracy level of 86.7% (26 out of 30 trials correct) or more on a block, then each frame for the next block was presented 20 ms faster. Likewise, if the subject did not reach an accuracy level of at least 73.3% (22/30) on a block, each frame for the next block was presented 20 ms slower. Otherwise, the speed remained the same as in the previous block of trials. The speed for the next training session for each subject was based on the frame speed and accuracy of the last block of the previous training session. The adaptive element of the multiple-frame procedure allowed accuracy to stabilize at approximately 80% within four sessions.

Frame times for the transfer sessions were at three fixed frame speeds (180 ms, 220 ms, and 260 ms) to assess difficult, moderate, and easy task situations.

Procedure. During the first session, subjects completed 90 task orientation trials. Stimuli for the orientation trials were categories other than those used in the actual experiment. The orientation allowed the subjects to become familiar with the requirements of the task and the experimental environment.

Each trial consisted of the following sequence: First, the memory set of either two, three, or six category labels, depending on the condition (e.g., fruits and occupations), appeared on the left side of the screen in a column. After studying the category names for up to 30 s, the subject pressed the space bar to initiate the trial. Following trial initiation, three plus signs were presented in the center of the screen for 500 ms to allow the subject to focus on the area of the screen in which the category exemplars would appear. Next, a sequence of eight frames appeared on the screen in rapid succession. A frame consisted of three exemplars presented in a column on the screen followed by a column of X's to mask the presentation of the previous words.

On a positive trial only one of the 24 exemplars was from one of the target categories held in memory (e.g., apple); the other exemplars were distractors. The position of the target within a frame (top, middle, or bottom of the column) was selected randomly. Likewise, the target occurred on a randomly determined frame with the restriction that it occur between frames two and seven, inclusive.

The subjects were required to note either the position of the target within the column of words (i.e., top, middle, or bottom) by pressing the corresponding key (labeled T, M, or B), or that no target was present by pressing the "no" key (labeled N). The top, middle, bottom, and no keys corresponded to the 7, 4, 1, and 5 keys on the number pad, respectively. Subjects could respond at any point during the eight frames and for up to 4 s after the final frame was presented.

A subject was provided with feedback after each trial and each block of trials. On correct trials the words "correct response" appeared. If the incorrect position was given, a tone sounded and the words "incorrect, target was in [top, middle, or bottom] position" appeared. If a position was selected and no target was presented, "no target present" appeared. After each block of trials the percentage of correct trials and frame times were shown to the subject for all completed blocks during the session.

Design. The 15 sessions of Experiment 1A were divided into five different stages: Subject Orientation (one session), Training I (six sessions), Transfer I (one session), Training II (six sessions), and Transfer II (one session). Memory set size (two, three, or six categories) during training was manipulated between subjects. The dependent variables recorded during training were frame time (speed) and detection accuracy. During transfer, frame times were held constant, and accuracy was the dependent variable.

Training sessions consisted of 10 blocks of 30 trials per block. Subjects completed 1800 trials prior to the first transfer session and another 1800 trials prior to the second transfer session. An average of 20% of the trials were negative (target absent) during each session. The number of negative trials varied between 5 and 7 out of 30 trials on any given block, with the mean being 6 negative trials per block for each session.

Three training conditions were manipulated between subjects (see Figure 1 for an outline of the category training sequence for each training condition): part-task training with two categories (PT2; two categories were in the memory set), part-task training with three categories (PT3; three categories

were in the memory set), and whole-task training with all six categories (WT6; all six categories were in the memory set). In condition PT2 subjects were trained with two categories during each training session. In condition PT3 subjects were trained with three categories during each training session. For both of these conditions subjects received equal training on each of the six categories after six sessions of training. Condition WT6 differed from the other two conditions in that all six categories were presented in the memory set as potential targets throughout the six sessions of training. For conditions PT2 and PT3 the assignment of the categories to search days was counterbalanced across subjects by a partial Latin square.

In both Transfer I and Transfer II all subjects completed 270 trials with all six training categories in the memory set (a total of 540 transfer trials). Three blocks (30 trials per block) were run at each of the following speeds: 180 ms, 220 ms, and 260 ms, for a total of nine blocks. The order of frame speed for each block was counterbalanced within and across the training conditions. There were six negative trials per block (20%). The same target and distractor categories from training were used for the transfer sessions.

Results: Experiment 1A

Training. Mean frame times and accuracies for each training session were aggregated across subjects. Accuracy stabilized close to 80% after four sessions of training as a result of the adaptive procedure used. Mean frame times for all three conditions decreased over training sessions according to a normal power function (see Figure 2). A fit of the power law function (Newell and Rosenbloom, 1981) to each of the training conditions yielded an $r^2 = 0.96$ for PT2, $r^2 = 0.98$ for PT3, and $r^2 = 0.96$ for WT6. Subjects' average frame time decreased from 879 ms after the

	TRAINING						TRANSFER
Day	**1/8**	**2/9**	**3/10**	**4/11**	**5/12**	**6/13**	**7/14**
PT2	Category 1, Category 2	Category 1, Category 2	Category 3, Category 4	Category 3, Category 4	Category 5, Category 6	Category 5, Category 6	Category 1, Category 2, Category 3, Category 4, Category 5, Category 6
PT3	Category 1, Category 2, Category 3	Category 1, Category 2, Category 3	Category 1, Category 2, Category 3	Category 4, Category 5, Category 6	Category 4, Category 5, Category 6	Category 4, Category 5, Category 6	Category 1, Category 2, Category 3, Category 4, Category 5, Category 6
WT6	Category 1, Category 2, Category 3, Category 4, Category 5, Category 6	Category 1, Category 2, Category 3, Category 4, Category 5, Category 6	Category 1, Category 2, Category 3, Category 4, Category 5, Category 6	Category 1, Category 2, Category 3, Category 4, Category 5, Category 6	Category 1, Category 2, Category 3, Category 4, Category 5, Category 6	Category 1, Category 2, Category 3, Category 4, Category 5, Category 6	Category 1, Category 2, Category 3, Category 4, Category 5, Category 6

Figure 1. *Training sequence for Experiment 1. Day refers to the session number (e.g., 1/8 means days, or sessions, 1 and 8). PT2 and PT3 mean part-task training with two and three categories, respectively. WT6 means whole-task training with all six categories.*

first session to 216 ms in the last session of training.

Transfer. Mean accuracy data are presented in Table 1 for each training condition (PT2, PT3, WT6) across three different frame speeds (180 ms, 220 ms, 260 ms), and the two transfer sessions (Transfers I and II). A $3 \times 2 \times 3$ (Training Condition $\times$ Transfer Session $\times$ Frame Speed) ANOVA was performed on the accuracy data. The main effect of transfer session was significant, $F(1,15) = 13.67$, $p < 0.0022$, $MS_e = 0.0327$, reflecting the improvement in accuracy after six additional days of consistent training. Also, the effect of frame speed was significant, $F(2,30) = 33.05$, $p < 0.0001$, $MS_e = 0.0020$. However, neither the main effect of training condition nor the higher-order interactions reached signifi-

cance. A Newman-Keuls test (alpha $= 0.05$) showed significant differences among all three frame speeds.

For the part-task training groups, analyses were conducted to determine whether categories learned at different points in the training led to significantly different performance levels at transfer. The PT2 condition included three different sets of two categories, and the PT3 condition included two different sets of three categories during each phase of training. The Training Sequence $\times$ Frame Speed ANOVA showed that the main effect of training sequence did not have a significant effect on either the PT2 or PT3 conditions, F's < 1. Similarly, the Frame Speed $\times$ Training Sequence interaction failed to reach significance for either the PT2, $F < 1$, or the PT3

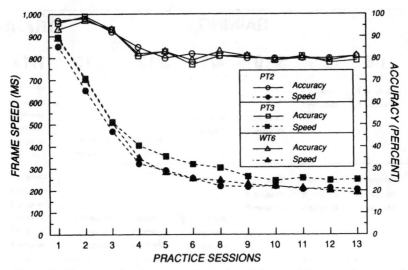

Figure 2. *Frame speed and accuracy for each training condition as a function of practice session (Experiment 1). PT2 and PT3 mean part-task training with two and three categories, respectively. WT6 means whole-task training with all six categories.*

conditions, $F(2,10) = 1.51, p = 0.2660, MS_e = 0.0036$.

Retention Performance

Experiment 1B method. Following Experiment 1A, subjects participated in an additional transfer session (Transfer III) and were asked to return 30 days later for another session (Transfer IV). Subjects participated in

an unrelated task for one day after they had completed Experiment 1A. The additional transfer test was included to ensure that experience had not disrupted performance on the search task.

The data in Table 1 show that their performance had not been disrupted; indeed, there was no difference between Transfers II and III. Transfers III and IV consisted of exactly

TABLE 1

Mean Accuracy for Transfer and Retention Sessions (Experiment 1)

Frame Speed (ms)	PT2			PT3			WT6		
	180	220	260	180	220	260	180	220	260
Transfer Session I	70	77	82	64	73	76	70	79	82
Transfer Session II	75	84	83	71	75	77	79	82	85
Transfer Session III	77	82	85	71	77	81	78	84	85
Retention	75	79	84	68	70	77	73	80	81

Note. Mean accuracies are expressed in percentages. PT2 = part-task training with two categories in the memory set; PT3 = part-task training with three categories; WT6 = whole-task training with all six categories.

the same categories and frame speeds as those of the two transfer sessions of Experiment 1A. One subject (PT2 condition) did not return for the 30-day session, and that subject's data were eliminated from the analyses.

Experiment 1B results. A summary of the mean accuracy data 30 days prior to and at retention testing is presented in Table 1. A 3 × 2 × 3 (Training Condition × Transfer Session × Frame Speed) repeated-measures ANOVA was performed on those accuracy data. A main effect of session was found, $F(1,14) = 11.87$, $p < 0.0039$, $MS_e = 0.0024$, reflecting the small performance decline (4%) over the 30-day interval. The effect of frame speed was significant, as found in Experiment 1A, $F(2,28) = 32.24$, $p < 0.0001$, $MS_e = 0.0020$, but the interaction of Frame Speed × Training Session did not reach significance, $F < 1$. As indicated by the ANOVA, the decline in performance was relatively stable across frame speeds.

The percentage decline of accuracy was greatest for the 220-ms frame speed (4.7%), followed by the 180-ms (3.3%) and finally by the 260-ms (3%) speed. Again, no differences were found among training conditions, $F(2,14) = 1.12$, $p < 0.3537$, $MS_e = 0.0420$, replicating the finding of Experiment 1A. None of the higher-order interactions reached significance (all F's < 1).

Experiment 1 Discussion

In this experiment no difference was found among training conditions. This finding is important because it suggests that consistent memory sets can be unitized—at least within the training context—even if part-task training is used. Although there is no deficit for learning only a portion of the six categories during a training session, the present experiment also demonstrated no advantage. This finding is consistent with previous experiments that have reported no benefit for part-

task training (e.g., Adams, 1987; Adams and Hufford, 1961; Briggs and Brogden, 1954; Briggs, Naylor, and Fuchs, 1962; Briggs and Waters, 1958; McGuigan and MacCaslin, 1955).

The retention results provide two important pieces of information. First, accuracy levels across training conditions remained equal after the retention interval. Second, performance levels after 30 days declined but remained higher than those of Transfer I. These data further suggest that part-task training is as effective as whole-task training for this class of tasks.

EXPERIMENT 2

In Experiment 1 no significant difference was found between part- and whole-task training. In Experiment 1 both target and distractor learning could benefit performance at transfer (because the same distractors were used during training and transfer). Hence, the goal of Experiment 2 was to explore the effects of part-task training on transfer and retention when target learning, independent of distractor learning, was the major contributor to performance improvement. In this experiment full-task performance was tested during transfer by introducing new distractor categories, isolating target set learning.

Experiment 2A Method

Subjects. Eight undergraduate females and ten undergraduate males were paid for their participation in the experiment. All subjects had at least 20/40 near and far vision. The averaged WAIS-R scaled scores were slightly higher than those of the average population: Vocabulary = 15.50 (range: 9 to 19), Digit Span = 12.67 (range: 9 to 17), and Digit-Symbol Substitution = 13.61 (range: 9 to 18).

Design and procedure. This experiment was identical to Experiment 1, except the distractor categories were changed at transfer.

Experiment 2A Results

Training. Mean frame times and accuracies for each training session were aggregated across subjects. Accuracy followed the same pattern as in the results of Experiment 1A and stabilized at approximately 80% after four sessions of training. As before, mean frame times decreased for all three training conditions according to a standard power function (see Figure 3). A fit of the power law function to each of the training conditions yielded an $r^2 = 0.97$ for PT2, $r^2 = 0.98$ for PT3, and $r^2 = 0.96$ for WT6. Subjects' average frame time decreased from 872 ms after the first session to 219 ms in the final session of training.

Transfer. Mean accuracy was determined for each training condition across frame speeds and transfer sessions and appears in Table 2. A 3 × 2 × 3 (Training Condition × Transfer Session × Frame Speed) ANOVA was performed on the accuracy data. The main effect of training session was signifi-

cant, $F(1,15) = 30.95$, $p < 0.0001$, $MS_e = 0.0022$, reflecting the improvement in accuracy after six additional days of CM practice. The main effect of frame speed also reached significance, $F(2,30) = 58.37$, $p < 0.0001$, $MS_e = 0.0016$. A Newman-Keuls test (alpha = 0.05) showed significant differences among all three frame speeds. No difference was found among training conditions, $F(2,15) = 1.24$, $MS_e = 0.0399$, replicating the finding of Experiment 1A. None of the higher-order interactions reached significance (all F's < 1). As in Experiment 1A, training sequence analyses indicated no effect on transfer performance of whether a category was learned early or late in training.

A 3 × 2 × 3 × 2 (Training Condition × Experiment Number × Frame Speed × Transfer Session) repeated-measures ANOVA was performed on the accuracy data from Experiments 1A and 2A. Results showed main effects for frame speed, $F(2,30) = 93.84$, $p < 0.0001$, $MS_e = 0.0019$; training session, $F(1,30) = 40.77$, $p < 0.0001$, $MS_e = 0.0027$;

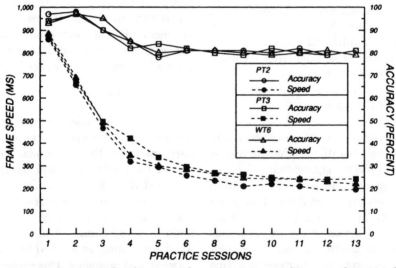

Figure 3. *Frame speed and accuracy for each training condition as a function of practice session (Experiment 2). PT2 and PT3 mean part-task training with two and three categories, respectively. WT6 means whole-task training with all six categories.*

TABLE 2

Mean Accuracy for Transfer and Retention Sessions (Experiment 2)

Frame Speed (ms)	PT2			PT3			WT6		
	180	220	260	180	220	260	180	220	260
Transfer Session I	57	63	66	56	65	66	64	68	73
Transfer Session II	62	68	73	59	66	74	70	73	78
Retention	66	68	73	65	72	74	68	69	75

Note. Mean accuracies are expressed in percentages. PT2 = part-task training with two categories in the memory set; PT3 = part-task training with three categories; WT6 = whole-task training with all six categories.

and experiment number, $F(1,33) = 14.08, p < 0.0007$, $MS_e = 0.0053$. The main effect of training condition was not significant, $F(2,30) = 2.04$, $p = 0.1477$, $MS_e = 0.0388$. None of the interactions reached significance. Transfer accuracy was higher for Experiment 1A most likely because of the ability to use both target and distractor learning at transfer in Experiment 1A but only target learning in Experiment 2A (Dumais, 1979; Rogers, 1992).

Experiment 2A Discussion

The pattern of results from Experiment 2A replicates those of Experiment 1A. Training performance (frame speed and accuracy during 12 sessions of adaptive training) was almost identical between the experiments. Comparing transfer performance for the two experiments, mean transfer accuracy (aggregated across frame speeds) was 74.8% versus 64.2% during Transfer I and 79.0% versus 69.2% during Transfer II, for Experiments 1A and 2A, respectively. The 10.6% difference for Transfer I and 9.8% difference for Transfer II was likely the result of switching distractor categories at transfer. This difference in transfer accuracy supports previous findings of distractor learning (Dumais, 1979; Rogers, 1992). During CM practice subjects strengthen consistent target categories as

well as weaken consistent distractor categories (i.e., subjects learn both target and distractor sets). The results suggest that part-task training is as good as whole-task training for tasks that involve only target learning (as in Experiment 2A) and for tasks that involve both target and distractor learning (as in Experiment 1A).

Retention Performance

All groups demonstrated high levels of retention during Experiment 1B. However, retention may have been influenced by target learning, distractor learning, or an interaction of the two. Next we tested performance after 30 days of inactivity to examine retention of search skill attributable to target learning.

Experiment 2B method. The same subjects participated in one retention session 30 days following the end of Experiment 2A. The retention session was identical to the previous transfer sessions.

Experiment 2B results and discussion. The mean accuracy data are reported in Table 2. A 3 × 2 × 3 (Training Condition × Transfer Session × Frame Speed) ANOVA was performed on the accuracy data. The effect of frame speed was significant, as found in Experiment 2A, $F(2,30) = 41.20, p < 0.0001, MS_e = 0.0021$, but the interaction of Frame Speed

× Training Session did not reach significance, $F(4,30) = 0.75, p = 0.5647, MS_e = 0.0020$.

However, an important finding was the significant interaction between training condition and training session, $F(2,15) = 3.77, p < 0.05, MS_e = 0.0030$. The interaction was the result of stable performance for the part-task training groups and declining accuracy for the whole-task group.

The interaction between training condition and training session suggests that when only target learning is important in the transfer/retention task, part-task training may facilitate the learning and/or retention of the target categories. More attention devoted to the task to be learned does result in better learning (Fisk and Schneider, 1984). Hence the superior retention performance of the part-task training groups in this experiment can be understood because the part-task training procedure allowed more attentional resources to be allocated to target categories during training.

In Experiment 1B, in which retention performance was based on two factors (target and distractor learning), degradation of one component (e.g., target learning) may have been compensated for by the other component (distractor learning). The Experiment 1B results may be completely attributable to the one additional session (Transfer III). This seems unlikely, however, because the additional session occurred after 15 other sessions and when performance appeared to have reached asymptote. In that experiment there were no differences between Transfers II and III (0.6%, 2%, and 0.3% differences for PT2, PT3, and WT6, respectively).

Regardless of the reason for stability of the retention performance in Experiment 1B, the results of the present retention experiment stand on their own. They demonstrate that part-task training may facilitate retention in situations in which only target learning can be used at retention.

CONCLUSIONS

The present series of experiments was designed to address how part-task training influences skill involving memory set unitization. The important results are as follows:

1. Part-task training and whole-task training did not lead to differences in *patterns* of transfer whether both target and distractor learning (Experiment 1A) or only target learning (Experiment 2A) was assessed.
2. However, a comparison of Experiments 1A and 2A suggested that distractor learning had a large effect on transfer *performance* for both part- and whole-task learning.
3. Performance of the three training conditions did not differ after a 30-day retention interval if retention performance was based on both target and distractor learning (Experiment 1B).
4. However, when retention of only target learning was assessed (Experiment 2B), part-task training was slightly but significantly superior to whole-task training.

The present laboratory task represents an abstraction of important real-world task characteristics, particularly for event detection tasks. Abstracting the important psychological characteristics of a targeted complex task has proven to be a successful and important strategy for understanding performance in complex task environments within the context of laboratory control (e.g., Brunswik, 1956; Hammond, 1966). The present laboratory task was designed in the spirit of Brunswik's representative design. Of course, one must have knowledge of the psychological characteristics of the complex task for the laboratory abstractions to be meaningful. Whether or not the present task captures those essential characteristics is still an empirical question, though we think the present task succeeds in that regard. With such a caveat in mind, we believe the implications of the present findings for training program design are straightforward and important.

The present data suggest that, within reason, tasks requiring memory set unitization for skillful performance may be efficiently trained by segmenting the to-be-unitized memory set elements. In fact, when target learning alone is the critical component of the search/detection skill, then such part-task training may lead to superior learning, at least as assessed by performance requiring retention of the skill. Such training will be important when it is more cost-effective to break a task into subcomponents (Adams, 1987) or when high initial memory load would impede learning (Fisk and Schneider, 1984).

These results also suggest that part-task training may be beneficial in refresher courses for tasks involving combined memory and visual search (e.g., air traffic control, computer operation). Refresher courses could provide a greater amount of practice on individual groups of subtasks without showing a deficit when the tasks are reintegrated. Concentration on the more important subtasks would allow more cost-effective refresher training to be developed (Wightman and Lintern, 1985).

ACKNOWLEDGMENT

This research was sponsored by the Air Force Human Resources Laboratory, Logistics and Human Factors Division (AFHRL/LRG), Wright-Patterson Air Force Base, Ohio (Contract F33615-88-C-0015). Beverley Gabel was the AFHRL contract monitor.

REFERENCES

Adams, J. A. (1987). Historical review and appraisal of research on the learning, retention, and transfer of human motor skills. *Psychological Bulletin, 101*, 41–74.

Adams, J. A., and Hufford, L. E. (1961). *Effects of programmed perceptual training on the learning of contact landing skills* (Tech. Report NAVTRADEVCEN 247-3). Port Washington, NY: U.S. Naval Training Device Center.

Anderson, J. R. (1982). Acquisition of cognitive skill. *Psychological Review, 89*, 369–406.

Anderson, J. R. (1983). *The architecture of cognition.* Cambridge, MA: Harvard University Press.

Bahrick, H. P. (1979). Maintenance of knowledge: Questions about memory we forgot to ask, *Journal of Experimental Psychology: General, 108*, 296–308.

Bahrick, H. P. (1984). Semantic memory in permastore: Fifty years of memory for Spanish learned in school. *Journal of Experimental Psychology: General, 113*, 1–29.

Battig, W. F., and Montague, W. E. (1969). Category norms for verbal items in 56 categories: A replication and extension of the Connecticut category norms. *Journal of Experimental Psychology Monographs, 80*(3, Pt. 2).

Briggs, G. E., and Brogden, W. J. (1954). The effect of component practice on performance of a lever-positioning skill. *Journal of Experimental Psychology, 48*, 375–380.

Briggs, G. E., Naylor, J. C., and Fuchs, A. H. (1962). *Whole versus part training as a function of task dimensions* (Tech. Report NAVTRADEVCEN 950-2). Port Washington, NY: U.S. Naval Training Device Center.

Briggs, G. E., and Waters, L. K. (1958). Training and transfer as a function of component interaction. *Journal of Experimental Psychology, 56*, 492–500.

Brunswik, E. (1956). *Perception and the representative design of psychological experiments.* Berkeley: University of California Press.

Collen, A., Wickens, D. D., and Daniele, L. (1975). The interrelationship of taxonomic categories. *Journal of Experimental Psychology: Human Learning and Memory, 1*, 629–633.

Dumais, S. T. (1979). *Perceptual learning in automatic detection: Processes and mechanisms.* Unpublished doctoral dissertation, Indiana University, Bloomington, IN.

Eggemeier, F. T., Fisk, A. D., Robbins, R. J., and Lawless, M. T. (1988). Application of automatic/controlled processing theory to training tactical command and control skills: II. Evaluation of a task-analytic methodology. In *Proceedings of the Human Factors Society 32nd Annual Meeting* (pp. 1232–1236). Santa Monica, CA: Human Factors and Ergonomics Society.

Eggemeier, F. T., Fisk, A. D., Robbins, R. J., Lawless, M. T., and Spaeth, R. (1988). *High-performance skills task analysis methodology: An automatic human information-processing theory approach* (Final Tech. Report AFHRL-TP-88-32, AD-B128 366). Wright-Patterson Air Force Base, OH: Air Force Human Resources Laboratory, Logistics and Human Factors Division.

Fisher, D. L., and Tanner, N. S. (1992). Optimal symbol set selection: A semiautomated procedure. *Human Factors, 34*, 79–95.

Fisk, A. D., and Eboch, M. (1989). Application of automatic/controlled processing theory to training component map-reading skills. *Applied Ergonomics, 20*, 2–8.

Fisk, A. D., and Eggemeier, F. T. (1988). Application of automatic/controlled processing theory to training tactical command and control skills: 1. Background and task-analytic methodology. In *Proceedings of the Human Factors Society 32nd Annual Meeting* (pp. 1227–1231). Santa Monica, CA: Human Factors and Ergonomics Society.

Fisk, A. D., and Gallini, J. K. (1989). Training consistent components of tasks: Developing an instructional system based on automatic/controlled processing principles. *Human Factors, 31*, 453–463.

Fisk, A. D., and Hodge, K. A. (1992). Retention of trained performance in consistent mapping search after extended delay. *Human Factors, 34*, 147–164.

Fisk, A. D., Lee, M. D., and Rogers, W. A. (1991). Recombination of automatic processing components: The effects of transfer, reversal, and conflict situations. *Human Factors, 33*, 267–280.

Fisk, A. D., and Rogers, W. A. (1992). The application of consistency principles for the assessment of skill development. In W. Regian and V. Shute (Eds.), *Cognitive approaches to automated instruction* (pp. 171–194). Hillsdale, NJ: Erlbaum.

Fisk, A. D., and Schneider, W. (1984). Memory as a function of attention, level of processing, and automatization. *Journal of Experimental Psychology: Learning, Memory, and Cognition, 10,* 181–197.

Hammond, K. R. (1966). *The psychology of Egon Brunswik.* New York: Holt, Rinehart & Wilson.

Healy, A. F., Fendrich, D. W., and Proctor, J. D. (1990). Acquisition and retention of a letter-detection skill. *Journal of Experimental Psychology: Learning, Memory, and Cognition, 16,* 270–281.

Kolers, P. A. (1976). Reading a year later. *Journal of Experimental Psychology: Learning, Memory, and Cognition, 5,* 554–565.

LaBerge, D., and Samuels, S. J. (1974). Toward a theory of automatic information processing in reading. *Cognitive Psychology, 6,* 293–323.

Logan, G. D. (1985). Skill and automaticity: Relations, implications and future directions. *Canadian Journal of Psychology, 39,* 367–386.

Logan, G. D. (1988). Toward an instance theory of automatization. *Psychological Review, 95,* 492–527.

MacKay, D. G. (1982). The problem of flexibility, fluency, and speed-accuracy trade-off in skilled behavior. *Psychological Review, 89,* 483–506.

McClelland, J. L., Rumelhart, D. E., and Hinton, G. E. (1986). The appeal of parallel distributed processing. In D. E. Rumelhart and J. L. McClelland (Eds.), *Parallel distributed processing: Explorations in the microstructure of cognition* (Vol. 1, pp. 3–44). Cambridge, MA: MIT Press.

McGuigan, F. J., and MacCaslin, E. F. (1955). Whole and part methods in learning a perceptual-motor skill. *American Journal of Psychology, 68,* 658–661.

Newell, A., and Rosenbloom, P. S. (1981). Mechanisms of skill acquisition and the law of practice. In J. R. Anderson (Ed.), *Cognitive skills and their acquisition* (pp. 1–55). Hillsdale, NJ: Erlbaum.

Prinz, W. (1979). Locus of the effect of specific practice in continuous visual search. *Perception and Psychophysics, 25,* 137–142.

Rabbitt, P. M. A., Cumming, G., and Vyas, S. (1979). Improvement, learning, and retention of skill at visual search. *Quarterly Journal of Experimental Psychology, 31,* 441–459.

Regian, J. W., and Schneider, W. (1990). Assessment procedures for predicting and optimizing skill acquisition after extensive practice. In N. Frederiksen, R. Glaser, A. Lesgold, and M. Shafto (Eds.), *Diagnostic monitoring of skill and knowledge acquisition* (pp. 297–323). Hillsdale, NJ: Erlbaum.

Rogers, W. A. (1992). Age differences in visual search: Target and distractor learning. *Psychology and Aging, 7,* 526–535.

Salasoo, A., Shiffrin, R. M., and Feustel, T. C. (1985). Building permanent memory codes: Codification and repetition effects in word identification. *Journal of Experimental Psychology: General, 114,* 50–77.

Schendel, J. D., and Hagman, J. D. (1991). Long-term retention of motor skills. In J. E. Morrison (Ed.), *Training for performance: Principles of applied human learning* (pp. 53–92). New York: Wiley.

Schneider, W. (1985). Training high-performance skills: Fallacies and guidelines. *Human Factors, 27,* 285–300.

Schneider, W. (1988). Micro Experimental Laboratory: An integrated system for IBM PC compatibles. *Behavior Research Methods, Instruments, & Computers, 20,* 206–217.

Schneider, W., and Detweiler, M. (1987). A connectionist/control architecture for working memory. In G. H. Bower (Ed.), *The psychology of learning and motivation* (Vol. 21, pp. 53–118). New York: Academic.

Schneider, W., and Fisk, A. D. (1984). Automatic category search and its transfer. *Journal of Experimental Psychology: Learning, Memory, and Cognition, 10,* 1–15.

Schneider, W., and Shiffrin, R. M. (1977). Controlled and automatic human information processing: I. Detection, search, and attention. *Psychological Review, 84,* 1–66.

Shiffrin, R. M. (1988). Attention. In R. C. Atkinson, R. J. Herrnstein, G. Lindzey, and R. D. Luce (Eds.), *Steven's handbook of experimental psychology* (2nd ed., pp. 739–811). New York: Wiley.

Shiffrin, R. M., and Czerwinski, M. P. (1988). A model of automatic attention attraction when mapping is partially consistent. *Journal of Experimental Psychology: Learning, Memory, and Cognition, 14,* 562–569.

Shiffrin, R. M., and Schneider, W. (1977). Controlled and automatic human information processing: II. Perceptual learning, automatic attending, and a general theory. *Psychological Review, 84,* 127–190.

Sperling, G., Budiansky, J., Spivak, J. G., and Johnson, M. C. (1971). Extremely rapid visual search: The maximum rate of scanning letter for the presence of a numeral. *Science, 174,* 307–311.

Wechsler, D. (1981). *Wechsler Adult Intelligence Scale: Revised.* New York: Psychological Corp.

Wightman, D. C., and Lintern, G. (1985). Part-task training for tracking and manual control. *Human Factors, 27,* 267–284.

The Impact of Cross-Training on Team Functioning: An Empirical Investigation

CATHERINE E. VOLPE,[1] JANIS A. CANNON-BOWERS,[2] *and* EDUARDO SALAS, *Naval Air Warfare Center, Orlando, Florida, and* PAUL E. SPECTOR, *University of South Florida, Tampa, Florida*

The effects of cross-training (presence vs. absence) and workload (high vs. low) on team processes, communication, and task performance were examined. Eighty male undergraduate students were randomly assigned to one of four training conditions: cross-training, low workload; cross-training, high workload; no cross-training, low workload; and no cross-training, high workload. Results indicated that cross-training was an important determinant of effective teamwork process, communication, and performance. Predicted interactions between cross-training and workload were not supported. Implications for the design and implementation of cross-training as a means to improve team functioning are discussed.

INTRODUCTION

Team effectiveness in the workplace is of vital concern to organizations that depend on teams to perform important tasks. Recently research on work teams has begun to examine several factors that foster effective team performance, including team composition, organizational context, and task demands (Hackman, 1986; Sundstrom, De Meuse, & Futrell, 1990; Swezey & Salas, 1992). However, despite the obvious importance of training as a means to enhance team performance, relatively little research can be found on team instructional strategies (Cannon-Bowers, Tannenbaum, Salas, & Volpe, 1995; Salas, Dickinson, Converse, & Tannenbaum, 1992). One particularly promising team training strategy in need of investigation involves the cross-training of team members to enhance their knowledge of one another's tasks (Cannon-Bowers & Salas, 1990).

[1] Now at Electronic Selection Systems, Orlando, Florida.
[2] Requests for reprints should be sent to Janis A. Cannon-Bowers, Naval Air Warfare Center, Training Systems Division, 12350 Research Pkwy., Orlando, FL 32826-3224.

Although *cross-training* is a familiar term to most people, it remains poorly defined, scarcely researched, and rarely mentioned in the literature. In this investigation we offer a definition of cross-training, explain the mechanisms by which we believe it can improve team functioning, and test its efficacy as a team training strategy.

Defining the Mechanisms of Cross-Training

Cross-training refers to a strategy in which each team member is trained on the tasks, duties, and responsibilities of his or her fellow team members. The goal of this type of training is to provide team members with a clear understanding of the entire team function and how one's particular tasks and responsibilities interrelate with those of the other team members (Baker, Salas, Cannon-Bowers, & Spector, 1992). The type of knowledge that individuals acquire through cross-training is referred to as *interpositional knowledge* (IPK).

IPK can be described as a type of "role"

knowledge held by team members. It is information that each team member holds regarding the appropriate task behavior of each of his or her interdependent teammates (Cream & Lambertson, 1975; Hemphill & Rush, 1952; Kahn, Wolfe, Quinn, Snoek, & Rosenthal, 1964). This knowledge pertains specifically to a teammate's individual job function requirements, including equipment operation, action-outcome contingencies, and task dynamics. It also includes context-dependent information pertaining to both temporal relationships and cause-and-effect associations within the task. In summary, IPK refers to the body of knowledge that a team member holds about the tasks, roles, and appropriate behavioral responses required of his or her teammates in various situations.

We hypothesize that IPK is crucial to team functioning because it allows team members to anticipate the task needs of fellow team members, thus allowing enhanced coordination with a minimal communication requirement. That is, team members can acquire information and other resources from one another without the necessity for explanation (e.g., when one team member provides information to another without being asked to do so). This enhanced ability to coordinate with reduced communication can be particularly important during periods of high workload, when it is difficult to overcome a lack of IPK via explicit communication among team members (i.e., because they are too busy attending to other task demands).

We maintained that the acquisition of IPK is the mechanism by which cross-training works. That is, cross-training improves team members' ability to predict, anticipate, and thus coordinate their activities by increasing the accuracy of their knowledge regarding the roles, tasks, and information needs of their teammates.

Cross-Training and Team Performance

Given the potential importance of IPK to team coordination, it is not surprising that several of the researchers who have sought an understanding of coordinated behavior have indirectly supported the use of cross-training as a team training strategy. As early as 1952, Hemphill and Rush touched on the potential utility of cross-training as a way to increase IPK. They proposed that effective coordination among members of a B-29 aircrew depended on every member having knowledge of the duties of the other crew members. They designed a study to examine the extent to which each crew member's understanding of the duties and responsibilities of his teammates would affect individual and crew effectiveness.

The participants in Hemphill and Rush's experiment were 364 team members who constituted 37 B-29 aircrew teams. Within each team, a cross-training examination was administered to each team member which contained questions about the tasks performed by seven crew positions (pilot, navigator, bombardier, radar observer, flight engineer, radio operator, and gunner). Findings indicated that crews that had higher levels of IPK (as measured by an "index of overlap" in knowledge among crew members) performed better than others in training missions, as rated by instructors. Specific findings included significant correlations between the cross-training exam scores and crew coordination (Pearson $r = .41, p < .05$), crew initiative ($r = .33, p < .05$), and crew leadership ($r = .34, p < .05$).

In a further test of the hypotheses, a longer version of the cross-training examination (lengthened to increase reliability) was administered to 122 crew members. Scores were then correlated with supervisor ratings on competence on arrival in the unit, competence at present, effectiveness in working with others, conformity with standard operating procedures, performance under stress, attitude and motivation, and overall effectiveness. Results indicated that cross-training exam scores correlated significantly with all criteria (Pearson rs ranged from .19 to .29; all were significant at the .05 level). The authors concluded that further research was needed to expand these results.

Overall, the Hemphill and Rush (1952) investigation lends considerable support to the notion expressed here: that effective team coordination is related to the amount of IPK possessed by the members of a team. In particular we were interested in testing cross-training as a mechanism to enhance shared knowledge among team members.

Additional support for the potential benefits of cross-training was provided in a coordination analysis conducted by Cream and Lambertson (1975). They demonstrated that each member of an aircrew must possess accurate expectations regarding the appropriate functional requirements of the other crew members in order for the team to perform effectively. Their research suggested that in order for these expectations to develop accurately, team members needed to familiarize themselves with the operational demands (e.g., the pilot hits this switch to lower the landing gear) and interactional demands (e.g., the pilot provides current altitude information to the copilot) of other team members.

These results are consistent with predictions made recently by team theorists regarding the ability of team members to anticipate and predict the resource and informational needs of their teammates, given some set of environmental demands. For example, Kleinman and Serfaty (1987) utilized a similar explanation to describe the findings of a study by Kohn, Kleinman, and Serfaty (1987) examining the effects of workload on decision-making teams.

Specifically, Kleinman and Serfaty (1987) argued that the reason some teams were able to maintain a high level of performance under heavy workload conditions (i.e., restricted communication) was that successful teams were able to compensate for their inability to communicate via the exercise of a common understanding of the situation. The investigators contended that this knowledge permitted the team members to use an "implicit coordination strategy" in which the need for overt communication is minimized. This strategy is one in which team members anticipate the actions and resource demands of environmental stimuli on other team members, thus reducing the need for explicit communications. Under conditions of high workload (as when an emergency or some other

event causes team members to become overloaded by the demands of their individual tasks), this type of coordination strategy appears to be critical for successful team performance.

Likewise, Orasanu (1990) inferred that high-performing aircrews were effective because the captains of successful crews used periods of low workload to "prepare" for crises (i.e., heavy-workload periods). Orasanu maintained that the successful teams used this downtime to share information regarding the situation and each member's role responsibilities, so that during periods of restricted communication, there was no need for overt strategizing. In contrast, the poorly performing crews were seen as attempting to handle the task in an inefficient manner. The author stated that unsuccessful teams were "continuously playing catch-up" during the high-workload situations, which were characterized by strained, ineffective communication. Orasanu reasoned that this less-than-efficient coordination pattern was evident because these teams apparently failed to share information during low-workload conditions, when complex, detailed information and assumptions about environmental conditions could be discussed. This is evidenced by the relatively low communication frequency recorded during low-workload periods for these teams.

Similarly, Cannon-Bowers, Salas, and Converse (1993) recently suggested that team members must hold shared or compatible mental models of the task and team to perform effectively. In this formulation IPK is seen as an important part of shared knowledge structures, which allow team members to form accurate expectations and explanations for the performance of the task and team. Team members can then anticipate the needs of teammates and quickly adapt to changing task conditions (particularly high workload) with little need for overt strategizing.

THE STUDY

Although the research cited in the previous section points to the potential benefits of cross-training, there has been no attempt to test these propositions empirically. The current investigation addressed this problem by using a controlled laboratory setting in which the amount of IPK presented to the team members was manipulated via training. In the cross-training condition, IPK was provided in an effort to illustrate the beneficial influence of cross-training on the team's ability to perform necessary teamwork behaviors. In the no-cross-training condition, IPK was experimentally restricted in order to examine such detrimental effects as the inability of team members to predict the behavior of other members, the inability to anticipate the needs and actions of other members, and the inability to function as team players. This was predicted to occur even though they were trained to proficiency in their own individual job functions.

Hypotheses

On the basis of past research, the following hypotheses regarding cross-training were tested:

1. Teams that received cross-training were predicted to exhibit significantly better teamwork behavior than teams that were not cross-trained. For the purpose of this study, *teamwork* was defined as a combination of technical coordination, team spirit, interpersonal cooperation, and cross-monitoring. This measurement scheme is described in more detail in a later section.
2. Teams that received cross-training were predicted to communicate more appropriately (i.e., to exhibit more volunteering of information, acknowledging a teammate's comment, and agreeing with a teammate, and to exhibit less requesting of information and providing task-irrelevant remarks) than teams that were not cross-trained.
3. Teams that received cross-training were predicted to perform significantly better on the team task than teams that were not cross-trained.
4. Under high-workload conditions, teams that received cross-training were expected to exhibit significantly better teamwork behavior, communicate more appropriately, and perform significantly better on the team task than teams that were not cross-trained; however, significantly smaller differences were expected between cross-trained and non-cross-trained teams under low-workload conditions.

The rationale was that during high-workload conditions, team members who were cross-trained would be able to anticipate the informa-

tional resource needs of their teammates, based on situational demands. This would be done without the need to coordinate orally through the use of role-clarifying communications (i.e., discussing exactly what needed to be done next in order to accomplish the task). In contrast, under low-workload conditions, cross-training was not predicted to have as beneficial an effect on team coordination and task performance because the team should have been able to communicate information regarding individual role demands and resource needs as required and, in this way, explicitly coordinate their actions to reach the desired objective.

METHOD

Participants

The participants were 80 male undergraduate psychology students who ranged in age from 18 to 45 years (mean = 23 years). These participants were taken from the available pool of students taking introductory psychology courses at a Southeastern university and received class credit for taking part in the study.

Students who possessed prior flight simulator experience were screened from the study. The remaining students were assigned to one of 40 two-person teams according to the time slot for which they volunteered. Each time slot was then randomly assigned to one of the four experimental conditions.

Design

The experiment was a 2 × 2 factorial between-subjects design. A laboratory paradigm was used to manipulate the variables of interest. These consisted of cross-training (presence vs. absence) and workload (high vs. low). Cross-training and workload were controlled via the experimental team task. The dependent measures consisted of three team-level variables: teamwork process, communication, and task performance. These measures consisted of objective as well as subjective observer ratings, which were collected during and after the experimental session with the aid of videotaped recordings of each team.

Team Task

Performing as two-person teams, the participants were required to fly a PC-based F-16 aircraft simulation called Falcon 2.0 (Louie, 1987) and "shoot down" enemy aircraft. The software package offers a simulated display of the cockpit with maps, gauges, and a standard instrument panel. It is designed as a task to be performed by one person, but it was split for the purpose of this study so that two people could perform it together. The students were randomly assigned to the joystick (Member A) or keyboard (Member B) position. The task responsibilities were assigned to the positions based on a team task analysis of the Falcon task.

The functions were divided so that each position had specific tasks for which they were responsible; Table 1 provides a list of task functions associated with each position. The task was structured so that neither team member was able to accomplish the mission objective without the sequenced task input of the other position. However, the mission objective could be reached in the no-cross-training condition because both team members were capable of orally communicating with each other via intercom headsets and microphones. Therefore, the question of interest was whether cross-training would allow teams to perform the task more *effectively* in terms of teamwork, communication, and task performance. Finally, no leadership designations were made by the trainers. As a result, both team functions were considered to possess equal authority (Brannick, Roach, & Salas, 1993).

This task was selected for a number of reasons. First, it requires coordination between the team members (i.e., interdependency). In fact, it is similar to PC-based simulators used to train actual pilots in teamwork skills (see Prince & Salas, 1991). Second, the task allowed either cross-training or positional training (no cross-training), depending on the condition. As noted, the task could be completed in the no-cross-training condition, but (we predicted) not as effectively. Third, the task is more representative

of an actual teamwork setting than are the group problem-solving tasks traditionally used (see Bowers, Salas, Price, & Brannick, 1992, for a thorough discussion). Fourth, these types of simulation tasks have been used by a number of researchers to reliably elicit the coordination behaviors of interest, and these behaviors have been found to be significantly related to mission performance (e.g., Bowers et al., 1992; Brannick et al., 1993; Lassiter, Vaughn, Smaltz, Morgan, & Salas, 1990).

Manipulations

Cross-training. Students in the cross-training condition were informed about, and received practice on, all operationally relevant tasks pertaining to their own functional responsibilities as well as those of the other position. The cross-training intervention that was designed to increase IPK consisted of a 10-min audiotaped explanation of how the joystick and keyboard were operated, accompanied by complementary written instructions to accomplish the tasks shown in Table 1. The purpose of each activity was then described and demonstrated by the instructor. In addition, the roles and responsibilities of the other team member were described. Overall, this intervention was designed to increase the participants' level of IPK by describing important aspects of the other team member's task and role.

Students in the no-cross-training condition received training identical to that of the cross-trained students, except that they received only the portion that was relevant to their own functional responsibilities. In both training conditions the students were told what their mission objective was and that it would require the input of their teammate.

Team workload. Objective team workload was manipulated by requiring both team members to perform a number of information-seeking behaviors in addition to those normally required by their primary flight duties. Specifically, in the high-workload condition, students were required to respond to prerecorded inquiries from

TABLE 1

Functions Performed by Team Members in Simulated Air Combat Task

Joystick Position	Keyboard Position
Direction: Maintains and controls current direction	Speed: Adjusts and maintains speed of aircraft
Altitude: Maintains current altitude and heading	Mach indicator: Monitors current airspeed information
Flight path ladder: Maintains and controls stable altitude	G-force indicator: Monitors g force of gravity
Looping maneuver: Performs maneuver to unlock enemy radar	Select missile: Weapon selection
Radar lock: Sustains radar lock on enemy aircraft	Shoot missile: Weapon firing
Intercept radar: Monitors the range, locking and shooting of enemy aircraft	Combined map/electronic display (COMED): Monitors current location, landmarks, and radar information of enemy aircraft
	Monitors screen: Monitors view for enemy aircraft and compass heading
	Threat indicator: Monitors and tracks location of enemy aircraft in range

"base" regarding a variety of variables. This task consisted of each member locating information on his monitor and relaying it orally via his headset at fixed 10-s intervals. An equal number of information requests were randomly provided every 10 s. The students were advised that this task was an essential part of their flight duties. In contrast, in the low-workload condition, students were responsible only for their primary flight duties.

Procedure

The teammates were first asked to complete a demographic information sheet and to indicate their level of familiarity with each other. Team member familiarity has been linked positively to teamwork and performance outcomes in prior research (Brannick et al., 1993; Stout, Cannon-Bowers, Salas, & Morgan, 1990). In the present research team member familiarity was included as a covariate in hypothesis testing so as to avoid having to screen for familiarity in selecting experimental participants. In this manner the impact of familiarity on team performance could be controlled statistically.

After completing the demographic information sheet, the students were taken to separate training areas and individually trained in one of the two training conditions. In both the cross-training and no-cross-training condition, an in-

structional booklet and an audiotape were used. In addition, both training conditions included a practice session in which each of the training components was practiced to proficiency as determined by an instructor. Because the aim of the study was to train each participant to proficiency on either his individual tasks alone (no cross-training) or both his task and his teammate's task (cross-training), the length of the training sessions varied. As a result the training intervention itself took approximately 30 to 45 min, depending on the participant's aptitude. However, the students were provided with the same amount of practice time (60 min) on the simulator in order to eliminate the possibility that the cross-trained teams performed better simply because they had more training time on the simulation. After training was completed, the students were seated at the simulator and instructed in the use of their headsets.

At this point the students in the high-workload condition were instructed for 5 min in the performance of the information-reporting task described earlier. This task was to be completed in addition to their primary flight activities. The experimenter then began the simulation, in which several enemy aircraft approached. In both workload conditions, participants were given 30 min to shoot down the enemy aircraft.

Measures

Three team-level dependent variables were measured: teamwork process ratings, task performance ratings, and communication frequencies. Task performance was rated on site during each session by experimenters. In addition, the experimenters had an opportunity later to review videotapes and audiotapes of each team in order to consider and possibly revise their task performance ratings, record communication frequencies, and generate the team process scores.

Teamwork scale. The Teamwork Rating Scale (TRS) was adapted for the present study to measure team processes specifically relevant to the team task of interest. It was a modified version of the rating form used by Brannick et al. (1993). Several teamwork process dimensions were selected for the TRS based on the results of past work (see Cannon-Bowers et al., 1995, for a full summary). These included technical coordination, interpersonal cooperation, team spirit, and cross-monitoring. These four aspects of teamwork process were selected for measurement on the basis of a factor analysis of observational items conducted by Brannick et al. (1993) indicating that similar categories comprised significant factors (though these labels are ours). Moreover, the dimensions have been shown to possess adequate internal consistencies (alphas range from .71 to .90) and interrater agreement (correlations between raters range from .57 to .87).

In order to test primary hypotheses, an overall teamwork process rating was derived by summing the dimension scores, given that no data exist to suggest that differential weighting is necessary. Other studies have successfully employed similar overall teamwork process measures (Stout, Salas, & Carson, 1994).

Communication. The frequency and pattern of intermember communication factors thought to be relevant to team functioning were collected using a communication content scheme that was based on a number of previously published classification systems. In order to investigate whether cross-training increased communica-

tion quality, the content of each message was categorized into one of the following categories: requesting information, volunteering information (observation, command, or suggestion), indicating agreement or compliance, task-irrelevant remarks, and acknowledgments. These dimensions were based on an instrument developed by Krumm and Farina (1962) that was designed specifically for aircrew communication.

A study conducted by Brannick et al. (1993) demonstrated that for the majority of communication behaviors, there existed adequate interjudge agreement. This finding suggested that observers are capable of reliably recording communication frequencies. It should also be noted that in the Brannick et al. study, these behaviors were recorded at the time the team was performing the task. It was predicted that higher rater agreement and better-quality measures would be obtained through the use of this scale in conjunction with the review of team audiotapes so that raters would be able to stop and replay tapes if a communication occurred too quickly or was missed.

Task performance. Five objective measures of task performance were collected from the simulation: (1) the time it took the team to shoot down the first enemy target, (2) the number of times the enemy was able to lock the team's aircraft with its radar (indicating poor positioning by the team), (3) the number of times the team's aircraft had the enemy in range (thereby threatening it), (4) the number of times the team had its radar locked on to the enemy, and (5) the total number of enemy aircraft destroyed by the team.

These five measures are task dependent and were used to provide objective indices of team performance. The measures have been shown to be reasonably correlated ($r = .53$; Brannick et al., 1993) and were directly related to the team mission as described in a team task analysis conducted on the Falcon simulation. In addition, subjective overall team quality and technical competency ratings were made independently by two observers (who were blind to experimental conditions). These types of ratings have had a reported interrater reliability of .83 (Brannick et al., 1993).

RESULTS

Table 2 presents means, standard deviations, interrater reliabilities, and internal consistency estimates for the dependent measures. Acceptable internal consistency estimates were achieved, with alphas ranging from .82 to .95. The agreement among the observers for all but one of the variables was reasonably high, with correlations ranging from .74 to .98. There was significantly less agreement regarding the acknowledgments variable (.55) than with other variables. On closer examination, it appeared that this may have been attributable to the infrequency with which the category was used by the raters. As shown in Table 2, the teams rarely exhibited this type of communication behavior.

Tests of Hypotheses

A series of 2 × 2 between-subjects analyses of covariance (ANCOVAs) was used to test the hypothesized main effect of training type and interactions of training and team workload.

ANCOVA was used based on the Brannick et al. (1993) report, which suggested that the level of team member familiarity (i.e., acquaintance) may affect a participant's task performance. The analysis used familiarity as a covariate to increase the precision of the experimental findings by removing a potential source of bias.

Preliminary analyses revealed that familiarity did not vary significantly among any of the conditions. It was a significant covariate in two analyses—teamwork process and overall communication frequency—indicating that teammates who were familiar with each other had higher teamwork process ratings and communication frequencies than did the others.

Hypothesis 1. Hypothesis 1 predicted that teams that received cross-training would exhibit significantly higher teamwork process ratings than would teams that were not cross-trained. This hypothesis was supported, as shown in Table 3. The first row of Table 3 presents the adjusted cell means, standard deviations, and ANCOVA results for teamwork process as a function of cross-training. As is evident, teams that were cross-trained were rated significantly higher in teamwork process than teams that were not, $F(1, 39) = 4.40, p < .05$.

TABLE 2

Means, Standard Deviations, Interrater Reliabilities (IRs) and Internal Consistency Estimates for Dependent Measures

Variable	Mean	SD	IR	Alpha
Teamwork process	97.75	17.7	82	.95
Communication variables				
Requests information	27.35	15.8	96	NA
Volunteers information	117.65	66.15	97	NA
Agreements	26.75	19.05	93	NA
Irrelevant remarks	21.13	23.4	95	NA
Acknowledgments	4.45	3.25	55	NA
Team task performance variables				
Enemy in range	13.73	6.45	NA	NA
Number of enemy aircraft destroyed	2.95	2.82	NA	NA
Time to destroy first enemy	17.2	10.70	NA	NA
Number of enemies locked on by the team	5.35	5.13	NA	NA
Number of times team locked on by enemy	3.25	3.18	NA	NA
Team competency	3.82	1.24	81	NA
Overall team quality	3.66	1.34	83	NA

NA: not applicable.

TABLE 3

The Impact of Cross-Training on Team Functioning

Variable	Cross-Training		No Cross-Training				
	Mean	SD	Mean	SD	MS	F	Ω
Teamwork process	102.95	13.87	92.73	15.79	967.52	4.40*	.09
Communication variables							
Requests information	24.89	14.80	29.35	12.90	222.56	1.37	.02
Volunteers information	137.52	49.53	97.86	50.19	14 475.66	7.46*	.09
Agreements	30.74	16.14	22.83	14.67	591.60	3.15	.04
Acknowledgments	4.66	2.39	4.24	2.50	1.73	.27	.01
Irrelevant remarks	28.51	21.92	14.06	9.71	194.69	5.27*	.05
Team task performance variables							
Enemy in range	17.63	7.14	9.83	5.77	556.63	12.57*	.25
Number of enemy aircraft destroyed	4.15	3.46	1.77	2.16	51.31	5.77*	.13
Time to destroy first enemy	12.40	11.48	22.00	9.93	844.31	7.36*	.17
Number of enemies locked on by the team	7.14	5.52	3.57	4.76	116.93	4.39*	.11
Number of times team locked on by enemy	2.58	2.00	3.94	1.38	116.93	4.39*	.11
Team competency	4.37	1.25	3.28	1.23	11.03	6.98*	.15
Overall team quality	4.24	1.30	3.09	1.38	12.19	6.63*	.15

Note. Degrees of freedom = (1,39) for all sources.
* $p < .05$.

Hypothesis 2. Hypothesis 2 predicted that teams that were cross-trained would communicate more appropriately than teams that were not. Specifically, it predicted that cross-trained teams would volunteer more information without being asked, acknowledge teammates more often, and agree with teammates more often but would request less information from teammates and provide fewer task-irrelevant remarks than would the non-cross-trained teams.

This hypothesis received mixed support. Turning to Table 3, it can be seen that cross-trained teams volunteered significantly more information, $F(1, 39) = 7.46, p < .05$, as expected. However, cross-training did not have the predicted effect on acknowledgments, agreements, or information requests. Further, cross-trained teams showed a significant increase in irrelevant remarks, $F(1, 39) = 5.27, p < .05$, the opposite of what was expected.

Hypothesis 3. Hypothesis 3 predicted that team task performance would be positively influenced by cross-training. This hypothesis was supported. As indicated in Table 3, cross-trained teams had significantly more targets in range, $F(1, 39) = 12.57, p < .05$, targets destroyed, $F(1, 39) = 5.77, p < .05$, and "lock ons" of the enemy

than did non-cross-trained teams, $F(1,39) = 4.39, p < .05$. Cross-training was also associated with less time to destroy the first enemy aircraft, $F(1, 39) = 7.36, p < .05$, and higher team task competency and overall quality ratings, $F(1, 39) = 6.98$ and 6.63, respectively ($p < .05$). The only measure that did not differ significantly across conditions was the number of times the enemy locked on to the team's aircraft.

Hypothesis 4. Hypothesis 4 predicted an interaction between cross-training and workload such that the difference between cross-trained and non-cross-trained teams would be more pronounced under high-workload conditions. This hypothesis received little support. To begin with, Table 4 shows the main effect of workload on teamwork process, communication, and task performance behaviors. As indicated, workload significantly degraded teamwork process, $F(1, 39) = 4.47, p < .05$. Workload also significantly affected communication variables—all communication frequencies decreased under high workload. However, workload had no impact on any of the team task performance measures.

With respect to the predicted interaction, in only two cases did cross-training and workload

TABLE 4

The Impact of Workload on Team Functioning

Variable	Low Workload		High Workload		MS	F	Ω
	Mean	SD	Mean	SD			
Teamwork process	102.79	13.43	92.88	16.26	983.04	4.47*	.09
Communication variables							
Requests information	34.49	13.80	19.75	13.90	2321.18	14.31**	.24
Volunteers information	148.46	56.99	86.95	42.73	37831.25	19.50***	.23
Agreements	34.81	19.93	18.67	10.89	2605.64	13.89**	.19
Acknowledgments	5.63	2.77	3.28	2.12	55.26	8.59**	.17
Irrelevant remarks	30.15	22.29	12.43	9.34	3141.68	8.53**	.15
Task performance variables							
Enemy in range	14.41	7.34	13.05	5.57	18.24	.41	.01
Number of enemy aircraft destroyed	3.66	3.3	2.25	2.35	19.65	2.21	.05
Time to destroy first enemy	16.34	11.05	18.07	10.37	29.90	.26	.01
Number of enemies locked on by the team	5.76	5.42	4.95	4.86	6.60	.25	.01
Number of times team locked on by enemy	3.55	4.12	2.95	2.24	2.74	1.49	.00
Team competency	4.09	1.4	3.56	1.08	2.73	1.73	.04
Overall team quality	3.93	1.46	3.4	1.22	2.74	1.49	.00

Note. Degrees of freedom = (1, 39) for all sources.
* $p < .05$, ** $p < .01$, *** $p < .001$.

interact significantly. First, for irrelevant remarks, workload interacted with cross-training such that under high workload the difference between cross-trained and non-cross-trained teams was significantly less than under low workload, $F(1, 39) = 4.68, p < .05$, which is the opposite of what was predicted. In the second case workload and cross-training interacted to affect acknowledgments, $F(1, 39) = 5.30, p < .05$, but the nature of this interaction was also not as predicted.

DISCUSSION

The aim of this investigation was to demonstrate the beneficial effects of cross-training on team functioning as it pertains to teamwork process, communication, and task performance. Also, the effects of workload and its interaction with the cross-training manipulation were investigated.

Effectiveness of Cross-Training

Team process variables. Results of the teamwork process analyses supported the first hypothesis, that cross-training would positively

affect overall teamwork process. That is, teammates who received interpositional knowledge through the cross-training manipulation interacted more effectively with each other, as determined by a measure of overall teamwork, than did teams that were not provided with this knowledge.

These findings are consistent with the arguments made by early team researchers who, though never testing the notion directly, proposed that an incomplete understanding of the duties of fellow team members would decrease team members' ability to effectively coordinate their individual activities with those of the other team members (Cream & Lambertson, 1975; Hemphill & Rush, 1952). These results are also consistent with the predictions made more recently by several team theorists who suggested that effective teamwork processes depend on shared knowledge among team members (Cannon-Bowers et al., 1993; Kleinman & Serfaty, 1987; Kohn et al., 1987). The unique contribution of the present report is that we demonstrated that a fairly simple cross-training intervention was potent enough to improve the teams' ability to coordinate.

Communication variables. The findings of the communication analyses were somewhat supportive of the second hypothesis. In general we found that, as predicted, cross-trained teams used more efficient communication strategies than did teams in the no-cross-training condition.

Specifically, although the number of requests for information and responses to those requests remained constant across training conditions, the number of times information was volunteered (i.e., without the other team member explicitly requesting it) was significantly greater in the cross-training condition. This finding may indicate that the cross-trained members were able to anticipate and predict the information needs of their teammates. This is consistent with early studies of team communication in which teams that received team training showed an increase in the number of voluntary messages over teams whose members had been individually trained (Krumm & Farina, 1962). For example, Krumm and Farina suggested that voluntary messages are strongly associated with team coordination and that their results pointed to the success of the training. They contended that in effective, coordinated teams, information needs will be anticipated and thus information will be volunteered in a timely manner.

It should also be noted that the cross-trained teams made more irrelevant remarks (i.e., not directly task related) than did teams that were not cross-trained. This is contrary to what we hypothesized but consistent with the findings of a study by Urban, Bowers, Morgan, Braun, and Kline (1992). Specifically, they found that teams working under high-workload conditions were able to maintain their performance while making more task-irrelevant remarks than their low-workload counterparts made.

Taken together, these findings appear to be consistent with the assertion of Kleinman and Serfaty (1987) that effective teams adjust their strategy in response to workload by relying on implicit coordination strategies and, hence, require less explicit (task-related) communication. In the current case cross-training may have in- creased the team's ability to engage in implicit coordination, thereby requiring less task-relevant communication and allowing more irrelevant remarks.

The findings of the communication analyses also offered strong evidence that the workload manipulation was detrimental to overall communication volume. In fact, the results of the content analysis indicated that every one of the communication variables was noted less frequently in the high-workload condition than in the low-workload condition, thus replicating the general findings in this regard (e.g., Briggs & Johnston, 1967; Briggs & Naylor, 1965; Johnston, 1966; Kidd, 1961; Kohn et al., 1987; Orasanu, 1990; Williges, Johnston, & Briggs, 1966).

Team outcome variables. The results of the team outcome analyses supported the first hypothesis in that cross-training produced significant improvements in team task performance. That is, cross-trained teams were more effective in the performance of the team task (as measured by objective performance indices) than were teams that were not cross-trained. Specifically, cross-trained teams shot down the first enemy aircraft in a shorter amount of time, destroyed a greater number of enemy aircraft, maneuvered the target into range more frequently, and got a greater number of locks onto the enemy aircraft than did the non-cross-trained teams. The only objective performance measure not influenced by training was the number of times the enemy locked onto the Falcon. This may be because this measure was the only one of a defensive rather than offensive nature (i.e., it involved avoiding being shot by the enemy rather than shooting down the enemy targets) and was therefore not emphasized in the training.

Overall, these findings provide additional support for the validity of cross-training as operationalized in the current investigation. In addition, the results of the present study represent one of the first attempts to manipulate shared knowledge among team members via cross-training and offer much-needed empirical support for the idea that team performance and functioning can be influenced by it. Further research is needed to examine the exact nature of the most effective cross-training for different types of tasks and teams.

Interaction Effects of Cross-Training and Workload

The current study also investigated whether, during high-workload conditions, teams that were cross-trained would be able to respond appropriately without unnecessary strategizing and thus have a significant advantage over teams that did not receive cross-training; further, we predicted that this difference would be less pronounced under low-workload conditions.

Results of the ANCOVAs for the three categories of dependent variables offer very little support for this proposition. Only two cases of significant Training × Workload interactions were found, and these were in contrast to what was expected. This may be explained by the fact that the workload manipulation produced such a strong effect on the amount of intrateam communication that the differential effects of training type on explicit communication frequency were diminished. Therefore, the reason for the lack of significant interactions for the majority of the dependent variables may lie in the fact that, even under low workload, the task was rather demanding. It is possible that under low-workload conditions, teams were still so busy that they were unable to compensate for the lack of cross-training through explicit communications.

IMPLICATIONS FOR TEAM TRAINING

To illustrate the applicability of these findings to a real-world setting, consider the case in which individuals who are technically proficient on their own job function are grouped to form an interdependent team. The nature of this type of team dictates that members will have a role in the task that requires some level of coordinated input and information exchange. Without spe-

cific training, it may be safe to assume that members' understanding of the task and team structure will be in some way deficient, even if they are generally familiar with the responsibilities of their (own) new position. The findings of the current study suggest that this lack of interpositional knowledge may degrade coordinated task activity, team interactions, and communication effectiveness. Cross-training team members on the team's role structure and each member's task responsibilities may be one way to circumvent these problems.

Unfortunately, the literature on training methods offers little guidance in selecting, designing, or applying appropriate training techniques (Hinrichs, 1976; McNelis, Salas, & Coovert, 1989). For example, interpositional knowledge does not necessarily have to be delivered with the type of cross-training in which each member is provided with a *complete overlap* in task knowledge and functional proficiency. In fact, it is likely that this type of strategy would be impractical and overly expensive—if not harmful—for teams of a highly technical nature (e.g., aircrews) or that require each member to possess specialized physical characteristics (e.g, football teams). To illustrate, a surgical team could not possibly function if a complete overlap in expertise (e.g., between the nurse and surgeon) was required. Indeed, this would be a ridiculous waste of resources; a more informational and less "hands-on" intervention seems appropriate.

For these reasons, research is needed to determine the specific nature of cross-training required for a particular task and team. The approach and format for conducting cross-training may vary considerably, depending on the degree of team involvement, the context, the level of technical expertise required, and a host of other factors. Practical considerations must also be considered in developing specific cross-training interventions. A further treatment of the factors involved in designing training for team competencies can be found in Cannon-Bowers et al. (1995).

CONCLUSION

The current investigation yielded important information regarding the utility of cross-training as a means to improve team performance. Results indicated that cross-training was a significant determinant of effective teamwork process and task performance and replicated the general finding that workload is damaging to team communications. To our knowledge, this study is the first to demonstrate the impact of cross-training on team outcomes in this manner.

Perhaps the most limiting characteristic of this investigation is that it was conducted in a laboratory setting using a college population. However, the task had fairly high real-work validity (as noted, it is used to train real pilots), thereby increasing somewhat the generalizability of obtained results. Furthermore, this experiment can be considered a first step in designing theory-based training as prescribed by Cannon-Bowers, Tannenbaum, Salas, and Converse (1991). Follow-up research is required in the field so that training interventions can be tested and refined in actual task settings.

ACKNOWLEDGMENTS

An abridged version of this paper was presented at the Seventh Annual Meeting of the Society for Industrial and Organizational Psychology. The views herein are those of the authors and do not reflect the official position of the organizations with which they are affiliated. We would like to thank Kim Travillian, Dawn Riddle, Kerry Burgess, and Tony Hubert for their assistance in conducting this research and Jennifer Greenis and Heriberto Velez for reviewing the manuscript.

REFERENCES

Baker, C. V., Salas, E., Cannon-Bowers, J. A., & Spector, P. (1992, April). *The effects of inter-positional uncertainty and workload on teamwork and task performance.* Paper presented at the annual meeting of the Society for Industrial and Organizational Psychology, Montreal, Canada.

Bowers, C., Salas, E., Price, C., & Brannick, M. (1992). Games teams play: A methodology for investigating team coordination and performance. *Behavior Research Methods, Instruments and Computers, 24,* 503–506.

Brannick, M. T., Roach, P. M., & Salas, E. (1993). Understanding team performance: A multimethod study. *Human Performance, 6,* 287–308.

Briggs, G. E., & Johnston, W. A. (1967). *Team training* (NTDC Tech. Report 1327-4). Orlando, FL: Naval Training Device Center.

Briggs, G. E., & Naylor, J. C. (1965). Team versus individual training, training task fidelity and task orientation effects on transfer performed by three-man teams. *Journal of Applied Psychology, 49,* 387–392.

Cannon-Bowers, J. A., & Salas, E. (1990, April). *Cognitive psychology and team training: Shared mental models in complex systems.* Paper presented at the annual meeting of the Society for Industrial and Organizational Psychology, Miami, FL.

Cannon-Bowers, J. A., Salas, E., & Converse, S. A. (1993). Shared mental models in expert team decision-making. In N. J. Castellan, Jr. (Ed.), *Current issues in individual and group decision making* (pp. 355–377). Hillsdale, NJ: Erlbaum.

Cannon-Bowers, J. A., Tannenbaum, S. I., Salas, E., & Converse, S. A. (1991). Toward an integration of training theory and technique. *Human Factors, 33,* 281–292.

Cannon-Bowers, J. A., Tannenbaum, S. I., Salas, E., & Volpe, C. E. (1995). Defining team competencies and establishing training requirements. In R. Guzzo & E. Salas (Eds.), *Team effectiveness and decision making in organizations* (pp. 333–380). San Francisco, CA: Jossey-Bass.

Cream, B. W., & Lambertson, D. C. (1975). *A functional integrated systems trainer: Technical design and operation* (AF-HRL-TR-75-6[II]). Brooks Air Force Base, TX: U.S. Air Force Human Resources Laboratory.

Hackman, J. R. (1986). The psychology of self-management in organizations. In M. S. Pallak & R. Perloff (Eds.), *Psychology and work* (pp. 89–136). Washington, DC: American Psychological Association.

Hemphill, J. K., & Rush, C. H. (1952). *Studies in aircrew composition: Measurement of cross-training in B-29 aircrews* (AD B958347). Columbus: Ohio State University, Columbus Personnel Research Board.

Hinrichs, J. R. (1976). Personal training. In M. D. Dunnette (Ed.), *Handbook of industrial and organizational psychology* (pp. 829–860). Chicago: Rand McNally.

Johnston, W. A. (1966). Transfer of team skills as a function of type of training. *Journal of Applied Psychology, 50,* 102–108.

Kahn, R. L., Wolfe, D. M., Quinn, R. P., Snoek, J. D., & Rosenthal, R. A. (1964). *Organizational stress: Studies in role conflict and ambiguity.* New York: Wiley.

Kidd, J. S. (1961). A comparison of one-, two-, and three-man work units under various conditions of workload. *Journal of Applied Psychology, 45,* 195–200.

Kleinman, D. L., & Serfaty, D. (1987). Team performance assessment in distributed decision making. In R. Gilson, J. P. Kincaid, & B. Goldiez (Eds.), *Proceedings of the Symposium on Interactive Networked Simulation for Training* (pp. 22–27). Orlando: University of Central Florida.

Kohn, C., Kleinman, D. L., & Serfaty, D. (1987). Distributed resource allocation in a team. In *Proceedings Joint Director of Laboratories Symposium on Command and Control Research* (pp. 221–233). Washington, DC: National Defense University.

Krumm, R. L., & Farina, A. J. (1962). *Effectiveness of integrated flight simulator training in promoting B-52 crew coordination* (MRL Tech. Documentary Report 62-1). Wright-Patterson Air Force Base, OH: Aerospace Medical Research Laboratories.

Lassiter, D. L., Vaughn, J. S., Smaltz, V. E., Morgan, B. B., & Salas, E. (1990). A comparison of two types of training interventions on team communication performance. In *Proceedings of the Human Factors Society 34th Annual Meeting* (pp. 1372–1376). Santa Monica, CA: Human Factors and Ergonomics Society.

Louie, G. (1987). Falcon 2.0 [Computer program]. Alameda, CA: Spectrum Holobyte.

McNelis, K., Salas, E., & Coovert, M. D. (1989). El Papel del modelaje de la conducta en el entrenamiento organizacional [The role of behavioral modeling in organizational training]. *Revista Interamericana de Psicologia Ocupacional, 8,* 7–19.

Orasanu, J. (1990, October). *Shared mental models and crew performance* (Tech. Report 46). Princeton, NJ: Princeton University, Cognitive Science Laboratories.

Prince, C., & Salas, E. (1991). The utility of low fidelity simulation for training aircrew coordination skills. In *Second International Training Equipment Conference and Exhibition Proceedings* (pp. 87–91). Wiltshire, England: International Training Equipment Conference, Ltd.

Salas, E., Dickinson, T. L., Converse, S. A., & Tannenbaum, S. I. (1992). Toward an understanding of team performance and training. In R. W. Swezey & E. Salas (Eds.), *Teams: Their training and performance* (pp. 3–29). Norwood, NJ: Ablex.

Stout, R., Cannon-Bowers, J. A., Salas, E., & Morgan, B. B. (1990). Does crew coordination behavior impact performance? In *Proceedings of the Human Factors Society 34th Annual Meeting* (pp. 1382–1386). Santa Monica, CA: Human Factors and Ergonomics Society.

Stout, R., Salas, E., & Carson, R. (1994). Individual task proficiency and team process behavior: What's important for team functioning. *Military Psychology, 6,* 77–192.

Sundstrom, E., De Meuse, K. P., & Futrell, D. (1990). Work teams: Applications and effectiveness. *American Psychologist, 45,* 120–133.

Swezey, R. W., & Salas, E. (1992). Guidelines for use in team-training development. In R. W. Swezey & E. Salas (Eds.), *Teams: Their training and performance* (pp. 219–245). Norwood, NJ: Ablex.

Urban, J. M., Bowers, C. A., Morgan, B. B., Jr., Braun, C. C., & Kline, P. B. (1992). The effects of hierarchical structure and workload on the performance of team and individual tasks. In *Proceedings of the Human Factors Society 36th Annual Meeting* (pp. 829–833). Santa Monica, CA: Human Factors and Ergonomics Society.

Williges, R. C., Johnston, W. A., & Briggs, G. E. (1966). Role of verbal communication in teamwork. *Journal of Applied Psychology, 50,* 473–478.

Catherine E. Volpe is senior staff consultant for Electronic Selection Systems Corporation in Orlando, Florida. She received her Ph.D. (1991) in industrial and organizational psychology from the University of South Florida.

Janis A. Cannon-Bowers is a senior research psychologist in the Training Technology Development Branch of the Naval Air Warfare Center Training Systems Division. She holds an M.A. and a Ph.D. in industrial/organizational psychology from the University of South Florida.

Eduardo Salas in a senior research psychologist and head of the Training Technology Development Branch of the Naval Air Warfare Training Systems Division. He received his Ph.D. (1984) in industrial and organizational psychology from Old Dominion University.

Paul E. Spector received his Ph.D. in industrial/organizational psychology from the University of South Florida in 1975. He has been a professor in that department since 1982.

Date received: September 20, 1993
Date accepted: June 14, 1995

Continuous Concurrent Feedback Degrades Skill Learning: Implications for Training and Simulation

RICHARD A. SCHMIDT,[1] *Failure Analysis Associates, Inc., Los Angeles, California, and* GABRIELE WULF, *Max-Planck-Institüt für psychologische Forschung, Munich, Germany*

In two experiments we investigated the role of continuous concurrent visual feedback in the learning of discrete movement tasks. During practice the learner's actions either were or were not displayed on-line during the action; in both conditions the participant received kinematic feedback about errors afterward. Learning was evaluated in retention tests on the following day. We separated (a) errors in the fundamental spatial-temporal pattern controlled by the generalized motor program from (b) errors in scaling controlled by parameterization processes. During practice concurrent feedback improved parameterization but tended to decrease program stability. Based on retention tests, earlier practice with continuous feedback generally interfered with the learning of an accurate motor program and reduced the stability of time parameterization. Continuous feedback during acquisition degrades the learning of not only closed-loop processes in slower movements (as has been found in earlier studies) but also motor programs and their parameterization in more rapid tasks. Implications for feedback in training and simulation are discussed.

INTRODUCTION

Over the past several decades, researchers have examined the role of numerous types of augmented (extrinsic, or supplemental) information feedback for the learning of movement skills. (For reviews, see Magill, 1993; Salmoni, Schmidt, & Walter, 1984; Schmidt, 1991.) One type of augmented feedback that has received both practical and theoretical attention in training and simulation settings is termed *concurrent feedback:* supplementary information presented to the learner during the actual action. This feedback

can be presented discontinuously to signal that the performer is on target at that moment or that a certain level of performance is being achieved. This feedback can also be presented continuously (on-line, with essentially no delay) to indicate the level of momentary performance error (Karlin & Mortimer, 1963; Kohl & Shea, 1995; Phillips & Berkhout, 1978), deviations from a goal movement pattern (Van der Linden, Cauraugh, & Greene, 1993), the pattern of ongoing electromyographic activity (Mulder & Hulstijn, 1985), or other sources of biofeedback. Continuous concurrent augmented feedback using vision is the focus of the present article.

Variants of this type of feedback are used in many practical situations of interest to the

[1] Requests for reprints should be sent to Richard A. Schmidt, Failure Analysis Associates, Inc., 11777 Mississippi Ave., Los Angeles, CA 90025; rschmidt@fail.com.

human factors/ergonomics field, such as simulators, in which a feature to be controlled (e.g., the location of an aircraft with respect to a glide path during landing; Lintern, 1980) is fed back to the learner concurrently and continuously with the action. In physical therapy, continuous feedback has been used for decades as a way to train weight-bearing in injured limbs (Winstein, 1991).

On the surface, continuous feedback appears to be effective for learning because it guides the learner powerfully to the correct response, minimizes errors, and holds behavior on target. The problem is that the performance gains during practice are seldom carried over to retention or transfer tests in which the augmented feedback is withdrawn. The usual finding is that people who have practiced with concurrent continuous feedback often perform worse on no-feedback retention tests than do people who have practiced without such feedback. In other words, continuous concurrent feedback appears to enhance performance during practice when the feedback is operating, but it does not contribute to learning and may even degrade learning, as measured on retention and transfer tests (see Annett, 1959, 1969; Karlin & Mortimer, 1963; Kohl & Shea, 1995; Patrick & Mutlusoy, 1982; Van der Linden et al., 1993).

If augmented continuous feedback generally degrades learning, it should be avoided in the design of learning settings. Nevertheless, augmented continuous feedback has been deliberately included in both simulators and training programs, often with impressive realism and fidelity but usually at considerable expense. Its inclusion is understandable in that it does generally improve performance when it is present, and it seems obvious at first glance that such information must contribute to learning as well. However, recent views of training (e.g., Schmidt & Bjork, 1992) emphasize that measures of performance during training are not generally good predictors of longer-term learning, which is usually best evaluated on retention or transfer tests that are separated from the training setting. Even more critical, there is considerable evidence that, relative to some so-called standard practice con-

dition, several factors that facilitate performance during practice are detrimental to retention and/ or transfer performance, which is widely considered important in the success of training and simulation programs.

Hypotheses for Feedback Processing

The puzzling and counterintuitive failure of continuous concurrent feedback to contribute to learning (even though it contributes strongly to performance during practice) has been explained mainly by the notion that concurrent feedback is overly guiding during practice. Originally suggested by Annett (1959, 1969), this idea has been formalized recently to form what is termed the *guidance hypothesis* for feedback (Salmoni et al., 1984; Schmidt, 1991; Schmidt, Young, Swinnen, & Shapiro, 1989). In this view the learner uses the guidance of concurrent feedback as a crutch to produce the correct action during practice. However, this reliance on feedback prevents the learner from exercising the capabilities that are needed on the retention or transfer test, on which augmented feedback is typically not available, so performance falls below even that of learners who did not have continuous feedback during practice. These overly guiding properties of concurrent feedback can be thought of in two ways.

Guidance properties. Concurrent feedback seems to be most commonly used to guide behavior. This process is probably most apparent in tasks for which the movement duration is relatively long (greater than several seconds). Most writers would agree that in such tasks, feedback processing is a critical part of the action. The augmented concurrent feedback indicates the nature and direction of needed moment-to-moment corrections and provides additional ways in which the information can be used for action (perhaps more directly than intrinsic feedback), facilitating momentary performance. Tasks in which this seems common are slow lever-positioning responses (Fox & Levy, 1969; Kohl & Shea, 1995), movement-patterning tasks (Van der Linden et al., 1993), and complex pursuit or compensatory tracking activities. In such tasks open-loop programming processes seem to be of minimal importance to the movement's control.

Attentional processes. A second, related view is that concurrent feedback during practice distracts attention from the intrinsic feedback that is naturally present. Attending to· these augmented sources prevents the learner from acquiring the capability to deal with the intrinsic information that will be present in the retention test and on which his or her performance will presumably depend when the augmented feedback is withdrawn. Most, if not all, experiments on continuous concurrent feedback have used such tasks, and their results can usually be interpreted in this general way.

A surprising counterexample is from Swinnen, Lee, Verschueren, and Serrien (in press), in which concurrent feedback facilitated both the performance and learning of a relatively difficult, long-duration bimanual coordination pattern. Lintern (1980) also showed some benefits of concurrent feedback in an aircraft simulator, in which a simple ("skeletal") display depicting a runway approach was augmented in various ways (e.g., the desired flight path information). The reasons for these contradictory patterns of results are unclear, and further work seems warranted to understand the nature of these effects.

Feedback in Motor Program Learning

An omission in the human factors literature concerns tasks for which open-loop programming processes are strongly involved—chiefly tasks that are performed relatively rapidly in stable, predictable environments (e.g., hammering, tossing, rapid control adjustments; see Schmidt, 1988). In such situations the preprogramming of an action would have the largest possibility of being successful, as environmental variations necessitating feedback-based modifications would be minimal or absent. Most researchers agree that such tasks depend little, if at all, on feedback processes. Therefore, if concurrent continuous feedback were presented during practice of these actions, would practice performance be enhanced and retention performance be degraded, as it is in tasks for which feedback

processing is important? Alternatively, because these tasks are thought to be performed largely on the basis of programming control, would these degrading effects of concurrent feedback be absent for this class of tasks, or would this feedback even facilitate learning?

To our knowledge there have been no previous attempts to evaluate the role of concurrent feedback on rapid, programmed activities; this formed a major goal of the present experiment. As such actions are frequently involved in many real-world training tasks, such information should be of value to trainers in the human factors field.

The additional goal of these experiments was to evaluate the nature of the effects, if any, of concurrent continuous feedback on the acquisition of several component processes of actions. In our previous work (Wulf & Schmidt, 1994; Wulf, Schmidt, & Deubel, 1993) we used a task and paradigm in which the learners acquire novel, relatively brief (<1 s) movement patterns (see also Wulf, Lee, & Schmidt, 1994, for similar procedures). In such tasks we have been able to separate the capability to produce an effective action into two components. One part involves the fundamental spatial-temporal patterning (or basic form) of the action and, in our terms, is based on the acquisition of a generalized motor program (GMP). The other part involves processes responsible for scaling the action properly in order to meet various environmental demands (e.g., size and speed) or, in our terms, the capability to parameterize the GMP.

Earlier we presented evidence that various experimental factors contribute differentially to these processes, adding to our confidence that these components of responding are psychologically and behaviorally distinct. Thus our second goal was to examine whether augmented concurrent feedback influences the acquisition of a GMP and/or the capability to parameterize the action to meet environmental demands.

EXPERIMENT 1

In Experiment 1 we examined the role of concurrent augmented visual feedback in learning a relatively complex arm action. Concurrent feedback was either provided or absent, and learning was evaluated on a delayed retention test without such augmented feedback.

Method

Participants. Students at the University of Munich ($N = 30$) participated in exchange for DM 15 (about US$9.00) for their services. The participants had no prior experience with the task, and they were not informed about the purposes of the experiment.

Apparatus

The apparatus consisted of a horizontal lever attached at one end to a vertical axle that turned almost frictionlessly in ball bearing supports. The supports were mounted to the side of a table, allowing the lever to move in the horizontal plane over the table surface. A vertical handle was attached at the other end of the lever; its position could be adjusted so that when the student's forearm rested on the lever, the elbow was aligned over the axis of rotation while the student grasped the handle comfortably. A potentiometer was attached to the lower end of the axle to record position; its output was sampled at 200 Hz by a Hewlett Packard Vectra QS/20 computer. A wooden cover over the tabletop prevented the seated participant from seeing the lever.

Task

Participants were asked to produce a right-arm lever movement with certain spatiotemporal goal movement patterns. Four goal movement patterns were used in the experiment, each with the same fundamental pattern (i.e., with the same relative timing and relative amplitudes). These figures were linearly scaled to different absolute amplitudes to create the four task versions, but the rate scaling was constant to produce a duration of 937 ms for each version (see Figure 1).

Before each trial the goal movement pattern was displayed for 1.5 s on a computer screen (EIZO Flexscan 9060S). A letter (*A, B,* or *C*) denoting a specific pattern was displayed in the upper left corner of the screen, together with the

goal movement pattern, to facilitate discrimination between the different patterns. The goal pattern and letter were then removed and replaced by two vertical cursors, one in the center representing the starting position and another representing the lever's present position. The participant then moved the lever to the starting position, indicated by the alignment of the two cursors. The two cursors then disappeared; immediately following was a tone indicating that the movement could begin. The participant's task was to make a lever movement that corresponded in space and time to the goal pattern presented earlier. Once the participant began moving the lever, the movement was recorded for 1.6 s.

After a 3-s interval, knowledge of results (KR) was provided for 5 s by superimposing the spatiotemporal trace of the participant's movement over that of the goal pattern. The goal pattern was displayed in white (on a black background); the participant-produced pattern was in yellow. In the lower right corner, *Vorgabe* (template) and *Reproduktion* (reproduction) were written in the same colors as the respective traces. Also, as the movement was recorded for 1.6 s, the subject's trace extended farther to the right than the goal pattern, allowing the two curves to be distinguished easily. A perfect movement was indicated by exact superimposition of the goal trace and the participant's trace (until the end of the goal trace), and deviations between the two traces indicated errors.

In addition to the two traces, the root-mean-square error (RMSE) between the participant's movement and the goal movement (for the duration of the goal pattern) was displayed in the upper right corner on the screen. Also, the trial number of the just-performed movement appeared in the upper left corner of the screen.

Procedures and Experimental Conditions

Participants were randomly assigned to one of two groups: one that received concurrent feedback (ConFB) and one that did not (No-ConFB). Both groups received KR (as described in the previous section) after each practice trial, but in addition the ConFB group received concurrent

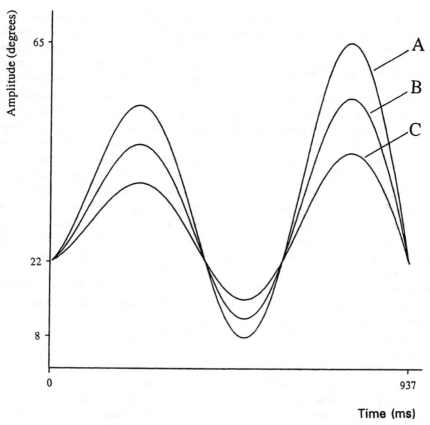

Figure 1. Spatiotemporal functions of the goal movement patterns.

feedback; that is, they saw their trace being drawn on the screen in real time as they produced the movement. For the ConFB group, once the lever was in the start position, the goal pattern appeared on the screen. As soon as the ConFB participant began moving the lever from the start position, the pattern he or she produced was drawn in real time over the goal pattern throughout the next 1.6 s, after which the screen turned blank. (For the No-ConFB group, the screen remained blank during the movement.) After 3 s, KR was provided to both groups.

Thus the only difference between the No-ConFB and the ConFB groups was that the latter was shown the position-time curve of their movement in relation to that of the goal pattern while the movement was being produced, whereas the former did not receive this continuous feedback during the movement. Note that both groups re-

ceived identical postmovement feedback information.

After the procedures had been explained and before the practice phase began, participants performed three practice trials under their respective practice conditions using Pattern B (see Figure 1). Their performance was discussed after each trial to ensure that they understood how to interpret the feedback. All participants then performed 90 practice trials using Patterns A, B, and C (see Figure 1) presented in random order, with the restriction that each pattern appeared 30 times.

One day after the practice phase, all participants performed a retention test consisting of 12 trials without either concurrent feedback or KR; the three practice-task versions (A, B, and C) were presented in random order (four trials per version). Note that the concurrent feedback that had been received by the ConFB group was now removed, so that the retention conditions were identical across groups.

Statistical Analyses and Dependent Measures

The analysis and computation of dependent measures focused on separating the errors in the underlying pattern (based on the GMP) from the errors in scaling the pattern (based on parameterization processes). These methods have been described previously (e.g., Wulf & Schmidt, 1994; Wulf et al., 1993), and we will not repeat the full description here. However, a brief summary of this method, with the interpretation of the various scores it generates, is given next.

Root-mean-squared error (RMSE). First, as a measure of overall error, we calculated the root-mean-squared deviations between the participant's pattern and the template across the 937 ms of the template. This measure is sensitive to both constant errors (the average deviation of the participant's movements from the goal pattern) and within-subject variability (Schmidt, 1988). It is important to note that RMSE is sensitive to both errors in GMP production and errors in parameterization, which are separated using the measures described next.

Scaling factors. Each participant-produced trajectory and its goal pattern were first synchronized at the moment that a displacement of 0.65° from the baseline was detected. Next, a computer program linearly scaled (stretched or compressed) the participant-produced trajectory in time by various factors. For each temporal rescaling factor, the within-subject correlation between the participant's rescaled trajectory and the goal pattern was calculated. This process was repeated with scaling factors ranging from 0.2 to 2.0 times the actual rate, which easily spanned the range of overall movement duration errors. We then selected the scaling factor for which the correlation with the goal pattern was maximized. This process did not introduce phase shifts between the two functions, only proportional changes in rate, with the start of the movement fixed.

The scaling rate that yielded the maximum cor-

relation between the participant's trace and the goal pattern was taken as a temporal scaling factor, or time factor, indicating the error in temporal parameterization. Thus the temporal scaling factor indicated the amount by which the participant's trajectory had to be linearly expanded or compressed in time so that it correlated maximally with the goal pattern. A correct temporal scaling had a value of 1.0, and values larger (or smaller) than 1.0 indicated movements with rates that were too slow (or fast). These were expressed as absolute constant error (|CE|) in time scaling for each participant (the absolute value of the mean across trials, indicating systematic bias) and as variable error (VE) in time scaling (the standard deviation [SD] across trials, indicating variability).

Next, an amplitude factor was determined as the variance in amplitudes (computed over the 193 samples in a trial) of the optimally time-scaled movement trajectory divided by the variance of the goal pattern. This ratio was taken as a measure of error in amplitude parameterization and indicated the amount by which the subject's trajectory had to be linearly scaled in amplitude so that it fit the goal pattern optimally. Again, the correct value for amplitude scaling was 1.0, and values larger (or smaller) than 1.0 indicated amplitudes that were too large (or small). These were expressed as |CE| and VE in a manner analogous to the temporal scaling factors.

Residual RMSE. For each trial the remaining (residual) RMSE between the optimally time- and amplitude-scaled trajectory and the goal pattern was calculated. We interpreted this residual RMSE as the inaccuracy of the GMP, as it represented the disagreement of the movement and the goal pattern when the errors in amplitude and temporal parameterization had been removed. Two measures were derived from the residual RMSE calculations.

First, the mean residual RMSE over a series of trials gives an estimate of the deviation in the participant's average trajectory from the goal pattern. It is mathematically equivalent to measuring the average movement trajectory for a series of trials and then computing the RMS deviation between this average trajectory and the goal pattern. It represents the deviation of the participant's average, or typical, trajectory from the goal pattern and is taken as a measure of GMP accuracy.

Second, residual RMSE-VE (a variable error) was calculated as follows. Within a block of trials, and using the rescaled templates, the within-subject (over trials) SDs of the amplitudes of the first through the last samples were computed (193 samples/movement were used). This provided estimates of the amplitude variability of the participant-produced templates. The mean of these 193 separate estimates provided a kind of average variability of the GMP's amplitude and was the residual RMSE-VE reported here.

As with many global performance scores, such a measure is insensitive to the fact that some portions of the movement are inherently more or less variable than others, and it contains several additional sources of variability (recording errors, neuromuscular variability); as a result, this measure will not be a very pure estimate of the extent to which the GMP is unstable. Within these limitations, we interpreted it as a measure that was at least sensitive to the instability of the amplitude characteristics of the GMP.

Interpretations. These scaling processes may be more clearly understood from Figure 2. The solid trace in each panel represents the goal movement pattern, parameterized properly. The participant's movement trajectory is indicated by the dashed trace. The dotted trace is the participant's movement rescaled in time and amplitude via the procedures described earlier. The deviation between the dotted trace and the goal pattern was the basis for the residual RMSE. In the upper panel the residual RMSE is relatively small after rescaling, indicating a fairly accurate GMP; the lower panel shows a much larger residual, indicative of a less accurate GMP.

Results: Practice

Overall performance. As can be seen from Table 1, both groups reduced their RMSEs over the course of practice. Throughout the entire prac-

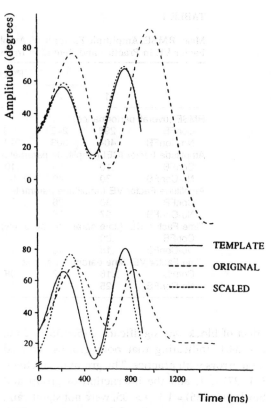

Figure 2. Examples of the effect of scaling movement trajectories; the solid trace is the goal movement pattern, the dashed trace is the participant's trajectory, and the dotted trace is the scaled version of the participant's trace.

tice phase, the ConFB group showed a greater overall accuracy in producing the goal pattern than did the No-ConFB group; continuous feedback was clearly effective in facilitating performance. The main effects of both block, $F(5, 135)$ = 31.5, $p < .001$, and group, $F(1, 27)$ = 12.4, $p <$ .01, were significant. There was no Group × Block interaction, $F(5, 135) < 1$. (The practice data of one participant in the No-ConFB group were accidentally deleted.)

Motor-program measures. Contrary to the global performance measure (RMSE), there were essentially no differences between groups during practice with regard to residual RMSE (Figure 3, left). The ConFB and the No-ConFB groups *seemed* similarly effective in producing the fundamental movement pattern, or GMP. The main

TABLE 1

Mean RMSE, Amplitude Factor |CE|, Amplitude Factor VE, Time Factor |CE|, and Time Factor VE in Practice and Retention in Experiment 1

	Practice Blocks						
	1	2	3	4	5	6	Retention
RMSE (overall performance)							
ConFB	296	242	209	189	178	162	380
No-ConFB	404	303	321	265	274	246	359
Amplitude Factor |CE| (amplitude parameter accuracy)							
ConFB	.10	.10	.10	.11	.09	.08	.42
No-ConFB	.30	.40	.35	.21	.19	.16	.32
Amplitude Factor VE (amplitude parameter stability)							
ConFB	.30	.26	.23	.21	.19	.18	.27
No-ConFB	.57	.76	.57	.19	.19	.23	.25
Time Factor |CE| (time parameter accuracy)							
ConFB	.09	.08	.04	.03	.03	.02	.20
No-ConFB	.18	.10	.11	.08	.08	.07	.14
Time Factor VE (time parameter stability)							
ConFB	.16	.12	.08	.08	.07	.06	.16
No-ConFB	.25	.18	.17	.13	.13	.12	.10

effect of block was significant, $F(5, 135) = 23.0$, $p < .001$, indicating that both groups reduced their errors with practice. The group main effect, $F(1, 27) < 1$, and the interaction of group and block, $F(5, 135) = 1.1$, $p > .05$, were not significant.

Participants also reduced their residual RMSE-VEs across practice blocks (Figure 4, left), indicating that they generally became more consistent in producing the fundamental movement structure. However, the ConFB group was clearly more variable in the fundamental pattern than the No-ConFB group. The main effects of both

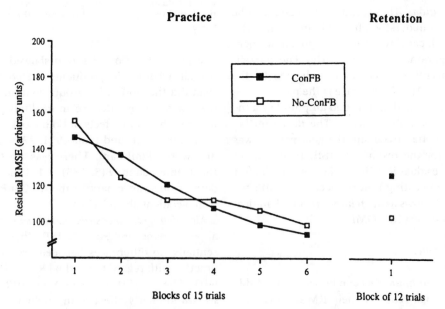

Figure 3. Residual RMSE in practice and retention in Experiment 1.

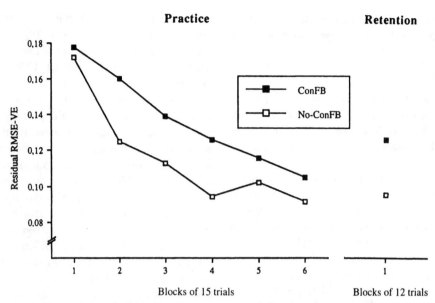

Figure 4. Residual RMSE-VE in practice and retention in Experiment 1.

block, $F(5, 135) = 26.9$, $p < .001$, and group, $F(1, 27) = 4.3$, $p < .05$, were significant. The Group × Block interaction was not significant, $F(5, 135) = 1.1$, $p > .05$.

Parameterization measures. Amplitude factor |CE|s are shown in Table 1. The No-ConFB group had comparatively large |CE|s in amplitude factors in the first half of practice and smaller errors in the second half, whereas the ConFB group was far more accurate in parameterization throughout practice. Thus continuous feedback clearly facilitated parameterization accuracy, relative to the KR-only condition. The main effect of group was significant, $F(1, 27) = 8.6$, $p < .01$, but the block main effect, $F(3, 135) = 2.1$, $p = .07$, and the interaction of group and block, $F(3, 135) = 1.7$, $p = .15$, did not quite reach conventional levels of significance.

For amplitude factor VEs, the ConFB group demonstrated relatively small and stable values throughout practice, whereas the No-ConFB group produced larger VEs during the first half of practice, with reduced errors to the level of the ConFB group in the second half (see Table 1). However, the main effect of group, $F(1, 27) = 3.5$, and block, $F(5, 135) = 1.7$, and the Group × Block

interaction, $F(5, 135) < 1$, were nonsignificant, all $ps > .05$.

For time parameterization, both groups became more accurate over practice, with the ConFB group demonstrating clearly smaller time factor |CE|s than the No-ConFB group throughout the whole practice phase (see Table 1). The main effect of group, $F(1, 27) = 14.0$, $p < .001$, and block, $F(5, 135) = 9.8$, $p < .001$, were significant, but there was no interaction of group and block, $F(3, 135) = 1.1$, $p > .05$. Both groups became more consistent in time parameterization across practice, with the ConFB group having considerably smaller time factor VEs than the No-ConFB group throughout the practice phase. The main effects of group, $F(1, 27) = 16.8$, and block, $F(5, 135) = 16.0$, $ps < .001$, were significant. There was no interaction of group and block, $F(5, 135) < 1$.

Summary. Both groups demonstrated similar performances with regard to the accuracy of the fundamental movement pattern (GMP) during practice, but continuous feedback led to decreased stability in the GMP as compared with the No-ConFB condition. In terms of parameterization measures, however, continuous feedback facilitated accuracy and stability in both amplitude and time parameterization (though the effect on amplitude factor VE was only marginally significant).

Results: Retention

One day after the practice phase, all participants returned to perform a retention test without feedback. Task Versions A, B, and C were presented in randomized order (four trials per version).

Overall performance. RMSEs in retention are shown at the top of Table 1. Errors were generally larger than those at the end of practice, with both groups showing similar error scores. The group effect was not significant, $F(1, 28) < 1$.

Motor-program measures. With regard to residual RMSE, the measure of GMP accuracy, the No-ConFB group demonstrated almost no retention loss as compared with performance at the end of practice, whereas the ConFB group showed a clear performance decrement (see Figure 3, right panel), to the point that the ConFB group had greater errors in the retention test. This group effect was significant, $F(1, 28) = 4.3, p < .05$. Even though both groups showed similar residual RMSEs during practice, the ConFB condition was less effective for learning of the GMP than the No-ConFB condition, at least as evaluated by this retention test.

For residual RMSE-VE, our measure of motor-program stability (Figure 4, right panel), the ConFB group was more variable in the fundamental movement pattern than the No-ConFB group. This effect was reliable, $F(1, 28) = 6.7, p < .05$. In addition to degrading the accuracy of the GMP, continuous feedback also degraded its stability, compared with the No-ConFB condition.

Parameterization measures. For accuracy of amplitude parameterization, both groups showed a performance decrement relative to performance at the end of practice (see Table 1). However, contrary to the situation in the practice phase, the ConFB group now had numerically larger errors than the No-ConFB group, but this difference was not significant, $F(1, 28) < 1$. For the variable errors in amplitude parameterization, the ConFB and No-ConFB groups showed

very similar values (see Table 1), and the effect of group was not significant, $F(1, 28) < 1$.

For time factor measures, the two groups tended to produce very similar |CE|s in time parameterization (see Table 1), and this group effect was not reliable, $F(1, 28) = 1.9, p > .05$. However, even though the ConFB participants were more stable in time parameterization than the No-ConFB participants during practice, they were less stable (greater time-factor VEs) in retention (see Table 1). This group effect in retention was significant, $F(1, 28) = 6.6, p < .05$.

Summary. For retention performance, measures of overall accuracy showed no effect of concurrent feedback during practice. However, measures of GMP proficiency indicated that concurrent feedback during practice degraded the learning of both the accuracy and stability of the motor-program representation. The measures of parameterization accuracy and stability in retention were generally degraded by concurrent feedback during practice, but these effects were reliable only for the time factor variability measure. Thus the chief effect of concurrent feedback seemed to be a degrading influence on motor-program accuracy and stability and on time parameterization stability.

EXPERIMENT 2

Even though Experiment 1 was fairly clear in demonstrating degrading effects of concurrent feedback on program and parameter learning, one concern needed to be defused in a second experiment. We were concerned because in the ConFB condition, the participant saw the template over which his or her action was drawn in addition to the feedback, whereas the No-ConFB group received neither the feedback nor the template. It was therefore possible that the results we obtained could have been attributable to the inclusion of the template rather than to the addition of the feedback. Therefore, to hold the factor of the template constant, in Experiment 2 both ConFB and No-ConFB groups were shown the template during movement execution.

TABLE 2

Mean RMSE, Amplitude Factor |CE|, Amplitude Factor VE, Time Factor |CE|, and Time Factor VE in Practice and Retention in Experiment 2

| | Practice Blocks | | | | | | |
	1	2	3	4	5	6	Retention
RMSE (overall performance)							
ConFB	241	212	177	167	162	162	322
No-ConFB	303	239	242	233	224	226	283
Amplitude Factor |CE| (amplitude parameter accuracy)							
ConFB	.08	.13	.10	.09	.07	.09	.23
No-ConFB	.17	.13	.14	.15	.12	.12	.23
Amplitude Factor VE (amplitude parameter stability)							
ConFB	.24	.22	.21	.19	.17	.19	.25
No-ConFB	.24	.20	.21	.21	.22	.23	.20
Time Factor |CE| (time parameter accuracy)							
ConFB	.08	.05	.04	.05	.03	.04	.42
No-ConFB	.11	.07	.05	.06	.05	.07	.23
Time Factor VE (time parameter stability)							
ConFB	.11	.08	.06	.08	.06	.08	.23
No-ConFB	.21	.14	.15	.11	.10	.10	.15

Method

Participants. Students from the University of Munich ($N = 32$) participated in exchange for DM 15 (about $9.00 U.S.). They had not served in Experiment 1 and had no prior experience with the task.

Apparatus, task, and procedure. For Experiment 2, the apparatus, task, and procedures were identical to those for Experiment 1, except that during movement execution, the template was displayed not only for the ConFB group, as in the previous experiment, but also for the No-ConFB group. Participants were randomly assigned to one of the two conditions (16 per group).

Results: Practice

Overall performance. As can be seen from Table 2, both groups became more accurate across practice. As in Experiment 1, overall the ConFB group demonstrated more effective performance than the No-ConFB group. The main effects of block, $F(5, 150) = 12.7$, $p < .001$, and group, $F(1,30) = 7.2$, $p < .05$, were significant. There was no interaction of group and block, $F(5, 150) < 1$.

Motor-program measures. Again, similar to Experiment 1, the No-ConFB and the ConFB groups showed similar residual RMSEs during practice (see Figure 5). The main effect of group was not significant, $F(1, 30) < 1$. Only the block main effect was significant, $F(5, 150) = 5.3$, $p < .001$, indicating that both groups became more accurate with regard to the fundamental movement pattern, or GMP, across practice. The interaction of group and block was not significant, $F(5, 150) = 1.3$, $p > .05$.

Both groups also became more consistent in producing the fundamental pattern across practice blocks (see Figure 6). The main effect of block was significant for residual RMSE-VE, $F(5, 150) = 8.9$, $p < .001$. However, even though the ConFB group tended to be more variable than the No-ConFB group, the group main effect was not significant, $F(1, 30) < 1$. Also, the interaction of group and block was not significant, $F(5, 150) < 1$.

Parameterization measures. Amplitude factor |CE|s can be seen in Table 2. There was no clear error reduction across practice, and the main effect of block was not significant, $F(5, 150) < 1$. The ConFB group tended to be more accurate in amplitude parameterization than the No-ConFB group, but, contrary to Experiment 1, the group effect was not significant here, $F(1, 30) = 3.0$,

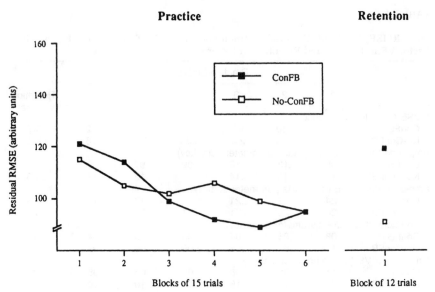

Figure 5. Residual RMSE in practice and retention in Experiment 2.

$p = .09$. There was no Group × Block interaction, $F(5, 150) = 1.2, p > .05$. In terms of the variability in amplitude parameterization, both groups tended to reduce their amplitude factor VEs across the practice phase (see Table 2), yet the main effect of block was only marginally signifi-cant, $F(5, 150) = 2.2, p = .06$. The main effect of group, $F(1, 30) > 1$, and the Group × Block interaction, $F(5, 150) = 1.5, p > .05$, were not significant.

In time parameterization, the No-ConFB group tended to have lower |CE|s than the ConFB group

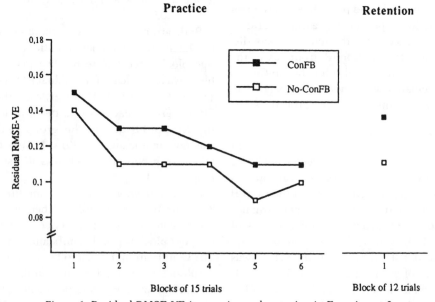

Figure 6. Residual RMSE-VE in practice and retention in Experiment 2.

(see Table 2). However, the main effect of group just failed to reach significance, $F(1, 30) = 3.3$, $p = .08$. Both groups became more accurate in time parameterization across practice blocks, $F(5, 150) = 4.0$, $p < .01$. The interaction of group and block was not significant, $F(5, 150) < 1$.

As in Experiment 1, the ConFB group was clearly more stable in time parameterization than the No-ConFB group (see Table 2). The group main effect, $F(1, 30) = 10.4$, $p < .01$, was significant. Also, the main effect of block was significant, indicating that both groups reduced their time factor VEs across practice, $F(5, 150) = 8.3$, $p < .001$. However, the No-ConFB group had clearly larger VEs than the ConFB group at the beginning of practice and showed a greater improvement across practice blocks. The Group × Block interaction was significant, $F(5, 150) = 2.4$, $p < .05$.

Summary. As in Experiment 1, the ConFB condition facilitated overall performance (RMSE) compared with the No-ConFB group; also, both groups demonstrated similar performance with regard to the fundamental movement pattern, or GMP, during practice. Concurrent feedback (ConFB) also tended to make GMP production more variable than postresponse feedback (No-ConFB), but this effect was not significant in the present experiment. In terms of parameterization, concurrent feedback again clearly facilitated stability in time parameterization, compared with the No-ConFB condition.

Results: Retention

Overall performance. As occurred in Experiment 1, even though the ConFB group was more effective than the No-ConFB group during practice, there were no significant group differences in RMSE in retention, $F(1, 30) = 1.6$, $p > .05$.

Motor-program measures. As to the accuracy of the GMP, the No-ConFB group demonstrated clearly lower residual RMSEs than the ConFB group (see Figure 4). The group effect was significant, $F(1, 30) = 7.7$, $p < .05$. Thus as in Experiment 1, concurrent feedback enhanced the learning of an accurate GMP. The ConFB group also tended to be more variable in the production of

the fundamental movement pattern (see Figure 5). However, the main effect of group failed to reach significance for residual RMSE-VE in this case, $F(1, 30) = 2.0$, $p > .05$.

Parameterization measures. Similar to Experiment 1, there were no group differences in terms of amplitude factor |CE|, $F(1, 30) < 1$, and VE, $F(1, 30) = 2.7$, $p > .05$. With regard to time factor |CE|, the ConFB group was less effective than the No-ConFB group. The group effect was significant, $F(1, 30) = 8.4$, $p < .01$. Also, as in Experiment 1, the ConFB group had clearly larger VEs in time parameterization than the No-ConFB group. This effect was significant, $F(1, 30) = 4.5$, $p < .05$.

Summary. The ConFB condition again degraded the learning of an accurate fundamental movement pattern (GMP), compared with the No-ConFB condition. In addition, stability in time parameterization was learned less effectively with the provision of concurrent feedback during practice (ConFB). These results were similar to those of Experiment 1 and indicate that the potential confound of having the template absent in the No-ConFB group was not a factor in the outcome.

DISCUSSION

We attempted to gain further insight into the functions of continuous, concurrent feedback for motor skill learning. Many investigators have found that concurrent feedback usually facilitates performance when it is present during practice but that it is detrimental to performance when measured in retention tests (or sometimes transfer tests). In this sense, concurrent feedback generally degrades learning. However, to our knowledge, there have been no previous experimental investigations of concurrent feedback effects in actions governed by primarily open-loop (or motor-program) control, and adding such information on this critical class of tasks seemed essential for a full understanding of the effect of feedback on learning. However, we also sought to examine the effects of concurrent feedback in our tasks and paradigm, such that the effects on motor-program learning could be separated from the effects on parameterization learning (e.g., Wulf & Schmidt, 1994, 1997; Wulf et al., 1993).

The two experiments presented here differed with respect to only one methodological detail: Whereas in Experiment 1 the No-ConFB group did not see the template during movement production in practice (as did the ConFB condition), in Experiment 2 both groups could see the template while they were executing the movement (to make sure that any obtained effects were attributable to the addition of concurrent feedback, not to the inclusion of the template). The results of both experiments were relatively consistent, indicating that the effects we found for Experiment 1 resulted mainly from the provision of concurrent feedback.

Feedback, Programming, and Parameterization

In both experiments, from the global performance measure (RMSE) it was obvious that ConFB condition facilitated performance during practice, as compared with the No-ConFB condition. However, overall learning, as measured by performances in retention, tended to be degraded by the concurrent feedback, though these effects were not reliable. There were certainly no tendencies toward facilitation of learning by the concurrent feedback condition. But when the various components of error (associated with proficiency in programming versus proficiency in parameterization) were separated mathematically, several clear effects emerged.

Programming. In terms of the accuracy in motor programming, indexed here by residual RMSE, continuous feedback had essentially no effect during the practice phase. However, the effects on residual RMSE-VE tended to be detrimental (this effect was significant only in Experiment 1, though), in that concurrent feedback seemed to decrease the stability of the GMP representations. On measures of learning (seen in the retention test), concurrent feedback during practice generally degraded performance in terms of the accuracy of the GMP in both experiments. Also, with regard to the stability of the GMP, the ConFB group showed less effective learning than the No-ConFB group (even though

this effect was again significant only in Experiment 1). We interpret these findings as showing that concurrent feedback was detrimental for the learning of the motor-program representation, particularly with respect to its accuracy, and somewhat less clearly with regard to its stability.

Parameterization. For parameterization performance, a different picture emerged. During practice, concurrent feedback generally facilitated performance of most of the measures of parameterization. These effects began very early in practice and generally persisted throughout the practice phase (exception: amplitude factor VEs; see Tables 1 and 2). However, in the retention tests these trends were reversed, with the ConFB group usually showing poorer retention performance than the No-ConFB group. With the exception of time factor VE, these detrimental effects were varied; however, they were not always statistically reliable and were sometimes essentially absent numerically (e.g., amplitude factor VE).

Interestingly, with regard to variability in time parameterization (time factor VE), the group differences were consistent and significant in both experiments. Whereas the ConFB group was clearly less variable in absolute timing than the No-ConFB group during practice, these differences were reversed in retention. That is, concurrent feedback interfered with learning to time the whole action consistently. Overall, there was no tendency for concurrent feedback to benefit any of these measures of learning, and our interpretation is that this form of feedback is probably detrimental—and certainly is not beneficial—for parameterization learning.

Mechanisms of Continuous Feedback

These findings extend the existing work on concurrent feedback in several ways. First, most earlier studies have used tasks with long-duration movements (tracking, positioning, etc.) for which feedback processes tend to dominate response production. The present findings generalize these principles to more rapid actions, in which open-loop programming processes are more strongly involved. Our tasks were not extremely rapid and ballistic (movement times of 937 ms), however, so it is difficult to be confident about the exact contributions of open- and closed-loop processes in these actions. The duration of the task is only one of the factors that determine the extent to which actions are structured in advance, however, as feedback processes can in some cases be far shorter than this 900-ms value. Tasks performed in closed, stable, and predictable environments generally tend to be structured in advance (clutch-accelerator movements in a car, operating a doorknob to open a door, etc.), which tends to free attentional processes from the analysis of feedback information for the detection of errors (Schmidt, 1988). It seems reasonable to suggest that, in any case, programming processes were considerably more involved in our task than they were in the earlier experiments on this question.

To the extent that programming and open-loop processes are involved in the present task, then, these data suggest that concurrent feedback degrades such processes during learning. Prior to this the detrimental effects of concurrent feedback (using mainly tasks with closed-loop control) had been explained by the general idea that concurrent feedback attracted attention and processing during performance, which interfered with the processing of relevant (i.e., intrinsic) sources of feedback. This generally facilitates performance temporarily, especially if the concurrent feedback provides information about movement accuracy, but it degrades learning, as seen in tests in which concurrent feedback has been removed. This is the most common finding when augmented concurrent feedback is used in simulators or other real-world learning settings in attempts to facilitate the learning of the tasks with the concurrent feedback removed in the transfer test.

Hypotheses that explain these results in terms of the substitution among sensory information sources do not account for our data very well, however, for several reasons. First, the pattern of results seen here appears to involve learning decrements in the open-loop (programmed) aspects of the action, which are presumably not dominated by feedback during performance. Also, our measures of GMP proficiency were generally not facilitated by concurrent feedback during practice (as many closed-loop tasks seem to be) but were degraded only when measures were taken on retention tests. Such results present a challenge to understanding how the learning of open-loop responses, involving processes believed to be largely free from feedback influences, can be degraded by manipulations of the within-response sensory information.

A second reason that such a sensory substitution hypothesis appears to be inadequate is that our measures of parameterization accuracy were generally facilitated by concurrent feedback during practice, but the long-term effects on retention were spotty and unreliable (with the exception of stability in time parameterization, which was clearly degraded by concurrent feedback). Our measures of parameterization accuracy might be influenced by closed-loop processes, however, in which errors in the initially programmed parameterization could be adjusted easily when the learner sees that the movement is too fast (or slow) or too small (or large). If so, then one would expect the learning of such parameterization processes to be degraded by concurrent feedback, which was not the case here. There were tendencies in this direction, but the effects were not strong, reliable, or consistent (again, with the exception of time factor VE). However, there were no indications that concurrent feedback facilitated parameterization.

Perhaps the reason for these patterns of effects with this task can be thought of in terms of what can be called an *incomplete programming hypothesis*. According to this general view, prior to each performance, participants presumably must retrieve the GMP from memory and parameterize it so that it meets the specified amplitude and timing demands. However, when concurrent feedback is present during practice—and when the participant has learned that such feedback is generally effective for performance—these programming processes might be less completely or effectively done than would be the case if the participant did not expect such augmented feedback. Programming could be considered less complete in several ways.

One way that programming could be less complete is based on the idea that some actions are thought to be composed of separate, sequentially organized units of control (e.g., Schneider & Schmidt, 1995; Young & Schmidt, 1990). If so, the initial segments of the action could be preprogrammed, with later segments of the action left unprogrammed. During practice, concurrent feedback during the final portions of the action could allow one to practice the last segments without preprogramming them. If augmented concurrent feedback is provided in practice, such information could induce a greater tendency to use such a mode of control. The disadvantage is that if concurrent feedback is removed in retention, there would no longer be a learned basis for control of the last part of the action, and performance would suffer. Such a hypothesis might be tested in future work by examining the locus of decrements in the action (initial vs. final segments) when augmented feedback is withdrawn in retention or transfer tests.

A second way that incomplete programming could be conceptualized is that the whole action might be programmed—but less completely so, in some sense—prior to response initiation. The general plan of the entire action could be specified, perhaps as Greene (1972) has suggested, with his notion of the "ballpark response," in which programming is sufficient only to bring the action "into the ballpark" in terms of its precision. However, when the response is then later generated, the precise details are added based on the feedback provided during the action. If so, then augmented concurrent feedback during practice could provide an increased reliance on the later, feedback-based control processes, and require less effort in terms of generating the precise details of control before the action is initiated. Again, such a mode of practice would benefit performance in acquisition, but it would be inadequate if the retention or transfer tests do not have the augmented feedback to provide the basis for adding fine control. These two subhypotheses are not exhaustive, and other analogous accounts could probably be formulated to account for the general pattern of results.

Training and Simulation

Whatever mechanisms are involved, our results join many others from a variety of tasks and paradigms in showing that when learning is measured on delayed retention and transfer tests, continuous concurrent augmented feedback is detrimental for learning. There seems to be little support for using these kinds of procedures in training, and in fact there is much support for the notion that such concurrent feedback should be avoided. This general notion has not, in our view, been readily accepted by trainers in many real-world training situations, considering that many such programs continue to promote continuous feedback as a way to facilitate performance and, ultimately, learning.

Probably one reason that the training difficulties with concurrent feedback have not been well accepted is because such feedback is usually a powerful variable for facilitating performance while it is being presented. Trainers are well aware of its effects, and it is understandable that well-motivated trainers would want to use whatever means they can to facilitate performance in practice. Furthermore, it seems natural to assume that any variable that is so effective for performance would also have powerful learning effects.

However, as we (Schmidt & Bjork, 1992) have pointed out with several experimental variables (including the use of various feedback manipulations), this assumption is not necessarily correct, and in many cases it appears to be wrong. Indeed, for several variables, deliberately organizing practice so that performance is "difficult" appears to provide decrements in performance during practice but sizable gains in performance on tests of learning conducted later. Assessing the effectiveness of variables operating in training appears to require a careful examination of the effects seen on long-term retention and/or transfer tests.

However, there are a few troubling exceptions to this generalization (e.g., Lintern, 1980; Swinnen et al., in press) that indicate that augmented concurrent feedback could be effective for learning after all. In most demonstrations of the degrading effects of concurrent feedback on learning, the training tasks have been relatively simple laboratory activities (though this is not universally the case). One suggestion, therefore, is that task complexity might somehow prove to be a moderating variable in these situations, although more evidence is needed to be certain. However, for the majority of tasks reported in the human factors literature, concurrent feedback apparently degrades the learning of processes leading to the development of effective retention and/or transfer performance, which is critical to the success of real-world training and simulation programs.

REFERENCES

Annett, J. (1959). Learning a pressure under conditions of immediate and delayed knowledge of results. *Quarterly Journal of Experimental Psychology, 11*, 3–15.

Annett, J. (1969). *Feedback and human behavior*. Baltimore: Penguin.

Fox, P. W., & Levy, C. M. (1969). Acquisition of a simple motor response as influenced by the presence or absence of action visual feedback. *Journal of Motor Behavior, 1*, 169–180.

Greene, P. H. (1972). Problems of organization of motor systems. In R. Rosen & F. M. Snell (Eds.), *Progress in theoretical biology* (Vol. 2, pp. 303–338). New York: Academic.

Karlin, L., & Mortimer, R. G. (1963). Effect of verbal, visual, and auditory augmenting cues on learning a complex motor skill. *Journal of Experimental Psychology, 65*, 75–79.

Kohl, R. M., & Shea, C. H. (1995). Augmenting motor responses with auditory information: Guidance hypothesis implications. *Human Performance, 8*, 327–343.

Lintern, G. (1980). Transfer of landing skill after training with supplementary visual clues. *Human Factors, 22*, 81–88.

Magill, R. A. (1993). Augmented feedback in skill acquisition. In R. N. Singer, M. Murphy, & L. K. Tennant (Eds.), *Handbook of research on sport psychology* (pp. 193–212). New York: Macmillan.

Mulder, T., & Hulstijn, W. (1985). Sensory feedback in the learning of a novel motor task. *Journal of Motor Behavior, 17*, 110–128.

Patrick, J., & Mutlusoy, F. (1982). The relationship between types of feedback, gain of a display, and feedback precision in acquisition of a simple motor task. *Quarterly Journal of Experimental Psychology, 34A*, 171–182.

Phillips, J. R., & Berkhout, J. (1978). Uses of feedback in computer-assisted instruction in developing psychomotor skills related to heavy machinery operation. *Human Factors, 20*, 415–423.

Salmoni, A. W., Schmidt, R. A, & Walter, C. B. (1984). Knowledge of results and motor learning: A review and critical reappraisal. *Psychological Bulletin, 95*, 355–386.

Schmidt, R. A. (1988). *Motor control and learning: A behavioral emphasis* (2nd ed.). Champaign, IL: Human Kinetics.

Schmidt, R. A. (1991). Frequent augmented feedback can degrade learning: Evidence and interpretations. In J. Requin & G. E. Stelmach (Eds.), *Tutorials in motor neuroscience* (pp. 59–75). Dordrecht, Netherlands: Kluwer Academic.

Schmidt, R. A., & Bjork, R. A. (1992). New conceptualizations of practice: Common principles in three paradigms suggest new concepts for training. *Psychological Science, 3*, 207–217.

Schmidt, R. A., Young, D. E., Swinnen, S., & Shapiro, D. E. (1989). Summary knowledge of results for skill acquisition: Support for the guidance hypothesis. *Journal of Experimental Psychology: Learning, Memory, and Cognition, 15*, 352–359.

Schneider, D. M., & Schmidt, R. A. (1995). Units of action in motor control: Role of response complexity and target speed. *Human Performance, 8*, 27–49.

Swinnen, S. P., Lee, T. D., Verschueren, S., & Serrien, D. J. (in press). Intrinsic and extrinsic information feedback for interlimb coordination: Evidence against the specificity hypothesis. *Human Movement Science.*

Van der Linden, D. W., Cauraugh, J. H., & Greene, T. A. (1993). The effect of frequency of kinetic feedback on learning an isometric force production task in nondisabled subjects. *Physical Therapy, 73*, 79–87.

Winstein, C. J. (1991). Knowledge of results and motor learning—Implications for physical therapy. *Physical Therapy, 71*, 140–149.

Wulf, G., Lee, T. D., & Schmidt, R. A. (1994). Reducing knowledge of results about relative versus absolute timing: Differential effects on learning. *Journal of Motor Behavior, 26*, 362–369.

Wulf, G., & Schmidt, R. A. (1994). Feedback-induced variability and the learning of generalized motor programs. *Journal of Motor Behavior, 26*, 348–361.

Wulf, G., & Schmidt, R. A. (1997). Variability of practice and implicit motor learning. *Journal of Experimental Psychology: Learning, Memory, and Cognition, 23*, 987–1006.

Wulf, G., Schmidt, R. A., & Deubel, H. (1993). Reduced feedback frequency enhances generalized motor program learning but not parameterization learning. *Journal of Experimental Psychology: Learning, Memory, and Cognition, 19*, 1134–1150.

Young, D. E., & Schmidt, R. A. (1990). Units of motor behavior: Modifications with practice and feedback. In M. Jeannerod (Ed.), *Attention and performance XIII* (pp. 763–795). Hillsdale, NJ: Erlbaum.

Richard A. Schmidt is a principal scientist in the Human Performance Group at Failure Analysis Associates, Inc., at the Los Angeles office. He is also a professor in the Department of Psychology at UCLA and director of the Motor Control Laboratory. He received a Ph.D. in physical education (human performance, experimental psychology) from the University of Illinois in 1967.

Gabriele Wulf is a research scientist at the Max Planck Institute for Psychological Research in Munich, Germany. She received her doctorate in sport sciences at the Deutsche Sporthochschule in Cologne, Germany, in 1984, specializing in human motor learning and movement control.

Date received: March 9, 1995
Date accepted: April 30, 1997

Principles of Educational Multimedia User Interface Design

Lawrence J. Najjar, Georgia Tech Research Institute, Atlanta, Georgia

This paper discusses principles of educational multimedia user interface design. The purpose of these principles is to maximize the learning effectiveness of multimedia applications. The principles are based on the results of studies in psychology, computer science, instructional design, and graphics design. The principles help user interface designers make decisions about the learning materials, learners, tasks that the learners perform, and tests for measuring learning performance.

INTRODUCTION

Multimedia user interfaces combine various media, such as text, graphics, sound, and video, to present information. Because of improvements in technology and decreases in costs, many human factors engineers will soon be designing user interfaces that include multimedia. Many educators, parents, and students believe that multimedia helps people to learn, so one popular application of this technology will be the field of education.

Unfortunately, the existing educational multimedia user interface design guidelines are based almost entirely on the opinions of experts (e.g., Allen, 1974; Arens, Hovy, & Vossers, 1993; Feiner & McKeown, 1990, 1991; Reiser & Gagné, 1982) rather than on the results of empirical research. This provides a weak foundation on which to make design decisions and slows progress in making educational multimedia user interfaces more effective.

The purpose of this paper is to describe empirically based principles that multimedia user interface designers can employ to create applications that improve the likelihood that people will learn. The principles are derived from studies conducted in a wide variety of fields, including psychology, computer science, instructional design, and graphics design. The principles focus on educational multimedia applications. Other sources (e.g., Mayhew, 1992; Smith & Mosier, 1986) provide more general user interface design principles and guidelines.

In any learning situation, four basic factors should be considered when evaluating learning (Bransford, 1978; Jenkins, 1979): the characteristics of (a) the materials, (b) the learner, (c) the learning task, and (d) the test of learning.

CHARACTERISTICS OF THE MATERIALS

The characteristics of the learning materials can significantly affect learning. Learning material characteristics include the medium, physical structure, psychological structure, conceptual difficulty, and sequence (Bransford, 1978). The following principles suggest ways to design the learning materials to improve learning.

Use the Medium That Best Communicates the Information

Although opinions differ (e.g., Clark, 1983; Mayer, 1997), limited evidence suggests that some media are better than others at communicating certain kinds of information (e.g., Najjar, 1996b). For example, when a learner needs to remember a small amount of verbal information for a short period of time, information that is presented via the auditory

Requests for reprints should be sent to Lawrence J. Najjar, Georgia Institute of Technology, Georgia Tech Research Institute, GTRI/EOEML/MARC Rm. 335, Atlanta, GA 30332-0823; http://mime1.marc.gatech.edu/imb/people/larry. html. **HUMAN FACTORS,** Vol. 40, No. 2, June 1998, pp. 311–323. Copyright © 1998, Human Factors and Ergonomics Society. All rights reserved.

medium is generally remembered better than information that is presented via text. In one study (Murdock, 1968), learners recalled and recognized 10 items from a list better when they were presented using sound than when using text. This result is consistent (Penney, 1975; Watkins & Watkins, 1980). Studies that found conflicting results (e.g., Marcer, 1967; Sherman & Turvey, 1969) used long retention intervals or inappropriate instructions or scoring methods.

For retaining information over longer periods, text appears to be better than sound for communicating verbal information. Text was superior to sound when the verbal information was a list of words (Severin, 1967), instructions (Sewell & Moore, 1980), four-line poems (Menne & Menne, 1972), and nonsense syllables (Chan, Travers, & Van Mondfrans, 1965; Van Mondfrans & Travers, 1964). However, one study (Van Mondfrans & Travers, 1964) found no learning differences between auditory and textual words. Also, if the learner's visual channel is already occupied, then it may be more appropriate to use audio verbal information than textual information. This situation occurs, for example, when pictorial animations and auditory verbal information are presented together (e.g., Baggett & Ehrenfeucht, 1983; Mayer & Anderson, 1992).

A picture, it is commonly said, can be worth a thousand words. Pictures seem to help people learn information more effectively than text. This picture superiority effect appears to be strong. For example, pictures of common objects were recalled and recognized better than their textual names (e.g., Lieberman & Culpepper, 1965; Nelson, Reed, & Walling, 1976; Paivio & Csapo, 1969, 1973; Paivio, Rogers, & Smythe, 1968). Exceptions seem to occur when the items are conceptually similar (e.g., all animals or all tools), causing the pictures to be easily confused (Nelson et al., 1976), or when the items are presented so quickly that learners cannot create verbal labels for the pictures (Paivio & Csapo, 1969). Also, pictures cannot be used to communicate abstract concepts, such as "freedom" and "amount."

Pictures also seem to be better than text or auditory instructions for communicating spatial information. For example, pictures helped people to draw and label the human heart (Dwyer, 1967a, 1967b), recall and recognize spatial relationships in a story (Garrison, 1978), and solve bus route problems (Bartram, 1980). To communicate motion-based information that changes continuously over time, when it is important to show how the information changes over time, animation and video appear to be best (Baek & Layne, 1988; Rieber, 1990a, 1990b; Rigney & Lutz, 1976; Roshal, 1961; Spangenberg, 1973). These studies used knot-tying tasks, assembly or disassembly tasks, and interactive, explanatory computer animations. Animation and video do not seem to be helpful when the information to be learned is difficult for learners to understand (e.g., Rieber, 1989), when the information does not need visual support (e.g., Carabello, 1985), and when the learners do not practice with an interactive animation (e.g., Rieber, 1990b).

When pictures are used in more complex ways, their benefits are less strong. One study (Baggett, 1979) found that on an immediate test, people recalled the structure of a story (exposition, complication, resolution) equally well whether the story was presented via a silent movie or via closely matched text. Another study (Nugent, 1982) also found no differences in immediate recall of story content when matched information was presented via silent video, text, or narration. However, when one of the studies (Baggett, 1979) tested recall performance after a week, performance was better for the pictorial story than the textual story.

Thus some media appear to communicate specific kinds of information better than other media. For communicating verbal information, text is better than auditory narration. For recalling and recognizing items, pictures are better than text. Pictures are also better than text or narration for communicating spatial information.

Use Multimedia in a Supportive, Not a Decorative, Way

There is strong empirical support for this design principle, especially for the use of supportive pictures with verbal information. Other multimedia combinations are not as well supported. The information being presented in one medium needs to support, relate to, or extend the information presented in the other medium. Several studies show that adding closely related, supportive illustrations to textual or auditory verbal information improves learning performance. For example, pictures improved recall of textual words (Paivio & Csapo, 1973), recall and comprehension of textual passages (Levie & Lentz, 1982), recall of auditory passages (Levin & Lesgold, 1978), and comprehension of auditory passages (Bransford & Johnson, 1972).

Some multimedia application designers apparently believe that pictures improve learner interest, motivation, and, therefore, learning. This does not appear to be the case. Adding unrelated illustrations does not improve learning, and in fact it may actually decrease learning. Unrelated illustrations did not improve comprehension and recall of textual material (Levie & Lentz, 1982; Sewell & Moore, 1980) or recall of illustration captions (Bahrick & Gharrity, 1976; Evans & Denny, 1978). One investigator (Peeck, 1974) found that adding supportive illustrations to text helped fourth-grade children retain verbal information. However, unrelated illustrations (Peeck, 1985, as cited in Winn, 1993) made it harder for learners to comprehend the text.

These results suggest that the mere presence of illustrations does not improve the learning of verbal information. The illustrations must help explain information that is presented by the verbal medium. It appears that supportive illustrations allow learners to build cognitive connections between the verbal and pictorial information (Clark & Paivio, 1991; Paivio, 1971, 1986, 1991). This dual-coded information leads to improved learning (Mayer & Anderson, 1991; Najjar, 1995a; Paivio & Csapo, 1973).

It is clear that supportive illustrations help people to learn verbal information. A small number of studies suggest that animations (e.g., Mayer & Gallini, 1990; Park & Hopkins, 1993) and videos (Nugent, 1982; Spangenberg, 1973) may also improve verbal information learning. However, additional studies must be performed before this principle can be extended to media combinations other than illustrations with textual or auditory verbal information.

Present Multimedia Synchronously

There is strong support for the idea that verbal-pictorial information should be presented together. For example, college students performed better on problem-solving transfer tests when textual annotations were integrated into explanative drawings than when the experimenters presented the text and drawings sequentially (Mayer, 1989a, 1989b; Mayer & Gallini, 1990) or simultaneously in time but physically spaced apart (Mayer, Steinhoff, Bower, & Mars, 1995). Creative problem solving and recognition were also higher when an auditory, explanative narration was synchronized with an explanative animation or movie than when the narration preceded or followed the animation (Baggett, 1984; Baggett & Ehrenfeucht, 1983; Mayer & Anderson, 1991, 1992).

Exceptions to the advantage of simultaneous presentation of related verbal and pictorial information occurred when the learners were very knowledgeable about the domain being studied (Mayer et al., 1995); when verbal recall, rather than problem-solving ability, was measured (Mayer, 1989a, 1989b; Mayer & Gallini, 1990); and when learning was measured a week later, after the learned information faded in both conditions (Baggett & Ehrenfeucht, 1983).

Synchronized presentation of verbal-pictorial information appears to improve learning better than sequential presentations. The synchronized presentation may help learners to use dual (verbal + pictorial) coding (Clark & Paivio, 1991; Paivio, 1971, 1986, 1991) to increase cognitive interconnections between the two forms of studied information and to prior knowledge.

Use Elaborative Media

There is limited, somewhat indirect evidence that the media themselves may encourage elaborative processing. Elaborative processing is extra cognitive processing of material that helps to integrate the material with prior knowledge. Elaborative processing often leads to improvements in learning performance (Anderson, 1980, 1983; Anderson & Reder, 1979; Palmere, Benton, Glover, & Ronning, 1983; Pressley, McDaniel, Turnure, Wood, & Ahmad, 1987; Reder, 1979; Stein & Bransford, 1979). Some media may encourage spontaneous elaborative processing of information more than other media (Najjar, 1996a). For example, pictures may be more elaborative than text. This appears to be the case when learning is measured using recognition (e.g., Hartman, 1961; Nelson et al., 1976; Read & Barnsley, 1977; Shepard, 1967) or recall (e.g., Paivio, 1975; Paivio & Csapo, 1973; Paivio et al., 1968). One study (Read & Barnsley, 1977) obtained this result even when the researchers used a study-test interval of 20 years.

Although there are some exceptions (see the previous section on using the medium that best communicates the information), the learning advantage for pictures, compared with text, may occur because pictures have more features available for processing than do words, and pictures may help access meaning more quickly and completely than words (Nelson, 1979; Smith & Magee, 1980).

Text may also be more elaborative than audio verbal media. Several studies (Menne & Menne, 1972; Severin, 1967; Van Mondfrans & Travers, 1964) found that text-only conditions produced better learning than audio-only conditions. However, unlike audio conditions, text conditions also allow the learner to process the verbal information at his or her own pace.

Some multimedia combinations may be more elaborative than other multimedia combinations or single media because of the advantages of dual coding. Information that is processed through both verbal and pictorial channels appears to be learned better than information that is processed through either channel alone (Barrow & Westley, 1959; Levin, Bender, & Lesgold, 1976; Mayer & Anderson, 1991; Nugent, 1982; Paivio, 1975; Paivio & Csapo, 1973; Pezdek, Lehrer, & Simon, 1984; Stoneman & Brody, 1983; Wetstone & Friedlander, 1974). For example, Severin (1967) found that learning performance in a combined audio and pictures condition was better than in a combined audio

and text condition. Nugent (1982) obtained the highest learning levels when she presented information via combined text and pictures or combined audio and pictures compared with the same content presented via text alone, audio alone, or pictures alone.

It appears that elaborative media (e.g., pictures vs. text, text vs. audio narration) may improve learning performance more than media that may not be as elaborative. Multimedia that encourages the learner to use both verbal and pictorial channels to process the information also appears to be very effective.

Make the User Interface Interactive

This design principle is strongly supported by a variety of studies. Interaction is mutual action among the learner, the learning system, and the learning material (Fowler, 1980). An interactive user interface may allow learners to control, manipulate, and explore the material, or it may periodically ask learners to answer questions that integrate the material. An interactive user interface appears to have a significant positive effect on learning from multimedia (e.g., Bosco, 1986; Fletcher, 1989, 1990; Stafford, 1990; Verano, 1987). For example, one researcher (Stafford, 1990) statistically analyzed 96 learning studies and concluded that interaction was associated with learning achievement and retention of knowledge over time. Other researchers (Bosco, 1986; Fletcher, 1989, 1990) examined 75 learning studies and found that participants learned the material faster and had better attitudes toward learning the material when they learned in an interactive instructional environment.

Interaction may improve learning because it encourages learners to elaboratively process the learning material (e.g., Bower & Winzenz, 1970; Jacoby, Craik, & Begg, 1979; Kolers, 1979; Salomon, 1984; Walker, Jones, & Mar, 1983). The interaction must be cognitively engaging. Learners who read screen after screen of text or who receive only simple "right" and "wrong" feedback to their responses are unlikely to learn (e.g., Bosco, 1986; Fowler, 1980; Verano, 1987). Also, interactivity may have a stronger effect on immediate learning than on long-term retention of the information (Fletcher, 1989).

CHARACTERISTICS OF THE LEARNER

Characteristics of the learner can have an impact on learning. Characteristics of the learner include the learner's current skills, knowledge, and attitudes (Bransford, 1978). The following principles describe learner characteristics that are associated with learning from educational multimedia.

Use Educational Multimedia with Naive and Lower-Aptitude Learners

Because few studies on this topic exist, the evidence supporting this design principle is limited. Multimedia information appears to be more effective for learners with low prior knowledge or aptitude in the domain being learned. Regarding naive learners, Mayer and Gallini (1990) found that illustrations helped college students with low prior knowledge of automobile mechanics to recall textual explanatory information and to solve creative problems. Adding illustrations to the text did not generally affect the learning performance of students who had high prior knowledge of these devices. Other studies found similar effects for teaching natural science to fifth graders (Kraft, 1961), geology and meteorology to college students (Dean & Enemoh, 1983; Kunz, Drewniak, & Schott, 1989), and basic training information to army recruits (Kanner & Rosenstein, 1960; Kanner, Runyon, & Desiderato, 1954). However, although this effect occurred in several studies, it was not always consistent (e.g., Mayer & Gallini, 1990, Experiments 2 and 3).

Multimedia also appears to be more helpful for learners with low aptitude than for learners with high aptitude. For example, in one study (Blake, 1977), college students with low or high aptitude in spatial and mental abilities learned the pattern of movement of five chess pieces via moving pictures (film), static pictures with animated arrows, or static pictures alone. The students with low aptitude performed better in the conditions with motion than in the condition with static pictures alone. However, the students with high aptitude performed similarly with all three kinds of pictures. Wardle (1977, as cited in Levie & Lentz, 1982) gave 800-word textual passages on various science topics to seventh-grade students. Some of the passages included supportive illustrations. Poor readers performed better on a comprehension test when the passages included illustrations. For good readers, the illustrations had no effect.

Although only a handful of studies examined this principle, their results suggest that multimedia is most effective for people with low prior knowledge or aptitude in the domain being learned. This may be because experts have prior knowledge that can be used to understand and integrate the new information, but novices lack this advantage. Also, novices may not know which information is important and on which information they should focus their attention. Learners with high aptitude appear to be able to learn from relatively nonelaborative media such as text, but low-aptitude learners benefit most from the elaborative and explanatory advantages offered by multimedia. High-aptitude learners may be good learners regardless of the medium used to present the information (e.g., Kanner & Rosenstein, 1960; Kanner et al., 1954; Kraft, 1961).

Present Educational Multimedia to Motivated Learners

A variety of studies provide moderately strong support for this design principle. Using external rewards, such as points or grades, to improve motivation does not appear to improve learning (Anderson, 1994; Harley, 1965; Loftus, 1972). For example, Loftus (1972) found that when he increased the number of points for recognizing certain pictures from a large set of pictures, people spent more time looking at the more rewarding pictures and recognized those pictures better in the recognition test. However, when he controlled for the amount of time spent looking at each picture, the reward had no effect. The reward affected what people learned but not how well they learned. Other researchers (Condry, 1977; Entin, 1974; Lepper & Greene, 1978; Lepper, Greene, & Nisbett, 1973) found that adding external rewards to a task may actually decrease learners' intrinsic motivation and cause them to spend less time on the task than those who are not externally motivated.

Intrinsic motivation, however, does appear to improve learning. An intrinsically motivated learner tends to learn more than an unmotivated learner. For example, Bickford (1989) found that high school students learned more from paper-based materials that were designed to improve intrinsic motivation than from materials that were not designed this way. Entin (1974) found that students who scored higher on an achievement motivation questionnaire tried more math problems and scored higher on an achievement test than did students who scored lower on the motivation questionnaire. Another study (Raynor, 1974) obtained similar results.

There are several ways to improve the learner's intrinsic motivation. The user interface designer can relate the content and objectives of the instructions to the needs and interests of the learner (Keller, 1983). This can be done by using familiar metaphors and analogies (Curtis & Reigeluth, 1984). For example, Ross (1983) found that learning performance in a statistics course improved when the exercises used contexts that were related to the students' majors rather than exercises that used generic contexts. Other studies (Flesch, 1948; Flesch & Lass, 1949; McConnell, 1978) suggest that instructional designs that use a personal style (e.g., personal pronouns, names of specific people, direct quotations, vignettes of famous people) rather than a formal style may stimulate learner interest. It appears that providing immediate, positive verbal praise and informative feedback in a context that does not control the consequences of the performance (e.g., does not have a direct impact on the student's grade) may improve intrinsic motivation (Bates, 1979; Condry, 1977; Deci, 1975; Keller, 1983). Also, general suggestions on how to improve learning performance should be given right before the next performance attempt (Keller, 1983; Tosti, 1978).

Multimedia material itself appears to offer motivational advantages because of its novelty, but these advantages (and the novelty) fade over time (Clark, 1983, 1985; Clark & Craig, 1992; Kulik, Bangert, & Williams, 1983). Finally, humor does not improve motivation because it can distract the learner from the instructional goals and interfere with comprehension (e.g., Markiewicz, 1974; Sternthal & Craig, 1973).

To Avoid Developmental Effects, Use Educational Multimedia with Adults and Older Children

Although it is difficult to establish specific age ranges, empirical studies provide moderate support for this design guideline. Multimedia appears to more effectively improve learning as children get older. On recognition and recall of information in films, older children did better than younger children and adults did better than older children (Stevenson & Siegel, 1969); the same result was obtained on recognition and recall of pictures (Dirks & Neisser, 1977; Hoffman & Dick, 1976), television commercials (Atkin, 1975a, 1975b, 1975c, 1975d; Rubin, 1972; Stoneman & Brody, 1983; Ward, 1972; Ward, Wackman, & Wartella, 1978), television programs (Leifer et al., 1971), and toy scenes (Dirks & Neisser, 1977).

Stoneman and Brody (1983) presented auditory-only, visual-only, or combined auditory-visual stories in which product advertisements were interspersed. Kindergarten children recognized more advertised products than did preschool children. Second-grade children recognized more advertised products than did kindergarten and preschool children. Hoffman and Dick (1976) showed several hundred pictures to three-year-old children, seven-year-old children, and adult college students. The seven-year-old children accurately recognized more pictures than did the three-year-old children. The adults recognized more pictures than did the seven-year-old children.

It appears that the younger children's processing occurs more at the perceptual level than the semantic level. Also, the ability to process auditory information seems to develop earlier than the ability to process visual information (Carterette & Jones, 1967; Stevenson & Siegel, 1969). With increasing experience and maturity, children appear to learn to process information at a deeper (e.g., Craik & Lockhart, 1972; Craik & Tulving, 1975), more semantic level and therefore improve their retention of information. This idea is supported by several studies (e.g., Ackerman, 1981; Hoffner, Cantor, & Thorson, 1989; Owings & Baumeister, 1979) that found that when presented with information using multimedia, younger children encoded more perceptual aspects of stimuli than did older children. Older children encoded more semantic information than did younger children. Rankin and Culhane (1970) obtained a similar effect when they compared the comprehension performance of sixth-graders and college students.

Older children and adults are more likely to be able to process the meaning of multimedia information than its appearance. Older children and adult learners should benefit from educational multimedia more than younger learners.

CHARACTERISTICS OF THE LEARNING TASK

The tasks that the learner performs with the learning materials can affect performance. Characteristics of the learning task include attending to the information, rehearsing it, and actively elaborating it. Educational multimedia user interface design principles for the learning task follow.

Use Multimedia to Focus the Learner's Attention

A small number of studies provide limited support for this design principle. Multimedia can help direct the learner's attention to relevant information and improve learning. For example, one study (Baxter, Quarles, & Kosak, 1978) asked adults in a shopping mall to "look over" a newspaper page that included a story with or without a large photograph. When asked questions about their recall of the newspaper story, participants remembered more information when they saw the story with the photograph than when they saw the story without the photograph. It appears that the photograph got the participants' attention and caused them to read the accompanying story. Other researchers successfully used drawings (e.g., Paradowski, 1967; Tennyson, 1978), motion (Baek & Layne, 1988; Park & Hopkins, 1993), small "chunks" of textual and graphical information (Rieber, 1990b), and

adjunct questions (e.g., McConkie, Rayner, & Wilson, 1973; Watts & Anderson, 1971) to focus the learner's attention.

However, getting a learner to pay attention to information does not necessarily mean that the learner will learn the information. For example, learners who are new to a field of knowledge may simply view a supplementary animation without trying to understand the information it shows (e.g., Reed, 1985). Also, irrelevant media such as unrelated pictures (e.g., Levie & Lentz, 1982) or motion (Park & Hopkins, 1993) may distract learners and actually decrease learning performance.

Encourage Learners to Actively Process the Information

A variety of multimedia studies were performed in this area, so there is strong empirical support for this design principle. Learning appears to improve when the learning task encourages the learner to actively process the information (e.g., Bobrow & Bower, 1969; Bower & Winzenz, 1970; Jacoby, 1978; Slamecka & Graf, 1978). For example, one study (Dean & Kulhavy, 1981) asked students to learn the features of a fictitious country. One group of students studied a map on which the features were labeled. Another group copied the features and labels onto a blank map. The students who were forced to actively process the spatial information by copying the map performed better on a free recall test of the map information.

Reading text may also cause the learner to more actively process the information than simply hearing verbal narration (e.g., Aldrich & Parkin, 1988; Baggett & Ehrenfeucht, 1983; Palmiter & Elkerton, 1991; Pezdek et al., 1984) or watching a silent movie (Salomon, 1984). Similarly, materials that force learners to figure out confusing information may cause them to more actively process the information, which thereby improves learning performance (e.g., Auble & Franks, 1978; Bock, 1978; Hunt & Elliot, 1980; Kolers, 1979; Sherman, 1976; Walker et al., 1983).

Simple repetition of the information does not encourage learners to actively process the information and does not necessarily improve learning. For example, before changing the frequency of its radio broadcast, the BBC advertised the new frequency via radio, television, newspaper, and direct mailings. Listeners received about 1000 exposures to the information about the new frequency. However, only 17% of the listeners learned the new frequency (Bekerian & Baddeley, 1980). To encourage listeners to actively process information, a different study (Thomson & Barnett, 1981) arranged for participants to hear 16 fake radio commercials. In one condition the listeners heard the product name (e.g., "Buy Brighto!") at the beginning and end of the commercial. In another condition listeners heard the product name at the beginning of the commercial (e.g., "Buy Brighto!"), but the product name was left unpronounced at the end of the commercial (e.g., "Buy!"). The final condition was the same as the previous condition, except listeners wrote down the name of the product that was left unpronounced at the end of the commercial. Fifteen minutes later, an unexpected test showed that recall accuracy improved across the groups from 16% to 29% to 46%. Extra processing appeared to improve learning.

In addition, the type of active processing is important. For example, Craik and Tulving (1975) found that processing the structural characteristics of each word in a list (e.g., "Is the word in capital letters?") was not as effective as processing the meaning of the word (e.g., "Would the word fit the sentence: 'He met a _____ in the street'?"). Other researchers (e.g., Craik & Watkins, 1973; Hyde & Jenkins, 1969; Parkin, 1984; Rundus, 1977) obtained similar results.

Processing tasks that encourage learners to integrate the information they are studying seem to improve learning. Several studies (e.g., Anderson & Biddle, 1975; Frase, 1975; Reder, 1979; Rothkopf, 1966) found that periodically asking learners to answer questions about the information they had just reviewed led to improvements in learning performance. Tasks that do not encourage learners to integrate the information may actually worsen learning performance (e.g., Stein & Bransford, 1979; Stein, Morris, & Bransford, 1978).

It is possible that tasks that encourage learners to actively process and integrate the information may focus their attention on the information and cause them to process the information more elaborately. This appears to be especially true when the processing focuses on the meaning of the information rather than its appearance, and when the processing integrates the information being studied. Information that is processed in this way is easier to connect with long-term memories, may improve retrieval, and may therefore result in improved learning (e.g., Anderson & Reder, 1979; Burns, 1992; Hirshman & Bjork, 1988; Reder, 1979).

CHARACTERISTICS OF THE TEST OF LEARNING

The characteristics of the test of learning can have a significant effect on learning performance (Najjar, 1995b). Tests can measure verbal, pictorial, semantic, or even procedural aspects of the learning materials. The types of tests can include recall, recognition, and problem-solving. A principle for designing a test of learning follows.

Match the Type of Information Tested to the Type of Information Learned

This design principle is moderately supported by the small number of studies performed on this topic. Scores on learning tests are higher when the kind of information (e.g., verbal, pictorial) that the learner needs to retrieve to complete the test matches the kind of information that he or she studied (e.g., Dwyer, 1967a, 1978; Morris, Bransford, & Franks, 1977; Samuels, 1967; Watkins, 1974). For example, on a verbal learning test, children in a verbal condition performed better than children in a verbal-pictorial condition. On a pictorial test, children in a verbal-pictorial condition performed better than children in a verbal condition (Beagles-Roos & Gat, 1983).

Studies also show (e.g., Frost, 1972; Leonard & Whitten, 1983; Stein, 1978) that learners perform better when they are given the same kind of test (e.g., recognition, recall) that they were told to expect when they began studying the learning material. Students who expected and got a recognition test performed better than students who expected a recognition test but got a recall test.

Learning performance improves when the way the learner stores the information (e.g., verbal, pictorial, or semantic, for recall or recognition) is similar to the way the information is tested (e.g., Morris et al., 1977; Tulving & Thomson, 1973). To improve student learning performance, the test should match the kind of information that was learned, and the given test should match the expected test.

CONCLUSIONS

These empirically validated principles will help educational multimedia user interface designers to build applications that improve learning. The most strongly supported principles suggest that designers should (a) use closely related verbal and pictorial information together and (b) build in tasks that encourage learners to elaboratively process the information. To make their educational multimedia applications even more effective, designers should also apply more general user interface design principles and guidelines (e.g., Mayhew, 1992; Smith & Mosier, 1986).

The design principles described in this paper are new, and user interface designers should use them with some caution. The number of studies supporting some of the principles (such as "Use educational multimedia with naive and lower aptitude learners") is limited, and some studies used narrow, somewhat artificial learning situations. Extending the results of these studies to other situations involves some risk. To more accurately evaluate the benefit of these design principles, studies are needed that use actual computer-based multimedia tutorials in realistic learning situations. Good candidates for these evaluations include the currently popular, compact-disk-based "edutainment" applications.

Designers need to understand not only when multimedia is effective but also why it is effective (e.g., Najjar, 1997). As more is learned about human perception, cognition, and learning, the existing educational multimedia design principles can be refined and new, more effective principles can be developed.

ACKNOWLEDGMENTS

The author thanks Laurie Najjar for editorial assistance and Chris Thompson of Georgia Tech Research Institute's Multimedia in Manufacturing Education laboratory for the use of support resources.

REFERENCES

Ackerman, B. P. (1981). Encoding specificity in the recall of pictures and words in children and adults. *Journal of Experimental Child Psychology, 31,* 193–211.

Aldrich, F. K., & Parkin, A. J. (1988). Improving the retention of aurally presented information. In M. M. Gruneberg, P. E. Morris, & R. N. Sykes (Eds.), *Practical aspects of memory* (pp. 490–493). Chichester, England: Wiley.

Allen, W. H. (1974). Media stimulus and types of learning. In H. Hitchens (Ed.), *Audiovisual instruction* (pp. 7–12). Washington, DC: Association for Educational Communications and Technology.

Anderson, J. R. (1980). *Cognitive psychology and its implications.* San Francisco: W. H. Freeman.

Anderson, J. R. (1983). *The architecture of cognition.* Cambridge, MA: Harvard University.

Anderson, J. R. (1994). *Learning and memory: An integrated approach.* New York: Wiley.

Anderson, J. R., & Reder, L. M. (1979). An elaborative processing explanation of depth of processing. In L. S. Cermak & F. I. M. Craik (Eds.), *Levels of processing in human memory* (pp. 385–403). Mahwah, NJ: Erlbaum.

Anderson, R. C., & Biddle, W. B. (1975). On asking people questions about what they are reading. In G. H. Bower (Ed.), *Psychology of learning and motivation* (Vol. 9, pp. 89–132). New York: Academic.

Arens, Y., Hovy, E. H., & Vossers, M. (1993). On the knowledge underlying multimedia presentations. In M. T. Maybury (Ed.), *Intelligent multimedia interfaces* (pp. 280–306). Menlo Park, CA: American Association for Artificial Intelligence.

Atkin, C. K. (1975a). *Effects of television advertising on children – First year experimental evidence* (Tech. Report). East Lansing: Michigan State University.

Atkin, C. K. (1975b). *Effects of television advertising on children – Second year experimental evidence* (Tech. Report). East Lansing: Michigan State University.

Atkin, C. K. (1975c). *Effects of television advertising on children – Survey of children's and mothers' responses to television commercials* (Tech. Report). East Lansing: Michigan State University.

Atkin, C. K. (1975d). *Effects of television advertising on children – Survey of preadolescents' responses to television commercials* (Tech. Report). East Lansing: Michigan State University.

Auble, P. M., & Franks, J. J. (1978). The effects of effort toward comprehension on recall. *Memory & Cognition, 6,* 20–25.

Baek, Y. K., & Layne, B. H. (1988). Color, graphics, and animation in a computer-assisted learning tutorial lesson. *Journal of Computer-Based Instruction, 15,* 131–135.

Baggett, P. (1979). Structurally equivalent stories in movie and text and the effect of the medium on recall. *Journal of Verbal Learning and Verbal Behavior, 18,* 333–356.

Baggett, P. (1984). Role of temporal overlap of visual and auditory material in forming dual media associations. *Journal of Educational Psychology, 76,* 408–417.

Baggett, P., & Ehrenfeucht, A. (1983). Encoding and retaining information in the visuals and verbals of an educational movie. *Educational Communication and Technology Journal, 31,* 23–32.

Bahrick, H. P., & Gharrity, K. (1976). Interaction among pictorial components in the recall of picture captions. *Journal of Experimental Psychology: Human Learning and Memory, 2,* 103–111.

Barrow, L. C., & Westley, B. H. (1959). Comparative teaching effectiveness of radio and television. *Audio Visual Communication Review, 7,* 14–23.

Bartram, D. J. (1980). Comprehending spatial information: The relative efficiency of different methods of presenting information about bus routes. *Journal of Applied Psychology, 65,* 103–110.

Bates, J. A. (1979). Extrinsic reward and intrinsic motivation: A review with implications for the classroom. *Review of Educational Research, 49,* 557–576.

Baxter, W. S., Quarles, R., & Kosak, H. (1978, August). *The effects of photographs and their size on reading and recall of news stories.* Presented at the Annual Meeting of the Association for Education in Journalism, Seattle, WA. (ERIC Document Reproduction Service No. ED 159 722)

Beagles-Roos, J., & Gat, I. (1983). Specific impact of radio and television on children's story comprehension. *Journal of Educational Psychology, 75,* 128–137.

Bekerian, D. A., & Baddeley, A. D. (1980). Saturation advertising and the repetition effect. *Journal of Verbal Learning and Verbal Behavior, 19,* 17–25.

Bickford, N. L. (1989). *The systematic application of principles of motivation to the design of printed instruction.* Unpublished doctoral dissertation, Florida State University, Tallahassee.

Blake, T. (1977). Motion in instructional media: Some subject-display mode interactions. *Perceptual and Motor Skills, 44,* 975–985.

Bobrow, D. G., & Bower, G. H. (1969). Comprehension and recall of sentences. *Journal of Experimental Psychology, 80,* 455–461.

Bock, M. (1978). Levels of processing of normal and ambiguous sentences in different contexts. *Psychological Research, 40,* 37–51.

Bosco, J. (1986). An analysis of evaluations of interactive video. *Educational Technology, 25,* 7–16.

Bower, G. H., & Winzenz, D. (1970). Comparison of associative learning strategies. *Psychonomic Science, 20,* 119–120.

Bransford, J. D. (1978). *Human cognition.* Belmont, CA: Wadsworth.

Bransford, J. D., & Johnson, M. K. (1972). Contextual prerequisites for understanding: Some investigations of comprehension and recall. *Journal of Verbal Learning and Verbal Behavior, 11,* 717–726.

Burns, D. J. (1992). The consequences of generation. *Journal of Memory and Language, 31,* 615–633.

Carabello, J. (1985). *The effect of various visual display modes of selected educational objectives.* Unpublished doctoral dissertation, Pennsylvania State University, University Park.

Carterette, E. C., & Jones, M. H. (1967). Visual and auditory information processing in children and adults. *Science, 156,* 986–988.

Chan, A., Travers, R. M. W., & Van Mondfrans, A. P. (1965). The effects of colored embellishments of a visual array on a simultaneously presented audio array. *Audio Visual Communication Review, 13,* 159–164.

Clark, J. M., & Paivio, A. (1991). Dual coding theory and education. *Educational Psychology Review, 3,* 149–210.

Clark, R. E. (1983). Reconsidering research on learning from media. *Review of Educational Research, 53,* 445–459.

Clark, R. E. (1985). Evidence for confounding in computer-based instruction studies: Analyzing the meta-analyses. *Educational Communication and Technology Journal, 33,* 249–263.

Clark, R. E., & Craig, T. G. (1992). Research and theory on multimedia learning effects. In M. Giardina (Ed.), *Interactive multimedia learning environments: Human factors and technical considerations on design issues* (pp. 19–30). New York: Springer-Verlag.

Condry, J. (1977). Enemies of exploration: Self-initiated versus other-initiated learning. *Journal of Personality and Social Psychology, 35,* 459–477.

Craik, F. I. M., & Lockhart, R. S. (1972). Levels of processing: A framework for memory research. *Journal of Verbal Learning and Verbal Behavior, 11,* 671–684.

Craik, F. I. M., & Tulving, E. (1975). Depth of processing and the retention of words in episodic memory. *Journal of Experimental Psychology: General, 104,* 268–294.

Craik, F. I. M., & Watkins, M. J. (1973). The role of rehearsal in short-term memory. *Journal of Verbal Learning and Verbal Behavior, 12,* 559–607.

Curtis, R. V., & Reigeluth, C. M. (1984). The use of analogies in written text. *Instructional Science, 13,* 99–117.

Dean, R. S., & Enemoh, P. A. C. (1983). Pictorial organization in prose learning. *Contemporary Educational Psychology, 8,* 20–27.

Dean, R. S., & Kulhavy, R. W. (1981). Influence of spatial organization on prose learning. *Journal of Educational Psychology, 73,* 57–64.

Deci, E. L. (1975). *Intrinsic motivation.* New York: Plenum.

Dirks, J., & Neisser, U. (1977). Memory for objects in real scenes: The development of recognition and recall. *Journal of Experimental Child Psychology, 23,* 315–328.

Dwyer, F. M. (1967a). Adapting visual illustrations for effective learning. *Harvard Educational Review, 37,* 250–263.

Dwyer, F. M. (1967b). The relative effectiveness of varied visual illustrations in complementing programmed instruction. *Journal of Experimental Education, 36,* 34–42.

Dwyer, F. M. (1978). *Strategies for improving visual learning.* State College, PA: Learning Sciences.

Entin, E. E. (1974). Effects of achievement-oriented and affiliative motives on private and public performance. In J. W. Atkinson & J. O. Raynor (Eds.), *Motivation and achievement* (pp. 219–236). Washington, DC: V. H. Winston.

Evans, T., & Denny, M. R. (1978). Emotionality of pictures and the retention of related and unrelated phrases. *Bulletin of the Psychonomic Society, 11,* 149–152.

Feiner, S. K., & McKeown, K. R. (1990). Coordinating text and graphics in explanation generation. In *AAAI-90: Proceedings, Eighth National Conference on Artificial Intelligence* (pp. 442–449). Menlo Park, CA: American Association for Artificial Intelligence.

Feiner, S. K., & McKeown, K. R. (1991). Automating the generation of coordinated multimedia explanations. *Computer, 24*(10), 33–41.

Flesch, R. (1948). A new readability yardstick. *Journal of Applied Psychology, 32,* 221–233.

Flesch, R., & Lass, A. (1949). *A new guide to better writing.* New York: Harper and Row.

Fletcher, D. (1989). The effectiveness and cost of interactive videodisc instruction. *Machine-Mediated Learning, 3,* 361–385.

Fletcher, D. (1990). *The effectiveness and cost of interactive videodisc instruction in defense training and education* (IDA Paper P-2372). Alexandria, VA: Institute for Defense Analyses.

Fowler, B. T. (1980). *The effectiveness of computer-controlled videodisc-based training.* Unpublished doctoral dissertation, University of Iowa, Iowa City.

Frase, L. T. (1975). Prose processing. In G. H. Bower (Ed.), *The psychology of learning and motivation* (Vol. 9, pp. 1–47). New York: Academic.

Frost, N. (1972). Encoding and retrieval in visual memory tasks. *Journal of Experimental Psychology, 95,* 317–326.

Garrison, W. T. (1978). *The context bound effects of picture-text amalgams: Two studies.* Doctoral dissertation, Cornell University. *(Dissertation Abstracts International, 39,* 4137A)

Harley, W. F., Jr. (1965). The effect of monetary incentive in paired-associate learning using a differential method. *Psychonomic Science, 2,* 377–378.

Hartman, F. R. (1961). Single and multiple channel communication: A review of research and a proposed model. *Audio Visual Communication Review, 9,* 235–262.

Hirshman, E., & Bjork, R. A. (1988). The generation effect: Support for a two-factor theory. *Journal of Experimental Psychology: Learning, Memory, and Cognition, 14,* 484–494.

Hoffman, C. D., & Dick, S. A. (1976). A developmental investigation of recognition memory. *Child Development, 47,* 794–799.

Hoffner, C., Cantor, J., & Thorson, E. (1989). Children's responses to conflicting auditory and visual features of a televised narrative. *Human Communications Research, 16,* 256–278.

Hunt, R. R., & Elliot, J. M. (1980). The role of nonsemantic information in memory: Orthographic distinctiveness effects on retention. *Journal of Experimental Psychology: General, 109,* 49–74.

Hyde, T. S., & Jenkins, J. J. (1969). The differential effects of incidental tasks on the organization of recall of a list of highly associated words. *Journal of Experimental Psychology, 82,* 472–481.

Jacoby, L. L. (1978). On interpreting the effects of repetition: Solving a problem versus remembering a solution. *Journal of Verbal Learning and Verbal Behavior, 17,* 649–667.

Jacoby, L. L., Craik, F. I. M., & Begg, I. (1979). Effects of decision difficulty on recognition and recall. *Journal of Verbal Learning and Verbal Behavior, 18,* 585–600.

Jenkins, J. J. (1979). Four points to remember: A tetrahedral model of memory experiments. In L. S. Cermak & F. I. M. Craik (Eds.), *Levels of processing and human memory* (pp. 429–446). Mahwah, NJ: Erlbaum.

Kanner, J. M., & Rosenstein, A. J. (1960). Television in army training: Color vs. black and white. *Audio Visual Communication Review, 8,* 243–252.

Kanner, J. H., Runyon, R. P., & Desiderato, O. (1954). *Television in army training: Evaluation of television in army training* (Tech. Report No. 14). Washington, DC: George Washington University, Human Resources Research Office.

Keller, J. M. (1983). Motivational design of instruction. In C. M. Reigeluth (Ed.), *Instructional-design theories and models: An overview of their current status* (pp. 383–434). Mahwah, NJ: Erlbaum.

Kolers, P. A. (1979). A pattern analyzing basis of recognition. In L. S. Cermak & F. I. M. Craik (Eds.), *Levels of processing in human memory* (pp. 363–384). Mahwah, NJ: Erlbaum.

Kraft, M. E. (1961). *A study of information and vocabulary achievement from teaching of natural science by television in fifth grade.* Unpublished doctoral dissertation, Boston University.

Kulik, J. A., Bangert, R. L., & Williams, G. W. (1983). Effects of computer-based teaching on secondary school students. *Journal of Educational Psychology, 75,* 19–26.

Kunz, G. C., Drewniak, U., & Schott, F. (1989, April). *On-line and off-line assessment of self-regulation in learning from instructional text and picture.* Presented at the Annual Meeting of the American Educational Research Association, San Francisco, CA.

Leifer, A. D., Collins, A. W., Gross, B., Taylor, P., Andrews, L., & Blackmer, E. (1971). Developmental aspects of variables relevant to observational learning. *Child Development, 42,* 1509–1516.

Leonard, J. M., & Whitten, W. B. (1983). Information stored when expecting recall or recognition. *Journal of Experimental Psychology: Learning, Memory, and Cognition, 9,* 440–455.

Lepper, M. R., & Greene, D. (1978). *The hidden costs of reward.* Mahwah, NJ: Erlbaum.

Lepper, M. R., Greene, W., & Nisbett, R. E. (1973). Undermining children's intrinsic interest with extrinsic rewards: A test of the overjustification hypothesis. *Journal of Personality and Social Psychology, 28,* 129–137.

Levie, W. H., & Lentz, R. (1982). Effects of text illustrations: A review of research. *Educational Communication and Technology Journal, 30,* 195–232.

Levin, J. R., Bender, B. G., & Lesgold, A. M. (1976). Pictures, repetition, and young children's oral prose learning. *Audio Visual Communication Review, 24,* 367–380.

Levin, J. R., & Lesgold, A. M. (1978). On pictures in prose. *Educational Communication and Technology Journal, 26,* 233–243.

Lieberman, L. R., & Culpepper, J. T. (1965). Words versus objects: Comparison of free verbal recall. *Psychological Reports, 17,* 983–988.

Loftus, G. R. (1972). Eye fixations and recognition memory for pictures. *Cognitive Psychology, 3,* 525–551.

Marcer, D. (1967). The effect of presentation method on short-term recall of CCC trigrams. *Psychonomic Science, 8,* 335–336.

Markiewicz, D. (1974). Effects of humor on persuasion. *Sociometry, 37,* 407–422.

Mayer, R. E. (1989a). Models for understanding. *Review of Educational Research, 59,* 43–64.

Mayer, R. E. (1989b). Systematic thinking fostered by illustrations in scientific text. *Journal of Educational Psychology, 81,* 240–246.

Mayer, R. E. (1997). Multimedia learning: Are we asking the right questions? *Educational Psychologist, 32*(1), 1–19.

Mayer, R. E., & Anderson, R. B. (1991). Animations need narrations: An experimental test of a dual-coding hypothesis. *Journal of Educational Psychology, 83,* 484–490.

Mayer, R. E., & Anderson, R. B. (1992). The instructive animation: Helping students build connections between words and pictures in multimedia learning. *Journal of Educational Psychology, 84,* 444–452.

Mayer, R. E., & Gallini, J. K. (1990). When is an illustration worth ten thousand words? *Journal of Educational Psychology, 82,* 715–726.

Mayer, R. E., Steinhoff, K., Bower, G., & Mars, R. (1995). A generative theory of textbook design: Using annotated illustrations to foster meaningful learning of science text. *Educational Technology Research and Development, 43,* 31–43.

Mayhew, D. J. (1992). *Principles and guidelines in software user interface design.* Englewood Cliffs, NJ: Prentice-Hall.

McConkie, G. W., Rayner, K., & Wilson, S. J. (1973). Experimental manipulation of reading strategies. *Journal of Educational Psychology, 65,* 1–8.

McConnell, J. V. (1978). Confessions of a textbook writer. *American Psychologist, 33,* 159–169.

Menne, J. M., & Menne, J. W. (1972). The relative efficiency of bimodal presentation as an aid to learning. *Audio Visual Communication Review, 20,* 170–180.

Morris, C. D., Bransford, J. D., & Franks, J. J. (1977). Levels of processing versus transfer appropriate processing. *Journal of Verbal Learning and Verbal Behavior, 16,* 519–533.

Murdock, B. B., Jr. (1968). Modality effects in short-term memory: Storage or retrieval? *Journal of Experimental Psychology, 77,* 79–86.

Najjar, L. J. (1995a). *Dual coding as a possible explanation for the effects of multimedia on learning* (GIT-GVU-95-29). Atlanta: Georgia Institute of Technology, Graphics, Visualization and Usability Center. Also available at http://www.cc.gatech.edu/gvu/reports.

Najjar, L. J. (1995b). *A review of the fundamental effects of multimedia information presentation on learning* (GIT-GVU-95-20). Atlanta: Georgia Institute of Technology, Graphics, Visualization and Usability Center. Also available at http://www.cc.gatech.edu/gvu/reports.

Najjar, L. J. (1996a). *The effects of multimedia and elaborative encoding on learning* (GIT-GVU-96-05). Atlanta: Georgia Institute of Technology, Graphics, Visualization and Usability Center. Also available at http://www.cc.gatech.edu/gvu/reports.

Najjar, L. J. (1996b). Multimedia information and learning. *Journal of Educational Multimedia and Hypermedia, 5,* 129–150.

Najjar, L. J. (1997). *A framework for learning from media: The effects of materials, tasks, and tests on performance* (GIT-GVU-97-21). Atlanta: Georgia Institute of Technology, Graphics, Visualization and Usability Center. Also available at http://www.cc.gatech.edu/gvu/reports.

Nelson, D. L. (1979). Remembering pictures and words: Appearance, significance, and name. In L. S. Cermak & F. I. M. Craik (Eds.), *Levels of processing in human memory* (pp. 45–76). Mahwah, NJ: Erlbaum.

Nelson, D. L., Reed, V. S., & Walling, J. R. (1976). Pictorial superiority effect. *Journal of Experimental Psychology: Human Learning and Memory, 2,* 523–528.

Nugent, G. C. (1982). Pictures, audio, and print: Symbolic representation and effect on learning. *Educational Communication and Technology Journal, 30,* 163–174.

Owings, R. A., & Baumeister, A. A. (1979). Levels of processing, encoding strategies, and memory development. *Journal of Experimental Child Psychology, 28,* 100–118.

Paivio, A. (1971). *Imagery and verbal processes.* New York: Holt, Rinehart & Winston.

Paivio, A. (1975). Coding distinctions and repetition effects in memory. In G. H. Bower (Ed.), *The psychology of learning and motivation* (Vol. 9, pp. 179–214). New York: Academic.

Paivio, A. (1986). *Mental representations: A dual-coding approach.* New York: Oxford University Press.

Paivio, A. (1991). Dual coding theory: Retrospect and current status. *Canadian Journal of Psychology, 45*, 255–287.

Paivio, A., & Csapo, K. (1969). Concrete-image and verbal memory codes. *Journal of Experimental Psychology, 80*, 279–285.

Paivio, A., & Csapo, K. (1973). Picture superiority in free recall: Imagery or dual coding? *Cognitive Psychology, 5*, 176–206.

Paivio, A., Rogers, T. B., & Smythe, P. C. (1968). Why are pictures easier to recall than words? *Psychonomic Science, 11*, 137–138.

Palmere, M., Benton, S. L., Glover, J. A., & Ronning, R. (1983). Elaboration and recall of main ideas in prose. *Journal of Educational Psychology, 75*, 898–907.

Palmiter, S., & Elkerton, J. (1991). An evaluation of animated demonstrations for learning computer-based tasks. In *Human Factors in Computing Systems: CHI '91 Conference Proceedings* (pp. 257–263). Reading, MA: Addison-Wesley.

Paradowski, W. (1967). Effect of curiosity on incidental learning. *Journal of Educational Psychology, 58*, 50–55.

Park, O., & Hopkins, R. (1993). Instructional conditions for using dynamic visual displays: A review. *Instructional Science, 21*, 427–449.

Parkin, A. J. (1984). Levels of processing, context, and facilitation of pronunciation. *Acta Psychologia, 55*, 19–29.

Peeck, J. (1974). Retention of pictorial and verbal content of a text with illustrations. *Journal of Educational Psychology, 66*, 880–888.

Peeck, J. (1985, March). *Effects of mismatched pictures on retention of illustrated prose.* Paper presented at the Annual Meeting of the American Educational Research Association, Chicago, IL.

Penney, C. G. (1975). Modality effects in short-term verbal memory. *Psychological Bulletin, 82*, 68–84.

Pezdek, K., Lehrer, A., & Simon, S. (1984). The relationship between reading and cognitive processing of television and radio. *Child Development, 55*, 2072–2082.

Pressley, M., McDaniel, M. A., Turnure, J. E., Wood, E., & Ahmad, M. (1987). Generation and precision of elaboration: Effects on intentional and incidental learning. *Journal of Experimental Psychology: Learning, Memory, and Cognition, 13*, 291–300.

Rankin, E. F., & Culhane, J. W. (1970). One picture equals 1,000 words? *Reading Improvement, 7*(2), 37–40.

Raynor, J. O. (1974). Relationships between achievement-related motives, future orientation, and academic performance. In J. W. Atkinson & J. O. Raynor (Eds.), *Motivation and achievement* (pp. 173–187). Washington, DC: Winston.

Read, J. D., & Barnsley, R. H. (1977). Remember Dick and Jane? Memory for elementary school readers. *Canadian Journal of Behavioral Science, 9*, 361–370.

Reder, L. M. (1979). The role of elaborations in memory for prose. *Cognitive Psychology, 11*, 221–234.

Reed, S. (1985). Effect of computer graphics on improving estimates to algebra word problems. *Journal of Educational Psychology, 77*, 285–298.

Reiser, R. A., & Gagné, R. M. (1982). Characteristics of media selection models. *Review of Educational Research, 52*, 499–512.

Rieber, L. P. (1989). The effects of computer animated elaboration strategies and practice on factual and application learning in an elementary science lesson. *Journal of Educational Computing Research, 5*, 431–444.

Rieber, L. P. (1990a). Animation in computer-based instruction. *Educational Technology Research and Development, 38*, 77–86.

Rieber, L. P. (1990b). Using computer animated graphics in science instruction with children. *Journal of Educational Psychology, 82*, 135–140.

Rigney, J. W., & Lutz, K. A. (1976). Effect of graphic analogies of concepts in chemistry on learning and attitude. *Journal of Educational Psychology, 68*, 305–311.

Roshal, S. M. (1961). Film-mediated learning with varying representations of the task: Viewing angle, portrayal of demonstration, motion, and student participation. In A. A. Lumsdaine (Ed.), *Student response in programmed instruction* (pp. 155–175). Washington, DC: National Research Council.

Ross, S. M. (1983). Increasing the meaningfulness of quantitative material by adapting context to student background. *Journal of Educational Psychology, 75*, 519–529.

Rothkopf, E. Z. (1966). Learning from written instructive materials: An exploration of the control of inspection behavior by test-like events. *American Educational Research Journal, 3*, 241–249.

Rubin, R. S. (1972). *An exploratory investigation of children's responses to commercial content of television advertising in relation to their stages of cognitive development.* Unpublished doctoral dissertation, University of Massachusetts, Amherst.

Rundus, D. (1977). Maintenance rehearsal and single-level processing. *Journal of Verbal Learning and Verbal Behavior, 16*, 665–681.

Salomon, G. (1984). Television is "easy" and print is "tough": The differential investment of mental effort in learning as a function of perceptions and attributions. *Journal of Educational Psychology, 76*, 647–658.

Samuels, S. J. (1967). Attentional processes in reading: The effect of pictures in the acquisition of reading responses. *Journal of Educational Psychology, 58*, 337–342.

Severin, W. J. (1967). The effectiveness of relevant pictures in multiple-channel communications. *Audio Visual Communication Review, 15*, 386–401.

Sewell, E. H., Jr., & Moore, R. L. (1980). Cartoon embellishments in informative presentations. *Educational Communication and Technology Journal, 28*, 39–46.

Shepard, R. N. (1967). Recognition memory for words, sentences, and pictures. *Journal of Verbal Learning and Verbal Behavior, 6*, 156–163.

Sherman, J. L. (1976). Contextual information and prose comprehension. *Journal of Reading Behavior, 8*, 369–379.

Sherman, M. F., & Turvey, M. T. (1969). Modality differences in short-term serial memory as a function of presentation rate. *Journal of Experimental Psychology, 80*, 335–338.

Slamecka, N. J., & Graf, P. (1978). The generation effect: Delineation of a phenomenon. *Journal of Experimental Psychology: Human Learning and Memory, 4*, 592–604.

Smith, M. C., & Magee, L. E. (1980). Tracing the time course of picture-word processing. *Journal of Experimental Psychology: General, 109*, 373–392.

Smith, S. L., & Mosier, J. N. (1986). *Guidelines for designing user interface software* (ESD-TR-86-278; MTR 10090). Bedford, MA: MITRE Corp.

Spangenberg, R. W. (1973). The motion variable in procedural learning. *Audio Visual Communication Review, 21*, 419–436.

Stafford, J. Y. (1990). *Effects of active learning with computer-assisted or interactive video instruction.* Unpublished doctoral dissertation, Wayne State University, Detroit, MI.

Stein, B. S. (1978). Depth of processing re-examined: The effects of precision of encoding and test appropriateness. *Journal of Verbal Learning and Verbal Behavior, 17*, 165–174.

Stein, B. S., & Bransford, J. D. (1979). Constraints on effective elaboration: Effects of precision and subject generation. *Journal of Verbal Learning and Verbal Behavior, 18*, 769–777.

Stein, B. S., Morris, C. D., & Bransford, J. D. (1978). Constraints on effective elaboration. *Journal of Verbal Learning and Verbal Behavior, 17*, 707–714.

Sternthal, B., & Craig, C. S. (1973). Humor in advertising. *Journal of Marketing, 37*, 12–18.

Stevenson, H. W., & Siegel, A. (1969). Effects of instructions and age on retention of filmed content. *Journal of Educational Psychology, 60*, 71–74.

Stoneman, Z., & Brody, G. H. (1983). Immediate and long-term recognition and generalization of advertised products as a function of age and presentation mode. *Developmental Psychology, 19*, 56–61.

Tennyson, R. D. (1978). Pictorial support and specific instructions as design variables for children's concept and rule learning. *Educational Communication and Technology Journal, 26*, 291–299.

Thomson, C. P., & Barnett, C. (1981). Memory for product names: The generation effect. *Bulletin of the Psychonomic Society, 18*, 241–243.

Tosti, D. T. (1978, October). Formative feedback. *National Society for Performance and Instruction Journal, 17*, 19–21.

Tulving, E., & Thomson, D. M. (1973). Encoding specificity and retrieval processes in episodic memory. *Psychological Review, 80*, 352–373.

Van Mondfrans, A. P., & Travers, R. M. W. (1964). Learning of redundant material presented through two sensory modalities. *Perceptual and Motor Skills, 19*, 743–751.

Verano, M. (1987). *Achievement and retention of Spanish presented via videodisc in linear, segmented and interactive modes.* Unpublished doctoral dissertation, University of Texas, Austin.

Walker, N., Jones, J. P., & Mar, H. H. (1983). Encoding processes and the recall of text. *Memory and Cognition, 11*, 275–282.

Ward, S. (1972). Children's reactions to commercials. *Journal of Advertising Research, 12*, 37–45.

Ward, S., Wackman, D. B., & Wartella, E. (1978). *How children learn to buy.* Newbury Park, CA: Sage.

Wardle, K. F. (1977, August). *Textbook illustrations: Do they aid reading comprehension?* Presented at the Annual Meeting of the American Psychological Association, San Francisco, CA.

Watkins, M. J. (1974). When is recall spectacularly higher than recognition? *Journal of Experimental Psychology, 102*, 161–163.

Watkins, O. C., & Watkins, M. J. (1980). The modality effect and echoic persistence. *Journal of Experimental Psychology: General, 109*, 251–278.

Watts, G. H., & Anderson, R. C. (1971). Effects of three types of inserted questions on learning from prose. *Journal of Educational Psychology, 62*, 387–394.

Wetstone, H. S., & Friedlander, B. Z. (1974). The effect of live, TV, and audio story narration on primary grade children's listening comprehension. *Journal of Educational Research, 68*, 32–35.

Winn, W. (1993). Perception principles. In M. Fleming & W. H. Levie (Eds.), *Instructional message design: Principles from the behavioral and cognitive sciences* (pp. 55–126). Englewood Cliffs, NJ: Educational Technology Publications.

Lawrence J. Najjar is a graduate research assistant at the Georgia Institute of Technology, where he expects to receive his Ph.D. in engineering psychology in 1998. He received an M.S. in engineering psychology from Georgia Tech in 1983.

Date received: May 17, 1996
Date accepted: December 12, 1997

A Methodology for Enhancing Crew Resource Management Training

Eduardo Salas and **Carolyn Prince**, Naval Air Warfare Center Training Systems Division, Orlando, Florida, **Clint A. Bowers,** University of Central Florida, Orlando, Florida, and **Renée J. Stout, Randall L. Oser,** and **Janis A. Cannon-Bowers,** Naval Air Warfare Center Training Systems Division, Orlando, Florida

Human error is an ever-present threat to the safe conduct of flight. Recently, applied psychologists have developed an intervention, crew resource management (CRM) training, designed to help prevent human error in the cockpit. However, as it is commonly applied within the aviation community, CRM lacks standardization in content, design, delivery, and evaluation. This paper presents a discussion of an applied program of research aimed at developing a methodology for the design and delivery of CRM training within the Navy. This long-term, theoretically based program of aviation team research included identification of skills to be trained, development of performance measures, application of instructional design principles, and evaluation of the training delivery. Our conclusion indicates that a systematic methodology for developing CRM training can result in better performance in the cockpit. Actual or potential applications of this research include any task environment in which teams are interdependent.

INTRODUCTION

In the 1970s, hundreds of airline passengers on routine, scheduled flights lost their lives because each of three aircrews committed an error. In one incident, the crew failed to take fuel levels into consideration during problem solving; in the second incident, the crew did not monitor the altitude; and in the third incident, the crew misinterpreted an air traffic control communication. The crew members who committed the errors had tens of thousands of hours of flight experience, yet the errors committed should have been avoided by even the most inexperienced pilots. The crews were not members of a country in which standards of pilot training and certification were questionable, and each of these crews worked for a major air carrier.

Two of the crews were flying domestic operations within the United States. In both domestic accidents, the crew members committed the errors while responding to a potentially unsafe problem with the plane by taking extra time and care to troubleshoot or prepare (or both) for this unplanned circumstance. As a result of their lapses, their planes ended the flight in one case by crashing into a stand of trees in Oregon, and in the other case by crashing into the Florida Everglades. The third plane, on an international flight, had results so catastrophic that it sent shock waves throughout the world. This plane collided with another aircraft, immediately ending the lives of everyone aboard both planes.

After 20 years, the aviation industry is still challenged by a haunting question: Why is the number of take-offs not equal to the number of safe landings? In the past 20 years, it has been commonly acknowledged that almost 60% to 80% of aviation incidents and accidents were attributable to human error in the cockpit (Foushee, 1984). This recognition led a number of applied psychologists to suggest an intervention aimed at improving human

Address correspondence to Eduardo Salas, Code 4961, Naval Air Warfare Center Training Systems Division, 12350 Research Parkway, Orlando, FL 32826-3275; salasea@navair.navy.mil. **HUMAN FACTORS**, Vol. 41, No. 1, March 1999, pp. 161–172.

performance and, in particular, teamwork in the cockpit. This intervention, commonly referred to as *crew resource management* (CRM) training, now has a long history of research and practice in the air carrier industry (Wiener, Kanki, & Helmreich, 1993).

On the military side, CRM training developments have also emerged. (The military labeled this team training intervention *aircrew coordination training*, but we will use the term CRM in this paper because it is most common in the airline industry and government regulatory agencies.) The U.S. Navy (in particular, the Marine Corps) enlisted the help of the Naval Air Warfare Center Training Systems Division about 10 years ago in improving the safety of its rotary wing fleet. Our response was to design and conduct a long-term program of research that began with theory building and moved into development of measures of performance, design of instruction, empirical testing, and evaluation. The purpose of this paper is to describe our efforts in this regard. To do this, we organized the presentation around the critical questions that have guided our research:

1. What is CRM, and, more specifically, what is CRM training?
2. Which theories provide a basis to develop CRM training?
3. Which skills underlie effective CRM?
4. Which instructional approaches and strategies are appropriate to impart CRM skills?
5. What evidence exists to support the effectiveness of CRM training?

We conclude with a presentation of our methodology for developing CRM training and a word about reciprocity between training science and practice.

What Is CRM and CRM Training?

The three accidents discussed in the previous section help illustrate a persistent threat to safe aviation: human error caused by inadequate coordination among team members. Although there were minor mechanical failures in two of the accidents, each occurred because of the crew's error. It is clear that the errors made in these accidents were not the result of inadequate technical training. All of those involved knew how to read the gauges and

how to conduct emergency procedures. It was also clear that on its own, human redundancy in the cockpit had not worked to make the system more secure; three people together in a single cockpit still overlooked a basic flight parameter. This led researchers to expand their view of what was required for effective aviation beyond technical aspects of the task (i.e., flying the aircraft) to include nontraditional competencies such as teamwork.

According to Lauber (1984), this new way of thinking (labeled first as *cockpit resource management* and later as crew resource management) is defined as "using all available resources – information, equipment, and people – to achieve safe and efficient flight operations" (p. 20). Foushee and Helmreich (1988) added that

> CRM includes optimizing not only the person-machine interface and the acquisition of timely, appropriate information, but also interpersonal activities including leadership, effective team formation and maintenance, problem-solving, decision-making, and maintaining situation awareness. (p. 4)

Given this definition, researchers began to examine what training for CRM might include. According to Foushee and Helmreich (1988, p. 4), such training involves "communicating basic knowledge of human factors concepts that relate to aviation and providing the tools necessary to apply these concepts operationally." At the time, CRM represented "a new focus on crew-level (as opposed to individual-level) aspects of training and operations" (p. 4). The major goal of the resulting training was to "help stem the tide of accidents caused by so-called human error" (Stone & Babcock, 1988, p. 553) by addressing "situational, sociopsychological and other factors that influence aircrew performance" (Caro, 1988, p. 258).

When it came to actually developing training to address CRM issues, early attempts held almost exclusively that CRM training should target crew members' attitudes toward teamwork (Chidester & Foushee, 1988; Helmreich & Wilhelm, 1991). In keeping with the definition of CRM, this approach to CRM training emphasized the social-psychological factors that influence crew performance. In fact, this work by Foushee, Helmreich, and others paved the way for the aviation industry to consider "softer" human performance and teamwork issues to be legitimate concerns. This was no small accomplishment. The success of these scientists was to shift the emphasis in aviation training onto social interactions in the cockpit (i.e., crew coordination). However, early CRM training practices did not go far enough; we believe that there were two fundamental problems.

First, initial CRM training overemphasized the affective, personality, and attitudinal aspects of crew coordination (i.e., the "right stuff") at the expense of the behavioral aspects of the problem of crew coordination. Second, little guidance was available regarding how to train crew coordination skills. So, although these early efforts had the effect of enlightening the aviation community, they did not address the CRM training problem sufficiently.

In the 10 years or so that have followed, the lack of a standardized methodology for developing CRM training (and the associated problem of disagreement about exactly what needed to be trained) caused a host of diverse CRM training programs to be developed. Whereas early programs emphasized attitudes toward teamwork, as described earlier, others were based on personality or skill, and still others on a combination of these (Byrnes & Black, 1993; Helmreich & Foushee, 1993; Yamamori & Mito, 1993). Furthermore, programs have varied in length from 1 h to 2 weeks. Training has been given in lectures, discussions, videotape observations, game-playing, classroom roleplay, mishap analyses, and both low- and high-fidelity simulations. Some courses have used only one technique (e.g., lecture) and others have used several. Content has shown diversity as well and has included topics such as interaction styles, stress reduction, and automation issues, in addition to the subjects of workload management, advocacy, and situation awareness that are included in many programs. Some programs are clearly based on attitude change or skill development, but others, for which training has been composed of elements copied from the programs of other organizations, have no discernible basis.

In sum, the lack of agreement regarding what CRM training should include and how it should be accomplished has led to confusion and, in many cases, adoption of atheoretical and suboptimal programs. To address these deficiencies, our work for the U.S. Navy adopted a more systematic approach to developing CRM training that was consistent with our team training work in similar domains.

As a starting point, we clarified the concept of CRM by extracting the notion of teamwork (or crew coordination) from the definitions offered earlier as the essential ingredient of CRM. We selected this focus because the evidence – gleaned mostly from accident reports – suggested that it was the lack of coordination of team member resources or the underutilization of team member resources (or both) that accounted for most of the errors (Foushee, 1984). Hence, we define CRM as being a set of teamwork competencies that allow the crew to cope with situational demands that would overwhelm any individual crew member (we elaborate on these competencies in a later section). Flowing from this definition, we consider CRM training to be a family of instructional strategies designed to improve teamwork in the cockpit by applying well-tested training tools (e.g., performance measures, exercises, feedback mechanisms) and appropriate training methods (e.g., simulators, lectures, videos) targeted at specific content (i.e., teamwork knowledge, skills, and attitudes).

It should be noted that our initial emphasis in CRM training was on skills (in part because other programs emphasized attitudes). More recently, our work has been focused more directly on the knowledge underlying effective teamwork (see, e.g., Stout, Salas, & Kraiger, 1997), but this work is not covered here. In the following sections, we describe the program of research that we conducted to enhance the training of CRM skills and the resulting methodology that we developed. As noted, we do this by addressing a number of crucial questions that guided our work.

Which Theories Provide a Basis for CRM Training Development?

The problem of improving teamwork in a task as complex as aviation provides a formidable challenge. Clearly, no single approach is likely to address the entire problem. Moreover, without a strong theoretical foundation –

one that rests on an understanding of individual and team performance, human learning and skill acquisition, and pedagogy – it is not likely that effective training will be developed or fielded. In fact, we submit as others have that there is nothing more practical than a good theory (Lewin, as cited in Marrow, 1969). This statement embraces the philosophy behind the design of all of the team training strategies that we have developed and tested (Salas & Cannon-Bowers, 1997).

Fortunately, there are a number of theoretical perspectives that can provide a strong foundation for CRM training development. For example, Tannenbaum, Beard, and Salas (1992) described a theoretical model of team performance that includes consideration of a variety of factors that might influence the performance of flight crews. This model has already been used as a foundation for research in CRM for automated cockpits (see Bowers, Thornton, Braun, & Salas, 1998).

Recent theoretical advances in cognitive psychology also offer promise in guiding what to train and how to train it. For example, Jentsch (1997) developed a metacognitive training strategy for junior first officers aimed at improving their ability to determine the appropriate time to act. The training focused on teaching the pilots monitoring skills and improving their ability to mentally simulate possible outcomes of either action or inaction in a particular situation. Pilots who had received the training performed significantly better in those situations that required judgment (whether to act or not) than those who did not receive the training. Similarly, using shared mental model theory (Cannon-Bowers, Tannenbaum, Salas, & Volpe, 1995) as a foundation, Stout, Salas, and Fowlkes (1997) found that participants who received CRM-type training improved their knowledge structures related to teamwork as compared with a control group that did not receive training. In addition, the trained group demonstrated 8% more teamwork skills when it was appropriate to do so during a simulated mission. Thus, training positively influenced aviators' knowledge structures and performance.

Although our work in CRM training has been influenced in one way or another by all of the theories just described, we based our notions about training teamwork in the cockpit largely on the body of work conducted in the 1980s by Salas, Morgan, Glickman, and colleagues (see, e.g., Morgan, Salas, & Glickman, 1994). Briefly, these researchers studied military command and control teams and found behaviors that were consistent across effective teams. Specifically, they found that using closed-loop communication, predicting each other's behavior, performing self-correction, and providing motivational and task reinforcement led to better team performance (for details, see McIntyre & Salas, 1995).

Which Skills Underlie Effective CRM?

We often found that aviators who attended early CRM training commented that the course was interesting and that they learned a lot about themselves, but they did not know what to do to change the way they operated when they got back in the cockpit. As Ryle (1949) pointed out, there is a difference between knowing that something should be done and knowing how to do it. Therefore, based on our notions about teamwork, we adopted a behavioral perspective to complement the attitude and personality-based work that was ongoing in the commercial sector in the early 1990s. This meant that from a learning standpoint, we needed to identify the requisite skills that enable effective performance in naval aviation. This training was designed to provide aviators with behaviors that they could take back to the cockpit to improve their performance. Thus, our vision for CRM training emphasizes what to do rather than how to feel.

In order to fulfill this vision, we synthesized the scientific literature (Salas, Dickinson, Converse, & Tannenbaum, 1992), made multiple observations of crews from several communities performing in full-mission simulations (Prince & Salas, 1993), and conducted structured interviews with over 200 aviators from a variety of operational communities. Through an integration of information gained from these three divergent sources, we identified a set of teamwork constructs with associated behaviors that we predicted would lead to effective teamwork in the cockpit (Prince & Salas, 1993). Along the way, we developed and tested a variety of approaches to identifying fleet-specific coordination behaviors (see Bowers, Baker, & Salas, 1994; Bowers, Morgan, Salas, & Prince, 1993). These tools exist and can be used to develop skills-based training today (Salas & Cannon-Bowers, 1997; Salas & Cannon-Bowers, in press).

In sum, by integrating theoretical models of teamwork and human learning with training needs-analysis tools, we addressed the first deficiency of CRM training noted previously by identifying the specific skills and associated behaviors that should improve coordination in the cockpit. The emphasis on specific skills offers several advantages from the theoretical and evaluation perspective (described in a later section). However, training success is also dependent upon one's ability to train those skills, which brings us back to the issue of how to train CRM skills.

How Can CRM Skills Best Be Trained?

Pertinent to the question of how best to train teamwork in the cockpit, we contend that the failure to create opportunities to practice newly acquired skills was the greatest shortcoming of early CRM training. In essence, early approaches failed to take advantage of literature that relates to human learning and skill acquisition (Kolb, 1984). Consequently, our efforts have endeavored to improve upon the traditional simulator-based practice experience, which typically lacked specific measurement and feedback, to more directly reinforce the targeted teamwork skills. For example, a program of research on expanded opportunities for teamwork skill practice (and feedback) has shown that CRM skills trained in one situation transfer to another (Brannick, Prince, Salas, & Stout, 1997; Prince, Brannick, Prince, & Salas, 1997). In addition, we have shown that this method of instruction provides enhanced training transfer as compared with game-playing, a technique currently used in many CRM programs (Brannick, Prince, et al., 1997; see Jentsch & Bowers, 1998 for a thorough discussion). Once our instruction was developed, it was necessary to assess the effectiveness of the training. The methods we employed to do this are described next.

What Evidence Exists to Suggest That CRM Training Is Effective?

As with any other complex human performance problem, the issue of establishing a criterion against which to evaluate the success of CRM training is difficult. At first glance, it is tempting to use broad performance indices to evaluate CRM effectiveness. Common examples include the use of accident rates or the number of mishaps attributed to CRM errors. However, there are several concerns regarding these types of indices: (a) CRM has become part of aviation vernacular (perhaps accident investigators are citing it as a cause more frequently); (b) aircraft are becoming increasingly automated, which places considerable additional coordination demands on crews (Bowers, Deaton, Oser, Prince, & Kolb, 1995; Thornton, Braun, Bowers, & Morgan, 1992); and (c) accidents that happened are a poor index of accidents that did not happen. There are numerous anecdotal accounts of accidents that were prevented with the use of CRM skills, but do the accidents that did happen negate these reported benefits?

A better approach is one that is multifaceted and considers several levels of evaluation, including trainee reactions, extent of learning, extent of performance change, and impact on organizational effectiveness (Cannon-Bowers et al., 1989; Ford, Kozlowski, Kraiger, Salas, & Teachout, 1997). This type of approach includes a variety of data points for consideration, such as trainee evaluation, specific indices of learning, evidence of application, and so forth. A multifaceted approach is not only more thorough than an approach focusing only on reactions or attitudes, but it also provides an improved ability to assess which aspects of training work and which do not work. Hence, it is also useful in identifying specific areas in need of remediation, because it assesses multiple aspects of learning/skill-acquisition.

Using a multifaceted approach, we conducted formal evaluations of our CRM training interventions in four aircraft communities with a total of 55 crews. In each of the evaluations we used a control group as a comparison so that we could actually assess the value of our training. We also used raters or assessors of performance who were blind to conditions. Finally, we used a measure of performance that was designed to identify whether aviators were demonstrating behaviors (identified through task analysis and assessment of training manuals and operating procedures) that had been delineated as more effective by fleet subject-matter experts (as described in Fowlkes, Lane, Salas, Franz, & Oser, 1994).

We obtained evidence that our variety of CRM training is effective, as evidenced by average performance improvements of 8% to 20% (for empirical evidence see Salas, Fowlkes, Stout, Milanovich, & Prince, in press; Smith-Jentsch, Salas, & Baker, 1996; Stout, Salas, & Fowlkes, 1997). That is, on average we observed 8% to 20% more teamwork behaviors exhibited in the cockpit by crews that were trained than by crews that were not trained. This may not seem like much, but it is actually impressive considering that many of the participants in these studies were experienced crews who actually changed their cockpit performance based on a relatively short training intervention. Furthermore, we submit that performing even one more teamwork behavior (when required) can make a difference between mission success and failure (and sometimes between life and death). Similar efforts in other military communities have also shown positive results (Cannon-Bowers & Salas, 1998; Leedom & Simon, 1995; Salas, Cannon-Bowers, & Johnston, 1997).

In addition, we have conducted studies to test specific components of the training. For example, Smith-Jentsch et al. (1996), using a variety of measurement approaches, demonstrated the effectiveness (74% performance improvement for the trained group) of skills-based assertiveness training, which is a component of our CRM program. Jentsch (1997) also used a multidimensional approach in assessing the effectiveness of metacognitive training for junior first officers.

Thus, the data are encouraging. We have shown that even experienced aviators can learn new ways to behave in the cockpit and improve their performance. In the future, we hope that others will also attempt to evaluate the CRM training they develop; in fact, this should be an ongoing process. Given the diffi-

culties noted earlier, it is the responsibility of all of the experts (researchers and practitioners), the airlines, the fleets, and cognizant agencies to devise acceptable methods for evaluating CRM training effectiveness and then to make the necessary commitment to, and investment in, conducting these studies.

A METHODOLOGY FOR DEVELOPING CRM TRAINING

Table 1 shows an overview of the methodology that we employed for developing effective CRM training. It is based on the work we conducted to develop CRM training as well as guidelines from the literature (Cannon-Bowers & Salas, 1998; Salas & Cannon-Bowers, in press; Salas, Cannon-Bowers, & Blickensderfer, 1997; Swezey & Salas, 1992). It is similar to the event-based approach to training that we have applied in other domains (Dwyer, Fowlkes, Oser, Salas, & Lane, 1997; Johnston, Cannon-Bowers, & Smith-Jentsch, 1995; Oser, Dwyer, & Fowlkes, 1995). For any particular community that we worked with, we followed the steps illustrated in Table 1 to develop theoretically rooted, behaviorally based, community-specific CRM training (for examples, see Prince & Salas, 1989; Stout, Salas, & Fowlkes, 1997).

The process delineated in Table 1 began with a thorough analysis of the aircraft mission and procedures and their impact on crew coordination demands. The second step was to assess the coordination demands directly (for details, see Bowers et al., 1993). This analysis allowed us to identify specific tasks that had a team component so that we could focus attention on these tasks in training. Next, we employed our theoretical notions regarding the nature of teamwork competencies to arrive at a set of coordination skills to be trained. Each skill was then translated into a training objective and used as the basis to develop scenarios or exercises that would allow the skill to be practiced. At the same time, we developed measures of performance and associated measurement tools that were linked to each of the training objectives (Baker, Prince, Shrestha, Oser, & Salas, 1993; Dwyer et al., 1997; Prince, Oser, Salas, &

Woodruff, 1993). Our goal was to focus on observable aspects of teamwork behavior so that raters could readily assess whether the behavior was demonstrated sufficiently (Fowlkes et al., 1994).

We then used the performance measurement data as a basis for developing feedback. Because feedback mechanisms were crucial to the training, we spent considerable effort tailoring our assessment tools so that they could form the basis of effective feedback (Fowlkes et al., 1994). Finally, whenever possible, we conducted training-effectiveness evaluations to determine the impact of our training on cockpit performance.

Description of the Training

Employing the methodology just described, we developed an instructional strategy for CRM training with behavioral modeling as its basis (Salas et al., in press; Stout, Salas, & Fowlkes, 1997). The training sessions followed an information-demonstration-practice-feedback sequence. The training began with a lecture (delivered live) explaining important teamwork skills. Next, videotaped models of effective and ineffective behavior were shown to the trainees. Following this demonstration phase, trainees were provided with the opportunity to practice on carefully crafted scenarios (as described earlier), and their performance was tracked. Finally, we provided feedback that was relevant to the behaviors that we sought to train.

Using this general strategy, we were able to train the repertoire behaviors that allowed crew members to exhibit effective teamwork in the cockpit, as evidenced by the effectiveness data described earlier. It should be noted that the methodology presented here should also be useful in developing other types of CRM training strategies. In other words, there may be other ways to train various aspects of teamwork in the cockpit; our argument is that it is the systematic process for devising it that helps ensure success. Further, we are aware that many CRM training programs have been developed that are consistent with one or even several of the steps shown in Table 1. Again, we argue that in order to develop and evaluate effective CRM training, all of the activities in Table 1 must be complet-

ed. Otherwise, there is a risk of suboptimizing; that is, there is a risk of developing training that addresses only parts of the problem or lacks a sound theoretical foundation. We also submit that this methodology may be useful in guiding the development of team training for domains other than aviation.

RECIPROCITY BETWEEN TRAINING RESEARCH AND PRACTICE

Given what we have presented thus far, it is clear that we have made progress in understanding CRM and applying sound training principles to improve it. However, as with other areas of training, the results of training research often do not influence training practice as much as they could. We suggest that there is still difficulty infiltrating operational communities with the results of team performance and training efforts. This problem is certainly not limited to CRM training; we have maintained that the application of research results to practice is a problem confronted by many areas of applied psychology (Cannon-Bowers, Tannenbaum, Salas, & Converse, 1991; Salas, Cannon-Bowers, & Blickensderfer, 1997).

Too often we find that training developers fail to make use of the results of training research. There may be many reasons for this. In the past, we have contended that it is attributable to a lack of translation mechanisms. That is, researchers and practitioners lack a common forum and language for communicating. Moreover, given that our work is conducted in a military context, it is sometimes dismissed as not generalizing to many commercial tasks, even though it could offer a substantial reduction in costs incurred in "re-inventing the wheel." Furthermore, by taking advantage of the research done with military participants – who often become commercial pilots after their military careers – it is likely that the state of the art will improve more quickly than it has in the past decade.

CONCLUDING REMARKS

In conclusion, the science and practice of CRM training has evolved considerably over the past 10 years. There are now viable (and tested)

TABLE 1: A Methodology to Design and Deliver CRM Training

Step	Description	Products	Sources
1. Identify operational/mission requirement	Review existing training curriculum, including course master material lists, instructor guides, standard operating procedures (SOPs); interview aviation subject matter experts (SMEs); observe crews performing missions; review relevant mishap/accident reports.	-Mission specific context/examples -General understanding of coordination demands within the task	Fowlkes, Lane, Salas, Franz, & Oser (1994); Hartel, Smith, & Prince (1991); Prince & Salas (1989)
2. Assess team training needs and coordination demand	Use same data sources as in Step 1, with emphasis on identifying deficiencies in existing team training and specifying all tasks required to perform missions that involve a teamwork element.	-Coordination demand analyses -List of tasks requiring coordination	Bowers, Morgan, & Salas (1991); Bowers, Morgan, Salas, & Prince (1993)
3. Identify teamwork competencies and knowledge, skills, and attitudes (KSAs)	Link team training needs to a theory of team performance that allows delineation of competencies (our emphasis initially was on skills) required to perform each of the team tasks identified in Step 2.	-Set of knowledge, skills, and attitudes	Cannon-Bowers & Salas (1997); Cannon-Bowers, Tannenbaum, Salas, & Volpe (1995); Prince & Salas (1993); Stout, Salas, & Fowlkes (1997)
4. Determine team training objectives	For each teamwork KSA, develop a training objective that can be empirically evaluated to determine whether or not it was accomplished.	-List of targets for training -Full list of generic and task-specific training objectives	Salas & Cannon-Bowers (1997); Stout, Salas, & Kraiger (1997); Swezey, Llaneras, Prince, & Salas (1991)

Step	Description	Methods/Tools	References
5. Determine instructional delivery method	The method for accomplishing the instruction should be specified (e.g., information, demonstration, or practice and feedback, or all) in this step. Consideration should be given to costs and availability of simulators.	-Lectures -Videos -Role-play exercises -PC-based methodologies -High-fidelity simulators -Accident reviews as analysis -Training curriculum	Baker, Prince, Shrestha, Oser, & Salas (1993); Beard, Salas, & Prince (1995)
6. Design scenario exercises and create opportunities for practice	Design scenarios or exercises in which events are embedded to provide trainees an opportunity to demonstrate each of the required KSAs identified in objectives in which accomplishment requires practice and feedback.	-Valid realistic scenario(s), including all relevant peripheral support	Dwyer, Fowlkes, Oser, Salas, & Lane (1997); Fowlkes, Dwyer, Oser, & Salas (1998); Prince, Oser, Salas, & Woodruff (1993)
7. Develop performance assessment/measurement tools	In conjunction with scenario design, develop measures that can reliably assess whether each of the KSAs was demonstrated at an observable behavioral level.	-Behaviorally-based checklists -Subjective evaluation forms -Outcome metrics and criteria	Baker & Salas (1992); Brannick, Salas, & Prince (1997); Fowlkes et al. (1994)
8. Design and tailor tools for feedback	Design or tailor (or both) measurement tools for use in debrief, in which trainees are made aware of those required team behaviors that they did and those that they did not perform successfully. This tool should also help instructors diagnose the causes of poor performance and provide guidance for specific improvement in future operations.	-Instructor training -Debriefing checklists and guides	Salas & Cannon-Bowers (1997); Smith-Jentsch, Zeisig, Acton, & McPherson (1998)
9. Evaluate the extent of improved teamwork in the cockpit	Design experiments to assess the effectiveness of the training. Because of operational constraints, quasi-research methods may need to be applied.	-Reaction data -Learning data -Knowledge-acquisition data -Transfer data -Performance data	Fowlkes et al. (1994); Salas et al. (1998); Stout, Salas, & Fowlkes (1997)

theories, methods, principles, approaches, and content available to shape the design and delivery of CRM training. Much has been done and much remains to be done. There is little doubt that our colleagues performing research in decision making (Orasanu, 1993), situation awareness (Endsley, 1995), leadership (Pettit & Dunlap, 1997), and other topics will contribute useful elements to CRM training as well. In fact, our own interests have now turned toward training the cognitive components of teamwork in the cockpit (Salas, Bowers, & Cannon-Bowers, 1995; Salas, Prince, Baker, & Shrestha, 1995; Shrestha, Prince, Baker, & Salas, 1995; Stout, Cannon-Bowers, & Salas, 1996). We believe that by exploiting advances in training technology and methods – and, perhaps more important, adopting a systematic methodology for developing training – teamwork in the cockpit will improve. It is up to all of us – scientists and practitioners – to see that the potential benefits of CRM training are realized in the aviation community.

ACKNOWLEDGMENTS

The views expressed here are those of the authors and do not reflect the official position of the organization with which the authors are affiliated. Several colleagues throughout the years have contributed to this program of research: David Baker, Rebecca Beard, Maureen Bergondy, Mike Brannick, Jim Driskell, Jennifer Fowlkes, Florian Jentsch, Danielle Merket, Dana Milanovich, Ben Morgan, Jr., Elizabeth Muñiz, Jerry Owens, Ashley Prince, Lisa Shrestha, Bob Swezey, Kimberly Smith-Jentsch, Mike Lilienthal, and Scott Tannenbaum. We appreciate their efforts in making this program of research offer solutions to the aviation community. We would also like to thank the 1400 or so aviators who have donated their time over the years to help us conduct this research.

REFERENCES

Baker, D. P., Prince, C., Shrestha, L., Oser, R. L., & Salas, E. (1993). Aviation computer games for crew resource management training. *International Journal of Aviation Psychology, 3*(2), 143–156.

Baker, D. P., & Salas, E. (1992). Principles for measuring teamwork skills. *Human Factors, 34,* 469–475.

Beard, R. L., Salas, E., & Prince, C. (1995). Enhancing transfer of training: Using role-play to foster teamwork in the cockpit. *International Journal of Aviation Psychology, 5*(2), 131–143.

Bowers, C. A., Baker, D. P., & Salas, E. (1994). Measuring the importance of teamwork: The reliability and validity of job/task analysis indices for team-training design. *Military Psychology, 6,* 205–214.

Bowers, C. A., Deaton, J., Oser, R. L., Prince, C., & Kolb, M. (1995). Impact of automation on aircrew communication and decision-making performance. *International Journal of Aviation Psychology, 5*(2), 145–168.

Bowers, C. A., Morgan, B. B., Jr., & Salas, E. (1991). The assessment of aircrew coordination demand for helicopter flight requirements. In *Proceedings of the 6th International Symposium on Aviation Psychology* (pp. 308–313). Mahwah, NJ: Erlbaum.

Bowers, C. A., Morgan, B. B., Jr., Salas, E., & Prince, C. (1993). Assessment of coordination demand for aircrew coordination training. *Military Psychology, 5,* 95–112.

Bowers, C. A., Thornton, C., Braun, C. C., & Salas, E. (1998). Automation, task difficulty, and aircrew performance. *Military Psychology, 10,* 259–274.

Brannick, M. T., Prince, C., Salas, E., & Stout, R. J. (1997). *Development and evaluation of a team training tool.* Unpublished manuscript.

Brannick, M. T., Salas, E., & Prince, C. (Eds.). (1997). *Team performance assessment and measurement: Theory, methods, and applications.* Mahwah, NJ: Erlbaum.

Byrnes, R. E., & Black, R. (1993). Developing and implementing CRM programs: The Delta experience. In E. L. Wiener, B. G. Kanki, & R. L. Helmreich (Eds.), *Cockpit resource management* (pp. 421–443). San Diego: Academic.

Cannon-Bowers, J. A., Prince, C., Salas, E., Owens, J. M., Morgan, B. B., Jr., & Gonos, G. H. (1989). Determining aircrew coordination effectiveness. In *Proceedings of the 11th Interservice/Industry Training Systems Conference* (pp. 128–136). Arlington, VA: National Defense Industrial Association.

Cannon-Bowers, J. A., & Salas, E. (1997). Teamwork competencies: The intersection of team member knowledge, skills, and attitudes. In H. F. O'Neil (Ed.), *Workforce readiness: Competencies and assessment.* Mahwah, NJ: Erlbaum.

Cannon-Bowers, J. A., & Salas, E. (1998). Individual and team decision making under stress: Theoretical underpinnings. In J. A. Cannon-Bowers & E. Salas (Eds.), *Making decisions under stress: Implications for individual and team training* (pp. 17–38). Washington, DC: American Psychological Association.

Cannon-Bowers, J. A., Tannenbaum, S. I., Salas, E., & Converse, S. A. (1991). Toward an integration of training theory and technique. *Human Factors, 33,* 281–292.

Cannon-Bowers, J. A., Tannenbaum, S. I., Salas, E., & Volpe, C. E. (1995). Defining team competencies and establishing team training requirements. In R. Guzzo & E. Salas (Eds.), *Team effectiveness and decision-making in organizations* (pp. 333–380). San Francisco: Jossey-Bass.

Caro, P. W. (1988). Flight training and simulation. In E. L. Wiener & D. C. Nagel (Eds.), *Human factors in aviation* (pp. 229–261). San Diego: Academic.

Chidester, T. R., & Foushee, H. C. (1988). Leader personality and crew effectiveness: Factors influencing performance in full-mission air transport simulation. In *Proceedings of the 66th Meeting of the Aerospace Medical Panel on Human Stress Situations in Aerospace Operations* (pp. 7-1–7-9). The Hague, Netherlands: Advisory Group for Aerospace Research and Development.

Dwyer, D. J., Fowlkes, J. E., Oser, R. L., Salas, E., & Lane, N. E. (1997). Team performance measurement in distributed environments: The TARGETs methodology. In M. T. Brannick, E. Salas, & C. Prince (Eds.), *Team performance assessment and measurement: Theory, methods, and applications* (pp. 137–153). Mahwah, NJ: Erlbaum.

Endsley, M. R. (1995). Measurement of situation awareness in dynamic systems. *Human Factors, 37,* 65–84.

Ford, J. K., Kozlowski, S. W. J., Kraiger, K., Salas, E., & Teachout, M. S. (Eds.). (1997). *Improving training effectiveness in work organizations.* Mahwah, NJ: Erlbaum.

Foushee, H. C. (1984). Dyads and triads at 35,000 feet: Factors affecting group process and aircrew performance. *American Psychologist, 39,* 885–893.

Foushee, H. C., & Helmreich, R. L. (1988). Group interaction and flight crew performance. In E. L. Wiener & D. C. Nagel (Eds.), *Human factors in aviation* (pp. 189–227). San Diego: Academic.

Fowlkes, J. E., Dwyer, D. J., Oser, R. L., & Salas, E. (1998). Event-based approach to training. *International Journal of Aviation Psychology, 8*(3), 209–221.

Fowlkes, J. E., Lane, N. E., Salas, E., Franz, T., & Oser, R. L. (1994). Improving the measurement of team performance: The TARGETs methodology. *Military Psychology, 6*, 47–61.

Hartel, C., Smith, K. A., & Prince, C. (1991). *Defining aircrew coordination: Searching mishaps for meaning.* Paper presented at the 6th International Symposium on Aviation Psychology, Columbus, OH.

Helmreich, R. L., & Foushee, H. C. (1993). Why crew resource management? Empirical and theoretical bases of human factors training in aviation. In E. L. Wiener, B. G. Kanki, & R. L. Helmreich (Eds.), *Cockpit resource management* (pp. 3–45). San Diego: Academic.

Helmreich, R. L., & Wilhelm, J. A. (1991). Outcomes of crew resource management training. *International Journal of Aviation Psychology, 1*(4), 287–300.

Jentsch, F. (1997). *Meta-cognitive training for junior team members: Solving the copilots catch-22.* Unpublished doctoral dissertation, University of Central Florida, Orlando.

Jentsch, F., & Bowers, C. A. (1998). Evidence of validity of low-fidelity simulations in studying aircrew coordination. *International Journal of Aviation Psychology, 8*(3), 243–260.

Johnston, J. H., Cannon-Bowers, J. A., & Smith-Jentsch, K. A. (1995). Event-based performance measurement system for shipboard command teams. In *Proceedings of the 1st International Symposium on Command and Control Research and Technology* (pp. 274–276). Washington, DC: Department of Defense.

Kolb, D. A. (1984). *Experiential learning: Experience as the source of learning and development.* Englewood Cliffs, NJ: Prentice-Hall.

Lauber, J. K. (1984). Resource management in the cockpit. *Air Line Pilot, 53*, 20–23.

Leedom, D. K., & Simon, R. (1995). Improving team coordination: A case for behavior based training. *Military Psychology, 7*, 109–122.

Marrow, A. J. (1969). *The practical theorist: The life and work of Kurt Lewin.* New York: Basic Books.

McIntyre, R. M., & Salas, E. (1995). Measuring and managing for team performance: Emerging principles from complex environments. In R. Guzzo & E. Salas (Eds.), *Team effectiveness and decision making in organizations* (pp. 149–203). San Francisco: Jossey-Bass.

Morgan, B. B., Jr., Salas, E., & Glickman, A. S. (1994). An analysis of team evolution and maturation. *Journal of General Psychology, 120*, 277–291.

Orasanu, J. L. (1993). Decision-making in the cockpit. In E. L. Wiener, B. G. Kanki, & R. L. Helmreich (Eds.), *Cockpit resource management* (pp. 137–172). San Diego: Academic.

Oser, R. L., Dwyer, D. J., & Folkes, J. E. (1995). Team performance in multi-service distributed interactive simulation exercises: Initial results. In *Proceedings of the 17th Interservice/Industry Training Systems and Education Conference* (pp. 163–171). Arlington, VA: National Defense Industrial Organization.

Pettit, M. A., & Dunlap, J. H. (1997). *Cockpit leadership and followership skills: Theoretical perspectives and training guidelines* (Tech. Report). Washington, DC: Federal Aeronautics Administration.

Prince, A., Brannick, M. T., Prince, C., & Salas, E. (1997). The measurement of team process behaviors in the cockpit: Lessons learned. In M. T. Brannick, E. Salas, & C. Prince (Eds.), *Team performance assessment and measurement* (pp. 289–310). Mahwah, NJ: Erlbaum.

Prince, C., Oser, R., Salas, E., & Woodruff, W. (1993). Increasing hits and reducing misses in CRM/LOS scenarios: Guidelines for simulator scenario development. *International Journal of Aviation Psychology, 3*(1), 69–82.

Prince, C., & Salas, E. (1989). Aircrew performance: Coordination and skill development. In D. E. Daniel, E. Salas,

& D. M. Kotick (Eds.), *Independent research and independent exploratory development (IR/IED) programs: Annual report for fiscal year 1988* (NTSC Tech. Report 89-009). Orlando, FL: Naval Training Systems Center.

Prince, C., & Salas, E. (1993). Training and research for teamwork in the military aircrew. In E. L. Wiener, B. G. Kanki, & R. L. Helmreich (Eds.), *Cockpit resource management* (pp. 337–366). San Diego: Academic.

Ryle, G. (1949). *The concept of mind.* Chicago: University of Chicago Press.

Salas, E., Bowers, C. A., & Cannon-Bowers, J. A. (1995). Military team research: Ten years of progress. *Military Psychology, 7*, 55–75.

Salas, E., & Cannon-Bowers, J. A. (1997). Methods, tools, and strategies for team training. In M. A. Quiñones & A. Ehrenstein (Eds.), *Training for a rapidly changing workplace: Applications of psychological research* (pp. 249–279). Washington, DC: American Psychological Association.

Salas, E., & Cannon-Bowers, J. A. (in press). The anatomy of team training. In L. Tobias & D. Fletcher (Eds.), *Handbook on research in training.* New York: Macmillan.

Salas, E., Cannon-Bowers, J. A., & Blickensderfer, E. L. (1997). Enhancing reciprocity between training theory and practice: Principles, guidelines, and specifications. In J. K. Ford, S. W. J. Kozlowski, K. Kraiger, E. Salas, & M. S. Teachout (Eds.), *Improving training effectiveness in work organizations* (pp. 291–322). Mahwah, NJ: Erlbaum.

Salas, E., Cannon-Bowers, J. A., & Johnston, J. H. (1997). How can you turn a team of experts into an expert team? Emerging training strategies (pp. 359–370). In C. Zsambok & G. Klein (Eds.), *Naturalistic decision making.* Mahwah, NJ: Erlbaum.

Salas, E., Dickinson, T. L., Converse, S. A., & Tannenbaum, S. I. (1992). Toward an understanding of team performance and training. In R. W. Swezey & E. Salas (Eds.), *Teams: Their training and performance* (pp. 3–29). Norwood, NJ: Ablex.

Salas, E., Fowlkes, J. E., Stout, R. J., Milanovich, D., & Prince, C. (in press). Does CRM training improve teamwork skills in the cockpit?: Two evaluation studies. *Human Factors.*

Salas, E., Prince, C., Baker, D. P., & Shrestha, L. (1995). Situation awareness in team performance: Implications for measurement and training. *Human Factors, 37*, 123–136.

Shrestha, L. B., Prince, C., Baker, D. P., & Salas, E. (1995). Understanding situation awareness: Concepts, methods, and training. In W. B. Rouse (Ed.), *Human/technology interaction in complex systems* (Vol. 7, pp. 45–83). Greenwich, CT: JAI Press.

Smith-Jentsch, K. A., Salas, E., & Baker, D. (1996). Training team performance-related assertiveness. *Personnel Psychology, 49*, 909–936.

Smith-Jentsch, K., Zeisig, R. L., Acton, B., & McPherson, J. (1998). Team dimensional training: A strategy for guided team self-correction. In J. A. Cannon-Bowers & E. Salas (Eds.), *Making decisions under stress: Implications for individual and team training* (pp. 271–297). Washington, DC: American Psychological Association.

Stone, R. B., & Babcock, G. L. (1988). Airline pilots' perspective. In E. L. Wiener & D. C. Nagel (Eds.), *Human factors in aviation* (pp. 529–560). San Diego: Academic.

Stout, R. J., Cannon-Bowers, J. A., & Salas, E. (1996). The role of shared mental models in developing team situational awareness: Implications for training. *Training Research Journal, 2*, 85–116.

Stout, R. J., Salas, E., & Fowlkes, J. E. (1997). Enhancing teamwork in complex environments through team training. *Group Dynamics: Theory, Research, & Practice, 1*, 169–182.

Stout, R. J., Salas, E., & Kraiger, K. (1997). The role of trainee knowledge structures in aviation team environments. *International Journal of Aviation Psychology, 7*(3), 235–250.

Swezey, R., Llaneras, R., Prince, C., & Salas, E. (1991). Instructional strategy for aircrew coordination training. In *Proceedings of the 6th International Symposium on Aviation Psychology* (pp. 302–307). Mahwah, NJ: Erlbaum.

Swezey, R. W., & Salas, E. (1992). Guidelines for use in team-training development. In R. W. Swezey & E. Salas (Eds.), *Teams: Their training and performance* (pp. 219–245). Norwood, NJ: Ablex.

Tannenbaum, S. I., Beard, R. L., & Salas, E. (1992). Team building and its influence on team effectiveness: An examination of
conceptual and empirical developments. In K. Kelley (Ed.),
*Issue, theory, and research in industrial/organizational psy-
chology* (pp. 117–153). Amsterdam: Elsevier.

Thornton, C., Braun, C. C., Bowers, C. A., & Morgan, B. B., Jr.
(1992). Automation effects in the cockpit: A low-fidelity
investigation. In *Proceedings of the Human Factors and
Ergonomics Society 36th Annual Meeting* (pp. 30–34). Santa
Monica, CA: Human Factors and Ergonomics Society.

Wiener, E. L., Kanki, B. G., & Helmreich, R. L. (Eds.). (1993).
Cockpit resource management. San Diego: Academic.

Yamamori, H., & Mito, T. (1993). Keeping CRM is keeping the flight
safe. In E. L. Wiener, B. G. Kanki, & R. L. Helmreich, *Cockpit
resource management* (pp. 399–420). San Diego: Academic.

Eduardo Salas is a senior research psychologist and
head of the Training Technology Development
branch of the Naval Air Warfare Center Training
Systems Division (NAWCTSD). He received his
Ph.D. in industrial/organizational psychology from
Old Dominion University, Norfolk, Virginia in 1984.

Carolyn Prince is a former research psychologist at
NAWCTSD. She holds a Ph.D. in industrial/organi-
zational psychology from the University of South
Florida, Tampa, which she received in 1984.

Clint A. Bowers is an associate professor of psychology
and director of the Team Performance Laboratory
at the University of Central Florida. He received his
Ph.D. in psychology from the University of South
Florida in 1987.

Renée J. Stout is a research psychologist who works
in the Aviation Team Training Lab at NAWCTSD.
She received her Ph.D. in human factors psychology
from the University of Central Florida in 1994.

Randall L. Oser is a research psychologist at NAW-
CTSD. He received his M.S. degree in industrial/
organizational psychology from the University of
Central Florida in 1990.

Janis A. Cannon-Bowers is a research psychologist
at NAWCTSD. She received her Ph.D. in 1988
from the University of South Florida in industrial/
organizational psychology.

Date received: November 5, 1997
Date accepted: July 13, 1998

Part 3: Training Devices and Simulators

The growth of simulation-based training has been dramatic in the last 30 years. We define simulation-based training as any instruction that uses a simulation of the real world as a major medium of instruction. The most prominent example of simulation-based training is operator simulation training (e.g., flight, driver, ship-helmsman). However, the capability to generate simulation-based graphics at significantly lower costs (cost reductions have been driven by the games industry) has made it possible for training to use simulation in ways that were not imagined in the past (e.g., cognitive training for decision making).

We define training devices as any instructional medium that allows trainees to learn knowledge and skills through interaction with cues that will be found on the actual job. These cues may be real (via training on actual equipment) or simulated (via a simulator). Training devices usually include instructional features that aid the learning and instruction process and include performance measurement capabilities, freeze and replay capabilities, cue enhancement capabilities, and aids for instructors.

Simulators are systems that replicate real-world equipment. They may be used for training or other purposes, such as engineering development and testing. A simulator may be a training device, but it must have instructional features like those mentioned earlier in order to be called a training device. A simulator's fidelity (physical and functional) may or may not be a factor in learning. That is, depending on the learning task, higher levels of fidelity might produce a more effective training environment, but it is also possible that high levels of fidelity might actually interfere with the learning process, especially when a novice learner is presented with real-world cues that are too fast, too subtle, or too complex for his or her level of expertise.

Although simulation technology has become ubiquitous in training because of technological improvements and cost reductions, many questions remain about how best to use it. For example, more than 40 years ago, Gagne (1954) wrote a classic paper that examined many research issues on training device and simulator development and use. It is informative to examine some of the issues he raised at that time to determine how far the research base has come in successfully addressing those issues. Gagne divided his issues into three main parts. We list only a few of the issues he discussed here. The articles included in this chapter address a number of them.

Analysis of the Task

"Since the building of a training device or simulator is undertaken in the first place because of a need to represent the actual job (either more simply, less expensive with smaller involvement of danger to the operator, etc.), it is obvious that some decisions must be made, at the very beginning, about what are the essential aspects of the job to represent. Although such decisions are always made when a training device is designed they are not always made in a systematic manner. And sometimes they seem to be perverted, rather than clarified, by attempts to follow the principle: 'Make the device as nearly like the actual equipment as possible'" (Gagne, 1954, p. 81).

Performance Measurement

"Research in this area [i.e., measurement reliability in a simulator] should have the aim of formulating a set of principles relevant to the arrangement of conditions in a trainer situation so as to produce maximum reliability of performance measurement with the sacrifice of as little validity as possible. The general question is: How must the characteristics of an operator task be deliberately altered in order to make possible, adequately reliable measurement of performance highly related to the operational task?" (Gagne, 1954, p. 85).

Effectiveness for Training

"Research should tell us how the physical characteristics of a trainer may be designed to bring about the most rapid acquisition and the highest possible level of performance in the operational skills for which training is required. The question can even be expressed in this way: How must the characteristics of an operational task be deliberately altered in order to insure the most effective training by means of a training device?" (Gagne, 1954, p. 86).

The four papers in this section address some key issues concerning training devices and simulators, including some of Gagne's issues. Williges, Roscoe, and Williges (1973) examine the degree of simulation and the fidelity of simulation as these factors relate to the design of simulators. They considered learning measurement, transfer, and retention in a synthetic flight trainer and then examined what were relatively new innovations in 1973 and their effect on synthetic flight simulation. The authors contend that transfer and cost-effectiveness are critical factors in synthetic flight training evaluation.

Carter and Laughery (1985) addressed the important and ever-present issue of fidelity in simulation. The question, "How much fidelity is enough for proper learning to occur?" has been asked by simulator designers, manufacturers, buyers, users, and researchers for many years. Psychological research has answered the question at least partially, but many arguments still occur when the topic is raised. Carter and Laughery's paper is especially interesting because they examined the fidelity question in the context of nuclear power plant simulation, in which the fidelity of simulation is not merely a matter of training or research interest but has profound regulatory implications.

Alluisi (1991) wrote a detailed account of the development of the SIMNET (simulator networking). This project aimed to produce a series of relatively low-cost armored vehicle simulators that could take advantage of then-novel technologies for linking the simulators. The intent was to enable operators to be trained as collective teams. SIMNET was important in the history of training devices and simulators because it was the first in a widespread effort in the military to develop low-cost, networkable trainers that allowed teams to learn and practice tactical and procedural skills. Because these networked trainers are focused not on procedural but on decision-making training for tactical operations, the question of how much fidelity to provide is critical. Alluisi explored this issue in detail.

Finally, Mackenzie, Harper, and Xiao (1996) present a practical examination of how simulator deficiencies can affect decision-making training for medical trainees. Technology limitations and/or cost efficiencies often require simulator designers to produce simulators that have known limitations. Although it is understandable that many, if not most, simulators will be affected by such limitations, our experience is that too often, instructors and trainees are not completely informed of these limitations and therefore are not able to make accommodations. We have included this paper because it explores how simulator limitations can have a profound impact on the resulting training.

Reference

Gagne, R.M. (1954). Training devices and simulators: Some research issues. *American Psychologist, 9*, 95–107.

Synthetic Flight Training Revisited

BEVERLY H. WILLIGES, STANLEY N. ROSCOE, *and* ROBERT C. WILLIGES,
University of Illinois at Urbana-Champaign

Critical issues in the development and use of synthetic flight trainers are reviewed. Degree of simulation and fidelity of simulation are discussed as key design considerations. Problems in measurement of original learning, transfer, and retention are presented. Both transfer effectiveness and cost effectiveness are described as critical factors in the evaluation of flight trainers. Recent training innovations, such as automatically adaptive training, computer-assisted instruction, cross-adaptive measurement of residual attention, computer graphics, incremental transfer effectiveness measurement, and response surface methodology, are discussed as potential techniques for improving synthetic flight training. It is concluded that broader application of simulation is necessary to meet the new demands of pilot training, certification, and currency assurance in air transportation.

BACKGROUND

Flight training at the close of World War I was a haphazard process at best. Basically, a new pilot was trained solely by demonstration and exhortation improvised by his particular flight instructor. The few flight-trainer type devices in use were short-winged aircraft, incapable of flight, known as "stub-winged Jennies" or "grass-cutters." Students would run these early trainers up and down a large field in an attempt to learn to control the craft.

Since those early days of flying, ground-based flight simulators and trainers have evolved from the famous Link "Blue Boxes" of World War II into precisely engineered devices capable of accurately computing the aerodynamic responses of an airplane to control inputs, and of reproducing realistic cockpit instrument indications for all flight situations. But, despite the sophistication of contemporary simulators and the longevity of their use, many research issues concerning ground-based flight simulators and trainers remain unanswered. This paper is a review of trainer-related research with an emphasis on these unresolved questions.

Why Simulate?

Obviously, the initial question is why use a simulator or trainer at all. According to Gagne (1962), the major difference between a simulator and the operational situation is that the simulator provides its users with greater control over ambient conditions. Whereas the real world is subject to unpredictable variations, a simulator provides planned variation of various elements of the real situation with unessential variables in the real situation omitted. The essential condition for effective training is that the simulator be procedurally faithful to the aircraft it is designed to represent. Determining which aspects of the operational situation can reasonably be left out is a central aspect of simulator design process.

The second major advantage offered by simulators is that dangerous elements in the operational situation may be represented safely. For example, an aircraft simulator might represent an engine on fire by a flashing red light rather than by an actual flame-producing fire. Furthermore, emergency procedures that would be too dangerous to teach in the air may be taught safely in ground-based devices.

A third advantage of simulators is their low operating cost, in comparison with the cost of operating counterpart aircraft. Simulator use is independent of weather or time of day, and the performance of individual flight tasks and procedures can be interrupted and repeated, thereby allowing errors to be corrected immediately and the distribution of practice on sequentially dependent flight tasks to be optimized. For these reasons, a major portion of any flight curriculum can be taught at a fraction of the cost of training in the actual aircraft.

How Can Simulation Be Used?

Flight simulators and trainers have several uses. Initially, performance in a simulator or trainer can be used in pilot selection as a predictor of future success in training and operations. Second, a simulator that reproduces the aerodynamic responses of an aircraft with good fidelity is valuable for teaching new psychomotor skills required for operating an aircraft. Furthermore, the training functions of simulators are not limited solely to initial acquisition of flying skills. Trainers can be used effectively to familiarize an experienced pilot with the operating procedures and characteristics of an aircraft to which he is newly assigned. Such training should reduce the transitioning pilot's initial erroneous responses in the air which result from negative transfer associated with the need to make different responses to highly similar stimuli. In addition, simulators can be used both to reassess and to maintain the proficiency of licensed pilots.

Proficiency assessment in a flight simulator is more economical and more readily controlled than a similar evaluation in the air. In fact, these devices have proven to be so useful that virtually all check rides for airline pilots are given in simulators rather than in their counterpart aircraft. Recently, commercial airline companies have conducted research to determine the feasibility of increasing the percentage of training in simulators for pilots of large jet aircraft. Results of studies conducted by both Trans World Airlines (1969) and American Airlines (1969) indicate that experienced pilots can be trained to type-rating proficiency entirely in a flight simulator. In addition, performance evaluations in the simulator accurately predicted performance in the corresponding aircraft. These studies suggest that the Federal Aviation Administration could modify pilot certification requirements to allow increased use of simulation equipment of proven effectiveness.

FLIGHT SIMULATOR DESIGN ISSUES

Smode, Hall, and Meyer (1966) have compiled a relatively comprehensive review of research studies using flight simulators and trainers for pilot training and have indicated areas in which additional research is needed. Although the terms *simulator* and *trainer* are sometimes used interchangeably, a distinction should be made: a simulator is designed to represent a specific counterpart vehicle or operational situation; a trainer is intended to represent a class of vehicles in various situations. Much of the research literature on simulation in pilot training can be subdivided into two areas: degree of simulation and fidelity of simulation. Degree of simulation refers to the inclusion of design features such as motion, extracockpit visual cues, and part-task *vs.* whole-task representation. Fidelity of simulation refers to the accuracy with which design features represent or duplicate their real world counterparts. The usual reason for striving for high fidelity of simulation is to maximize transfer of training to performance in the operational situation (Muckler, Nygaard, O'Kelly, and Williams, 1959). The following draws upon Smode's review of degree of simulation and Muckler's analysis of fidelity of

Degree of Simulation

Motion. Research findings dealing with motion simulation are as yet inconclusive. Results may be divided into three categories: those that support the value of motion, those that suggest that the value of motion depends upon the transfer task, and those that suggest that the value of motion is merely a transient effect.

Because motion provides the trainee with additional cues, many researchers have concluded that motion facilitates transfer performance (Besco, 1961; Buckhout, Sherman, Goldsmith, and Vitale, 1963; Ruocco, Vitale, and Benfari, 1965a; 1965b; Townsend, 1956). However, these studies evaluated learning with criterion trials in a simulator rather than in the air.

There has been some speculation that motion is necessary only in specific situations, where acceleration cues either improve performance by facilitating anticipatory responses or hinder performance by making it more difficult for the pilot to make necessary control adjustments (Rathert, Creer, and Douvillier, 1959; Rathert, Creer, and Sadoff, 1961). A body of evidence suggests that, although motion cues do seem to facilitate an initially higher level of performance, this effect rapidly fades with subsequent flight experience (Caro and Isley, 1966; Feddersen, 1961).

Several studies have linked the value of motion with the experience level of the pilot. Flexman (1966) and Briggs and Wiener (1959) noted that, because experienced pilots often rely on motion rather than instrument readings, motion becomes more important as experience level increases. On the other hand, Muckler, *et al.* (1959) suggested that motion combined with contact cues is more important during the initial stages of learning. When visual and vestibular cues are conflicting, pilots tend to rely more upon their vestibular cues as their confidence in the visual information decreases (Johnson and Williams, 1971).

Smode, *et al.* (1966) indicated a need for additional research concerning simulator motion, and we believe such research is still needed today. In fact, the question of whether or not motion cues influence transfer at all is, as yet, unanswered. Because of the possibility that erroneous motion cues might actually cause negative transfer, the issue of whether or not simulator motion is beneficial cannot be separated from a consideration of the fidelity of motion cues necessary to produce positive transfer and the relative transfer effectiveness and the cost-effectiveness tradeoff of increasing motion-cue fidelity. Studies are also needed to determine the relationship between cockpit workload and the value of motion cues of varying experience levels.

Extracockpit visual cues. Because only selected visual cues are used by pilots, and visual cues are not required to perform every flying maneuver, the complete external visual environment does not need to be reproduced in a flight simulator. The real problem is determining exactly which visual cues are necessary.

An early study at the University of Illinois, using a crude contact training device, confirmed the value of visual cues for training private pilots (Flexman, Matheny, and Brown, 1950). Interestingly, the same device was ineffective for training military pilots (Ornstein, Nichols, and Flexman, 1954). The explanation given for this difference was that the value of extra-cockpit cues is limited by the quality of the instruction associated with their use.

Studies using slightly more complex extra-cockpit visual devices have confirmed their value (Creelman, 1955; Payne, Dougherty, Hasler, Skeen, Brown, and Williams, 1954). However, the value of extracockpit visual simulation in the learning of perceptual responses in flying, in the absence of related psychomotor responses, has not been substantiated (Adams and Hufford, 1961; Creelman, 1955).

A second group of studies has been concerned with training pilots to divide their attention between external visual cues and the instruments within the aircraft. Pfeiffer, Clark, and Danaher (1963) concluded that training does improve a pilot's time-sharing ability. Further, such training can be given in relatively inexpensive training devices (Gabriel, Burrows, and Abbott, 1965). The questions of whether or not the time-sharing skills learned in such devices transfer to flight and are retained over extended periods remain to be answered.

Smode, *et al.* (1966) point out that information on the value of contact devices is muddled because the utility of the visual device is so intertangled with the fidelity of the particular simulator being used. They indicate that more information is needed to clarify the value of contact displays both as a part of, and independent of, the specific simulator used. In addition, satisfactory methods of presenting visual and motion cues simultaneously need to be explored. For example, is moving the visual display in response to the pilot's input equivalent, perceptually, to moving the trainer? A related issue is the value of open-loop training for closed-loop tasks. Again, the factor of level of pilot experience needs to be investigated. Finally, the effectiveness of inexpensive devices to teach time-sharing should be further explored, and the value of tachistoscopic training to increase instrument reading speeds needs to be determined.

Part-task vs. whole-task training. Although a great deal of research has been concerned with part-task *vs.* whole-task learning, the results of this research have limited applicability to pilot training. Traditionally, part-task training in verbal and simple motor tasks involves the development of component parts of a skill and subsequent practice on all parts concurrently. Part-task pilot training is typically molecular rather than atomic, often involving training on individual whole tasks that are later practiced in

series with different tasks rather than in parallel with other parts of the same task. An example of a part-task flight trainer is a device that simulates, with high fidelity, only the attack phase of an air-to-air intercept mission (Nygaard and Roscoe, 1953).

The results of several studies specifically concerned with part-task trainers for pilots have supported the utility of such devices (Dougherty, Houston, and Nicklas, 1957; Miller, 1960; Parker and Downs, 1961; Pomarolli, 1965). A frequent limitation of part-task trainers is that they fail to provide an opportunity for practice in time-sharing attention among tasks (Adams, Hufford, and Dunlop, 1960; Hufford and Adams, 1961). A subsequent period of integration is necessary to allow students to perform various subtasks on a time-shared basis. On the positive side, part-task training seems to require less relearning after a period of rest (Hufford and Adams, 1961).

Smode, *et al.* (1966) suggest two areas for future research. First, the relative contributions of part-task and whole-task trainers need to be determined so that less expensive trainers can be used whenever appropriate. In addition, information concerning how task integration proceeds is needed. Eventually a pilot must scan his instruments, tune radio receivers, navigate, and communicate—all while flying his airplane. Occasionally he may have to disarm a hijacker. Practice in doing all such things concurrently is required in a comprehensive training program.

Because almost any training must be part-task training to some degree, the real issue would appear to be the optimum size for each learning "chunk." Obviously, very few people get into a helicopter for the first time and solo; rather, training proceeds in steps. Optimum step size in pilot training is an open question.

Fidelity of Simulation

Muckler, *et al.* (1959) made a thorough review of early findings concerning the fidelity of simulation necessary for maximum transfer and found widely varying results. Several studies led to the conclusion that fidelity of simulation made little difference in the amount of transfer to the air. Mahler and Bennett (1950) found no differences in transfer among several training devices varying widely in fidelity. With the exception of performance on one maneuver (recovery from unusual attitudes), Wilcoxon, Davy, and Webster's (1954) results support Mahler and Bennett.

On the other hand, a study by Ornstein, Nichols, and Flexman (1954) isolated particular components of the pilot's task and found that training in a simulator of higher fidelity (Link P-1) consistently resulted in better transfer on each of 22 instrument maneuvers than training in either the Link AN-T-18 or C-8 trainers. Similarly, the results of Dougherty, *et al.* (1957) favor trainers of higher fidelity. They found better transfer to the SNJ aircraft when pilots were trained either in an SNJ operational flight trainer or a procedural trainer than with a photographic mockup. However, in this study the advantage enjoyed initially by trainers of higher fidelity was negligible by the sixth air trial.

Each of the preceding four studies was conducted under similar experimental conditions, but the results are irreconcilable. Muckler, *et al.* (1959) concluded that studies concerned with fidelity of simulation are plagued by a variety of problems such as lack of generalizability from oversimplified laboratory tasks and inadequate measurement techniques. Before any definite conclusions can be drawn about fidelity of simulation, more detailed information is needed to determine how such variables as instructor ability, variations in the difficulty of the training task, and pilot experience level affect transfer performance. As stated previously, there has been no experimental effort to determine the relationship between transfer of training and fidelity of simulated motion cues.

FLIGHT SIMULATOR EVALUATION ISSUES

What to Measure and How?

As noted by Blaiwes and Regan (1970) and more recently by Roscoe (in McGrath and Harris, 1971), three criteria must be considered in properly evaluating any training device: (1) efficiency of original learning, (2) transfer of what was learned in one situation to another, and (3) retention of what was once learned.

Original learning. To determine the effectiveness of synthetic flight trainers against any of these criteria, objective performance measures are necessary. One traditional measure of learning is instructor ratings. In general, such ratings tend to be subjective and, as such, are hampered by gross inconsistencies among independent observers. In an attempt to overcome many of the difficulties associated with subjective grading by check pilots, the development of objective flight inventories has been encouraged. One of the first of these was the Ohio State Flight Inventory which combined a series of five-point rating scales for each maneuver with some objectively scored items completed during flight. Ericksen (1952) summarized studies using this inventory.

In 1947, an extensive program to develop an objective checklist for pilot evaluation was begun under the sponsorship of the CAA through the National Research Council Committee on Aviation Psychology (Gordon, 1947; 1949). The decisions to include items were based upon critical incidents, accident reports, and job analyses. Evaluated tasks were arranged into a standard flight sequence, including both subjective and objective items. To maximize objectivity, graphics or pictures, quantitative data, and precise descriptions were used.

Another flight inventory, developed at the Human Resources Research Office, had, as a goal, the complete description of pilot performance (Smith, Flexman, and Houston, 1952). Based on reported critical behaviors in flying, the HumRRO inventory consisted of two types of items: scale items (whether or not within predetermined tolerances) and categorical items (whether or not completed). Although its use has been limited to research, the inventory has provided reliable normative data to set standards for pilot training.

More recently, the Illinois Private Pilot Performance Scale has been developed (Povenmire, Alvares, and Damos, 1970; Povenmire and Roscoe, 1973; Selzer, Hulin, Alvares, Swartzendruber, and Roscoe, 1972). This scale evaluates performance on each of 10 maneuvers from the FAA's private pilot flight test guide. Four to six quantitative variables for each maneuver are scored by marking the maximum deviation from desired performance on appropriate scales. Equal weights are given to all variables measured. Individual deviation scores are converted into standard scores based upon the observed variability among students tested for private pilot certification at the University of Illinois. Observer-observer reliability in excess of .80 has been found for this testing instrument (Selzer, *et al.*, 1972).

Despite the increasing objectivity of pilot performance grading, the reliability of the so-called objective checks has been disappointing in routine use. According to Smode, *et al.* (1966), several factors contribute to the limited capability of objective measures. They are: check pilot biases, inadequate descriptions of acceptable performance, low validity based upon the failure to define precisely the critical skills to be assessed, and the need to give special training to check pilots on the use of objective measurements.

The shortcomings of check pilot ratings have given impetus to the development of automatic recording devices built into synthetic trainers. Danneskiold (1955) conducted a study to determine the feasibility of mechanical scoring devices. Although accuracy of measurement was an asset, the mechanical devices were limited by inflexibility, cumbersome size, and the failure to reflect meaningful aspects of

flight. Current research on automated performance measurement by Knoop (in McGrath and Harris, 1971) has thus far resulted only in tentative conclusions about requirements and feasibility. However, it is evident that semiautomatic performance assessment methods resulting from this research will receive considerable attention in connection with the evaluation of the effectiveness of new synthetic training devices.

The most elusive problem in the semiautomatic assessment of pilot performance is determining what ideal pilot behavior is. At present the "real world" criterion most often used seems to be expert judgments of what maneuvers are essential and what range of performance variation can be tolerated. Another approach is the collection of normative data from the performance of experienced pilots. However, Flexman (in McGrath and Harris, 1971) asserts that the variance among experienced pilots is greater than that among student pilots. Until some agreement is reached about what constitutes ideal pilot performance, evaluative techniques that measure deviations from a standard will be severely limited.

Transfer. A critical measure of the effectiveness of flight simulators is their transfer to performance in the air. Although early evaluations of flight trainers provided estimates of air time savings, many failed to include a control group, thus eliminating any objective measure of transfer of training (Conlon, 1939; Crannell, Greene, and Chamberlain, 1941; Greene, 1941).

A number of studies conducted at the University of Illinois were designed to measure the value of synthetic training in reducing the flight hours necessary to obtain a private pilot's license. Williams and Flexman (1949) measured the amount of flight time until students were judged ready to "solo". The results revealed no significant differences among groups of subjects having 0, 2, or 4 hours of experience in a C-3 Link trainer. The experimenters recognized that the amount of flight time until ready to solo was not a good criterion for the evaluation of an early Link trainer, because skills required in landing and other presolo maneuvers requiring visual cues other than a horizon line were not easily taught in this type of trainer.

In their second study, evaluating the School Link, Williams and Flexman (1949) used three errorless trials on selected maneuvers as the criterion measure. The experimental group that received both Link training and inflight training learned the maneuvers to criterion with 28% fewer air trials and 22% fewer errors in flight than a control group receiving only inflight training. The experimenters suggested that approximately 25% of beginning flight training could be accomplished on the ground.

A further study by Flexman, Matheny, and Brown (1950) compared two groups of student pilots after 10 hours of flight training. One group received no Link training and a second group received whatever Link training each individual student considered to be beneficial. Results indicated that the Link group was more proficient on a flight examination similar to the Private Pilot Performance Scale.

At about the same time that Link trainers were undergoing these evaluations, similar studies were conducted with the Link P-1 (SNJ) simulator which approximated a military aircraft, the T-6 (SNJ). In general, students receiving partial synthetic training performed as well or better than students trained solely in the air. Comparisons were based upon various criteria including number of flight failures and accidents, check flight grades, and total training hours (Flexman, Townsend, and Ornstein, 1954; Mahler and Bennett, 1950; Ornstein, *et al.*, 1954; and Wilcoxon, Davy, and Webster, 1954).

The first and only studies that have allowed an assessment of transfer effectiveness of a specific flight simulator to its counterpart airplane on a maneuver-by-maneuver basis were conducted in 1950 by Flexman and reported

22 years later (Flexman, Roscoe, Williams, and Williges, 1972).

In 1969, almost 20 years later, Povenmire and Roscoe (1971) measured the transfer effectiveness of the relatively new Link GAT-1 and the Link AN-T-18 of World War II as used by typical flight instructors in a routine private pilot training program. Ground training for 11 hours in the GAT-1 saved an average of 11 hours (34.5 *vs.* 45.5 for the control group) in the Piper Cherokee, thereby yielding a tranfer effectiveness ratio of 1.00. An equal amount of training in the famous and venerable AN-T-18 saved an average of nine hours in the Cherokee for a transfer effectiveness ratio of 0.82, thereby providing further justification for its continued widespread use more than 30 years after its invention.

Retention. Interestingly, the most common measure of training effectiveness, retention of material learned, has been generally ignored in the evaluation of simulators. Most studies fail to measure the permanence of simulator learning, despite the obvious importance of retaining flying skills. One notable exception is a study by Mengelkoch, Adams, and Gainer (1958) which measured simulator performance after a 4-month retention interval. Unfortunately, both training and retention trials were conducted solely in a trainer with no measure of performance in the air.

Other studies in pilot training have not been designed to use the retention scores obtained as a measure of the effectiveness of various types and amounts of original learning including simulator training (Seltzer, 1970). Measurement of retention is hindered by such problems as variations in the original training of subjects, difficulty of controlling the amount of flying experience each individual pilot receives during the retention period, and unavailability of subjects after a sufficiently long retention period. The lack of simulator studies using a retention measure reflects the general insufficiency of information relating to retention of pilot skills or, for that matter, retention of any complex motor skill.

Cost Effectiveness

A trend in simulator development has been to duplicate as closely as possible every detail of the operational aircraft. As hardware technology develops, new capabilities are added to flight simulators resulting in a rapid cost spiral. Unfortunately, the training value of each added capability is seldom assessed. With inflated equipment costs, the need to weigh the relative value of physical fidelity against its cost has become evident.

In most of the quantitative transfer studies with simulators, the speed of learning by an experimental group, previously trained to a specified level of proficiency in a simulator, has been compared with that of a control group receiving no simulator training. Transfer has been measured solely by the saving in flight time or reduction in errors with no regard for the amount of simulator training given members of the experimental group.

The study by Povenmire and Roscoe (1971), in which subjects were given a fixed amount of ground training, was an exception. It is doubtful that anyone would seriously propose replacing the GAT-1 with the AN-T-18 in a modern private pilot training program; nevertheless, a strong case could be made for doing so, as shown below in a cost effectiveness analysis based on the Povenmire and Roscoe data.

Assume the hourly cost of dual flight training to be $22 ($14 for the Cherokee plus $8 for the instructor) and the corresponding values for the GAT-1 and the AN-T-18 to be $16 ($8 + $8) and $10 ($2 + $8), respectively. In a flight course normally requiring 46 hours in the air, if 11 hours of training in the GAT-1, costing $176, save 11 hours in the Cherokee, costing $242, each $1 spent in the GAT-1 buys $1.38 worth of air training. Similarly, if 11 hours of training in the AN-T-18 costing $110, save 9

hours in the Cherokee, costing $198, each $1 spent in the AN-T-18 saves $1.80 in the air.

Determining Essential Realism

Several approaches to lowering equipment costs are possible. The first requires a realistic appraisal of the amount of realism essential for the training task. Too often factors adding realism to a simulator are evaluated strictly in a go/no-go fashion. For example, the research question generally has been whether or not to include extracockpit visual displays, rather than what visual cues are necessary to achieve high transfer to flight.

Payne, *et al.* (1954) used a relatively simple visual display, providing only a dynamic perspective outline of a runway on a screen in front of a 1-CA-2 (SNJ) simulator, to prepare a group of beginning students for solo flight in a T-6 aircraft. The transfer group reached proficiency in landing with 61% fewer air trials and 74% fewer errors in landing approaches than did a control group trained only in the aircraft.

The inclusion of motion in present-day simulators is another example of the realism-*vs.*-cost problem. Cohen (1970) estimated the cost of a three-degree-of-freedom motion system at about $100,000 and a six-degree-of-freedom system at $250,000. Such costs are not insignificant; nevertheless, most large-aircraft simulators have complex motion systems even though there is little evidence to indicate that such motion capability significantly improves ground-based training, and much of a pilot's training encourages him to disregard acceleration cues in flight.

Cohen (1970) indicated that a systematic research effort is needed to determine what kinds and what degrees of motion are essential for the flight training task. An initial effort in this regard might be to determine what aspects of motion a pilot can perceive and how acceleration thresholds vary under stress. Obviously, if certain types of motion cues cannot be perceived by the human operator, providing them is, at best, wasteful. In addition, if motion of some sort is included in a simulator, an effort should be made to avoid introducing misleading cues that hinder rather than facilitate transfer.

In view of the large sums invested in the design, development, and production of complex simulator motion systems, it is difficult to understand why there has been no objective, controlled experiment to assess their transfer effectiveness. An experiment by Matheny, Dougherty, and Willis (1963) showed that relatively faithful cockpit motion improves pilot performance in the simulator, presumably by providing alerting cues, and recent experiments at Ames Research Center (Guercio and Wall, 1972) and at the Aviation Research Laboratory of the University of Illinois (Jacobs, Williges, and Roscoe, 1973; Roscoe, Denney, and Johnson, 1971) support this finding. However, there is no evidence one way or the other to indicate that this improvement transfers to flight. The general experimental finding that relatively difficult training tasks yield higher transfer than easier ones suggests that transfer might be reduced as a consequence of adding motion cues that make the simulated flight task easier.

The evident reason that large sums are spent for simulator motion systems, with no evidence of their training value, is their high face validity. A high-fidelity motion system is a delight to any pilot; the illusion of flight is extemely realistic. The decision to include a complex motion system in a simulator is invariably determined by the enthusiasm of pilots, particularly ones in responsible positions.

Training Objectives and Low-Cost Trainers

A related approach to lowering equipment costs involves identifying aspects of flight training that can be taught in low-cost devices. When a less complex training device is appropriate, it should not be overlooked in favor of complex simulators with higher face validity.

The failure to consider low-cost devices when procuring flight training equipment was well illustrated by Prophet (1966). A procedures trainer that had cost over $100,000 was pitted against a plywood, photographic instrument-panel mockup costing less than $100. As predicted by Prophet, the static mockup fared as well as the costlier model for teaching cockpit procedures. Surprisingly, the mockup trainees also did as well as the simulator trainees on other tasks such as reading instruments and making precise control settings. Although the more expensive trainer had capabilities far beyond the scope of the mockup, training in the costlier device should have been devoted primarily to task elements that could not be taught effectively with less costly equipment.

The value of less than full simulation in a variety of flight training situations is obvious; cost reductions in training equipment may be quite large. However, the development and evaluation of simple training devices depend upon the imagination of the designer, the ingenuity of the instructor, and the financial support of the potential user.

INNOVATIONS IN SYNTHETIC FLIGHT TRAINING AND RESEARCH

Although the unresolved issues in synthetic flight training all have their origins in relative antiquity, progress toward resolution is evident. Terms such as "adaptive training", "computer-assisted instruction", "cross-adaptive measurement of residual attention", "computer graphics", "incremental transfer effectiveness", and "response surface methodology" have blossomed during the past decade. Progress in each of these areas shows positive acceleration.

Adaptive Training

Although all personalized instruction is, in a sense, adaptive, the term "adaptive training" has come to refer specifically to the automatic adjustment of the difficulty, complexity, or newness of a training task as a function of the individual student's progress. Examples of adaptive training include: automatically increasing the average amplitude of the forcing function for a tracking task as a student learns; requiring a student to handle more and more subtasks simultaneously in accordance with his immediately preceding performance; and introducing new and different tasks as old tasks are mastered.

Adaptive training employs predetermined decision rules for the adjustment of a training system to the requirements of the individual trainee. Subsequent system outputs are determined by the previous output from the student. In effect, the training task is programmed to advance appropriately with increasing student proficiency.

The first formal application of automatically adaptive logic to the training of pilots has been incorporated into the Synthetic Flight Training System (SFTS) developed by the Naval Training Device Center for helicopter pilot training by the United States Army. In this system, one central digital computer drives four cockpit simulators in which four pilots learn to fly simultaneously under the supervision of a single instructor. The difficulty of certain flight tasks adapts automatically to the individual student's continuously measured performance.

The application of automatic adaptation of task difficulty to the SFTS (Caro, 1969) was inspired mainly by the studies of Hudson (1964) and Kelley (1966). In a conference on adaptive training held at the University of Illinois in 1970 (McGrath and Harris, 1971), it became evident that a central issue was the nature of the adaptive logic employed. Specifically, should error limits be held constant as

skill increases and the task becomes more difficult, as advocated by Kelley, or should error limits vary as the individual's performance improves, as advocated by Hudson? This issue is currently under investigation at the University of Illinois (Crooks and Roscoe, 1971).

Computer-Assisted Instruction

Automated adaptive skill training is a form of computer-managed instruction, as is programmed cognitive training, which may or may not be adaptive. However, the term "computer-assisted instruction" (CAI) implies programmed cognitive learning in which an automatically branching logic allows each student to progress through a course at his own rate.

The application of CAI to the ground-school portion of the flight curriculum at the Institute of Aviation of the University of Illinois is currently in progress. Courses designed to prepare students for private, commercial, instrument, instructor, and airline transport pilot certificates and ratings will be programmed for the PLATO system which eventually will have terminals throughout the nation. PLATO is the acronym for Programmed Logic for Automatic Teaching Operations.

The PLATO system (Bitzer and Johnson, 1971) was designed to aid both student and instructor in the educational process through use of the capabilities of the modern digital computer. The PLATO computer interacts with each student by presenting information and reacting to student responses. The actions of the computer follow the instructor's rules which specify what is to be done in each and every possible situation. A lesson constructed of such a set of rules can have a flexibility approaching that possible when each student has a human tutor. In fact, the rules defining a useful tutorial lesson presented by computer are quite similar to those used implicitly by a human teacher. For example, areas in which a student has proven competence are given minimal coverage, whereas areas in which the student lacks competence are developed more thoroughly.

In contrast to a conventional classroom in which a teacher manages 20 to 30 students simultaneously and can seldom give special attention to individual students, PLATO appears to give each student undivided attention. This appearance results from the ability of the computer to identify and handle most student requests in a fraction of a second. When several students request material simultaneously, the PLATO system processes their requests in turn. However, the last processed student seldom has to wait more than one-tenth of a second for a reply from the computer. To most students, one-tenth of a second appears to be instantaneous. One aspect of individual attention is rapid feedback. The student receives immediate knowledge of the correctness of his responses.

The primary application of PLATO to the training, certification, and currency assurance of pilots will be in the cognitive domain, although PLATO is also capable of certain types of perceptual-motor training. The individual attention capability of PLATO, together with computational and graphic display abilities, allows authors of ground school courses to select and present stored material, such as special characters, photographic slides, and either printed or audio messages, and to construct geometric figures or graphs activated by instructions of either the author or the student. A constructed graphic display, for example, might be used to allow a student in an aviation course to specify the shape and construction of an airfoil. PLATO could then produce a cross-sectional view of the airfoil on the student's plasma display screen. Upon request, PLATO might also show the paths of air molecules flowing around the airfoil in flight.

As the number of terminals grows throughout the country, it will become increasingly possible and desirable to leave much of the certification and currency assurance testing to the PLATO system. Doing this would allow students to take FAA tests at their own convenience and would also free many FAA examiners for more important tasks. When legislation is passed requiring all pilots to undergo periodic recertification, the extra load on the present testing system is going to be enormous. Using CAI techniques to conduct these tests will provide great relief to the system.

Cross-Adaptive Measurement of Residual Attention

The automatically adaptive measurement of a pilot's "residual attention" while performing routine flight tasks can consist of anything from rhythmic tapping on a microswitch with a finger or foot (Michon, 1966) to complex information processing (Ekstrom, 1962; Knowles, 1963). Such tasks serve at least two functions. They provide an inferential measure of the pilot's mastery of the primary task, and they can realistically create elevated workload pressures typical of those encountered in flight emergencies. The demands imposed by such tasks can be made to cross-adapt automatically to the pilot's performance of his primary flight control task. The better he flys, the faster the information to be processed flows. In this way the pilot's total cockpit workload capacity can be measured as a function of his level of training or the decay in his proficiency following periods of inactivity.

The use of automatically adaptive and cross-adaptive secondary tasks for the measurement of residual attention has been applied both in the experimental study of flight display and control design variables (Kraus and Roscoe, 1972) and in the prediction of success in pilot training (Damos, 1972). From these experiments, it has become evident that the technique also can produce a powerful instructional effect in the important areas of attention sharing and decision making. Futhermore, it is well established that pilots show small decrements in flying skills over long periods of inactivity, but show large decrements quickly in procedural efficiency, particularly in situations requiring attention sharing and rapid decision making.

Thus, residual attention tasks provide not only a measure of the initial attainment of proficiency, but also a quick and reliable means of testing the currency of certificated pilots. Tasks similar to those already employed effectively in human engineering experiments can be integrated into either ground-based or airborne flight trainers, but new techniques will have to be developed for their routine use in pilot training, certification, and currency assurance.

Computer Graphics

The simulation of extracockpit visual cues is essential for training in ground-referenced maneuvers involving great danger in actual flight. Extremely costly visual systems have been employed for training in high-speed, low-altitude military operations and in emergency procedures, such as single-engine approaches and engine loss on takeoff in multiengine transport aircraft. High costs are justified in such cases. However, there is an urgent need for less costly but nonetheless effective visual systems for use in various phases of flight training. Perhaps the greatest payoff would be found in the initial training of pilots to land an airplane safely with a minimum of exposure to the hazards of pre-solo and early post-solo landing practice.

Valverde (1968) emphasizes the importance of understanding the capabilities and limitations of visual equipment in order to evaluate properly its use in meeting specific training requirements. He points out, for example, that a large generator is necessitated by the use of a large visual envelope. Therefore, if a small envelope can be used, the cost saving will be extended to other equipment dependent on it.

A computer-generated, line-drawing display system (LDS-1) developed by the Evans and Sutherland Computer Corporation (Ogden, 1970) fits into Valverde's small envelope category. This graphic display system allows automatic windowing and perspective projection of three-dimensional objects, such as an aircraft carrier or an airport with runways and hangars, and therefore lends itself to the simulation of approaches to landings and other contact flight operations requiring a limited field of view.

The advanced simulator for undergraduate pilot training (ASUPT) being developed for the Flying Training Division of the USAF Human Resources Laboratory will present an enormous computer-generated visual envelope around the simulated cockpit of a T-37 airplane (Gum, Knoop, Basinger, Guterman, and Foley, 1972; Smith, 1972). This application of computer graphics presents a somewhat less than literal black-and-white image of the outside world on seven 36-in. circular CRTs, each framed by a pentagonal display window, in a faceted arrangement covering the full forward, lateral, and vertical limits of the external visual field from the cockpit of the T-37. This colossal device is designed to allow the systematic experimental determination of the external visual cues that contribute to takeoff and landing training.

Another advanced application of computer graphic techniques, developed jointly by Hughes Aircraft Company and the University of Illinois for the Federal Aviation Administration, generates a moving-map display for cockpit presentation, continuously showing present position, heading, and area navigation guidance commands. Similar systems have been developed by several companies, including Boeing, Astronautics, and Sperry-Phoenix.

Incremental Transfer Effectiveness

To determine the relative value of simulator training, Roscoe (1971; 1972) proposed the concept of "incremental transfer effectiveness" which postulates a function found by comparing successive increments of time spent in one training task with successive increments of time saved in subsequent training. When the incremental transfer effectiveness ratio drops below the ratio of the hourly cost for ground trainers to that of training aircraft, continued ground training is not cost effective.

The incremental transfer effectiveness concept recognizes the decreasing value of successive increments of simulator training in terms of the time saved in generally more expensive equipment. Povenmire and Roscoe (1973) demonstrated the negatively decelerated relationship between hours saved in the Cherokee airplane and hours spent in the Link GAT-1 in the training of a private pilot. Comparison of the incremental transfer effectiveness functions of different training provides a rational basis for procurement and use in economic terms.

Response Surface Methodology

Previous research has concentrated on the separate effects of numerous variables important in simulator training, but little effort has been directed toward investigating the simultaneous effects of these variables. It is possible that important interactions may be present or that the effect of one variable may be so strong that it overrides other variables. Methodologically, however, it is extremely difficult to examine many variables at once without quickly approaching an unwieldy number of essential data points. For example, if three variables were observed at three levels in a traditional factorial analysis of variance design, 27 treatment combinations or data points would be required for each replication of the design. If seven variables were investigated at three levels each, 2187 data points would be required for each replication. Obviously, the latter experiment would not be conducted. It is also not surprising that such a methodological impass was quickly realized in early research on flight simulators (Williams and Adelson, 1954).

A research technique called response surface methodology (RSM) has been developed for investigating many variables simultaneously. Box and Wilson (1951) originally used RSM to determine the optimum combination of variables for producing the maximum yield of a chemical reaction. The RSM designs minimize the number of data collection points necessary to determine a multiple-regression prediction equation describing the relationship between a predicted score and the experimental variables. Details and examples of this technique are provided by Box and Hunter (1957) and Cochran and Cox (1957).

Recently, Williges and Simon (1971) discussed the utility of using RSM techniques in human performance research. In addition to the economy of the data collection, the designs are flexible and efficient. The designs are flexible in that the data can be collected in sequential order. At the end of each stage of data collection, the experimenter can analyze his results and decide on the appropriate data points to investigate during the next stage of experimentation. The designs are also efficient in that controls are readily available for undesirable fluctuations when the experiment is extended over time. However, certain design modifications are necessary before these techniques can be used successfully to assess human behavior. Some of these considerations are described by Clark and Williges (1973).

With the increased use of RSM in engineering research, it is surprising that limited applications have been made to behavioral research. Only two studies concerned with problems of human learning have used RSM. Meyer (1963) used RSM to study the effects of degree of original learning, time between interpolated and original learning, length of the interpolated list, and degree of interpolated learning on the amount of retroactive inhibition in verbal learning. He plotted a response surface relating the four independent variables to amount of recall. Williges and Baron (1973) used RSM to plot a transfer surface of trials to criterion in an epicycloid pursuit rotor task as a function of tracking speed during training, time between training trials, and number of training trials on a simple pursuit rotor task.

The RSM technique appears to be a viable procedure or model for systematically developing a training simulator. First, it allows for simultaneous investigation of many variables. Second, the sequential research strategy of RSM provides an orderly procedure for determining the variables of importance in simulation to maximize learning, transfer, and retention. Third, the resulting prediction equations can be used to determine tradeoffs among the various independent variables important in simulation to maintain a specific level of learning, transfer, and retention. Finally, the separate RSM prediction equations for level of learning, transfer, and retention can be compared to determine the necessary tradeoffs among the important simulation variables to optimize systematically the combined level of learning, transfer, and retention provided by a particular simulator.

One overall limitation of research on training simulators appears to be that simple piecemeal approaches are used to solve complex research problems. The potential power of RSM is that it allows the investigator to examine the problems of simulation research from a complex, multiparameter, yet systematic, point of view.

PROSPECTUS

User demand for air transportation, recreational flying, and an ever-increasing variety of agricultural, industrial, and scientific flight operations is placing unprecedented pressures on the National Aviation System (NAS). The rapidly increasing complexity of the system itself is demanding new levels of flying skill and knowledge to which few pilots have been trained, and training costs to prepare pilots to operate safely and effectively in the NAS are becoming prohibitive. Furthermore, there is inadequate assurance that those presently flying are qualified to do so, and this problem is growing.

What is needed is a scientifically rigorous investigation into fundamental flight requirements, including not only the perceptual, cognitive, and motor skills required of pilots, but also the attitudes and judgmental factors essential to safe flight. The investigation must start with the identification of the types of flight operations, or missions, that will be undertaken during the foreseeable future and the functions to be performed by pilots in such operations. From this functional analysis must be derived the minimum standards of skill, knowledge, and judgment required of all categories of pilots permitted to fly in the National Aviation System. Current pilot training and certification practices must be evaluated in this new context. Where existing requirements and methods are found to be deficient, new approaches must be devised to close the training and certification gaps at a bearable cost.

A new pilot training, certification, and currency assurance system is needed—one that will automatically qualify each pilot for his particular level of operation at a bearable cost to him as well as to the aviation community. Representative advances in training technology applicable to this objective include computer-assisted cognitive training and testing, automatically adaptive skill training and performance assessment, and the extended use of simulation to previously unexploited areas of pilot training, certification, and currency assurance.

ACKNOWLEDGMENTS

This paper is a report of progress on a continuing research project being conducted by the Aviation Research Laboratory of the Institute of Aviation, University of Illinois at Urbana-Champaign. The research is sponsored by the Life Sciences Program, Air Force Office

of Scientific Research. Dr. Glen Finch, Program
Manager, Life Sciences Directorate, directed the
project which includes eight tasks, four in the
area of pilot selection, training, and performance assessment, and four that deal with avionic
system design principles.

REFERENCES

Adams, J. A. and Hufford, L. E. Effects of
programmed perceptual training on the
learning of contact landing skills. (Tech. Rep.
NAVTRADEVCEN 297-3) Port Washington, N. Y.:
Office of Naval Research, Naval Training Device
Center, April, 1961. (AD 264 377)

Adams, J. A., Hufford, L. E., and Dunlop, J. M. Part-
versus whole-task learning of a flight maneuver.
(Tech. Rep. NAVTRADEVCEN 297-1) Port Washington, N. Y.: Office of Naval Research, Naval
Training Device Center, June, 1960. (AD 242 580)

American Airlines, Inc. Optimized flight crew training:
A step toward safer operations. Fort Worth, Tex.:
American Airlines, Flight Training Academy, April,
1969.

Besco, R. V. The effects of cockpit vertical accelerations on a simple piloted tracing task. (Rep. No.
N.A. 61-47) Los Angeles: North American Aviation, April, 1961.

Bitzer, D. L. and Johnson, R. L. PLATO: A computer-based system used in the engineering of
education. *Proceedings of the IEEE*, 1971, 59,
960-968.

Blaiwes, A. S. and Regan, J. J. An integrated approach
to the study of learning, retention, and transfer—A
key issue in training device research and development. (Final Rep. NAVTRADEVCEN-IH-178)
Orlando, Fla.: Office of Naval Research, Naval
Training Device Center, 1970.

Box, G. E. P. and Hunter, J. S. Multifactor experimental designs for exploring response surfaces.
Annals of Mathematical Statistics, 1957, 28,
195-241.

Box, G. E. P. and Wilson, K. B. On the experimental
attainment of optimum conditions. *Journal of the
Royal Statistical Society Series B (Methodological)*, 1951, 13, 1-45.

Briggs, G. E. and Wiener, E. L. Fidelity of simulation:
I. Time sharing requirements and control loading
as factors in transfer of training. (Tech. Rep.
NAVTRADEVCEN-TR-508-4) Port Washington,
N. Y.: Office of Naval Research, Naval Training
Device Center, 1959.

Buckhout, R., Sherman, H., Goldsmith, C. T., and
Vitale, D. A. The effect of variations in motion
fidelity during training on simulated low altitude
flight. (Tech. Rep. AMRL-TR-63-108) Wright-
Patterson Air Force Base, Ohio: Aerospace Medical
Research Laboratory, December, 1963.

Caro, P. W., Jr. Adaptive training—An application to
flight simulation. *Human Factors*, 1969, 11,
569-576.

Caro, P. W., Jr. and Isley, R. N. Changes in flight
trainee performance following synthetic helicopter
flight training. (Professional Paper 1-66) Alexandria, Va.: Human Resources Research Office,
April, 1966.

Clark, C. and Williges, R. C. Response surface methodology central-composite design modifications for
human performance research. *Human Factors*,
1973, 15, 295-310.

Cochran, W. G. and Cox, G. M. Some methods for the
study of response surfaces. *Experimental designs.*
New York: Wiley, 1957, 335-375.

Cohen, E. Is motion needed in flight simulators used
for training? *Human Factors*, 1970, 12, 75-79.

Conlon, E. W. Preliminary report on contact flight
instruction on a Link trainer. Washington, D. C.:
National Research Council, Committee on Aviation
Psychology, May, 1939.

Crannell, C. W., Greene, E. B., and Chamberlain, H. F.
The experimental study of the use of the Link
trainer in visual contact flight instruction. Washington, D. C.: National Research Council, Committee
on Aviation Psychology, 1941.

Creelman, J. A. Evaluation of approach training
procedures. (Rep. 2, Proj. NM 001-109-107)
Pensacola, Fla.: U. S. Naval School of Aviation
Medicine, 1955.

Crooks, W. H. and Roscoe, S. N. Varied and fixed
error limits in automated adaptive skill training. In
M. P. Ranc, Jr. and T. B. Malone (Eds.) *Pro-
ceedings of the seventeenth annual meeting of the
Human Factors Society.* Santa Monica, Calif.:
Human Factors Society, October 1973, 272-280.

Damos, D. L. Cross-adaptive measurement of residual
attention in pilots. Unpublished master's thesis,
University of Illinois at Urbana-Champaign, 1972.

Danneskiold, R. D. Objective scoring procedure for
operational flight trainer performance. (Tech. Rep.
SPECDEVCEN-TR-999-2-4) Port Washington, N.
Y.: Office of Naval Research, Special Devices
Center, February, 1955. (AD 110 925)

Dougherty, D. J., Houston, R. C., and Nicklas, D. R.
Transfer of training in flight procedures from
selected ground training devices to the aircraft.
(Tech. Rep. NAVTRADEVCEN-TR-71-16-16) Port
Washington, N. Y.: Office of Naval Research, Naval
Training Device Center, September, 1957. (AD 149
547)

Ekstrom, P. Analysis of pilot work loads in flight
control systems with different degrees of automation. Paper presented at the Institute of Radio
Engineers, International Congress on Human Factors in Electronics, Long Beach, Calif., May, 1962.

Ericksen, S. C. A review of the literature on methods
of measuring pilot proficiency. (Res. Bull. 52-25)

Lackland Air Force Base, Tex.: Human Resources Research Center, 1952.

Feddersen, W. E. The effect of simulator motion upon system and operator performance. Paper presented at Seventh Annual Army Human Factors Engineering Conference, Ann Arbor, Michigan, October, 1961.

Flexman, R. E. Man in motion. *The Connecting Link*, 1966, 3, 12-18.

Flexman, R. E., Matheny, W. G., and Brown, E. L. Evaluation of the School Link and special methods of instruction in a ten-hour private pilot flight-training program. *University of Illinois Bulletin*, 1950, 47 (80) (Aeronautics Bulletin 8).

Flexman, R. E., Roscoe, S. N., Williams, A. C., Jr., and Williges, B. H. Studies in pilot training: The anatomy of transfer. *Aviation Research Monographs*, 1972, 2 (1).

Flexman, R. E., Townsend, J. C., and Ornstein, G. N. Evaluation of a contact flight simulator when used in an Air Force primary pilot training program: Part I. Over-all effectiveness. (Tech. Rep. AFPTRC-TR-54-38) Lackland Air Force Base, Tex.: Air Force Personnel and Training Research Center, September, 1954. (AD 53730).

Gabriel, R. F., Burrows, A. A., and Abbott, P. E. Using a generalized contact flight simulator to improve visual time-sharing. (Tech. Rep. NAV-TRADEVCEN-TR-1428-1) Port Washington, N. Y.: Office of Naval Research, Naval Training Device Center, April, 1965. (AD 619 047).

Gagne, R. M. Simulators. In R. Glaser (Ed.) *Training research and education.* Pittsburgh: University of Pittsburgh Press, 1962, 223-246.

Gordon, T. A. The airline pilot: A survey of the critical requirements of his job and of pilot evaluation and selection procedures. (Rep. No. 73) Washington, D. C.: Civil Aeronautics Administration, Division of Research, 1947.

Gordon, T. A. The development of a standard flight-check for the airline transport rating based on the critical requirements of the airline pilot's job. (Report 85) Washington, D. C.: Civil Aeronautics Administration, Division of Research, 1949.

Greene, E. B. The Link trainer in visual contact flying. Washington, D. C.: National Research Council, Committee on Aviation Psychology, 1941.

Guercio, J. G. and Wall, R. L. Congruent and spurious motion in the learning and performance of a compensatory tracking task. *Human Factors*, 1972, 14, 259-269.

Gum, D. R., Knoop, P. A., Basinger, J. D., Guterman, I. M., and Foley, W. L. Development of an advanced training research simulation system. In E. B. Tebbs (Ed.) *Third annual symposium: Proceedings of psychology in the Air Force.* Colorado Springs, Colo.: USAF Academy, Department of Life and Behavioral Sciences, April, 1972.

Hudson, E. M. Adaptive training and nonverbal behavior. (Tech. Rep. NAVTRADEVCEN-

TR-1395-1) Port Washington, N. Y.: Office of Naval Research, Naval Training Device Center, July, 1964.

Hufford, L. E. and Adams, J. A. The contribution of part-task training to the relearning of a flight maneuver. (Tech. Rep. NAVTRADEVCEN-TR-297-2) Port Washington, N. Y.: Office of Naval Research, Naval Training Device Center, March, 1961. (AD 259 505)

Jacobs, R. S., Williges, R. C., and Roscoe, S. N. Simulator motion as a factor in flight director display evaluation. *Human Factors*, 1973, 15, 569-582.

Johnson, B. E. and Williams, A. C., Jr. Obedience to rotation-indicating visual displays as a function of confidence in the displays. *Aviation Research Monographs*, 1971, 1 (3), 11-25.

Kelley, C. R. Self-adjusting vehicle simulators. In C. R. Kelley (Ed.) *Adaptive simulation.* Arlington, Va.: Office of Naval Research, August, 1966. (AD 637 658).

Knowles, W. B. Operator loading tasks. *Human Factors*, 1963, 5, 155-161.

Kraus, E. F. and Roscoe, S. N. Reorganization of airplane manual flight control dynamics. In Knowles, W. B. and Sanders, M. S. (Eds.) *Proceedings of the 16th annual meeting of the Human Factors Society.* Los Angeles: Human Factors Society, October, 1972. [Also Tech. Rep. ARL-72-22/AFOSR-72-11. Savoy, Ill.: University of Illinois at Urbana-Champaign, Institute of Aviation, Aviation Research Laboratory, September, 1972.]

Mahler, W. R. and Bennett, G. K. Psychological studies of advanced Naval air training: Evaluation of operational flight trainers. (Tech. Rep. SPECDEVCEN-TR-999-1-1) Port Washington, N. Y.: Office of Naval Research, Special Devices Center, September, 1950. (AD 643 499).

Matheny, W. G., Dougherty, D. J., and Willis, J. M. Relative motion of elements in instrument displays. *Aerospace Medicine*, 1963, 34, 1041-1046.

McGrath, J. J. and Harris, D. H. Adaptive training. *Aviation Research Monographs*, 1971, 1.

Mengelkoch, R. F., Adams, J. A., and Gainer, C. A. The forgetting of instrument flying skills as a function of the level of initial proficiency. (Tech. Rep. NAVTRADEVCEN-71-16-18) Port Washington, N. Y.: Office of Naval Research, Naval Training Device Center, 1958.

Meyer, D. L. Response surface methodology in education and psychology. *The Journal of Experimental Education*, 1963, 31, 329-336.

Michon, J. A. Tapping regularity as a measure of perceptual motor load. *Ergonomics*, 1966, 9, 401-412.

Miller, R. B. Task and part-task trainers and training. (Tech. Rep. WADD-TR-60-469) Wright-Patterson Air Force Base, Ohio: Wright Air Development Division, Behavioral Sciences Laboratory, 1960.

Muckler, F. A., Nygaard, J. E., O'Kelly, L. I., and

Williams, A. C., Jr. Psychological variables in the design of flight simulators for training. (WADC Tech. Rep. 56-369) Wright-Patterson Air Force Base, Ohio: Wright Air Development Center, January, 1959. (AD 97130).

Nygaard, J. E. and Roscoe, S. N. Manual steering display studies: I. Display-control relationships and the configuration of the steering symbol. (Tech. Mem. 334) Culver City, Calif.: Hughes Aircraft Company, October, 1953.

Ogden, S. The Evans and Sutherland LDS-1 display system. *Proceedings of the Society for Information Display*, 1970, 11, 52-60.

Ornstein, G. N., Nichols, I. A., and Flexman, R. E. Evaluation of a contact flight simulator when used in an Air Force primary pilot training program. Part II. Effectiveness of training on component skills. (Tech. Rep. AFPTRC-TR-54-110) Lackland Air Force Base, Tex.: Air Force Personnel and Training Research Center, December, 1954. (AD 62373)

Parker, J. F., Jr. and Downs, J. E. Selection of training media. (Tech. Rep. ASD-TR-61-473) Wright-Patterson Air Force Base, Ohio: Aeronautical Systems Division, Aerospace Medical Laboratory, September, 1961. (AD 271 483)

Payne, T. A., Dougherty, D. J., Hasler, S. G., Skeen, J. R., Brown, E. L., and Williams, A. C., Jr. Improving landing performance using a contact landing trainer. (Tech. Rep. SPECDEVCEN 71-16-11) Port Washington, N. Y.: Office of Naval Research, Special Devices Center, March, 1954. (AD 121 200)

Pfeiffer, M. G., Clark, W. C., and Danaher, J. W. The pilot's visual task: A study of visual display requirements. (Tech. Rep. NAVTRADEVCEN-TR-783-1) Port Washington, N. Y.: Office of Naval Research, Naval Training Device Center, March, 1963. (AD 407 440)

Pomarolli, R. S. The effectiveness of the Naval air basic instrument trainer. (Special Rep. 65-7) Pensacola, Fla.: Naval Aviation Medical Center, November, 1965. (AD 627 218)

Povenmire, H. K., Alvares, K. M., and Damos, D. L. Observer-observer flight check reliability. (Tech. Rep. LF-70-2) Savoy, Ill.: University of Illinois at Urbana-Champaign, Institute of Aviation, Aviation Research Laboratory, October, 1970.

Povenmire, H. K. and Roscoe, S. N. An evaluation of ground-based flight trainers in routine primary flight training. *Human Factors*, 1971, 13, 109-116.

Povenmire, H. K. and Roscoe, S. N. The incremental transfer effectiveness of a ground-based general aviation trainer. *Human Factors*, 1973, 15, 534-542.

Prophet, W. W. The importance of training requirements information in the design and use of aviation training devices. (Professional Paper 8-66) Alexandria, Va.: Human Resources Research Center, December, 1966.

Rathert, G. A., Jr., Creer, B. Y., and Douvillier, J. G.,

Jr. Use of flight simulators for pilot control problems. (NASA Memorandum 3-6-59A) Washington, D. C.: National Aeronautics and Space Administration, February, 1959. (AD 210 526)

Rathert, G. A., Jr., Creer, B. Y., and Sadoff, M. The use of piloted flight simulators in general research. (AGARD Rep. 365) Paris, France: Advisory Group for Aeronautical Research and Development, North Atlantic Treaty Organization, April, 1961. (AD 404 196)

Roscoe, S. N. Incremental transfer effectiveness. *Human Factors*, 1971, 13, 561-567.

Roscoe, S. N. A little more on incremental transfer effectiveness. *Human Factors*, 1972, 14, 363-364.

Roscoe, S. N., Denney, D. C., and Johnson, S. L. The frequency-separated display principle: Phase III. (Annual Summary Rep. ARL-71-15/ONR-71-1) Savoy, Ill.: University of Illinois at Urbana-Champaign, Institute of Aviation, Aviation Research Laboratory, December, 1971. (AD 735 915)

Ruocco, J. N., Vitale, P. A., and Benfari, R. C. Kinetic cueing in simulated carrier approaches. (Tech. Rep. NAVTRADEVCEN-TR-1432-1) Port Washington, N. Y.: Office of Naval Research, Naval Training Device Center, April, 1965. (AD 617 689) (a).

Ruocco, J. N., Vitale, P. A., and Benfari, R. C. Kinetic cueing in simulated carrier approaches. Supplement I. Study details. (Tech. Rept. NAVTRADEVCEN-TR-1432-1-S1) Port Washington, N. Y.: Office of Naval Research, Naval Training Device Center, April, 1965. (AD 618 756) (b).

Seltzer, L. Z. A study of the effect of time on the instrument skill of the private and commercial pilot. (Tech. Rep. FAA-DS-70-12) Washington, D. C.: Federal Aviation Administration, April, 1970.

Selzer, U., Hulin, C. L., Alvares, K. M., Swartzendruber, L. E., and Roscoe, S. N. Predictive validity of ground-based flight checks. In W. B. Knowles, M. S. Sanders, and F. A. Muckler (Eds.) *Proceedings of the Sixteenth annual meeting of the Human Factors Society*. Santa Monica, Calif.: Human Factors Society, October 1972, 318.

Smith, J. F. Applications of the advanced simulation in undergraduate pilot training (ASUPT) research facility to pilot training programs. In E. B. Tebbs (Ed.) *Third annual symposium: Proceedings of psychology in the Air Force.* Colorado Springs, Colo.: USAF Academy, Department of Life and Behavioral Sciences, April, 1972.

Smith, J. F., Flexman, R. E., and Houston, R. C. Development of an objective method of recording flight performance. (Tech. Rep. 52-15) Lackland Air Force Base, Tex.: Human Resources Research Office, December, 1952.

Smode, A. F., Hall, E. R., and Meyer, D. E. An assessment of research relevant to pilot training. (Tech. Rep. AMRL-TR-66-196) Wright-Patterson Air Force Base, Ohio: Aerospace Medical Research Laboratory, November, 1966. (AD 804 600).

Townsend, J. C. Evaluation of the Link ME-1 basic instrument flight trainer. (Tech. Rep. AFPTRC-TR-

56-84) Lackland Air Force Base, Tex.: Air Force Personnel and Training Research Center, June, 1956. (AD 113 519).

Trans World Airlines. Flight simulator evaluation. Kansas City, Mo.: Trans World Airlines, Flight Operations Training Department, June, 1969.

Valverde, H. H. Flight simulators: A review of the research and development. (Tech. Rep. AMRL-TR-68-97) Wright-Patterson Air Force Base, Ohio: Aerospace Medical Research Laboratory, July, 1968. (AD 855 582).

Wilcoxon, H. C., Davy, E., and Webster, J. C. Evaluation of the SNJ operational flight trainer. (Tech. Rep. SPECDEVCEN-TR-999-2-1) Port Washington, N. Y.: Office of Naval Research, Special Devices Center, March, 1954. (AD 86988).

Williams, A. C., Jr. and Adelson, M. Some considerations in deciding about the complexity of flight simulators. (Tech. Rep. AFPTRC-TR-54-106) Lackland Air Force Base, Tex.: Air Force Personnel and Training Research Center, December, 1954. (AD 62986).

Williams, A. C., Jr. and Flexman, R. E. Evaluation of the School Link as an aid in primary flight instruction. *University of Illinois Bulletin*, 1949, 46 (71) (Aeronautics Bulletin 5).

Williges, R. C. and Baron, M. L. Transfer assessment using a between-subjects central-composite design. *Human Factors*, 1973, 15, 311-319.

Williges, R. C. and Simon, C. W. Applying response surface methodology to problems of target acquisition. *Human Factors*, 1971, 13, 511-519.

Nuclear Power Plant Training Simulator Fidelity Assessment*

Richard J. Carter, Oak Ridge National Laboratory, Oak Ridge, Tennessee, and **K. Ronald Laughery**, Micro Analysis and Design, Boulder, Colorado

The fidelity assessment portion of a methodology for evaluating nuclear power plant simulation facilities in regards to their appropriateness for conducting the Nuclear Regulatory Commission's operating test, was described. The need for fidelity assessment, data sources, and fidelity data which need to be collected were addressed. Fidelity data recording, collection, and analysis was discussed. The processes for drawing conclusions from the fidelity assessment and evaluating the adequacy of the simulator control-room layout were presented.

INTRODUCTION

The U.S. Nuclear Regulatory Commission (NRC) has proposed that revisions be made to Part 55 (Operating Tests) of Title 10 to the "Code of Federal Regulations" and to Regulatory Guide 1.149 (1984). If the rule changes are enacted, the operating test would be administered in a plant walk-through and in a simulation facility, which could be the plant, a plant-referenced simulator, or another simulation device, alone or in combination. During the simulation facility part of the operating test, reactor operators would be assessed on their ability to respond to normal plant operations and malfunctions in a realistic environment. The proposed modifications would require the facility licensee for each nuclear power unit to evaluate their simulation facility as to its appropriateness for the conduct of the operating test.

NRC's Office of Nuclear Regulatory Research contracted to Oak Ridge National Laboratory for the development of a methodology for performing the simulation facility evaluations. The methodology is to be utilized

*Research sponsored by the U.S. Nuclear Regulatory Commission Office of Nuclear Regulatory Research with Oak Ridge National Laboratory operated by Martin Marietta Energy Systems, Inc. under contract #DE-AC05-840R21400 with U.S. Department of Energy.

during two phases of the life-cycle, initial simulator acceptance and recurrent analysis. Initial evaluation is aimed at ensuring that the simulation facility provides an accurate representation of the reference plant. There are two components of initial simulator evaluation: fidelity assessment and a direct determination of the simulator's adequacy for operator testing. Recurrent evaluation is aimed at ensuring that the simulation facility continues to accurately represent the reference plant throughout the life of the simulator. It involves three components: monitoring reference plant changes, monitoring the simulator's hardware, and examining the data from actual plant transients as they occur.

This paper describes the methodology's fidelity assessment portion of initial simulation facility evaluation.

THE NEED FOR FIDELITY ASSESSMENT

The need for each simulation facility to realistically mimic the actual plant has become increasingly important. This is attributable to at least two factors. First, simulators have become a desired medium for providing nuclear power plant (NPP) operators with the skills required for plant operation. Second, it has become increasingly apparent that some simulators may not function in the same manner as

Reprinted from Proceedings of the Human Factors and Ergonomics Society 29th Annual Meeting (pp. 1017–1021).

the actual plant for some plant conditions. When opportunities have arisen to compare actual plant performance during a transient to a simulator's performance, the simulation facility has sometimes behaved very differently.

Fidelity is not the "bottom-line" of a simulator's performance. The true determination of a simulation facility's effectiveness is how the operators trained and tested in it can perform in the plant. However, since direct measurement of operator performance is difficult, impractical, or even impossible for many of the NPP tasks which are tested, the measurement of simulator fidelity is often the best measure than can reasonably be taken.

SOURCES OF FIDELITY DATA

To assess fidelity, two types of data have to be collected for each task for which the simulation facility will be used for operator testing, namely, simulator performance and baseline plant data. The simulator performance data will involve setting up a set of simulation facility starting conditions, developing a scenario of events which will occur, and then collecting data on the changes in values of selected operator display parameters for some period of time. Those parameters which are monitored will depend upon the particular task.

There are three primary sources of reference data on which simulation facility evaluations are based. First, there is actual plant data from the reference plant for which the simulator is being designed. This is clearly the best measure since it represents the ultimate goal of simulation facility performance. However, actual plant data cannot be obtained to represent all operator tasks which will be tested on the simulator. The situation may never have occurred in the plant, particularly for the relatively severe transients for which the simulator will be used. Power plants which are under construction will, obviously, not have any operation data. Even if the plant exists and the situations have occurred, the data collected on plant performance during the occurrence may not be of sufficient precision, accuracy, or completeness to use for simulation facility evaluation. If good plant data had been collected, it may have been used for developing the sim-

ulator mathematical models, in which case the data could not be used for simulation facility evaluation.

The second source of baseline data is from similar plants. The definition of what constitutes "similar" is not a simple issue. Some of the issues which must be considered are as follows:

1. The nuclear steam supply system including reactor type, number of coolant loops and the power rating.
2. The emergency core cooling system including system types, number of pumps, and automatic initiation conditions.
3. The arrangement of reactor auxiliaries.
4. The secondary plant.

If a plant exists that is sufficiently similar to the reference plant, then one should consider collecting baseline data from the similar plant for those tasks for which data are available. Of course if reference plant data exist for an operator task, then similar plant data need not be collected.

The same problems may arise for similar plant data as for reference plant data; the situation may never have arisen, the data may be inadequate, or the data may have been used during simulator design. In addition, there are obvious logistical problems in locating and securing the data from other NPPs.

The third potential source of data that is considered is plant performance data generated by the use of best-estimate engineering models. These models are generally more sophisticated than the mathematical models that are used in the simulator, primarily because the constraint for real-time model execution does not exist. Because these models can include more variables and interactions among variables, and can operate in smaller time increments, they are generally better predictors of plant performance than the simulator models. Therefore, if no actual plant data exist, the engineering models can provide a baseline for comparison. There are, however, drawbacks of these models including the time required to set up the model for any scenario, the computer costs of running the models, and some doubt as to these models' ability to predict plant operation during a scenario.

The selection of a baseline data source should be made individually for each operator task. The baseline selected should be the best possible. As was previously stated, reference plant data is far and away the preferred alternative with similar plant and engineering-model data as acceptable alternatives for the situations in which plant data do not exist and cannot be obtained. When deciding whether to use similar plant vs. engineering-model data, one should consider first, the degree of similarity to the plant as outlined above and, second, the expected accuracy of the engineering models which would be used.

THE FIDELITY DATA WHICH NEED TO BE COLLECTED

The assessment of the fidelity of the simulation is considered to be the same as assessment of the accuracy of the simulator's mathematical models. To do this, two approaches might be taken. One alternative would be to examine the simulator models directly. This would involve examining the lines of computer code and determining whether (1) the correct variables were included in each of the submodels, (2) the variable update frequency was sufficient, (3) the numerical approximation techniques were adequate, and (4) the functional relationships were correct. Unfortunately, the state-of-the-art in NPP modeling is not sufficiently well advanced to permit a determination of what constitutes "adequate," "sufficient," or "correct."

The second approach, and the one used in our methodology, is simply to observe the outputs of the simulator models at a level where they can be directly compared to the baseline data, i.e., display parameters. It can be argued that this is a better approach, regardless, since we are directly measuring the simulator's ability to correctly simulate the baseline plant performance data. The disadvantage of this approach is that it is far less clear when one has collected a sufficient amount of data to support the contention that the simulation facility does predict plant performance under conditions which differ from those under which the baseline data were collected.

Two factors are considered in determining the display parameters for each operator task: (1) those displays which operators rely on most in performing the task and (2) those displays parameters for which data have been or can be collected.

Determining the Display Parameters on Which Operators Rely in Performing the Task

What must be developed to answer this question is a "critical-display-parameters-by-operator-tasks" matrix. The operator tasks are the evolutions and malfunctions listed in American National Standard (ANS) 3.5 (American Nuclear Society, 1981). The critical displays for each task can include any of the displays which are available to the operator. The design of this matrix proceeds according to the following steps.

Develop a list of operator displays. To determine which displays are used by operators, one must first develop a list of all displays that are available. This includes all analog, digital, and binary displays. To facilitate development and use of this display listing, the displays should be grouped hierarchically by first, the plant system to which they pertain and then, within each system, the types of information which are presented on the physical status of the plant.

Obtain opinions from at least two plant operators on the ten most critical displays for each task. The first set of data is collected from experienced plant operators regarding which of the displays are used for each of the tasks. Each of the operators is given a series of forms, one for each operator task. On these forms is the following:

1. The title of the operator task.
2. A brief description of the task.
3. A set of instructions to the plant operator who will fill out the form.
4. A list of all operator displays.

The brief description of the task serves to clarify any ambiguity about the task which is not clarified in the task title. The instructions indicate to the plant operator filling out the form that he is to allocate a total of 100 points to no more than ten of the listed displays. The number of points allocated to any display should reflect the importance of that display to the operator when he performs the task. If less than ten displays are truly important, then points should only be allocated to those that are important. Then, the operator is to mark the appropriate number of points in the space provided by each of the displays listed.

Obtain opinions from at least two nuclear engineers/designers on the ten most critical displays for each task. In an identical manner to that used for plant operators, data is collected on the most critical displays from at least two nuclear engineers or designers. This step is intended to provide a check on the data obtained from the operators. The engineers/ designers should also know what parameters the operators should be considering in performing the task. Preferably, the engineers/ designers should be familiar with the reference plant.

Reconcile differences between plant operators and nuclear engineers, if necessary. Once these data are collected, the opinions from the individuals (operators and engineers/designers) are compiled onto one sheet for each operator task. Ideally, the two lists would be identical. However, since this is a subjective rating of relative importance, differences should be expected. If the operators and engineers agree on less than 50% of the critical displays, then that may be an indication that the operators and engineers have perceived the definition of the task differently. A reconciliation is attempted which can simply involve a meeting between the two groups to mutually determine which displays are truly important. At worst, more than ten parameters can be measured for a task during fidelity assessment, thereby reflecting the combination of the two groups' opinions on the most critical displays. If the two groups agree on more than 50% of the displays, then the total points for each display are summed. The ten displays receiving the highest total score are deemed as those for which measurements should be taken during fidelity assessment of the simulator for that task.

Determining the Display Parameters on Which Data Have Been or Can be Collected

At this point, one has determined what variables should be measured during the fidelity portion of initial simulator evaluation. Now practical considerations must come into play.

The question is whether what one has is what one wants and, if not, what can be done to rectify the shortcomings?

Operator tasks for which reference or similar plant data are to be used as a baseline. The rule for determining whether the plant data are satisfactory is if data were collected on 40% of the critical display parameters, then these data are sufficient. If one samples 40% of the critical displays and they are found to be satisfactory, then it can be assumed with some level of confidence that the other displays will function properly. However, if data for less than 40% of the critical displays can be assessed with plant data, then engineering models should be used instead to collect the baseline data.

Operator tasks for which engineering-model data are to be used as a baseline. The rule for determining whether the engineering-model data are satisfactory is if data were collected on 60% of the critical display parameters, then these data are sufficient. The reasons for setting a higher requirement for engineering-model data than for plant data are two-fold. First, the engineering-model data are probably less valid than plant data and, therefore, one should be more cautious in accepting the simulator when using engineering-model data as a baseline. Second, engineering models can usually be modified to provide output about the parameters of interest.

Absolute/Trend Parameters

Before one can develop the critical-display-parameters-by-operator-tasks matrix, he needs to know, for each variable, whether it is an absolute or trend parameter. An absolute parameter is one in which the absolute value of the parameter is important to the operator during performance of the task. Trend parameters, on the other hand, are only important with respect to the rate at which they are changing, not necessarily the absolute value of the parameter. For every operator task, each parameter is designated as either a trend or absolute parameter. This designation is made by a subject-matter expert. It is conceivable that display parameters may be trend parameters in some tasks and absolute parameters in others.

Once this analysis is completed, the matrix is constructed. The dimensions of the matrix are (1) the operator tasks which are to be tested in the simulation facility (rows of the matrix) and (2) the plant control-room displays (columns of the matrix). The cells of the matrix are either (1) blank indicating that the display is not critical for performance of the task, (2) a "T" indicating that the display presents a critical trend parameter for performing the operator task, or (3) an "A" indicating that the display presents a critical absolute parameter for performing the operator task.

COLLECTING AND RECORDING THE FIDELITY DATA

In order to minimize the overall data collection and analysis effort, careful attention should be given to the form and format of recording the data as well as the source and content of the data themselves. To conceptualize the problem, consider that during data analysis every baseline data point must have a corresponding simulator data point and the focus will be to determine whether the two numbers are nearly the same.

To facilitate this determination, the prime goal of data collection is to ensure that these pairs of points are truly comparable. This requires that the baseline and simulator data are synchronized and that any simulated operator actions or equipment malfunctions occur at the same relative time. A shift of even a few seconds can lead to the appearance of great differences between the simulator and the baseline when, in fact, the differences are simply due to a phase shift in the data collection timing.

The methods of collecting and recording the data have a significant impact upon the effort required in analyzing the data. With state-of-the-art NPP parameter recording systems and simulator performance monitoring systems, the data analysis requires little more than developing several computer programs to reformat the data. However, if all data must be collected manually, then hundreds of man-hours may be required to reduce the data. Even if the data are collected automatically, careful attention must be paid to ensure that the synchronization issues are adequately addressed.

ANALYZING THE FIDELITY DATA

Four summary descriptive statistics are computed for both absolute and trend parameters: (1) root mean squared (RMS) error, (2) percent error, (3) maximum error, and (4) error t-score. The first three statistics provide descriptive information about the simulator's fidelity. Each of these three statistics represents a different aspect of fidelity, each of which is important to human perception in a different way. The computation of an error t-score provides an inferential statistical test of the simulator's resemblance to the plant with respect to the observed parameters. The four statistics are computed for each of the critical displays on the tasks which are being evaluated.

DRAWING CONCLUSIONS

The fidelity assessment procedure does not result in a statement as to whether the simulator has adequate fidelity as a whole. Rather, the simulation facility is deemed acceptable or unacceptable for the testing of individual tasks as is evidenced by the simulator's performance during a scenario embodying that task. To assess the simulation facility one must first determine if the simulator sufficiently replicates each of the critical operator display parameters within each scenario. Then, based on the number of critical display parameters successfully simulated, the simulation facility's overall acceptability in simulating the scenario is decided. If the scenario can be faithfully replicated by the simulator, we assume that the simulation facility can simulate other scenarios of the same task with roughly equal success, and, hence, one should consider the simulator acceptable for testing of that task.

Two levels of acceptability were defined for each individual critical display parameter, fully acceptable or marginally acceptable. The criteria for each of these levels of acceptability with respect to each of the four statistical measures computed were also specified. All criteria are such that if the observed measure is less than the criteria, it is acceptable at the appropriate level. The selection of these criteria was based upon the recommendations of ANS 3.5.

To consider the simulator fidelity for testing

a particular task acceptable, at least 75% of the critical display parameters must be deemed fully acceptable with respect to all four criteria and at least 90% of the critical display parameters must be deemed either fully or marginally acceptable as they were measured during the performance of the scenario. If the displays are critical to task performance, then it is essential that they behave properly in the simulator.

ASSESSING THE ADEQUACY OF THE SIMULATOR CONTROL-ROOM LAYOUT

Up to this point, we have described procedures designed to investigate whether the simulation facility's representation of the reference plant's dynamics are adequate for operator testing. If these criteria are satisfied, then one must ensure that the control-room layout is a sufficient replication of the reference plant's control room.

The procedures for conducting this part of fidelity assessment are relatively straightforward. For each operator task for which the simulator is to be used for operator testing, a collection sheet is prepared. On this sheet is a list of the critical displays for the task. Data is collected by comparing the simulation facility's control-room layout with respect to each critical display. Each of the displays is compared with respect to the similarity of (1) display type, (2) minimum display value, (3) maximum display value, and (4) relative display location. To assess the similarity of relative display location, the following factors are considered:

1. The display should be located on a panel which is similarly located in the control room with respect to other control panels.
2. The display should be located in a similar location on the panel with respect to other displays on the panel.

It is not essential that the relative location of each of the displays be identical to the reference plant. Rather, if an operator could identify a display strictly by its location (i.e., without referring to any display markings or other identifiers), then it is considered as having the same relative location.

For a control-room layout to be considered adequate, all of the critical displays must be rated as the same in the simulator as in the plant on all of the about four dimensions.

REFERENCES

American Nuclear Society (1981). Nuclear power simulators for use in operator training. American National Standard 3.5.

U.S. Nuclear Regulatory Commission (1984). 10 CFR parts 50 and 55, operator's licenses and conforming amendment. *Federal Register*, 49, 46428-46440.

U.S. Nuclear Regulatory Commission (1984). Nuclear power plant simulation facilities for use in operator license examinations. Proposed Revision to Regulatory Guide 1.149.

NOTE

The methodology described in this paper has not been adopted (or rejected) by NRC. The paper in no way represents current or planned policy of NRC.

The Development of Technology for Collective Training: SIMNET, a Case History

EARL A. ALLUISI,[1] *Institute for Defense Analyses*

SIMNET, an acronym for *simulator networking*, was initiated in 1983 as a project on large-scale simulator networking by the Defense Advanced Research Projects Agency (DARPA; Order AO 4739, signed February 15, 1983). It is a proof-of-principle technology demonstration of interactive networking for real-time, person-in-the-loop battle engagement simulation and war-gaming. Intended for military collective training, SIMNET is also adaptable for training or exercising commanders and staffs at higher echelons, usable in the development of military concepts and doctrine, and suitable to the testing and evaluation of alternative weapon system concepts prior to acquisition decisions. This paper summarizes the technical history of SIMNET development and identifies lessons learned that could contribute to the success of future efforts to develop training technologies and systems, especially for collective training. It concludes with a discussion of the implications and challenges of SIMNET for the human factors and training technology communities.

INTRODUCTION

SIMNET is the first system to achieve true interactive simulator networking for the collective training of combat skills in military units from mechanized platoons to battalions. As of January 1, 1990, its components consisted of about 260 ground vehicle and aircraft simulators, communications networks, command posts, and data-processing facilities distributed among eleven sites—seven in the continental United States (Cambridge, Massachusetts; Fort Benning, Georgia; Fort Knox, Kentucky; Fort Rucker,

Alabama; Fort Stewart, Georgia; MacDill Air Force Base, Florida; and Arlington, Virginia) and four at U.S. Army locations in Europe (in West Germany at Grafenwühr, Friedberg, Schweinfurt, and Fulda.) The Fort Knox site is currently the largest SIMNET facility, with simulators for 44 M1 Abrams tanks, 28 M2/3 Bradley fighting vehicles, 2 scout/attack helicopters, 2 close air support fighter aircraft, a battalion task force tactical operations center, an administrative-logistics operating center, and other command and control, artillery and mortar fire, and close air support control elements—all fully interactive on a local area network.

In 1989 SIMNET technology was adopted into the army and became SIMNET-T, a collective or unit training capability that

[1] Requests for reprints should be sent to Earl A. Alluisi, Science and Technology Division, Institute for Defense Analyses, 1801 Beauregard St., STD, Alexandria, VA 22311.

the army is planning to extend through a large-scale, follow-on acquisition program. SIMNET-D at Fort Knox and AIRNET at Fort Rucker are other versions that provide developmental capabilities. They can be reconfigured to simulate new design concepts for evaluation in SIMNET trials and are the basis of a new joint U.S. Army/Defense Advanced Research Projects Agency (DARPA) initiative to demonstrate advanced distributed simulation technology for use in system development, studies, and analyses (Lunsford, 1989).

Technical History

The technical development of SIMNET is divided into five phases: (a) its origins in related efforts prior to 1979; (b) the gestation and planning that led to initiation of the SIMNET project in 1983; (c) component development and early demonstrations from 1983 to 1985; (d) prototyping, networking, and testing from 1985 to 1987; and (e) system development and field testing, culminating in transition to the army in 1989.

Origins

The late J. C. R. Licklider became the first director of DARPA's Information Processing Techniques Office in 1962. He influenced DARPA to support many developments and applications of the human-computer interactive technologies in areas of the behavioral sciences (Licklider, 1960, 1988). He also stimulated the establishment of DARPA's Cybernetics Technology Office (CTO; initially called the Behavioral Science Research Office and then the Human Resources Research Office), of which he was the initial director, and through which the SIMNET project was executed (discussion with A. W. Kibler, Falls Church, Virginia, January 18, 1990).

Changes relevant to the later development of SIMNET were taking place at DARPA during the late 1970s. The CTO program had

been cut by half, all work in the behavioral sciences was being questioned, and internal support seemed increasingly limited. The trend throughout DARPA was away from efforts that produced only written reports and toward work that developed technologies or things of direct military use. CTO complied by reducing its academic or basic research and increasing its adaptations of new computer technologies in areas such as military decision making and training (telephone discussion with S. J. Andriole, George Mason University, February 7, 1990).

In 1978 a new CTO staff member was given a new Apple II personal computer by the Apple Education Foundation (discussion with J. D. Fletcher, Institute for Defense Analyses [IDA], January 4, 1990). The DARPA director and his deputy were invited to a demonstration of the new tabletop computer—an example of a new technology that showed promise of advancing military (and civilian) education and training in the future. Some of the microcomputer's potential applications to military training issues were discussed, and the CTO was encouraged to prepare an internal DARPA proposal to develop a tank gunnery trainer (TGT) based on personal computer technology.

The statement of military need for a TGT cited the high cost of tank gunnery practice in the field and the consequent severe limitations on the amount of practice that tank gunners acquired in the then-current training systems. It is important to note that low cost was a goal from the start, because system concepts developed with and without such a low-cost constraint are very likely to differ substantially (S. J. Andriole, February 7, 1990; J. D. Fletcher, January 4, 1990). The notion was to use an Apple II computer to drive a videodisc with computer graphics overlaying the display. The plan won support, and the TGT project was initiated in June 1979 (DARPA AO 3791, signed June 11, 1979).

Then, near the end of 1979, computer graphics were successfully overlaid on a videodisc image in what was probably the first demonstration of that new technology (i.e., the mixing on a single visual display of digital computer generated images with analog video images from a videodisc; see S. J. Andriole, February 7, 1990; J. D. Fletcher, January 4, 1990; telephone discussion with R. S. Jacobs, Illusion Engineering, Inc., March 21, 1990).

Computer image generation (CIG) had been demonstrated and was being exploited prior to this time by the training technology research and development (R&D) community. In 1976 the Advanced Simulator for Pilot Training (ASPT) had been installed at the Williams Air Force Base, Arizona, facilities of the Air Force Human Resources Laboratory. The visual system was a dodecahedron of seven CIG channels (limited to 2500 edges) displayed through special optics (pancake windows) by monochromatic video projectors, with three additional channels reserved for planned insertion of a high-resolution inset.

CIG was also used in the Visual Technology Research Simulator under development at the Naval Training Systems Center (NTSC; previously the Naval Training Equipment Center) in Orlando, Florida (Lintern, Wightman, and Westra, 1984). Its visual system projected a CIG image on the interior surface of a dome, with a high-resolution area-of-interest inset from a second projector slaved to the head and eye movements of the pilot (Balwin, 1981; Breglia, Spooner, and Lobb, 1981). The U.S. Army Project Manager for Training Devices (PM TRADE), collocated with NTSC in Orlando, also planned to use a CIG display in a full-mission tank simulator under development at Fort Knox—the Unit Conduct of Fire Trainer.

However, all of these systems were expensive. The cost of a single CIG channel during the late 1970s was $2 to $3 million, and the estimated procurement costs of CIG-based training simulators ranged from $5 million for the Unit Conduct of Fire Trainer to at least $30 million for ASPT. Concern was growing in the R&D community regarding the military's ability to acquire such systems in sufficient numbers for effective training. Efforts were under way to develop lower-cost systems. For example, during spring 1978 the NTSC converted its mainframe-based simulator computer system to a multi-minicomputer architecture, a precursor to the multi-microprocessor architecture eventually employed in the SIMNET system (personal correspondence with W. S. Chambers, NTSC, March 2, 1990). DARPA's concern was revealed through its emphasis on developing a low-cost TGT—specifically, a device that could be procured in quantity for $10 000 each at most.

Gestation and Planning

During the winter of 1979–80, the initial TGT objectives were expanded to include the idea of developing a tank team gunnery trainer (TTGT) by networking—quickly and at low cost—five TGTs together so that they could interact (J. D. Fletcher, January 4, 1990; discussion with A. Freedy, Perceptronics, at IDA, December 5, 1989). The objective was to have several TGTs show a single videodisc-generated scene, with the trainees competing to be the first to sight and fire at an enemy tank. This concept did not include an interactive engagement, and thus it lacked one of the central characteristics of the SIMNET system developed later.

Before the project manager left DARPA in fall 1980, he located an Air Force officer to continue the TGT development—Jack A. Thorpe. Thorpe was familiar with the use of computer image generation in flight simulators for training. While at the Air Force Human Resources Lab during the mid-1970s, he and several of his colleagues, including D.

Bustell, M. Cyrus, J. Fuller, L. E. Martin, G. Reid, R. Reis, and W. Waag, had often discussed the potential benefits of a network of low-cost simulators to provide combat skills training for pilots (telephone discussion with E. L. Martin, AFHRL/OT, March 14, 1990; telephone discussion with M. Cyrus, La Jara, Colorado, April 11, 1990). He was also familiar with the high costs of large aircraft simulators and appreciated the technologies that could contribute to the future development of simulators for collective as well as individual training (discussion with J. A. Thorpe, DARPA European Office, October 16, 1989). In 1978, while at the Air Force Office of Scientific Research, he had proposed research to develop a network of simulators that would provide large-scale battle engagement interactions for training air combat skills that pilots could not learn in peacetime (Thorpe, 1978).

Thorpe joined the Cybernetics Technology Office staff in January 1981. He managed the TGT project and pushed for the networking of several devices to provide a platoon-level, low-cost team device—the tank team gunnery trainer. He began exploring video game technology for application to tank gunnery training, telerobotics for maintenance training, and the feasibility of molding fiberglass to form a tank cockpit (E. L. Martin, March 14, 1990). He also began an expanded proposal on networking technology for use of simulators in collective training, an idea that later became the SIMNET project (A. Freedy, December 5, 1989). His concept was to demonstrate the utility of low-cost, large-scale battle engagement simulation technology for combat training, and DARPA contracts aimed toward that goal were awarded (e.g., Owens and Stalder, 1982).

Affordability had become a major issue in the military departments, and it had a substantial effect on the development and use of simulators for training. Because of excessive

cost (upward of $30 million), the Air Force had canceled Project 2360, an engineeering development effort to create a prototype of a high-resolution, high-brightness, full-field-of-view, CIG-based visual system for flight simulators. However, it had continued a less expensive advanced development effort, Project 2363 (Ferguson, 1984). The army had canceled a similar program for a full crew tank simulator (Project FCTS) because of excessive costs ($18 million). Therefore Thorpe established affordability as a core TTGT requirement: unit cost would have to be sufficiently low to permit the army to procure the thousands of copies needed for collective combat skills training (J. A. Thorpe, October 16, 1989).

During the summer of 1982, Thorpe asked a retired army colonel to help develop a network of tank simulators suitable for army unit training. The DARPA plans were reported to the deputy chief of staff for training at Headquarters, U.S. Army Training and Doctrine Command (TRADOC). The deputy chief of staff, with others, had authored a training study that identified army training needs, some of which DARPA was trying to address through its development of videodisc, microprocessor, and other emerging technologies in the TGT project (R. S. Jacobs, March 21, 1990). A TRADOC representative, sent to learn more about the proposal, reported that it seemed potentially able to meet some army training requirements, and TRADOC then indicated a willingness to support DARPA's project to the extent that it would address the army's collective training needs (telephone discussion with G. W. Bloedorn, February 20, 1990).

Component Development

The SIMNET project was approved in 1982 and initiated by DARPA in early spring 1983 (DARPA AO 4739, signed February 15, 1983; J. A. Thorpe, October 16, 1989). There were three initial contracts: (a) one to develop the training requirements and conceptual designs for the vehicle simulator hardware and system integration (telephone discussion with J. M. Levine, Northridge, California, March 6, 1990), (b) a second to develop the networking and graphics technology, and (c) a third to conduct a six-month "lessons learned" study of army field training experiences with armor and mechanized infantry units using the instrumented ranges of the National Training Center at Fort Irwin, California (G. W. Bloedorn, February 20, 1990; R. S. Jacobs, March 21, 1990; J. A. Thorpe, October 16, 1989).

The first volume of the data handbook for SIMNET development appeared in March 1983 (Bloedorn, 1983a; Bloedorn, Kaplan, and Jacobs, 1983), and the second volume in August 1983 (Bloedorn, 1983b). The two volumes provided initial data and background information on the M1 Abrams tank, the M2 Bradley fighting vehicle, and the M3 cavalry fighting vehicle. The design philosophy was highly innovative. At a time when simulators were designed to approach full physical fidelity (i.e., to emulate the vehicles they represented as closely as engineering technology and the available funds permitted), the SIMNET goal was selective functional fidelity. The procedure was first to learn what functions were needed to meet the training objectives and only then to specify simulator hardware requirements. Thus hardware items regarded as nonessential to the combat operations to be trained either were not included or were designated only by drawings or photographs in the simulator. This made it easier to design relatively low-cost devices and thereby to minimize simulator unit cost (G. W. Bloedorn, February 20, 1990; R. S. Jacobs, March 21, 1990; J. A. Thorpe, October 16, 1989).

An early low-cost prototype—a so-called 60% solution—was planned for field demonstration as a concrete device that would be modified and improved on the basis of informal tests and evaluations (A. Freedy, December 5, 1989; discussion with U. Helgesson, Los Angeles, California, April 12, 1990). In lieu of detailed engineering specifications for this prototype, DARPA agreed to provide a listing of the "minimum essential characteristics" for the simulators (J. A. Thorpe, October 16, 1989). To obtain this list, five missions were first identified as best for the training that SIMNET should address (hasty attack, deliberate attack, hasty defense, deliberate defense, and passage of lines). These missions were further broken down into 9 collective skills and 62 tasks (G. W. Bloedorn, February 20, 1990; J. A. Thorpe, October 16, 1989).

The developers had to know how armor units operated and trained, so during the fall of 1983, the design team's scientists and engineers went to the Armor School at Fort Knox for field training in close combat heavy skills with actual equipment. Design became behaviorally driven and began by identifying cues that crew members needed to learn their specific duties and tasks. Input came from a wide variety of persons: operators, trainers, engineers, and psychologists (J. A. Thorpe, October 16, 1989). The design did not concentrate on the armored vehicle per se; rather, the vehicle simulator was viewed as a tool that would enhance the training of the crews as a collective—a military unit—when networked with other simulators. The design goal was to make the crews and units (not the devices) the center of the simulations; the major interest was in collective, not individual, training (G. W. Bloedorn, February 20, 1990).

Where design options were to be exercised, the development team insisted that such decisions be based on the likelihood of obtaining the desired trainee behavior. Each active display and control in each vehicle simulator was tied to the performance of a specific task to be trained. No superfluous live displays or controls were inserted in the simulators merely because the real vehicles had them. Instead, the SIMNET simulators sought to achieve the design goal of selective functional fidelity (G. W. Bloedorn, February 20, 1990; U. Helgesson, April 12, 1990; R. S. Jacobs, March 21, 1990; J. A. Thorpe, October 16, 1989).

There were other design issues, of course, but they were always resolved in the direction of the behavioral goals. For example, some of the trainers wanted roles for instructors and designs for instructor operator stations, but the training issue was posed, "How does a commander at any level diagnose performance deficiencies of a unit and correct them while in contact with the enemy?" Because the answer did not provide for so-called third-party scoring, it was decided that neither should SIMNET. Instead, the chain of command controls its own training in SIMNET, with no instructors, no instructor operator stations, and no third parties scoring soldiers' performances (J. A. Thorpe, October 16, 1989).

The developmental process was to construct mock-ups of hardware elements designed to provide the proper touch of the cues required for training the desired performances, progressing from laboratory racks, to plywood foam-core mock-ups, to a fiberglass prototype. At each stage of the development, models of the product were demonstrated to soldiers, who worked very closely with the development team and gave them reactions and suggestions. These often led to improvements in the design of successive units. This process provided the necessary input for the approach—that is, for stopping short of a full-scale emulation of the vehicle with a less complex and less costly device that was good enough to achieve the training objectives desired (G. W. Bloedorn, February

20, 1990; U. Helgesson, April 12, 1990; R. S. Jacobs, March 21, 1990).

The first SIMNET foam-core mockup of the M1 tank was demonstrated at Fort Knox during the spring of 1984, about one year after project initiation—an extremely rapid prototyping. At about the same time, the SIMNET concept was demonstrated at the facilities of the army's principal acquisition agency for simulators and training equipment, PM TRADE (J. A. Thorpe, October 16, 1989).

However, a major crisis had arisen during the early months of 1984 when it appeared that the planned visual display and networking architectures would not satisfy the SIMNET system requirements. Specifically, analyses and expert judgments indicated that the planned use of off-the-shelf visual display technology would not provide the required scene complexity within the cost, computer, and communications constraints set by the SIMNET goals (G. W. Bloedorn, February 20, 1990; M. Cyrus, April 11, 1990; R. S. Jacobs, March 21, 1990; J. A. Thorpe, October 16, 1989). DARPA's management, projecting the technology to be inadequate and too costly, considered aborting the project (J. A. Thorpe, October 16, 1989).

Fortunately, development of a new low-cost microprocessor-based CIG technology for visual displays such as those envisioned for SIMNET had been proposed. The new technology appeared interesting: it promised to meet the scene complexity requirements at acceptably low dollar and computational costs. In addition, if it worked, it would permit use of a simpler networking architecture, one that would be less costly in both dollar and communications capacity requirements. The proposal was to use microprocessors in each tank simulator to compute the visual scene for that tank's virtual world, including the needed representations of other vehicles, both friendly and enemy. The network would not have to carry all the data in the visual scenes of all simulators; rather, network transmissions could be limited to relatively small packages of calibration and status-change information (G. W. Bloedorn, February 20, 1990; M. Cyrus, April 11, 1990; E. L. Martin, March 14, 1990; J. A. Thorpe, October 16, 1989).

Thorpe proposed that the contract for development of the networking and graphics technology be terminated or amended and that the new proposal be funded (G. W. Bloedorn, February 20, 1990; J. A. Thorpe, October 16, 1989). Some were skeptical: they judged the risk to develop the new technology as too high. Thorpe argued the case and won approval to proceed, provided the new technology was successfully demonstrated by the beginning of the next fiscal year (October 1984). If it was not successful, the project would be terminated. DARPA's early sign-off on the feasibility test of this technical approach is one of four critical milestones in SIMNET development identified by one of the contractors (personal memorandum from G. Weltman on "SIMNET Case History," March 2, 1990).

A "breadboard" demonstration of the graphics technology was made within the deadline, and in January 1985 a rack-mounted SIMNET system mock-up with graphics was put together in the contractor's office near the DARPA headquarters building to show the SIMNET concept: a network of low-cost armored vehicle simulators that could be used for collective training of armor and mechanized infantry units. The demonstration included not only the SIMNET ply-wood M1 tank simulator but also communications, logistics, and fire support and a representation of the tactical operations center. When this was shown to the secretary of the army, the army chief of staff, the vice chief of staff, and other high-level army offi-

cials that month, their reaction was positive—sufficiently so that the army began to provide additional funding and support for DARPA's further development of the SIMNET technology (G. W. Bloedorn, February 20, 1990).

Prototyping, Networking, and Testing

The SIMNET design concept was well established by 1985. SIMNET would consist of local and long-haul nets of interactive simulators for manuevering armored vehicle combat elements (M1 tanks and M2/3 fighting vehicles), combat support elements (including artillery effects and close air support, with both rotary and fixed-wing aircraft), and all the necessary command and control, administrative, and logistics elements for both friendly and enemy forces. A distributed net architecture would be used, with no central computer exercising executive control or major computations, but rather with essentially similar (and all the necessary) computation power resident in each vehicle simulator or center-nodal representation (see Thorpe, 1987).

The terrains for the battle engagements would be simulations of actual places, 50 × 50 km initially but eventually expandable by an order of magnitude in depth and width. Battles would be fought in real time, with each simulated element—vehicle, command post, administrative and logistics centers, and so on—being operated by its assigned crew members. Scoring would be recorded on combat events such as movements, firings, hits, and outcomes, but actions during the simulated battle engagements would be completely under the control of the personnel who were fighting the battle (i.e., no third-party interventions). Training would occur as a function of the intrinsic feedback and lessons learned from the relevant battle engagement experiences. Development would proceed in steps, first to demonstrate platoon-level networking, then to company and battalion levels, and later perhaps to even higher levels (Thorpe, 1987).

The system would be developed by applications from three technology domains: the simulators; the computational hardware, software, and networking; and the graphics for the visual displays (DARPA AO 5608 and AO 5835, amendments signed July 11 and 28, 1986, respectively). The development team consisted of contractor personnel from each of the three areas. Thorpe, as project manager, served essentially as the program's chief executive officer, "matching the contractors' capabilities with the users' interests and needs throughout the program, to the advantage of all" (A. Freedy, December 5, 1989). The subdivision of the program effort into the three technical areas (simulators, computer hardware/software, and vision systems) with a program structure "able to manage and integrate contributions of the three associated contractors" is the second of the four contractor-identified critical milestones (G. Weltman, March 2, 1990).

Each simulator was to be a self-contained unit, with its own host microprocessor, terrain data base, graphics and sound systems, cockpit with only task-training-justified controls and displays, and network plug-in capability. Each would generate all the battle engagement environment necessary for the combat mission training of its crew. For example, each tank crew member would see a part of the virtual world created by the graphics generator using the terrain data base and information arriving via the net regarding the movements and status of other simulated vehicles and battle effects—the precise part being defined by the crew member's line of sight: forward for the tank driver or from any of three viewing ports in a rotatable turret for the tank commander (Thorpe, 1987).

As developed, the visual display depends primarily on the graphics generator resident in each simulator. It is a computer image generation system that differs in several ways from earlier CIG systems such as the Advanced Simulator for Pilot Training and Visual Technology Research Simulator. First, it is based on microprocessors rather than minicomputers or a large mainframe, so it is relatively low in cost—less than $100 000 per simulator visual display subsystem as opposed to more than $1 million per ASPT visual channel. Second, it is high in visual complexity, with many moving models and special effects, but low in display complexity, with relatively few pixels, small viewing ports, and a relatively slow update rate of 15 frames/s. The development of this essentially unique graphics generator for SIMNET was a principal factor in permitting SIMNET to meet its low-cost-per-unit constraint (Ferguson, 1984; Thorpe, 1987).

SIMNET employs both local area and long-haul networks. Thus it supports a network distribution not only horizontally, among simulators or elements of the same or similar kinds, but also vertically, among simulated nodes representing the command and control, combat support, combat service support, and logistics elements that constitute an entire battle force (i.e., an entire military collective committed to engage an enemy force in battle). This kind of battle engagement simulation includes all the important elements of reality needed to support SIMNET's collective training objective. Furthermore, having been designed to use both local area networks (LANs) and long-haul networks (LHNs), SIMNET also permits geographically separated military elements to interact in a common battle engagement over the same simulated terrain. The LAN architecture in SIMNET is a relatively simple application of ether-net technology as shown in Figure 1.

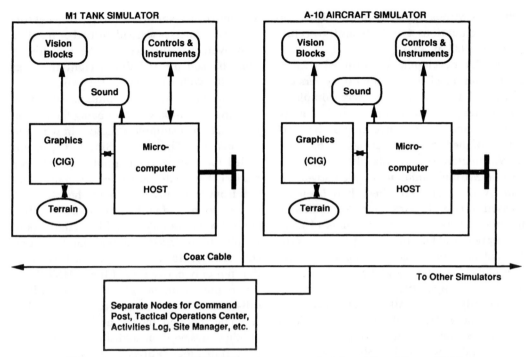

Figure 1. *Architecture of SIMNET local area network. Adapted from Thorpe (1987).*

Because the architecture of the microprocessor-based graphics generator permits any similarly equipped simulator to connect to the net, which is based on a distributed computing architecture, the entire SIMNET system is extremely powerful and robust. New or additional elements can be included simply by plugging them into the network. Once connected to the net, simulators transmit and receive data packets from other simulators or nodes, such as stations for combat support or logistics elements, and compute their visual scenes and other cues, such as the special effects produced by the sound system. Because the data packets need to convey only a relatively small amount of information (position coordinates, orientation, and unique events or changes in status), the communications load on the net is modest, as is the increase in load with the addition of another simulator.

The long-haul network architecture initially employed wideband land lines capable of joining separated LANs or individual simulators into a common net. Plans called for the LHN eventually to use satellite communications capabilities to expand the flexibility of SIMNET participation. Packet-switching protocols, previously developed as part of the ARPANET project, were adapted to provide for transmitting the data needed by the simulators and other SIMNET nodes to compute the actions taking place in their virtual-world battlefields (Thorpe, 1987). Where updating information is slow in coming from another simulator, its state is inferred, computed, and displayed. Then, when a new update is received, the actual-state data are used in the next frame, and any serious discontinuity is masked by the receiving simulator's automatic activation of a transition-smoothing algorithm. Should a simulator fail, the rest of the network simply continues without its contribution. Thus network degradations are "soft and graceful" (Thorpe, 1987, p. 495).

SIMNET development continued according to plan. The first two laboratory versions of the M1 tank simulators were completed and tested, then networked and demonstrated at the annual convention of the Association of the U.S. Army during October 1985 in Washington, D.C. By May 1986 two final-version M1 tank simulators were networked, and by October 1986 these had been increased to two platoons (eight M1 tank simulators)—including representations of combat support and logistics elements such as the tactical operations center, supporting artillery, close air support, and the administrative and logistics operations center—all housed temporarily in Hill Hall at Fort Knox (R. S. Jacobs, March 21, 1990). The test of the first prototypes and the networking of the first eight simulators in full public view at Fort Knox were the final two contractor-identified critical milestones; they showed "everybody that the rest was just a matter of getting the production cranking—and, of course, evaluating the full-scale war-fighting exercises" (G. Weltman, March 2, 1990).

The experiences gained through the platoon trials with armor troops using the available simulators provided valuable information that influenced development of the additional simulators required for the SIMNET goal of battalion-level training. Through design modifications supported by the iterative prototyping approach, the lessons learned from each new SIMNET field trial were incorporated into subsequent simulator units (G. W. Bloedorn, February 20, 1990; M. Cyrus, April 11, 1990; R. S. Jacobs, March 21, 1990; J. A. Thorpe, October 16, 1989). Thus with the flexibility permitted by developmental (as contrasted with production) prototyping, each new addition to the pool of available simulators was essentially

an improved version of its predecessor. Where weaknesses were identified or possible improvements discovered, the developers "just backed up and did it right the next time" (G. Weltman, March 2, 1990).

As development of the simulators was progressing, attention was also being focused on the design of a suitable structure in which to house them. The initial SIMNET concept called for trailers (two simulator units per trailer) that could be hauled by semitrailer rigs from one army post to another, there to be interconnected with one or more so-called control room trailers or buildings that would supply the necessary support facilities and connections. However, a solution was found that not only better served the SIMNET system requirements but also was less costly: a self-contained, preengineered facility that could be unbolted, moved, and reconstructed at other sites, with no additional costs except for suitable foundations (U. Helgesson, April 12, 1990). The structure has single connections for the water, electricity, and gas supplied by the host installation and is otherwise entirely self-contained, easily maintained, and efficient. The move into the SIMNET-T facility at Fort Knox took place in February 1987 (U. Helgesson, April 12, 1990; discussion with J. Owens, Fort Knox, Kentucky, April 12, 1990).

SIMNET development continued to progress rapidly during 1987. For example, in April 1987 a laboratory test of a simulator-satellite-simulator long-haul (64 000 km) network connection was successful, thus giving promise of eventual implementation of a satellite-based LHN as had been planned. With the addition of two attack (A-10) aircraft simulators for close air support in November 1987, the Fort Knox SIMNET-T facility consisted of 54 SIMNET ground vehicle and aircraft simulators, all fully interactive via the local area network. By the end of 1987, five SIMNET air defense system simulators and two scout-attack helicopter simulators had been added, a second structure at Fort Knox (the SIMNET-D facility) had been occupied, "time travel" had been demonstrated (i.e., SIMNET's ability to provide movement back and forward in time, as well as in space, to permit the viewing of any event in any area of the simulated battlefield using the digitized electronic record of a SIMNET engagement), and the first SIMNET battalion-level battle engagement training exercises had been conducted (G. W. Bloedorn, February 20, 1990; U. Helgesson, April 12, 1990; J. A. Thorpe, October 16, 1989).

System Development and Field Testing

As 1987 was drawing to a close, there were general discussions and fears throughout the Department of Defense regarding anticipated cuts in the defense budgets, including those for the joint army-DARPA funding of SIMNET development. Moreover, the SIMNET system, as then constituted, was addressing essentially only the training needs of active army heavy mechanized units. Thorpe realized that its potential was much greater, and he hoped to expand future SIMNET applications to include navy, marine, and air force units, reserve and active, in joint war-fighting training. Believing that he would benefit from advice on how to expand SIMNET applications, he asked General Paul F. Gorman to head a small group of retired senior officers consisting of Admiral S. Robert Foley, Jr., General Robert Dixon, and Major General George E. Coates to assess the potential of the SIMNET system, its then-available products, and the future developments planned for it (J. A. Thorpe, October 16, 1989).

Gorman and the others visited the SIMNET facilities at Fort Knox in December 1987. Their reactions were positive, and they provided advice regarding the directions in which they judged the technology should be

taken and how SIMNET could be expanded to address still more of the important combat-relevant tasks on which collectives needed to train to do well in battle. Gorman and a former army chief of staff (and chairman of the Joint Chiefs of Staff) carried their support of the SIMNET development to the secretary of defense (Gorman, 1987, p. 2).

Funding support for SIMNET development continued from both DARPA and the army, and as development proceeded from early 1988, SIMNET evolved more fully into a warfighting training system. The SIMNET sites were structured to provide the full table of organization and equipment capabilities of all three elements of combat (maneuver, combat support, and logistics), and the army began playing an increasingly active role in planning for the test and transition of SIMNET technology into use.

For example, the army's Combined Arms Center at Fort Leavenworth, Kansas, had formulated the Battle Command Integration Program (BCIP), a comprehensive program for training the use of automation applied to battlefield requirements for command and control. SIMNET and BCIP seemed to fit together nicely; SIMNET technology would serve as the basis for comprehensive military simulations that included command modules, a centralized world-class, opposing force (OPFOR) capability, manned SIMNET-level simulations, and automated workstations for semiautomated forces. The initial sites would be Forts Knox and Leavenworth, but the system would be capable of expansion by replication of capabilities at other sites. When it ultimately matured, SIMNET technology would transition to army use, and follow-on engineering development and procurement would "create a distributed simulation network capable of supporting a 20-corps exercise with an OPFOR of 40 corps-sized units"—a simulation network of roughly 200 divisions, 800 brigades, and 3000 battalions (Gilbert, Robbins, Downes-Martin, and Payne, 1989).

The vision of a matured and expanded SIMNET-supported BCIP depicted by Gilbert and his colleagues (1989) shows some elements (such as the self-contained sites at Forts Knox, Rucker, and Benning) completely in the "SIMNET world" and others partly out of it. Some elements would exist in SIMNET simulators, workstations, and networks, some in real-world tactical environments, and some in academic environments. This longer-term goal challenged the development to design the SIMNET command modules so that they would interface transparently with other, non-SIMNET elements—from academic classrooms, through SIMNET simulators (wherever located), to command posts in the field (Gilbert et al., 1989).

Development proceeded along those lines. by February 1988 the SIMNET facility at Fort Knox had grown to a total of 71 interactively networked simulators on the LAN, and in April 1988 battalion-level, force-on-force operational exercises with Forward Area Air Defense elements were conducted there. During the same month, the SIMNET attack helicopter simulators were updated to the next-generation aircraft, and a SIMNET site was established at Fort Benning, Georgia. During the next month a major milestone was reached with the successful operation of the long-haul net between Fort Knox and BBN Laboratories in Cambridge, Massachusetts. The two U.S. field sites (Forts Benning and Knox) were connected in June 1988, and the prime contractor and developer of the SIMNET-D capability at Fort Knox began to support the Armor Center in an evaluation of proposed improvements to the M1A1 tank (Garvey, 1989). Then, in July 1988, the third U.S. SIMNET site (and fourth total, including the one in West Germany) was activated at Fort Rucker and connected with the others.

During the same period, a semiautomated forces capability was developed to allow one person at a SIMNET command module to orchestrate interactively the battlefield movements of a collective (a unit such as a platoon or company, with all of its assigned vehicles) instead of being limited to control of a single vehicle. Three European-based SIMNET platoon sites became operational during summer and fall 1989 (at Schweinfurt in June, Fulda in August, and Friedberg in November), and work continued on development of a long-haul network connecting all four European sites to the three U.S. sites. Documentation grew from both the development team (e.g., Garvey and Radgowski, 1988; Miller, Pope, and Waters, 1988) and the army's test and evaluation teams (e.g., Black, 1988; Brown, Pishel, and Southard, 1988; Ground and Schwab, 1988; Kraemer and Bessemer, 1987; Pate, Lewis, and Wolf, 1988; Schwab, 1987; Walter, 1987).

SIMNET was completed and adopted into the army starting in October 1989. As of January 1990, the army's project manager for training devices was preparing the necessary documents for procurement of the first several hundred units of the Close Combat Tactical Trainer (CCTT) system, the production follow-on to SIMNET-T. This procurement is to be the first production purchase of several thousand planned units. If completed as planned, the system will provide a CCTT (SIMNET-type) training capability at each army battalion's home base at an estimated total cost of $1.5 billion. The October 1989 acquisition plan includes provisions for air and ground vehicle developmental test beds (AIRNET-D and SIMNET-D) intended to "provide materiel developers the ability to try out their ideas prior to issuing doctrinal changes or the bending of metal" (Lunsford, 1989, p. 14).

The successful development and demonstration of SIMNET's collective training capabilities have stimulated new plans to expand this type of computer-based interactive networking for real-time, person-in-the-loop battle engagement simulation and war-gaming. The SIMNET-D and AIRNET-D capabilities are to be expanded in a new joint army-DARPA initiative to demonstrate advanced distributed simulation technology for use in system development, studies, and analyses. This will provide a developmental capability that can be reconfigured to stimulate new design concepts for evaluation in SIMNET-like, person-in-the-loop battle engagement trials.

A second conference on communications and network standards for simulator interoperability was held in January 1990 at the Institute for Simulation and Training of the University of Central Florida in Orlando; the first conference had been held in August 1989 (Cadiz, Goldiez, and Thompson, 1989). Industry and armed services representatives at the conference agreed to use the SIMNET protocols as a starting point from which to develop further modifications to support the more rapid update rates and additional attributes needed for the net to handle other weapon systems adequately, such as faster-flying fighter aircraft and missiles (Danisas, Glasgow, Goldiez, and McDonald, 1990).

In a related effort, the armed services have agreed to provide continuing funding for the operation of a facility to produce a standardized digital data base of the earth's terrain, including cultural features. This production facility is now scheduled to open under the management of the Defense Mapping Agency in May 1991. The services are united with industry to develop the technology and standards to support a data base and communications net with protocols that will allow simulators of different types and architectures to interoperate, thus permitting person-in-the-loop simulations not only of army combined arms but also of joint services combat operations (Demers, 1989).

This cooperation among the armed services and American industry provides a credible prediction of SIMNET's eventual nature and scope: a future defense simulation net on which any simulator or war game can be connected to interoperate with any other simulator or war game in real time using accurate terrain data bases that include cultural features and up-to-date changes. This would be a valid and realistic tool not only for training but also for the development of concepts and doctrine, the analysis and evaluation of the effectiveness of proposed new weapon systems, and the rehearsal of combat missions before they are undertaken. With addition of a valid capability for asking "what if?" questions, it could even form the basis of an effective battle management aid.

OBSERVATIONS ON SUCCESS: LESSONS LEARNED

Contributions to the successful development of SIMNET can be identified in seven areas; recognized needs, realistic objectives, feasible technologies, a risk-tolerant research and development environment, an iterative approach, reliable user participation, and resourceful people.

Recognized Needs

The pressing need for improvement in collective combat skills training has been recognized by the Defense Science Board. In 1976, the board's Task Force on Training Technology concluded that insufficient attention had been given to collective training in defense technology base research and development. It suggested that increases in team training research and development could lead to substantial improvements in the efficiency and effectiveness of both the training and the performance of military units (Defense Science Board, 1976, pp. xi, 36–37). The DSB 1982 Summer Study on Training and Training

Technology reached essentially similar conclusions.

From its beginning, the SIMNET project sought to address recognized service needs for collective combat skills training. During the 1980s these needs became even greater, especially for the army, as more stringent restrictions were placed on opportunities for conducting large-scale operational exercises, the traditional means of military collective training. In addition, there was growing recognition and concern regarding unfilled needs for more frequent and effective combined arms and joint services training at all levels (Defense Science Board, 1988).

The recognition of these needs, and of SIMNET's potential for addressing them, was one element contributing to its success. Because SIMNET promised to provide the army with an effective capability for collective battle engagement training of heavy mechanized units at an affordable cost, its development was supported in substantial and critical ways by the army. Of the total of about $214 million spent through 1989 on SIMNET-T, AIRNET, and SIMNET-D, DARPA provided about $33 million, or slightly more than 15%, and the army provided the rest (discussion with Col. L. L. Mengel, April 12, 1990).

Realistic Objectives

A second area that contributed to the success of the SIMNET project was its promising goal of developing armored vehicle simulators that could be networked to provide a means for collective combat skills training, with sufficient visual scene capability to simulate the appropriate complexity of battle engagement scenarios—and at relatively low unit costs, to permit the acquisition of a sufficiently large number to affect training (and therefore readiness). The emphasis on affordability led to the objective of low unit cost being stated clearly and forcefully from inception. This in turn established the neces-

sary attitude for success—that is, it placed constraints on the system concept, approach, and design without which the SIMNET development very likely would have failed.

The establishment of realistic objectives, including the low-cost constraint to address the affordability issue, was a second element contributing to SIMNET's success. The objectives were easily understood, and, if attained, the technology promised to provide a realistic way for the army to address its collective training needs, at least in the area of its most difficult and expensive heavy mechanized units. The concept of a network of simulators that permitted entire units to train by experiencing the "fog of war" in realistically simulated interactive battle engagements with "live" opposition forces was both understandable and attractive; it generated army support for SIMNET, even with the understanding that the SIMNET solution would be achieved, not with a perfect simulation of vehicles but, rather, with a simulation that would be affordable and good enough.

Feasible Technologies

The critical computing, imaging, and networking technologies were actually or potentially available for further development and integration into the SIMNET system. Moreover, they were well fitted to the SIMNET goals. The concepts were not new; they had been articulated and discussed since at least the mid-1970s, but the critical enabling technologies were not then available. The SIMNET development team recognized that rapid advances in relevant technologies provided capabilities that would make the development of SIMNET feasible during the 1980s.

The identification of feasible technologies was a third element contributing to the successful development of SIMNET. It is true that once demonstrated, the SIMNET technology appears relatively simple and easy to understand. It appears innovative but not especially revolutionary; after all, there are many computer graphics video games in the marketplace. But there were considerable risks involved. For example, the development of the graphics generator—with sufficient visual field complexity to support SIMNET's battle engagement collective training mission, and to do so within the SIMNET dollar and time cost constraints—represented an extreme in technological risk, essentially equivalent to that of a critical enabling technology.

Risk-Tolerant Research and Development Environment

The R&D environment—including the then-current modus operandi at DARPA, where project managers were expected to know their research areas thoroughly, including all the players (the persons and firms that had the requisite expertise and capabilities)—certainly contributed to the project's successful execution. It was the standard operating procedure for project managers to work collegially with potential or actual contractors in the conduct of the work and, to some extent, in the development of the concept, the identification of the relevant enabling technologies, and the specific and innovative applications to which it could contribute. The typical operating mechanism was that of a development team, and many—if not most or nearly all—of the DARPA contracts at the time were sole source, a situation that ended after 1984 with passage of the Competition in Contracting Act.

This risk-tolerant environment at DARPA in the early to mid-1980s, including its rapid prototyping persuasion and pervasive scientific and management culture, was a fourth element contributing to SIMNET's successful development. In fact, much of the success of the SIMNET project can be attributed to the DARPA culture, which not only permitted support of high-risk projects that showed promise of compensation via large potential

payoffs, such as SIMNET, but also fostered rapid prototyping. The SIMNET project flourished in this environment, for it required not only the acceptance and support of its high-risk development but also a tolerance and supporting infrastructure for the rapid prototyping it needed for continuing army support.

Iterative Approach

Nothing in the design phase of SIMNET's rapid prototyping iterative approach was concretely defined; no detailed engineering specifications were written. Rather, goal-oriented functional specifications were used. These were based on thorough front-end analyses with considerable user impacts on the articulation of what had to be trained (i.e., on the military training requirement). Then the procedure was to (a) build a mock-up as rapidly as possible, expose it to many different people with relevant expertise—army armor personnel (both operational and training), psychologists and engineers, developers and users—and, on the basis of their reactions, comments, and advice, (b) modify the design before building the first prototype, (c) try it, and (d) revise the design for the next unit—in short, to iterate the design on the basis of experience and lessons learned from working with concrete objects, not merely concepts on paper, and to do so rapidly.

The rational iterative-prototyping developmental approach adopted was a fifth element contributing to the success of the SIMNET project. It was consistent with Thorpe's 60% solution and *not-forever* principles—that is, principles that recognized that trying to develop a perfect product would extend the development time substantially, probably fail, and, even if it were successful, leave the military user with very expensive, technologically outdated equipment that would last forever. Instead, the development approach employed in the SIMNET project was to buy, whenever possible, commercially available components that might not meet military specifications constraints or last as long but that could be obtained more quickly and much less expensively. This also ensured that SIMNET would have state-of-the-art, proven, high-technology components and parts that cost less than specially designed and manufactured equipment.

Reliable User Participation

In part because of the prompt demonstrations of concrete products instead of the more usual notional viewgraphs with artist's conceptions of a proposed product, and in part because of the iterative and rapid prototyping approach instead of the typical detailed engineering specifications and 10-year acquisition cycle, SIMNET promised early implementation of products that would address specific army collective training requirements with new (not outdated) technology. Consequently, at each stage of its development there were extensive SIMNET demonstrations to very high levels of the Department of Defense—the Office of the Secretary of Defense and all three military departments. High levels of the army hierarchy were convinced of SIMNET's potentially high payoff in readiness. As a result, the army provided user support from the very top—the secretary of the army as well as the army chief and vice chief of staff. This support resulted in broad and welcomed user participation and cooperation in the development of and total funding for SIMNET-T.

The reliable user support obtained was a sixth element contributing to SIMNET's successful development. SIMNET could not have been rapidly prototyped and transitioned without the kind of top-down user support it enjoyed. If it had obtained only the bottom-up advocacy typical of the Planned Program Budget System used by the armed services' laboratories and acquisition agencies,

SIMNET might never have been approved as a DARPA project.

Resourceful People

The SIMNET development was blessed with properly qualified people throughout, including, but not limited to, the members of the development team and others representing DARPA, the army, and various elements of the Department of Defense. The people were committed and dedicated to the successful development and implementation of SIMNET. They were resourceful and able to handle each of the many crises. They believed in the goal of an affordable, effective network of simulators for the collective battle engagement training of U.S. forces. They were confident that they could do it and committed to the task.

The resourceful people engaged in the SIMNET project constitute a seventh (and most important) element contributing to its successful execution. Of these, the predominant essential ingredient was the contribution of the SIMNET project manager, Jack Thorpe. Without his concept, commitment, determination, technical expertise, willingness to accept risk, and steadfast pursuit of the means to bring the SIMNET concept to concrete reality and operational test, the SIMNET project would have faltered many times and perhaps even failed. This assessment is a conclusion based on the independent statements of many persons (e.g., G. W. Bloedorn, J. D. Fletcher, A. Freedy, R. S. Jacobs, J. M. Levine, E. L. Martin). It is yet another instance of a lesson learned many times over: good people are the most important of the necessary ingredients to success in research and development.

IMPLICATIONS AND CHALLENGES FOR HUMAN FACTORS

SIMNET is among the largest and most successful developments in training technology since the RAND Corporation's and Systems Development Corporation's development and operation of the SAGE system in the 1950s. In potential impact, assessed in terms of the potential numbers of individual trainees and military units affected, it ranks along with the development of the National Training Center at Fort Irwin, California. The lessons learned are applicable even to the pursuit of research issues in academic settings and appear to be mandatory for training technology developments in applied settings. To be successful, one should

- address recognized, real, and substantial needs
- with realistic objectives
- using feasible enabling technologies
- applied in iterative, rapid prototyping, innovative approaches
- that make frequent use of concrete demonstrations
- with customer participation and high-level customer support
- in a risk-tolerant research and development environment
- with competent people
- organized into a development team with appropriate leadership.

Lacking any one of these ingredients raises the risk of failure.

One might ask, "Is it worth the trouble?" The answer is doubtless idiosyncratic. Each researcher or developer will have to decide for himself or herself. For the sponsor the question is basically the same, but phrased differently: "If the project's proposer is successful, what difference will it make in my operations, and what will that difference be worth?" A definitive answer to those questions at the beginning of a proposed development requires a front-end analysis—a type of analysis that the human factors and training technology communities are not particularly well trained to do. To learn how, and to develop new analytic techniques appropriate to the issue, is not a trivial challenge for the field.

That collective training is important to military readiness is embedded in military concepts and doctrine and based on observations "that small but well-trained and intelligently led forces can defeat larger but poorly trained and led forces" (Hart and Sulzen, 1988, p. 281; also see Dupuy, 1984, pp. 326–336). In their analysis of the data of 237 platoon-level, light-infantry simulated battles conducted over a period of 10 years as part of 11 separate field exercises, Hart and Sulzen found that the relative odds of winning an offense was "increased 30 to 1 by belonging to a high as opposed to a low training-ratio population" (p. 279).

That SIMNET can be employed to provide the same kind of effective crew, group, team, and unit (CGTU) battle engagement training is supported by at least one early datum: A platoon of four M1 tank simulators (with Range-301 data base) had been installed in March 1987 at Grafenwuhr, a site in West Germany. The army used it that spring for training, and in June 1987 a SIMNET-trained U.S. Army platoon won the competition for the Canadian Army Trophy—the first time a U.S. team had won the trophy (Kraemer and Bessemer, 1987).

In addition, Deitchman (1988) has used a warfare simulation model in an analytic study to examine the military value of training in terms of what would happen to the outcome of a war if unit training could lead to various assumed increases in the capability of armored ground forces and tactical air attack forces. He found that a factor-of-two improvement in certain parameters considered relevant to training could reverse the course of a war in central Europe, as portrayed by the model. In a second study that employed a sample of available data from various sources on platoon-level tank warfare and squadron-level tactical bombing, Deitchman (1990) found that, in the cases he explored, "either training or equipment improvement for specific military tasks improve force effectiveness by roughly comparable magnitudes. Depending on the cost elements included, training or equipment improvement may be either comparable in cost or else training tends to cost less—sometimes considerably less" (p. S-3). Establishment of the military value of training for a proposed development by suitable front-end analytic techniques remains a challenge, but one that could be met with use of person-in-the-loop simulations such as those available with SIMNET capabilities and appropriate analytic warfare models.

Another challenge deals with evaluation of the training effectiveness of large-scale training systems such as SIMNET, and of SIMNET-T itself. As indicated earlier, the army plans to conduct a large-scale field test of SIMNET's effectiveness in training many battalion-level units in different regiments and divisions at distributed locations. The army, like the other services, is well versed in conducting competent tests and evaluations of weapon systems and military equipment. For SIMNET to be viewed by the army as successful—that is, as an effective training system that is worth the expenditure of funds for additional procurement—the army will have to be convinced that it will make a difference in readiness. If the test results indicate that SIMNET does what it purports to do—that military units trained with its use are able to fight wars like combat veterans—the army will doubtless fulfill its plan to extend the validated SIMNET training capability army-wide.

The challenge for the human factors and training technology communities is to adapt the disciplines' traditional scoring and analytic techniques to the army's real training procedures. At the start the two appear to be incompatible; recall that the behaviorally driven training design approach adopted in the SIMNET development excluded instructors, instructor operator stations, and third

parties scoring the soldiers' performances in favor of having the unit's chain of command control its own training.

However, each battle engagement simulation in SIMNET produces a complete record of all the events that occurred with each participant in each vehicle in each unit involved in the simulation. That record constitutes a mountain of information that can be scored, analyzed, and interpreted, even in terms of classical learning curves. Input, process, and output variables abound, but the disciplines are not yet at ease in dealing with collectives or in transforming the discrete data of individual events, persons, and vehicles into meaningful, valid, and important collective measures and variables.

This challenge transcends the specific case of SIMNET's training effectiveness. It calls for development of a science of collective or CGTU performance measurement and evaluation. Progress in this direction is being made, especially in analysis of data from the army's National Training Center at the Monterey Field Unit of the Army Research Institute, and in the establishment of a comprehensive program of research and exploratory development on team training and performance at the Naval Training Systems Center (Glickman et al., 1987; Guerette et al., 1987; Morgan and Salas, 1988; Swezey and Salas, 1989, in press).

SIMNET is essentially an early practical application of virtual world technology. It has been initially directed at training issues. However, it is also applicable—perhaps with even greater potential value—to acquisition, tactical, and operational issues, all of which raise additional challenges for the human factors, training technology, and human performance communities. For example, SIMNET-D is configured especially to permit, merely through changes in software, the introduction of possible or conceivable weapon systems for test and evaluation with person-in-the-loop battle engagement simulations, even before the design, or construction of "breadboard" or "brassboard" prototypes. Thus with use of this kind of SIMNET technology, data regarding the potential military value of a proposed weapon system could be obtained and assessed prior to costly acquisition decisions.

Similarly, the capabilities of SIMNET technology can be used to propose, develop, test, and evaluate tactical concepts and doctrine with person-in-the-loop battle engagement simulations against any hypothesized enemy force, on any type of terrain, at any simulated site on earth, before any potentially battle-costly doctrinal decisions are institutionalized. Furthermore, with the addition of real-time updating of the terrain and cultural features, SIMNET technology could provide a powerful aid to situational awareness for combat pilots and ground troops. And if these capabilities were available to front-line units in actual war-fighting situations, they could be used with representations of the enemy forces, with their actual deployments and capabilities, not only for mission rehearsals but also to develop, test, evaluate, and practice the tactics to be employed in actual battle engagements. Were SIMNET technology then tied with networking to higher-level wargaming models, it could become a powerful and effective battle management aid.

The possible beneficial applications of virtual world technology, only partly demonstrated in SIMNET, are astounding. That they provide considerable challenges for the human factors and training technology communities is evident. The challenges span the range from definitions, taxonomies, and fundamentals of measurement to the formulation, empirical test, and validation of theories of collective human behaviors—that is, the behaviors in and of groups. A question that remains to be answered is, "Can the human factors and training technology communities

and related disciplines adapt to, cope with, and conquer the orders-of-magnitude increases in complexity over their currently typical studies?" Let us hope that they can.

REFERENCES

Balwin, D. M. (1981). Area of interest—Instantaneous field of view vision model. In E. G. Monroe (Ed.), *Proceedings of the 1981 IMAGE Generation/Display Conference II* (pp. 481–496). Williams Air Force Base, AZ: Air Force Human Resources Laboratory.

Black, B. A. (1988, August). *Review of activities supporting DCD block II trials* (Memo. Report PERI-IK: 70-1r). Fort Knox, KY: U.S. Army Research Institute for the Behavioral and Social Sciences, Fort Knox Field Unit.

Bloedorn, G. W. (1983a, March). *Large-scale simulation data package: Vol. I. M1 Abrams tank*. Woodland Hills, CA: Perceptronics.

Bloedorn, G. W. (1983b, August). *Large-scale simulation data package: Vol. II. M2 and M3 fighting vehicle*. Woodland Hills, CA: Perceptronics.

Bloedorn, G. W., Kaplan, R., and Jacobs, R. S. (1983, March). *Large-scale simulation data package*. Woodland Hills, CA: Perceptronics.

Breglia, D. R., Spooner, A. M., and Lobb, D. (1981). Helmet-mounted laser projector. In E. G. Monroe (Ed.), *Proceedings of the 1981 IMAGE Generation/Display Conference II* (pp. 241–258). Williams Air Force Base, AZ: Air Force Human Resources Laboratory.

Brown, R. E., Pishel, R. G., and Southard, L. D. (1988, April). *Simulator networking—Preliminary training developments study*. Fort Monroe, VA: TRADOC Analysis Command.

Cadiz, J., Goldiez, B., and Thompson, J. (Eds.). (1989, August). *Summary report: The First Conference on Standards for the Interoperability of Defense Simulations*. Orlando, FL: University of Central Florida, Institute for Simulation and Training.

Danisas, K., Glasgow, B., Goldiez, B., and McDonald, B. (Eds.). (1990, January). *Summary report: The Second Conference on Standards for the Interoperability of Defense Simulations* (Vols. I–III). Orlando, FL: University of Central Florida, Institute for Simulation and Training.

Defense Science Board. (1976, February). *Summary report of the DSB Task Force on Training Technology*. Washington, DC: Author.

Defense Science Board. (1988, May). *Report of the DSB Task Force on Computer Applications to Training and Wargaming*. Washington, DC: Author.

Deitchman, S. J. (1988, March). *Preliminary exploration of the use of a warfare simulation model to examine the military value of training* (IDA Paper P-2094). Alexandria, VA: Institute for Defense Analyses.

Deitchman, S. J. (1990, January). *Further explorations in estimating the military value of training* (IDA Paper P-2317). Alexandria, VA: Institute for Defense Analyses.

Demers, W. A. (1989). All together now. *Military Forum, 6* (November/December), 38–43.

Dupuy, T. N. (1984). *The evolution of weapons and warfare*. Fairfax, VA: Hero Books.

Ferguson, R. L. (1984). AVTS: A high-fidelity visual system. In E. G. Monroe (Ed.), *Proceedings of the 1984 IMAGE Conference III* (pp. 475–485). Williams Air

Force Base, AZ: Air Force Human Resources Laboratory.

Garvey, R. E., Jr. (1989, November). SIMNET-D: Extending simulation boundaries. *National Defense*, pp. 40–43.

Garvey, R. E., Jr., and Radgowski, T. (1988). Data collection and analysis: The keys for interactive training for combat readiness. In *Proceedings of the 10th Interservice/Industry Training Systems Conference* (pp. 572–576). Washington, DC: National Security Industrial Association.

Gilbert, A. L., Robbins, J., Downes-Martin, S., and Payne, W. (1989, February). *Aggregation issues for command modules in SIMNET*. Paper presented at the MORIMOC-II Workshop on Human Behavior and Performance as Essential Ingredients in Realistic Modeling of Combat, Alexandria, VA.

Glickman, A. S., Zimmer, S., Montero, R. C., et al. (1987, November). *The evolution of teamwork skills: An empirical assessment with implications for training* (Tech. Report 87-016). Orlando, FL: Naval Training Systems Center.

Gorman, P. F. (1987, December). *Battalion task force training with SIMNET* (Memorandum report for J. Thorpe, DARPA, December 15, 1987). (Available from the author)

Ground, D., and Schwab, J. R. (1988, March). *Concept evaluation program of simulation networking (SIMNET)* (Tech. Report 86-CEP345). Fort Knox, KY: Armor and Engineering Board.

Guerette, P. J., Miller, D. L., Glickman, A. S., et al. (1987, November). *Instructional processes and strategies in team training* (Tech. Report 87-017). Orlando, FL: Naval Training Systems Center.

Hart, R. J., and Sulzen, R. H. (1988). Comparing success rates in simulated combat: Intelligent tactics vs. force. *Armed Forces and Society, 14*, 273–286.

Kraemer, R. E., and Bessemer, D. W. (1987, October). *U.S. tank platoon training for the 1987 Canadian Army Trophy (CAT) competition using a simulation networking (SIMNET) system* (Research Report 1457). Fort Knox, KY: U.S. Army Research Institute for the Behavioral and Social Sciences, Fort Knox Field Unit.

Licklider, J. C. R. (1960). Man-computer symbiosis. *IRE Transactions on Human Factors in Electronics, 1*, 4–11.

Licklider, J. C. R. (1988). The early years: Founding IPTO. In T. C. Bartee (Ed.), *Expert systems and artificial intelligence* (pp. 219–227). Indianapolis: Howard Sams.

Lintern, G., Wightman, D. C., and Westra, D. P. (1984). An overview of the research program at the Visual Technology Research Simulator. In E. G. Monroe (Ed.), *Proceedings of the 1984 IMAGE Conference III* (pp. 205–221). Williams Air Force Base, AZ: Air Force Human Resources Laboratory.

Lunsford, R. J., Jr. (1989, October). *U.S. Army training systems forecast, FY 1990–1994*. Orlando, FL: Project Manager for Training Devices, U.S. Army Materiel Command.

Miller, D. C., Pope, A. R., and Waters, R. M. (1988). Long-haul networking of simulators. In *Proceedings of the 10th Interservice/Industry Training Systems Conference* (pp. 577–582). Washington, DC: National Security Industrial Association.

Morgan, B. B., Jr., and Salas, E. (1988). A research agenda for team training and performance: Issues, alternatives, and solutions. In *Proceedings of the 10th Interservice/Industry Training Systems conference* (pp. 550–565). Washington, DC: National Security Industrial Association.

Owens, A. J., and Stalder, R. F., Jr. (1982, May). *The adaptive maneuvering logic in tank warfare simulation* (Final Report DSI-82-413-F). San Diego, CA: Decision Science.

Pate, D. W., Lewis, B. D., and Wolf, G. F. (1988, June). *Innovative test of the simulator network (SIMNET) system*. Fort Bliss, TX: Air Defense Artillery Board.

Schwab, J. R. (1987, May). *Innovative test of simulation networking—Developmental (SIMNET-D)*. Fort Knox, KY: Armor and Engineering Board.

Swezey, R. W., and Salas, E. (1989). Development of instructional design guidelines for individual and team training systems. In *Proceedings of the 11th Interservice/Industry Training Systems Conference* (pp. 422–426). Washington, DC: National Security Industrial Association.

Swezey, R. W., and Salas, E. (in press). *Teams: Their training and performance*. Norwood, NJ: Ablex.

Thorpe, J. A. (1978, September 15). *Future views: Aircrew training 1980–2000*. Unpublished concept paper at the Air Force Office of Scientific Research. (Available from the author)

Thorpe, J. A. (1987). The new technology of large-scale simulator networking: Implications for mastering the art of warfighting. In *Proceedings of the 9th Interservice/Industry Training Systems Conference* (pp. 492–501). Washington, DC: American Defense Preparedness Association.

Walter, J. L. (1987, May). *Post-NTSC SIMNET training evaluation* (Memo.). Washington, DC: Department of the Army.

Simulator Limitations and Their Effects on Decision-Making

Colin F. Mackenzie, Ben D. Harper, and **Yan Xiao,** colin@anesthlab.ab.umd.edu, bharper@wam.umd.edu, yxiao@umabnet.ab.umd.edu, University of Maryland School of Medicine, Baltimore, Maryland

The use of simulators for skill evaluation is common in a number of fields, but the effect that various characteristics of simulators have on subject performance is not fully understood. Ten experienced trauma anesthesiologists participated in a full mission simulation of a real video taped trauma case management. This case contained critical events that require the participants to use clinical examinations and familiar and unfamiliar equipment contained in the simulator. Participants' performance differed from both the performance in the real case and ideal performance. The unfamiliarity with one complex piece of the equipment (ventilator function and controls) caused 4 out of 10 participants to be unable to finish the simulation. Surprisingly, unfamiliarity with another piece of equipment (new to all participants) did not have any dramatic impact. Anesthesia simulators have been successfully used in training anesthesiologists, but the demands of evaluation of clinical skills differ from training. Complex equipment and the limitation in realism of the mannequins in the anesthesia simulators prevent fair assessment of clinical diagnostic skills.

Full-mission simulators are used in training and research in aviation (Foushee & Helmreich, 1988), medicine (Howard, Gaba, Fish, et al., 1992) and other industries (Plott, Wachtel & Laughery , 1988). In aviation, simulators are an important part of the re-certification process. With the advances of simulator technologies we can only expect increasing usage of simulators, particularly in the sensitive area of certification and re-certification. However, there is little information available that critically evaluates factors that influence performance in the simulator that are a result of the simulator rather than the individual. For example, the implementation and design of a simulation may differ from reality in crucial but deceptively subtle ways, changing the ways in which decisions and diagnoses are made.

To examine the effect of familiarity with simulator equipment and realism of some aspects of the simulation on decision making when basic domain knowledge is similar, we constructed a high-fidelity full-mission medical simulation of real videotaped critical events.

The objectives of the study were 1) to determine the effects of familiarity with simulator equipment on the performance of participants, and 2) to identify how and why simulator participants' responses differed from those recorded on videotape from real life. In this paper we will concentrate specially on the changes in the decision making process when simulators provide clinical cues that are unfamiliar to simulation participants.

METHODS

The real videotaped critical events were recorded as part of a larger study examining real-life decision-making under stress and were previously extensively analyzed and reported by Mackenzie, Craig, Parr, Horst & LOTAS (1994). Figure 1 shows the flow of the critical events along with the ideal action sequence.

The videotape showed management of an unconscious patient after admission to a trauma center. During airway management two critical events occurred: insertion of the

Reprinted from Proceedings of the Human Factors and Ergonomics Society 40th Annual Meeting (pp. 747–751).

breathing tube too far into the airway so that only one lung is ventilated (endobronchial intubation) and a failure of mechanical ventilation of the patient caused by a faulty ventilator setting. Two full-mission anesthesia simulators (CAE Electronics Inc. and LORAL) were used to re-create the original case. The differences between these two simulators were transparent to the participants and were in computer models and other devices re-creating the pharmacological and physiological responses to interventions.

Six instructor/actors in the simulator were trained using a script of the real communications transcribed from the original videotape. A training session one month before the simulation discussed re-creation of the event and carried out practice on the CAE simulator. On the night before the simulation all six instructors/actors performed each scenario several times on both the CAE and LORAL simulators.

Participants voluntarily signing up to manage the critical incident were randomly assigned to the CAE or LORAL simulator. They were trau-

ma anesthesiologists familiar with management of unconscious trauma patients. However, four of the ten participants were unfamiliar with the mechanical ventilator used in the simulation and all the participants were unfamiliar with a particular manufacturer's end-tidal CO_2 ($ETCO_2$) monitor used in the simulator. They all, however, knew the technique of using $ETCO_2$ monitoring to detect ventilation and correct placement of the breathing tube in the airway. In addition all participants were unfamiliar with the two simulators.

Participants were given material describing the simulators and their functions. Then together, all participants received a verbal explanation of what was planned. They were given a written case summary providing a history and physical exam of the real patient. The real videotape of patient management was next shown and the supervisor whose role they would play was identified. The videotape was stopped before the critical events shown in Figure 1 occurred, but the case continued in the simulator with each participant individually managing the case scenario. The sound track of the real video was played as background throughout the continued simulation.

Each simulator was assigned three instructor/actors. After the simulation, participants and instructor/actors completed a questionnaire identifying their views on usefulness of simulator in training, research and evaluation of competence. Two hours later, when all participants had completed the simulation the participants were fully debriefed. The entire proceedings including introduction, all simulations and the debriefing were videotaped.

RESULTS

There were ten participants, five performed on each of the two simulators. The various paths taken during the simulated and real event are shown schematically in Figure 2.

An extensive questionnaire completed by participants and instructors gauged the fidelity of the simulator, the ability of the actors, the performance of the participant, and the usefulness of the process, measuring the usefulness of the simulators in training and research. The lowest score on the questionnaire by partici-

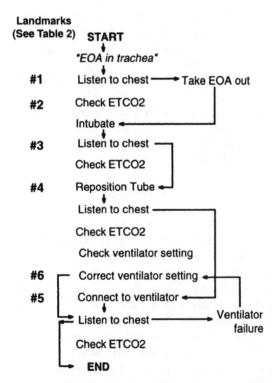

Figure 1. Ideal action sequence is listed in vertical order, whereas real actions are the path shown by arrows. Interesting deviation points from the ideal action sequence are marked by landmarks 1–6.

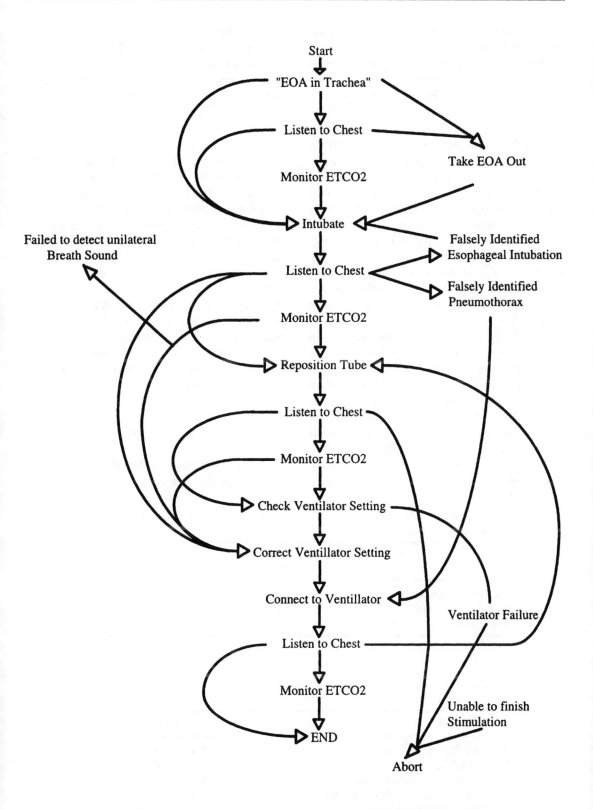

Figure 2. Various paths followed by simulator participants as compared to the path taken in the real event, and the ideal activity sequence.

pants was its usefulness in evaluating competence (Table 1).

The evaluation of the participants' performance on the simulator was carried out at six major landmarks (see Figure 1) in case management (Table 2).

Standard operating procedure in management of the esophageal obturator airway (EOA) recommends leaving it in position until the airway is secured with a conventional cuffed tracheal tube (ETT). Seven participants removed the EOA without first inserting the ETT. Four participants made an incorrect diagnosis that the ETT was in the esophagus (Figure 3) after observing the movements of the upper abdomen and examination of the chest. Only two participants used the available $ETCO_2$ analyzer to distinguish esophageal from tracheal placement of the ETT.

Nine of ten participants eventually recognized endobronchial intubation from the unilateral movements of the chest and absence of breath sounds. However, one participant confused this with pneumothorax which has similar clinical signs, with the main clinical differentiation being that percussion of the chest is hyperresonant in pneumothorax and dull in endobronchial intubation. Such differentiation is not realistically made in the simulator man-

nequin (although it is the state of the art). The other participants corrected the endobronchial intubation and one used the tube insertion depth marked on the ETT to confirm correct depth placement during manual ventilation with a resuscitator bag.

Four participants who were unfamiliar with the mechanical ventilator were unable to find the on/off switch (located at the back and to one side of the ventilator). They continued manual ventilation and called for help. Of the six participants familiar with the ventilator, one identified the incorrect control setting and rectified the problem before connecting the patient mannequin. A second participant connected the ventilator and patient mannequin, identified the failure to ventilate and corrected the problem without disconnecting the ventilator. The remaining four participants disconnected the non-functioning ventilator and reverted back to manual ventilation with the resuscitator bag while they examined the ventilator control settings and rectified the ventilator problem.

DISCUSSION

Our study indicates that clinical cues (as opposed to monitor cues) used by clinicians

TABLE 1: Aggregate responses of participants and instructors to questionnaire evaluation

	Mean for Participants $n = 6$	Mean for Instructors $n = 10$
Useful for Training	4.0	4.3
Useful for Research	4.2	4.5
Useful to evaluate competence	3.6	3.9

Note: Score 5 = Ideal and 1 = no use

TABLE 2: Evaluation parameters, and frequency of completion of simulator tasks among ten participants

	Landmark	Simulator Result
1.	Take EOA out/Leave in	7 out / 3 in
2.	Use $ETCO_2$ analysis	2 used / 8 no
3.	Recognize endobronchial intubation	4–9
4.	Correct endobronchial intubation	5–9
5.	Detect failure of ventilator	6
6.	Correct failure of ventilator	6

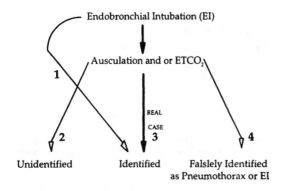

Figure 3. Different diagnoses made by simulator participants after endobronchial intubation (EI). The real case followed the dashed line.

change the decision making process. These cues are subtle and the provision of them is limited by the limitations in mannequin's technology. False positive information obtained from examination of the mannequin (e.g. movements of upper abdomen and examination of chest) can change performance dramatically compared to the real case management. Twice as many participants failed initially to diagnose endobronchial intubation as were successful. In the real event cues obtained from clinical examination lead immediately to accurate diagnosis. In the mannequin, movement of the upper abdomen made the diagnosis ambiguous. The results also suggest that clinicians use physical examination (inspection and auscultation particularly) but rarely confirm findings with technology. $ETCO_2$ monitoring provides clear cut data in ambiguous circumstances of EOA placement, endobronchial intubation and ventilator malfunction but was not often utilized in the simulator, or the real environment, because there was no connector for CO_2 sampling in the manual ventilation circuit. A procedural change in equipment preparation (placing the CO_2 connector in the set-up before cases begin) has eliminated this barrier to $ETCO_2$ monitoring in the real environment.

The approach of using simulation of real videotaped cases in which performance has been evaluated in detail has several advantages over the use of hypothetical or prototypical cases, including: comparison of simulator participants' performance and real events, re-creation of situational awareness and precise specification of evaluation criteria using the real video-

taped event, training of instructor/actors using the real communication cues so that variations in the presentation of the simulation to the participants are minimized, and selection of vignettes that allow very brief simulations with defined goals.

Unlike airline pilots, anesthesia care providers are not certified using specific equipment, rather they are expected to perform adequately with a variety of different possible equipment components. These components vary from hospital to hospital and frequently vary between anesthesia locations within the same hospital. The results of our study show the importance of simulator components when judging performance of management of real critical events in the simulator.

Traditionally medical evaluations occur during clinical practice in the real environment, or by use of hypothetical cases given using either written tests or by oral examinations. High-fidelity, full-mission simulators are potentially a better means of evaluating skills than other tests that do not assess a physician in the clinical environment. Certification of clinical competence and continued demonstration of qualification to practice medicine could be determined from such realistic simulations. However, evaluation of clinical competence has different requirements of a simulator and the simulated case compared to training. For evaluation of competence, simulator design should be optimized to allow assessment of the participants' clinical skills. Non-essential differences in generic simulator equipment should not bias candidates' performance. For training, however, some features of equipment must be exactly duplicated in the simulator. The participants rated these simulators to be useful for training and research but gave their lowest rating for evaluation of competence suggesting that they too had reservations about the adequacy of clinical skills assessment in the simulator. The limiting factor is the mannequin realism which provides degraded clinical cues, unlike real patients. We believe this was a major cause of the misdiagnoses made by the experienced trauma anesthesia providers.

The results of the current study on simulators show that evaluation of participants performance: a) is altered by unfamiliarity with complex equipment; b) is biased by mannequin realism which limits the availability of subtle cues; and c) allows only certain aspects of the clinician's skills (e.g. task prioritization and therapeutic interventions, but not physical diagnoses) to be tested in a simulator.

ACKNOWLEDGMENTS

Supported by the Office of Naval Research Grant #N- 000114-91-J-1540 and the National Aeronautics and Space Administration Grant #NCC 2-921. The opinions expressed in this paper are those of the author and do not reflect the opinion or policy of the US Navy or NASA.

We acknowledge with thanks the assistance of CAE Technologies Inc., J. Doyle, T. Votapka, C. Green MD, Loral Inc., B. Reuger, M. Good MD, K. Abrams MD, C. Reuger MD, ITACCS, the LOTAS Group, the simulator actors and participants. Additional support: D. Ovelgone, R. Kahn and D. Emge.

REFERENCES

Howard, S.K., Gaba, D.M., Fish, K.J. et al.(1992). Anesthesiology Crisis Resource Management Training: Teaching anesthesiologists to handle critical incidents. *Aviation Space Environmental Medicines*, 63; pp. 763.

Foushee, H.C. and Helmreich, R.L. (1999). Group interaction and flight crew performance. pp. 189 *in Weiner. R.L. and Nagel. D.C.(eds) Human Factors in Aviation.* Academic Press, San Diego.

Plott, C., Wachtel, J. and Laughery, K.R. (1988). Operational Assessment of Simulator Fidelity in the Nuclear Industry. *Proceedings of the Human Factors Society 32nd Annual Meeting,* Santa Monica, CA. pp. 705–709.

Mackenzie, C.F., Craig, G.R., Parr, M.I., Horst, R. and the LOTAS Group. (1994). Video Analysis of Two Emergency Tracheal Intubations Identifies Flawed Decision-making. *Anesthesiolgy.* 81; pp. 763–771.

Part 4: Application Areas

A variety of factors have accelerated the pace of change in the workplace since 1970. New management strategies, numerous corporate mergers, the end of the Cold War, a dazzling array of new technologies, changes in demographics that keep workers in the workforce longer, and the globalization of the world's economies are just a few of these factors. Many young people entering the workplace today can expect to change careers at least once and perhaps more during their careers. Training has become an essential area of focus for both organizations and individuals.

New technologies have also had a substantial effect on training. For example, there are now better training devices and simulation, a variety of new computer-assisted instruction tools, embedded training (which puts training capability on the actual workplace equipment), Web-based learning tools, and multiway video teleconferencing. Unfortunately – but predictably – as is often the case with new technologies, understanding of how best to use the technologies has not kept pace with their development. As we noted in the Introduction, research on education and training worldwide is funded at a minuscule level relative to the total education and training budget worldwide. This part in particular serves to demonstrate the level of quality of research in some of the key application areas: flight training, maintenance training, and cognitive skill training. By selecting these topics, we do not mean to imply that the other training application domains are not important or that high-quality research has not been done. Rather, these three areas have been a topic of consistent interest to human factors training researchers since the 1950s.

Flight Training

Part 3 contained a number of articles on training devices and simulators. Although these training tools are used in a variety of training applications, probably their greatest contribution has been to flight training. Perhaps this is because training devices and simulators offer

an effective training environment for a very dangerous and unforgiving training domain: flight. Also, the cost of flight is so expensive that these devices help to make flight training affordable. It is because of reasons like these that this subsection and the two that follow concern the use of training devices and simulators as components of quality training programs for flight.

Despite the many advantages that training devices and simulators offer the flight training community, we have experienced many instances in which they were not used as effectively as they might have been. Too often the feeling on the part of the simulator instructors seems to be, "As long as the simulator can replicate well the flight environment cues that a pilot will experience in the actual aircraft, we're satisfied that the trainees are getting the right kind of training." This attitude has not only led to ineffective and inefficient flight training systems, but it can lead to dangerous training systems. We have found it imperative that the simulator be viewed as only one part of a quality training system. Caro's article (1973) details some critical points that should be heeded if a simulator is to function as an effective part of a quality flight training system.

Caro followed his classic 1973 article with another (1979), which examined a controversial topic in flight training that is simulator-based: What contribution, if any, does motion cueing make to an effective simulator-based training program? Despite the widespread use of these programs in both military and civilian aviation training, the empirical research is somewhat equivocal about their effectiveness. Caro presents the evidence up to 1979 and provides some cogent recommendations about the use of motion. (Research performed on this topic since Caro's article has done very little to clear up the controversy he describes.)

James Kleiss' article (1995) addresses a crucial part of flight simulation: out-of-the cockpit visual imagery. As described in Part 3, a long-standing question regarding simulator-based training is, "How much fidelity is enough for learning to occur?" This question is especially important when applied to the visual cue issue, because pilots gain a great deal from out-of-the cockpit visual cues. Also, visual systems can account for a large portion of a flight simulator's total cost. Kleiss presents important information about the visual cues required when learning to fly at low levels.

Maintenance Training

Aircraft, trucks, factory production equipment, computers, power plants, and myriad other workplace equipment have become more reliable than in years past. They need regularly scheduled maintenance less often, but even regularly scheduled maintenance now requires more complex maintenance skills than in the past. When a piece of modern equipment does malfunction, the diagnostic and repair skills needed to fix it require significant amounts of training. Maintenance training, even in a more reliable world, has become an essential element of keeping the high-tech workplace running smoothly.

This part contains two important articles about maintenance training. Johnson and Rouse (1982) examine the effectiveness of maintenance simulation training against that of more traditional instructional methods and concluded that simulation training for maintenance can be as effective as more traditional instructional approaches. Allen, Hays, and Buffardi (1986) also examine the theme of simulator-based maintenance training, particularly the issues of physical fidelity (does the simulation look like the actual equipment?) and functional fidelity (does the simulation act like the actual equipment?). Their findings about the utility of these two types of fidelity have important implications not only for those who train maintenance troubleshooting skills but for anyone who uses simulation to train analytical skills.

Cognitive Skill Training

As we examined training research articles since 1970, we noticed a significant trend related to the training of higher-order, cognitive skills. The number of articles in these important areas increased significantly over the years. Whereas earlier articles concentrated mostly on the training of procedural skills, from about

1985 on, the number of articles focusing on cognitive skill training increased dramatically. We believe there are at least two reasons for this trend. First, because of the rapid increase in automated systems in the workplace, the role of many workers has been shifting from procedural tasks to tasks that require more complex analytical and decision-making skills. Second, understanding of the cognitive processes of workers, and tools for studying and analyzing those processes, have improved in quality and increased in quantity. We expect this trend to continue as the workplace demands more executive oversight of highly automated systems.

The first paper in this section (Schneider, 1985) examines the nature of expertise and explores the problems that occur when training methods for lower-performance skills are used to train these high-performance skills. The guidelines he provides for training such skills are well founded.

In his 1988 paper, Logan explores the nature of memory and automaticity. The two sides of the automaticity controversy he examined are cognitive processing resources and retrieval of information from memory. Logan's examination and guidelines are still important for training developers and deliverers today.

The need for workers to process information, make decisions, and perform actions in dual-task settings has grown as the complexity of the workplace has grown. Schneider and Detweiler (1988) investigate the theoretical constructs of dual-task performance and present a thoughtful architecture for both explaining and predicting dual-task performance.

Entin and Serfaty (1999) hypothesize that successful teams have better adaptive strategies than their less successful counterparts. They report on their development of a team training procedure based on concepts of shared mental models of team members and the ability of good teams to implicitly coordinate their activities instead of using a more explicit approach. The paper describes the success of their team training procedure and presents implications for training teams in an adaptive fashion. Many workplaces depend on quality team performance, and Entin and Serfaty's article should be carefully considered by team trainers.

Aircraft Simulators and Pilot Training

PAUL W. CARO, *Human Resources Research Organization, Fort Rucker, Alabama*

Flight simulators are built as realistically as possible, presumably to enhance their training value. Yet, their training value is determined by the way they are used. Traditionally, simulators have been less important for training than have aircraft, but they are currently emerging as primary pilot training vehicles. This new emphasis is an outgrowth of systems engineering of flight training programs, and a characteristic of the resultant training is the employment of techniques developed through applied research in a variety of training settings. These techniques include functional context training, minimizing over-training, effective utilization of personnel, use of incentive awards, peer training, and objective performance measurement. Programs employing these and other techniques, with training equipment ranging from highly-realistic simulators to reduced-scale paper mockups, have resulted in impressive transfer of training. The conclusion is drawn that a proper training program is essential to realizing the potential training value of a device, regardless of its realism.

INTRODUCTION

I would not consider the money being spent on flight simulators as staggering if we knew much about their training value, which we do not. We build flight simulators as realistic as possible, which is consistent with the identical elements theory of transfer of Thorndike, but the approach is also a cover-up for our ignorance about transfer because in our doubts we have made costly devices as realistic as we can in the hopes of gaining as much transfer as we can. In these affluent times, the users have been willing to pay the price, but the result has been an avoidance of the more challenging questions of how the transfer might be accomplished in other ways, or whether all that complexity is really necessary (Adams, 1972, pp. 616-617).

Personnel responsible for the design of flight simulators are almost exclusively engineers. Sometimes they are assisted by psychologists, but, as may be inferred from the above quotation, the influence of this latter group is minimal. In view of the identical elements orientation of most simulator designers and the large amounts of money available to satisfy their strivings for system identity and engineer-ing excellence, the results are as might well be expected: most aircraft simulators are land-locked duplicates of their flying counterparts.

THE ROLE OF SIMULATORS IN PILOT TRAINING

It is not at all surprising that flight simulators are built as realistically as possible. It is not surprising, either, that pilot-training program designers and administrators have tended to rely upon such realism to assure adequate pilot training. Too often many of these individuals appear to forget that the simulator does not train. *It is the manner in which the simulator is used that yields its benefit.*

Gagne (1962) pointed out that transfer of training is a function of factors such as training objectives and instructional quality as well as of the fidelity characteristics of synthetic training equipment, and Muckler, Nygaard, O'Kelly, and Williams (1959) identified instructional techniques and instructor ability as important variables involved in transfer of training in flight simulators. Prophet (1966) stated that the flight simulator is only the vehicle for the training program and is often less important than are the synthetic training instructor and the organization and content of the synthetic training program.

There probably has never been a serious challenge to the suggested importance of the manner in which simulators are used. Gagne, Muckler, Prophet, their associates, and many others who could be cited, have stated no more than that which is obvious to all. In spite of this apparent consensus, however, it is my observation that very little attention is devoted to simulator training programs in many pilot training organizations, certainly much less attention than is devoted to the design of the simulators themselves.

The Traditional Role

In many pilot training programs, simulators are used as an adjunct to training conducted in flight. Their use is intended principally to effect a reduction in the overall cost of flight training, but in many instances (in fact, in almost all military training programs) there is little evidence that simulators have led to reduced training costs. In one such program, synthetic training was shown to add to the cost of pilot training without demonstrable transfer of training benefits (Isley, Caro, and Jolley, 1968; Jolley and Caro, 1970).

In these traditional or adjunct programs, there is often a division of responsibility between aircraft and simulator instructors. The former are the *real* instructors, while the latter are second-class citizens and are sometimes known as "device operators" rather than instructors. Because of their lower status, communication between the device operators and the pilots who *really* teach flying is infrequent, and students soon learn to revere the real

instructors and tolerate the operators and the simulators they use.

Training tasks are also divided between the aircraft and the simulator in such programs along status lines. In spite of the sophisticated engineering features and dynamic realism of many modern simulators, they seldom are used to their full capabilities. A survey of simulator utilization in the Air Force (Hall, Parker, and Meyer, 1967) found that device instructors, probably because of their limited ability and a lack of command emphasis upon their jobs, tend to concentrate upon procedural tasks in simulator training and deemphasize or ignore completely the training value of simulators with respect to dynamic flight tasks. It appears that if a task can be taught in both a simulator and an aircraft, it will be taught in a simulator only if the flight instructor finds it boring to teach in the aircraft.

The Emerging Role

Fortunately, instances of the traditional role of simulator utilization are being encountered less frequently as economic pressures upon pilot training organizations are forcing management to be concerned over the relatively high costs of conducting training in aircraft that can be conducted in simulators. The airline industry has been a pace-setter for much of the new emphasis upon simulator training, possibly because of the high cost and adverse publicity associated with accidents during in-flight training activities. But, for whatever reasons, a new role is emerging for simulators in pilot training programs.

The new role is characterized by emphasis upon simulators as primary vehicles for pilot training and is a natural outgrowth of the application of systems engineering concepts to the design of total training systems (Hall and Caro, 1971; Prophet, Caro, and Hall, 1972). To an increasing extent, pilot training is being conducted in simulators with the exception of a few maneuvers that, because of engineering state-of-the-art limitations, cannot be performed in present-day simulators (American Airlines, 1969) and for the flying necessary to confidence-building or equipment-familiarization purposes (Caro, 1972).

The shift of training from the aircraft to the simulator, while in itself a major break with traditional pilot training programs, is not the most important aspect of the emerging role of simulators. It is the manner in which these devices are being used that makes the biggest difference. Training program content has begun to become more responsive to mission requirements; the instructor has become a training resource manager; and the goals of training are beginning to be viewed in objectively measurable performance terms, rather than primarily in terms of flight hours logged. It is becoming evident in these programs that the training vehicles—the simulators principally, but also the training aircraft—are less important in many respects than are the instructors and the organization and content of the training programs.

SIMULATOR TRAINING PROGRAMS

Some of the newer flight simulators have hardware features that are intended solely to enhance the training value of the equipment rather than to duplicate aircraft features (Caro and Prophet, 1971). In some instances, these devices incorporate deliberate deviations from realism in attempts to improve, from the transfer-of-training standpoint, upon the relatively poor learning environment of the design-basis aircraft. But, with or without such advanced design-for-training features, it is still necessary to have an appropriately designed training program for use with these simulators if we expect to make significant gains in pilot training efficiency and effectiveness.

Most readers are already familiar with such terms as "systems engineering of training", "task analysis", "specific behavioral objec-

tives", and "commonality analysis". Military and commercial pilot training programs have made much use in recent years of concepts underlying such terms in defining more objectively the required content of training. Because of the resulting critical training program content reviews, many programs now are devoted largely to "need to know" skills, rather than the mass of miscellaneous "nice to know" and curiosity information that still clutters up many traditional training programs.

Along with better training simulators and more clearly defined training program content has come new status for the simulator instructors. They no longer are viewed as second class citizens who use make-do equipment to accomplish uninteresting aspects of training. Instead, they are *the* instructor, often the best qualified personnel available, and they conduct or oversee all training received by their students. The resources these instructors need to attain their training objectives, *e.g.*, an aircraft, a simulator, programmed learning material, and personnel to assist as might be required, are all under their control.

These features of modern simulator training programs—better simulators, clearly defined content, and well-qualified instructors—provide the essential ingredients for effective and efficient training, but they are nothing more than that. They still do not constitute a training program. A training program is the manner in which the well-qualified instructor uses the appropriately designed simulator to establish the clearly defined course content within the skills repertoire of the trainee.

In our work in Army and Coast Guard aviation during the past decade, we at the HumRRO Aviation Division have devoted considerable effort to the methodology involved in the use of simulators and other synthetic flight training equipment in modern training programs. We have been involved in the full range of activities associated with pilot training, including definition of the training requirement itself (*e.g.*, Hall, Caro, Jolley, and Brown,

1969), design of aircraft simulators (*e.g.*, Caro, Hall, and Brown, 1969), development of simulator training programs (*e.g.*, Caro, 1971), evaluation of simulator training program transfer of training (*e.g.*, Caro, 1972), evaluation of off-the-shelf training devices and simulators (*e.g.*, Caro, Isley, and Jolley, 1968), and investigation of costs associated with simulator training programs operation (Jolley and Caro, 1970).

During these activities, our purpose has been to bring into pilot training programs the advances made through applied training research in a wide range of training settings. We believe we have been reasonably successful in our early efforts in this regard, and we believe our success has been largely due to our orientation that training is a technology which can be engaged in, after appropriate training, by reasonably bright and adaptable people, not an art which is an inherent characteristic of the "good instructor". We note also that our view of training is not unique. Training programs of several other pilot training agencies are employing many of the same techniques we are using.

Some of the training techniques currently being employed are described in the paragraphs below.

Functional context training. Pilot training programs have been organized around a functional context, *i.e.*, around sets of meaningful, purposeful, mission modules. Course content is taught within the context of the mission-oriented purpose it supports. For example, aircraft maneuvers such as descending turns are taught to undergraduate level instrument flight trainees within the functional context of a simulated instrument approach, rather than as an exercise, *per se*, as is done during early stages of some traditional instrument training programs.

Individualization of training. The pace and redundancy of training—all aspects of training, including supporting "academic" activities—are adapted to the rate of learning of each student. An individually-paced student, thus, is ad-

vanced to the next set of instructional content only after he has demonstrated to his instructor a specified level of mastery of an earlier set.

Sequencing of instruction. The order of instructional content is arranged so as to assure that students have been taught (and have mastered) prerequisite knowledges and skills before training in a new set is undertaken.

Minimizing over-training. Steps are taken to assure that training time is restricted to that time needed to bring a trainee to the required level of training and no more. In some cases, this means overriding an instructor who feels that a particular trainee can achieve higher skill levels even though his performance at the time has already reached the specified requirements for that phase of training.

Efficient utilization of personnel. Each instructor is optimally qualified for his task, is provided the tools he may require for efficient use of his time and talents, and is trained to administer the particular course of instruction in a standardized manner. In this regard, it should be noted that an optimally-qualified instructor in the aircraft is very likely to be optimally qualified to instruct in the simulator as well. Our most productive approach has been to assign both jobs to the same individual.

Use of incentive awards. Motivation to achieve in-flight training is largely a manipulable, rather than an inherited, characteristic. The behavior control techniques of "behavior modification" or "contingency management" have been found useful in flight training, as well as in other training situations. We have found, for example, that incentive awards such as free time for both the trainee and his instructor are effective "motivators" for the achievement of stated performance goals in less training time.

Crew training. Simulators lend themselves to simultaneous pilot and copilot training much more readily than do the aircraft they simulate because of the need for the instructor to occupy a pilot seat in the latter for safety reasons. By deviating from this real-world model and moving the instructor to another seat position, we have found that effective training can be given in both pilot and copilot tasks simultaneously, thus effectively increasing the availability of simulator seats for training.

Peer training. Trainees, themselves, are used to assist fellow trainees in many simulator training activities. This technique has been found particularly useful with respect to cognitive problem-solving activities such as those which occur during navigation problems. Simulators are particularly well suited for peer instruction because the instructor can be removed from the cockpit area without creating flight safety problems with relatively unskilled trainees.

Minimizing equipment costs. To the extent that it is efficient, medium- to low-fidelity training devices, or other less expensive equipment, can be substituted for the much more expensive training in simulators. Training tasks should be allocated among the various training vehicles principally on the basis of cost effectiveness.

Objective performance measurement. All training goals are stated in objective, measurable terms which relate to the performance of the trainee or the simulator (or aircraft) he controls. With objective data, the usefulness of observations does not depend upon who is doing the observing, and there can be assurance that the proficiency data obtained are a dependable measure of the performance in question rather than a reflection of personal or other factors in the evaluation situation. Reliable data obtained through objective performance measurement can provide a basis for the standardization of the products of training. In our pilot training programs, objective performance measurement is a technique employed throughout training, not just for checkrides.

The techniques described above can be employed with almost any training equipment from simple paper mockups to operational aircraft themselves. They are not limited in their applicability to simulators, *per se.* In contrast, there are other training techniques which can be used only in those cases where specific provision is made for them in simulator design. Such training techniques include automated instruction and performance monitoring, feedback augmentation through video and simulator performance recording techniques, modeling through simulator programing, and trainee-initiated and trainee-paced instruction. For a more systematic discussion of such design features, see Caro and Prophet (1971).

TRANSFER OF TRAINING EVIDENCE

The various pilot training programs in which we have employed the training techniques described above have been quite successful. For example, in an Army undergraduate helicopter instrument-pilot training program, in which a new and quite realistic simulator was used, all of the described training techniques were incorporated into training program design at the time the simulator was introduced. The result was a 90% reduction in the amount of aircraft time required to attain the course objectives (Caro, 1972).

In that particular instrument-training program, prior to introduction of the new simulator training program, 60 hours aircraft time and 26 hours training-device time, using a modified 1-CA-1 trainer, were devoted to instructing aviators in instrument flight techniques and procedures. Graduates of the course were awarded an Army Standard Instrument Card. When the new simulator, the 2-B-24, and its specially-designed training program were introduced, the same training goals were achieved after only 6½ hours aircraft time and just under 43 hours simulator time, on the average. In addition, the total calendar time required to accomplish the training was reduced from 12 to 8 weeks.

The introduction of new training equipment often provides an opportunity to introduce new training program concepts, as is illustrated by the above instance. A similar opportunity was provided when new trainers were obtained for a fixed-wing instrument course. The training device in that instance was a commercially-available instrument trainer, the GAT-2, which was modeled after a "generalized" light twin-engine aircraft, *i.e.*, it clearly was not a "simulator" and many training activities could not be conducted in it. Nevertheless, impressive transfer of training benefits were shown when the trainer was used in conjunction with a training program incorporating the training features described above (Caro, 1971).

The training goals of the fixed-wing course included transition to a twin-engine aircraft as well as qualification for a Standard Instrument Card. The programmed allocation of aircraft time between these two goals was 10 hours for twin-engine transition and 50 hours for instrument training. Additionally, 21 hours of training in a 1-CA-1 trainer were included in the course prior to the introduction of the new commercially-available trainer. Using the new trainer with the training program we developed for it, a total of 25 hours of instruction resulted in a reduction of the 60 hours training time in the aircraft to only 35 hours—approximately 5 hours for twin-engine transition and 30 hours for instrument training. In spite of the fact that substantial savings were realized with respect to the VFR transition training goals in this course, it should be noted that there was no synthetic visual display associated with the new trainer.

In another study where device realism might be considered exceptionally low, five instructional periods in a cockpit mockup made of plywood and photographs by unskilled labor (psychologists) at a cost of about $30 were found to be about as effective as five hours of instruction in the aircraft itself (Prophet and Boyd, 1970). The training task in that study consisted of aircraft pre-start, start, runup, and shutdown procedures for the OV-1 Mohawk aircraft. The training consisted of a highly-structured program which incorporated most of the techniques described above. The same training program was used in the mockup and

in the aircraft. For the tasks involved, pilots trained in the mockup were found to be as proficient in the aircraft as were pilots who received comparable training in the aircraft itself.

In another course, where a slightly more realistic mockup built by the Army at a cost of about $4,300 was introduced, again with a training program incorporating many of the techniques described above (Caro, Isley, Jolley, and Wright, 1972), the instructors reported impressive transfer-of-training results. The course was a transition course for the Army's U-21 aircraft, and it consisted of 25 hours instruction in the aircraft. When the mockup and its training program were added, without any change in the 25 flight hours, there was about a 10% increase in the amount of that time each trainee spent in learning to fly instead of sitting on the ramp learning procedures. Although no attempt was made to measure the increased pilot proficiency which presumably resulted, it is evident that they at least had 2½ hours more actual flight experience upon graduation with no increase in programmed flight hours.

To complete the description of instances of training device utilization, I shall mention one more item. We also have obtained substantial, demonstrable transfer of training using reduced-scale paper mockups when an appropriately-designed training program is used with them. Admittedly, the amount of training which can be undertaken with such simple devices is limited. On a cost-effectiveness basis, however, simple devices can often be much more efficient training vehicles for the tasks for which they were designed than more realistic simulators.

CONCLUSIONS

At this point it is appropriate to return to the quotation which introduced this paper. I am of the opinion that we know more about the training value of simulators than the quotation implies, although I do not suggest that we know very much. Perhaps we build simulators as realistically as possible because people who design them do not know much about training. Or, perhaps it is because those who design them know that those who use them do not know much about training, and the safest thing to do is to build simulators like aircraft. In that way, at least, instructor pilots will be able to get some training value out of them by using simulators just like they would aircraft.

It is true that the users have been willing to pay the price for simulator realism, although in some instances realism was bought for the sake of realism, not to meet known training goals. In spite of such affluence, the question of how transfer might be accomplished in less expensive ways is not being avoided altogether. It is receiving attention in research centers such as that which I represent. Even now, there is substantial applied research evidence that much of the training being conducted in expensive simulators could be accomplished in less expensive devices *if* the training programs used with them were properly designed and conducted.

Finally, let me acknowledge that the present state-of-the-training art is relatively primitive, and I do not suggest we should cancel all orders for realistic simulators. I do believe that in many cases we are paying for realism where it cannot be justified from a transfer-of-training standpoint. A proper training program can compensate for lack of physical similarity between the training device and the aircraft, but a realistic simulator is a poor substitute for competent training. Obviously, transfer of training from a device to an aircraft is limited to the tasks which can be performed in the device. But, whether that limit is reached is a function of the way in which the device is used. There probably would have been zero transfer, or even a great deal of negative transfer, in all the instances of device utilization I described above had they been used with inappropriate

training programs. The key is the program, not the hardware.

REFERENCES

Adams, J. A. Research and the future of engineering psychology. *American Psychologist*, 1972, 27(7), 615-622.

American Airlines, Inc. Flight Training Academy. Optimized flight crew training, a step toward safer operations. American Airlines, Inc., Fort Worth, Texas, 1969.

Caro, P. W. An innovative instrument flight training program. Paper No. 710480, Society of Automotive Engineers, New York, N. Y., 1971.

Caro, P. W. Transfer of instrument training and the synthetic flight training system. Proceedings of Fifth Naval Training Device Center and Industry Conference, Orlando, Florida, 1972.

Caro, P. W., Hall, E. R., and Brown, G. E., Jr. Design and procurement bases for Coast Guard aircraft simulators. Technical Report 69-103, Human Resources Research Organization, Alexandria, Virginia, 1969.

Caro, P. W., Jr., Isley, R. N., and Jolley, O. B. The captive helicopter as a training device: experimental evaluation of a concept. Technical Report 68-9, Human Resources Research Organization, Alexandria, Virginia, 1968.

Caro, P. W., Jolley, O. B., Isley, R. N., and Wright, R. H. Determining training device requirements in fixed wing aviator training. Technical Report 72-11, Human Resources Research Organization, Alexandria, Virginia, 1972.

Caro, P. W. and Prophet, W. W. Some considerations for the design of aircraft simulators for training. Proceedings of Second Annual Psychology in the Air Force Symposium, U. S. Air Force Academy, Colorado Springs, Colorado, 1971.

Gagne, R. M. (Ed.). *Psychological principles in system development.* New York: Holt, Rinehart and Winston, 1962.

Hall, E. R. and Caro, P. W. Systems engineering of Coast Guard aviator training. Proceedings of Second Annual Psychology in the Air Force Symposium, U. S. Air Force Academy, Colorado Springs, Colorado, 1971.

Hall, E. R., Caro, P. W., Jolley, O. B., and Brown, G. E., Jr. A study of U. S. Coast Guard aviator training requirements. Technical Report 69-102, Human Resources Research Organization, Alexandria, Virginia, 1969.

Hall, E. R., Parker, J. F., Jr., and Meyer, D. E. A study of Air Force flight simulator programs. USAF: AMRL-TR-67-111, Aerospace Medical Research Laboratories, Wright-Patterson AFB, Ohio, 1967.

Isley, R. N., Caro, P. W., and Jolley, O. B. Evaluation of synthetic instrument flight training in the officer/warrant officer rotary wing aviator course. Technical Report 68-14, Human Resources Research Organization, Alexandria, Virginia, 1968.

Jolley, O. B. and Caro, P. W., Jr. A determination of selected costs of flight and synthetic flight training. Technical Report 70-6, Human Resources Research Organization, Alexandria, Virginia, 1970.

Muckler, F. A, Nygaard, J. E., O'Kelly, L. I., and Williams, A. C., Jr. Psychological variables in the design of flight simulators for training. USAF: WADC-TR-56-369, Wright-Patterson AFB, Ohio, 1959.

Prophet, W. W. The importance of training require-ments information in the design and use of aviation training devices. Paper presented at 16th Annual International Air Safety Seminar, Athens, Greece, 1963.

Prophet, W. W. and Boyd, H. A. Device-task fidelity and transfer of training: aircraft cockpit procedures training. Technical Report 70-10, Human Resources Research Organization, Alexandria, Virginia, 1970.

Prophet, W. W., Caro, P. W., and Hall, E. R. Some current issues in the design of flight training devices. *25th Anniversary Commemorative Technical Journal*, U. S. Naval Training Device Center, Orlando, Florida, 1971.

The Relationship between Flight Simulator Motion and Training Requirements

PAUL W. CARO[1], *Seville Research Corporation, Pensacola, Florida*

Flight simulator motion has been demonstrated to affect performance in the simulator, but recent transfer of training studies have failed to demonstrate an effect upon in-flight performance. However, these transfer studies examined the effects of motion in experimental designs that did not permit a dependency relationship to be established between the characteristics of the motion simulated and the training objectives or the performance measured. Another investigator has suggested that motion cues which occur in flight can be dichotomized as maneuver and disturbance cues, i.e., as resulting from pilot control action or from external forces. This paper examines each type cue and relates it analytically to training requirements. The need to establish such relationships in simulator design is emphasized. Future transfer studies should examine specific training objectives that can be expected to be effected by motion.

INTRODUCTION

Transfer of training research related to flight simulator motion has addressed the general question, "Is motion needed for training?" The negative answer obtained during recent studies (Jacobs and Roscoe, 1975; Woodruff, Smith, Fuller, and Weyer, 1976; Gray and Fuller, 1977; Martin and Waag, 1978a; 1978b; Pohlmann and Reed, 1978) has been counter to the opinions of many pilots and has not been supported by evidence from research on the effects of motion on pilot performance in aircraft control tasks. A possible explanation for this discrepancy is that the question is too general and does not focus upon specific training objectives.

A more appropriately stated question is, "For what training is motion needed?" That

is, what are the specific training objectives that might possibly be met (or met in a shorter period of training) through the use of motion cues during training in a simulator? This paper addresses the latter question by examining aircraft motion and relating possible training needs to specific motion characteristics that might be expected to enhance transfer of training.

BACKGROUND

Edwin Link, the father of flight simulation, had no doubts concerning the need for motion in flight trainers. His devices all had roll, pitch, and 360° rotational motion cueing. In fact, it was not until the late 1940's and early 1950's that trainers appeared on fixed bases—a development based solely on cost reduction considerations.

In the ensuing years, flight simulators have been built with a variety of platform motion systems, ranging from very limited displace-

[1] Requests for reprints should be sent to Dr. Paul W. Caro, Seville Research Corporation, 400 Plaza Building, Pensacola, Florida 32505, U.S.A.

ment roll and pitch systems to massive synergistic platforms and large amplitude booms. With the exception of those very few devices built specifically for use in the conduct of research on motion, cost and a consensus that motion is important have been the bases for virtually all motion systems design and procurement decisions. For example, when it became possible to procure two- and three-axis systems at moderate expense, small platform motion systems regained their earlier popularity. Then, not too many years ago, relatively cost-efficient, six-axis systems were developed and became the frequently accepted solution to motion cueing requirements. More recently, the cost effectiveness of motion when used in conjunction with visual displays has been questioned, and interest has once more turned toward simulators with little or no platform movement (Scientific Advisory Board, 1978).

This vacillation over motion does not exist because of a lack of user concern or research endeavor. Most simulator users strongly endorse motion, and a substantial body of research has been devoted to motion cueing in simulators.

Yet, today, one is still unable to give a generally accepted answer to the question, "Is motion needed for training?" This is so, in part, because until about 1975, transfer of training studies examining the role of motion were not being conducted. Even today, the few such studies that have been done have tended to treat motion as an all-or-none feature without demonstrating a possible relationship between the characteristics of the motion simulated and the training objectives or the dependent variables employed to measure transfer. As a consequence of the possible absence of such a relationship, these studies have not demonstrated a useful effect of platform motion upon subsequent in-flight aircrew performance.

One might more readily accept these transfer studies, some of which have been in-

terpreted as "proving" null hypotheses, were it not for the numerous reports demonstrating that the presence or absence of certain kinds of motion has predictable effects upon pilot performance in simulators. For example, Perry and Naish (1964) found that pilots respond to external forcing functions such as side gusts more rapidly, with more authority, and in a more precise manner, in a simulator with motion and visual cues than when only visual cues are present. Rathert, Creer, and Sadoff (1961) found that the correlation between pilot performance in an aircraft and in a simulator increased with the addition of simulator motion cues where such cues helped the pilot cope with a highly damped or unstable vehicle or a sluggish control system, or under some circumstances, where the control system was too sensitive. When the aircraft was easy to fly, however, motion had no effect. Douvillier, Turner, McLean, and Heinle (1960) studied the effects of simulator motion on pilots' performance of flight tracking tasks. The results from a moving-base flight simulator resembled the results from flight much more than did those from a motionless simulator. Huddleston and Rolfe (1971) reported that the presence of simulator motion produced patterns of control response more closely related to those employed in flight. That is, using simulators without motion, experienced pilots were able to achieve acceptable levels of performance, but their patterns of control response showed that their performance was achieved using a strategy different from that used in a dynamic training environment. Ince, Williges, and Roscoe (1975) found differences in pilot responses to cockpit display types that were attributed to differences in simulated motion. In a transfer of training study in which criterion trials were conducted in a simulator with a visual system, not in flight, DeBerg, McFarland, and Showalter (1976) found that pilots trained with motion performed significantly better during criterion trials than pilots trained in the same simulator without motion.

These and numerous other studies (reports by Gundry, 1976 and 1977, and Levison and Junker, 1977, include reviews of such studies) provide evidence that the movement of the platform upon which the simulator cockpit rests does affect performance in the simulator; i.e., the performance of the pilot in the presence of motion is often different than it would be in the absence of motion. With motion, the pilot's simulator control responses to external forcing functions appear to be more rapid and accurate and more like responses used to control the aircraft in flight. These studies suggest that simulator motion can affect performance in the simulator. Such findings lead one to expect motion to affect transfer of training to the aircraft as well.

How can one account for this apparent discrepancy between the results of the transfer studies and those addressing performance in the simulator only? Some transfer studies may be suspect simply on the grounds of experimental design; others from grossness of airborne criterion data. It may be, however, that regardless of other merits, all such studies have been deficient in that they have not paid sufficient attention to the function and composition of the motion being provided. For example, it is more difficult to hypothesize a cue-response relationship between turbulence motion and loss of an outboard engine on takeoff than it would be between yaw motion and such an event. One must examine motion more closely to determine which motion element or cues should even be expected to affect operator performance and subsequent transfer of training in a particular study.

TWO KINDS OF MOTION CUES

Gundry (1976; 1977) made a distinction that can be helpful in analyzing the effects of motion. He distinguished between two kinds of motion cues and suggested that they might affect pilot performance differentially. He defined *maneuver motion* as that motion which arises within the control loop and results from pilot-initiated changes in the motion of the aircraft in order to change its heading, altitude, or attitude. *Disturbance motion*, on the other hand, arises outside the control loop and results from turbulence or from failure of some component of the airframe, equipment, or engines that causes a motion of the aircraft that is unexpected by the pilot.

In Gundry's view, platform motion appears to produce quicker, more accurate pilot control of a simulator because the disturbance components of motion resulting from simulated turbulence or equipment failure can provide more rapid and relevant alerting cues about forces acting upon the aircraft than can be obtained from other available cue sources. Maneuver motion, however, does not fulfill alerting functions, because it results from pilot-initiated control movements. Research reviewed by Gundry involving maneuver motion suggested that this component of platform motion has little effect upon the control of a simulated aircraft whose flight dynamics are stable. For unstable vehicles, however, the presence of maneuver motion will allow the pilot to maintain control even in flight regions where control by visual cues alone would be impossible. Thus, disturbance motion permits more rapid and accurate control under all simulated flight conditions in which such motion is appropriate. Maneuver motion is a useful control cue only when the simulated aircraft is unstable.

In recent transfer of training studies involving platform motion, emphasis was upon simulation of maneuver motion rather than disturbance motion (Jacobs and Roscoe, 1975; Woodruff, Smith, Fuller, and Weyer, 1976; Gray and Fuller, 1977; Martin and Waag, 1978a; 1978b; Pohlmann and Reed, 1978). These investigators found that the motion variable in their studies did not enhance transfer of training to the aircraft. Since maneuver motion is pilot-induced, and the aircraft involved in these studies were relatively stable, the most likely effect of motion was to provide feedback to the pilot concerning his control inputs. If sufficient feedback were available from other sources, such as the aircraft instruments or an extra-cockpit visual display, as likely was the case in those studies, the cues provided by maneuver motion could not be expected to have an appreciable effect upon simulator training effectiveness. Under such circumstances, the cues would be redundant and probably would be ignored altogether by the pilot. Had these transfer studies examined the influence of disturbance motion resulting from factors outside the pilot's control loop, one might predict on the basis of the distinction made by Gundry that the results would have been different.

It should be noted that Martin and Waag (1978a) did examine two flight maneuvers in which disturbance cues were provided. However, their data do not help establish the point made here. In the power-on stall and traffic pattern stall maneuvers examined in their study, disturbance motion cues to indicate stall onset were provided through a platform motion system to one group undergoing training in a simulator. In a group trained without platform motion, disturbance cues were provided, nevertheless, to indicate stall onset through a manual shaking of the control stick by the instructor. Both of these groups performed significantly better in the aircraft than a no-simulator training control group, but they were not significantly different from each other. Because a disturbance motion cue associated with stall onset was provided to both simulator groups, no conclusions can be drawn with respect to the value of disturbance motion per se. No significant differences between motion and

no-motion simulator trained groups during transfer trials were reported in this study for 14 other maneuvers that did not involve disturbance cues.

The distinction between maneuver and disturbance motion helps in our understanding of both the prior research on motion and the reactions of pilots to the motion component of aircraft simulators. In the cited transfer of training studies in which motion did not appear to influence subsequent pilot performance in the aircraft, the motion involved was predominantly, if not exclusively, of the maneuver variety. On the other hand, disturbance motion has been the predominant type of motion in nontransfer studies in which changes in pilot performance were related to motion simulation. Thus, the results of both sets of studies can be accepted and attributed to the nature of the motion simulation involved in each. Disturbance motion appeared to be important, at least in training situations where disturbance cues could be related to specific training objectives. On the other hand, maneuver motion appeared to be important when the aircraft simulated was unstable or was particularly sensitive to control input, but the research data available at this time have not shown that it contributes to transfer of training in stable, easy-to-fly aircraft.

Hence, in considering the performance characteristics of motion systems for a particular simulator to be used for training, it is important to distinguish between maneuver motion and disturbance motion, and to identify the training needs associated with each.

MANEUVER MOTION AND TRAINING

All currently operational aircraft are generally easy to control. Even vertical takeoff airplanes and helicopters are stable and easy to control during flight at moderate to high airspeeds. Attitude changes and other motions which occur in these aircraft during pilot-induced maneuvers such as steep turns, aerobatics, and in the case of helicopters, rapid deceleration and autorotation, are directly responsive to pilot inputs and present no unusual control problems. The cues associated with such motion probably do no more than confirm to the pilot what he already has learned from other sources, that is, that the aircraft has responded to his control input. It is likely that the pilot would frequently even be unaware of the presence (or absence) of maneuver motion cues during flight at maneuver airspeeds, since those cues would be compatible with, and redundant to, information he already has. In fact, there have been numerous anecdotal reports of pilots not knowing whether the simulator's motion platform was on or off or even that the direction of movement had been reversed during periods of training when only maneuver motion was being simulated.

Conditions of relative instability can exist, nevertheless, as during operation at near-stall airspeeds, or when in ground effect in the case of helicopters and other VTOL aircraft. In order to taxi a helicopter, for example, power must be applied to lighten pressure on the wheels. When this is done, the aircraft tends to yaw due to torque. The yaw motion will be felt by the pilot since it occurs rapidly, and a prompt correcting response is required in order to maintain directional control.

Furthermore, in the hover mode of VTOL operation, the pilot probably uses motion cues as the primary or initial source of information about changes in the position, movement, and attitude of the aircraft. Visual cues that would reflect these small, but rapidly occurring, changes tend to be noted by the pilot later than motion cues (Reed, 1977) and thus would possibly be inadequate for aircraft control. In fact, the pilot might even be unable to learn to hover in a VTOL simulator that lacked maneuver motion cues simulated through a platform motion system. Such a

learning task could be comparable to learning to ride a unicycle without being able to feel the onset of an imbalance condition. Visual cues alone would probably be insufficient for efficient learning to take place. Hence, appropriate maneuver motion cues for these simulator training tasks would probably enhance training.

Certain necessary characteristics of such motion cues can be anticipated, also. When operating a VTOL within ground effect, the sensitivity of the vehicle to control input is such that the onset of motion resulting from pilot control input is prompt, and motion acceleration is rapid, particularly with respect to rotational movements. The magnitude of displacement tends to be small, however, because large movements must be prevented to preclude contact with external objects and the ground. Consequently, in simulating maneuver motions in an unstable flight situation, particular attention should be directed to the rate of motion onset, but large displacement would not appear to be required.

In other words, the magnitude of excursion of motion in a simulator is probably of less importance than the promptness and correlatedness of such motion. Therefore, lags between pilot control input and simulator response that appreciably exceed corresponding lags in the aircraft would likely have an adverse effect upon pilot performance, since the consequence would be loss of the early alert to the pilot that the cues associated with such vehicle motion can provide.

It is not possible, on the basis of available behavioral and training research data, to quantify precisely for platform motion system design the excursions required in each degree of freedom in order to provide the maneuver motion cues appropriate to simulator training. Certain requirements for motion systems can be stated, however. First, time lags in a motion system between control input and the onset of simulator motion

should approximate those of the aircraft itself. Second, the rate of motion onset must be
sufficient to alert the pilot. That is, the rate
must be above his threshold of perception if it
is to provide a cue. Third, because rotational
and translational displacements are important in simulators insofar as they provide required alerts to the pilot, displacements that
permit only the required alerts, with washout effects, are appropriate and should be
sufficient.

DISTURBANCE MOTION AND TRAINING

There are a number of events outside the
pilot's control loop, that is, external forcing
functions, which result in motions of the aircraft that are unexpected by the pilot. These
motions provide a degree of realism to simulation of specific aircraft or aircraft types.
They also sometimes provide prompt cues to
the need for action to overcome the effects of
equipment failure, and they can influence
training problem difficulty. Examples of
these disturbance motions include the shakes
and vibrations that characterize helicopter
flight under normal conditions, the sudden
yaw that accompanies outboard engine failure, and simulated turbulence which can
make precise aircraft control difficult.

Two kinds of disturbance motion should be
provided in an aircraft simulator designed for
training. *Uncorrelated disturbance motion*, the
first kind, is low frequency motion that is not
correlated with pilot control movements.
Such motion appears to the aircrew to be
either irregular in occurrence and essentially random in frequency, direction, and
amplitude; or to be of a relatively fixed frequency, direction, and amplitude, but virtually always present. Turbulence, engine vibration, and oscillatory shakes are examples
of uncorrelated disturbance motion. Though
apparently important for motivation and
vigilance, these motions do not alert the pilot
to a disturbing event and thus do not cue him

to initiate a particular control input. For this
reason, simulation of uncorrelated motion
probably can be relatively gross with respect
to the magnitude and direction of corresponding motion in the aircraft, but it should
be simulated in the device under circumstances which characterize its presence in
the aircraft. (It is recognized that uncorrelated disturbance motion, such as certain vibrations, can be used by the pilot to detect
abnormal conditions or impending malfunctions. Such "troubleshooting" functions of
motion are not being addressed in this paper.)

Correlated motion, the second kind of disturbance motion of concern in aircraft simulation, is motion that is a consequence of
events that are of immediate interest to the
pilot and require his prompt attention. Motion that results from an equipment failure or
sudden (and unintended) change in configuration of the aircraft, such as asymmetrical
external stores hangup or jettison, is illustrative of a correlated disturbance motion. Its
characteristic is a rapid onset or jolt that has
a predictable effect on the performance of the
aircraft. The pilot must learn to respond to
such motion by rapidly identifying its probable cause in order to initiate an appropriate
corrective procedure, and he must rapidly
make an input to the aircraft's controls that
will allow him to maintain control over the
vehicle's flight. Hence, accurate simulation of
correlated disturbance motion cues is important to effective simulator training.

Correlated disturbance motion cues that
probably should be provided in a training
simulator include those that result from aircraft failures and malfunctions, as well as
motion cues correlated with disturbing
events such as buffets, touchdown impact,
stores release, weapons firing, projectile impacts on the airframe, and helicopter blade
stall and strikes. In addition, certain disturbance motion cues uncorrelated with specific
events might also be included in a simulator

because they can affect task difficulty and trainee motivation. These latter cues include those associated with turbulence and general vibrations and oscillations associated with routine operations.

Disturbance cues in the simulator that are correlated with specific events probably should reproduce as faithfully as possible the cues that are caused by similar events in the aircraft with respect to time of onset, and for the same reasons discussed above related to maneuver motion cues. Likewise, unless magnitude of excursion is a significant cue that enables the pilot to determine the event with which the disturbance is correlated, these cues can probably be of relatively low magnitude, since it is primarily the alerting acceleration or jolt that provides information to the pilot and is useful in training.

DETERMINING SIMULATOR MOTION SYSTEM DESIGN REQUIREMENTS

The simulator motion transfer of training studies conducted to date have not provided an empirical basis for determining whether a simulator for a particular aircraft should be procured with or without motion, and if so, what the principal characteristics of that motion should be. Until an empirical basis can be developed, the above rationale concerning the potential value of motion in training simulators can provide an analytic basis for such decisions. The following paragraphs discuss implications of that rationale for decisions concerning motion system design and procurement.

If the simulated aircraft is easy to control, i.e., is relatively stable, a platform motion system designed to provide maneuver motion cues would not normally be required. If, however, the aircraft tends to be unstable during certain flight modes (e.g., at airspeeds approaching stall), or if the occurrence of particular malfunctions could render the aircraft

unstable, maneuver motion cues might well be beneficial to pilots learning to fly the aircraft during such unstable conditions. An example of a malfunction that could render an otherwise stable aircraft relatively unstable, and thus would lead to a requirement for maneuver motion cues in training, would be the loss of an automatic stability augmentation system. If the aircraft is unstable, either by reason of its basic design or due to such an abnormal condition, the provision of maneuver motion cues should be important in simulator training, and a motion system capable of providing maneuver motion cues should be procured.

Likewise, the need for disturbance motion cues in a flight simulator can be determined by examining the role in training of turbulence, vibration, and other uncorrelated disturbance motion cues, as well as the need for disturbance motion cues that might be correlated with specific events and thus related to specific training objectives. The need for uncorrelated cues can be determined, on the one hand, by assessing their importance with respect to trainee expectations, motivation, and attitudes. However, except to the extent that uncorrelated disturbance motion cues may be useful in the manipulation of training problem difficulty, in the maintenance of vigilance, or in order to train personnel to operate the aircraft under conditions of severe turbulence, they will have little effect on the attainment of training objectives. If it is determined that such motion cues can have these roles in a particular simulator training application, then a motion system capable of providing those cues should be provided.

The need for correlated disturbance motion cues can be determined objectively through an analysis of the motions of the simulated aircraft when disturbing events occur. The aerodynamic effects upon aircraft motion of an abrupt change in aircraft configuration, for example, are precisely predictable. When those effects include a sudden change in the direction, attitude, or rate of motion of the aircraft, a motion cue that is potentially useful in training can be identified. It then must be determined whether the onset of the cue is above the threshold of perception of the pilot. It so, the pilot can use the motion-onset cue to initiate a control input that may be necessary to continue safe flight. If the analysis indicates that such cues do occur, particularly if the cues provide the first, and possibly only, timely alert to the need for a pilot control response, then a simulator motion system capable of providing such motion cues should be provided.

The amount of simulator motion associated with these correlated disturbances, i.e., the physical displacement of the pilot in the simulator, need not duplicate the displacement he would experience in the aircraft. Rather, it need only be great enough to serve an alerting function. In other words, motion in the simulator must be sufficiently intense and abrupt to provide a stimulus which will permit the pilot to detect reliably the need for a control response. If this requirement is met, he then can be trained to make the appropriate response. Thus, by identifying the disturbance cues that are correlated with sudden changes in aircraft configuration or systems failure, it should be possible to specify the alerting motion cues as initial accelerations to be provided in a simulator motion system.

CONCLUSION

The foregoing analysis focused upon the establishment of a logical relationship between simulator motion and training requirements. It emphasized the need to define the performance characteristics of a flight simulator motion system in response to specific training objectives or to motivational and attitudinal factors related to training. Recent empirical studies of the effects of simulator motion upon transfer of flight training have not ad-

dressed these important considerations. They did not establish that the performance of the motion platforms involved in the studies provided motion cues that were related to the training under study in a manner that would lead one to expect a transfer effect to be demonstrated.

Whether motion (or any other dimension of simulation) is needed in order to achieve a particular training objective remains an empirical question. However, empiricism must build upon logical analyses that establish relationships to be tested. In the case of flight simulation, it is suggested that the analysis should examine the relationship between the motion cues provided and the motion-related objective of the training to be conducted.

ACKNOWLEDGMENTS

This paper was prepared during research sponsored by the Air Force Office of Scientific Research (AFSC), United States Air Force, under Contract No. F49620-77-C-0112. Appreciation is expressed to William V. Hagin, Andrew M. Junker, John B. Sinacori, William D. Spears, and Wayne L. Waag for having reviewed and commented upon a draft of this paper.

REFERENCES

DeBerg, O. H., McFarland, B. P., and Showalter, T. W. The effect of simulator fidelity on engine failure training in the KC-135 aircraft. Paper presented at American Institute of Aeronautics and Astronautics Visual and Motion Simulation Conference, Dayton, Ohio, April 26-28, 1976.

Douvillier, J. G., Jr., Turner, H. L., McLean, J. D., and Heinle, D. R. Effects of flight simulator motion on pilots' performance of tracking tasks. Washington, D.C.: National Aeronautics and Space Administration, Technical Note NASA-TN-D-143, 1960.

Gray, T. H. and Fuller, R. R. Effects of simulator training and platform motion on air-to-surface weapons delivery training. Williams AFB, Arizona: Air Force Human Resources Laboratory, Technical Report AFHRL-TR-77-29, July, 1977.

Gundry, A. J. Man and motion cues. *Third Flight Simulation Symposium Proceedings.* London, England: Royal Aeronautical Society, April, 1976.

Gundry, A. J. Thresholds to roll motion in a flight simulator. *Journal of Aircraft,* 1977, *14,* 624-630.

Huddleston, H. F. and Rolfe, J. S. Behavioral factors influencing the use of flight simulators for training. *Applied Ergonomics,* 1971, *2.3,* 141-148.

Ince, F., Williges, R. C., and Roscoe, S. N. Aircraft simulator motion and the order of merit of flight attitude and steering guidance displays. *Human Factors,* 1975, *17,* 388-400.

Jacobs, R. S. and Roscoe, S. N. Simulator cockpit motion and the transfer of initial flight training. *Proceedings of the Human Factors Society 19th Annual Meeting.* Santa Monica, California: Human Factors Society, 1975.

Levison, W. H. and Junker A. M. A model for the pilot's use of motion cues in roll axis tracking tasks. Wright-Patterson AFB, Ohio: Aerospace Medical Research Laboratory, Technical Report AMRL-TR-77-40, June, 1977.

Martin, E. L. and Waag, W. L. Contribution of platform motion to simulator training effectiveness: Study I—Basic contact. Williams AFB, Arizona: Air Force Human Resources Laboratory, Interim Report AFHRL-TR-78-51, June, 1978. (a)

Martin, E. L. and Waag, W. L. Contribution of platform motion to simulator training effectiveness: Study II—Aerobatics. Williams AFB, Arizona: Air Force Human Resources Laboratory, Interim Report AFHRL-TR-78-52, September, 1978. (b)

Perry, D. H. and Naish, J. M. Flight simulation for research. *Journal of the Royal Aeronautical Society,* 1964, 68, 645-662.

Pohlmann, L. D. and Reed, J. C. Air-to-air combat skills: Contribution of platform motion to initial training. Williams AFB, Arizona: Air Force Human Resources Laboratory, Technical Report AFHRL-TR-78-53, October, 1978.

Rathert, G. A., Jr., Creer, B. Y., and Sadoff, M. The use of piloted flight simulators in general research. Paris, France: Advisory Group for Aeronautical Research and Development, North Atlantic Treaty Organization, Report 365, 1961.

Reed, L. E. Effects of visual-proprioceptive cue conflict on human tracking performance. Wright Patterson AFB, Ohio: Air Force Human Resources Laboratory, Technical Report AFHRL-TR-77-32, June, 1977.

Scientific Advisory Board. Report of the USAF Scientific Advisory Board ad hoc committee on simulation technology. Washington, D.C.: U.S. Air Force, April, 1978.

Woodruff, R. R., Smith, J. F., Fuller, J. R., and Weyer, D. C. Full mission simulation in undergraduate pilot training: An exploratory study. Williams AFB, Arizona: Air Force Human Resources Laboratory, Technical Report AFHRL-TR-76-84, December, 1976.

Visual Scene Properties Relevant for Simulating Low-Altitude Flight: A Multidimensional Scaling Approach

JAMES A. KLEISS,[1] *University of Dayton Research Institute, Dayton, Ohio*

In the present experiments I sought to identify the properties of visual scenes relevant for simulating low-altitude flight. The approach was first to identify the relevant properties of real-world scenes. The stimuli were videotape segments or still photographs of real-world scenes exhibiting a variety of scene properties. Ratings of similarity between stimulus pairs were submitted to multidimensional scaling analyses. Results using videotape segments provided consistent evidence for two relevant scene properties: variation in terrain shape and variation in object size or spacing. Results using still photographs were less interpretable, supporting the argument that motion information is important. Results suggest that designers of flight simulator visual scenes should focus specifically on rendering elements of terrain shape and objects in scenes.

INTRODUCTION

Vision provides important information for perceiving and controlling one's own motion within a fixed environment (see Warren, 1990, for an overview of visual self-motion perception). An important component of any vehicle simulator intended to teach vehicle control skills is, therefore, the artificial visual scene. The purpose of the present experiments is to identify scene properties relevant for simulating visual low-altitude flight, a task of interest to the military aviation community. The use of the term *scene* in this investigation is consistent with its use in reference to the computer-generated im-

age displayed to pilots in flight simulators (e.g., Lintern and Koonce, 1992; Lintern, Thomley-Yates, Nelson, and Roscoe, 1987; Lintern and Walker, 1991). The scene properties of interest are assumed to be environmental surfaces and objects that mediate optical transformations specifying one's position or change in position within the environment. This corresponds to Marr's (1982) 2½-dimensional sketch in a viewer-centered frame of reference.

It is important to distinguish between scene content defined this way and the semantic content of scenes reflected in meaningful relationships among objects, which influence the process of object identification (e.g., Biederman, Mezzanotte, and Rabinowitz, 1982; Biederman, Rabinowitz, Glass, and Stacy, 1974). This is not to deny a possible role for semantic factors in flight simulator visual scenes. For example, object identification is an important component of some flight tasks, such as aerial reconnaissance

[1] Requests for reprints should be sent to James A. Kleiss, Delco Electronics Corp., User Interface Development, 1800 E. Lincoln Rd., Mail Station E110, Kokomo, IN 46904-9005. A portion of these data were presented at the Human Factors Society 35th Annual Meeting, San Francisco, California, September 1991.

and target identification. Pilots also report a strategy of seeking out familiar objects in the environment, the apparent size of which serves as a cue for distance (162d Tactical Fighter Group, 1986). However, the dominant information for perceiving self-motion appears to be structural relations among environmental elements. For example, one important form of visual information for perceiving change in altitude is change in a perspective gradient (e.g., Flach, Hagen, and Larish, 1992; Wolpert, 1988; Wolpert, Owen, and Warren, 1983). In contrast, Biederman (1981) found that a background depth gradient did not facilitate perception of objects compared with a control condition in which there was no depth gradient. Biederman concluded that object identification is influenced by semantic relationships.

Computer-generated scenes commonly used in flight simulators comprise polygonal surfaces and, with more advanced technology, complex textured patterns mapped onto surfaces. Global scene detail increases as a function of the number of polygons and texture elements in scenes, and this has been shown to have a significant positive effect on performance of tasks involving flight near the terrain surface (e.g., Barfield, Rosenberg, and Kraft, 1989; Buckland, Edwards, and Stevens, 1981; Buckland, Monroe, and Mehrer, 1980; DeMaio, Rinalducci, Brooks, and Brunderman, 1983; Lintern and Koonce, 1992; Lintern et al., 1987; Lintern and Walker, 1991; Martin and Rinalducci, 1983; McCormick, Smith, Lewandowski, Preskar, and Martin, 1983).

Some have sought to relate scene quality to quantitative factors, such as polygon count and number of texture maps, with the important design question being the amount required to ensure effective training (see Padmos and Milders, 1992, for a recent discussion along these lines). However, concern for these factors alone fails to consider that available polygons and texture elements may be allocated disproportionately among various scene properties. In consequence, two scenes may be quantitatively similar but qualitatively dissimilar with respect to the specific scene properties represented. This is also an important consideration because various types of scene elements differ in their effectiveness for tasks involving perception or control of altitude in flight simulators (Flach et al., 1992; Kleiss and Hubbard, 1993; Wolpert, 1988; Wolpert et al., 1983). Efficient use of available computer image generator (CIG) processing resources therefore requires some consideration of which specific scene properties are relevant.

Evidence also suggests a classification of relevant visual scene content into various subtypes. For example, Kleiss and Hubbard (1993) found that detection of altitude change in a flight simulator improved significantly both when objects were added to scenes and when a complex, textured pattern was rendered on the terrain surface. However, each factor affected performance independently, suggesting that the presence of one does not compensate for the absence of the other. DeMaio et al. (1983) also manipulated objects and texture (texture consisted of a mosaic of geometric shapes) in flight simulator visual scenes and found that objects were effective for estimating altitude, whether scenes were presented dynamically or statically. In contrast, texture was effective only with dynamic presentation. These authors concluded that texture provided optic flow information, whereas objects provided a different type of information that was less dependent on motion. Findings such as these suggest that an adequate conceptualization of flight simulator visual scene content must take into account multiple types of relevant scene properties.

In seeking to identify relevant properties of flight simulator visual scenes, one might continue a process of experimental evaluation similar to that reflected in the studies cited. In so doing, one could draw on the body of evidence provided by basic research investigating perception and control of self-motion, particularly perception of altitude (e.g., Flach et al., 1992; Johnson, Tsang, Bennett, and Phatak, 1989; Wolpert, 1988). One is struck, however, by the relative simplicity of the stimulus displays used in such research (e.g., grids or random dot patterns on flat surfaces) compared with the apparent complexity of the real-world environment. Although it is possible that the real-world environment can ultimately be reduced to a few abstract stimulus properties, a question persists as to whether other potentially relevant scene properties remain to be identified.

Laboratory manipulations of computer-generated scenes are also potentially limited by available CIG technology. Hence some relevant scene properties may be excluded from experimental designs simply because technology is not sufficiently sophisticated to render them. As a result, significant gaps may exist between computer-generated scenes and the operational environments they are intended to simulate.

An alternative is to inquire, first, about the relevant properties of real-world scenes. It is safe to assume that all relevant scene properties are represented in the variety and complexity of the real-world flight environment. The problem is to isolate relevant scene properties, given the large number of plausible alternatives. Fender (1982) noted that pilots know a great deal about the visual environment in which they fly and that this knowledge would be of great value in designing flight simulators. Unfortunately, pilots are not necessarily consciously aware of all that they know and may not be able to reliably communicate that information verbally. Fender suggested that multidimensional scaling (MDS) would be well suited to tapping this type of knowledge.

MDS is a method for identifying the perceived structure in a set of stimuli by mapping stimuli within an *n*-dimensional spatial configuration. The mapping is based on measures of perceived similarity between stimuli, such that similar stimuli are positioned close to one another in multidimensional space and dissimilar stimuli are positioned far apart. Subsequent examination of relationships among stimuli within the spatial configuration is useful for revealing stimulus properties that affect subjects' perceptual judgments.

Kleiss (1990) undertook an MDS analysis using videotape segments depicting low-altitude, high-speed flight over a variety of real-world terrains. Pilots rated the degree of visual similarity between pairs of scenes with regard to properties important for visual low-altitude flight. A two-dimensional spatial configuration revealed orderly variation in two scene properties: terrain vertical development mediated by presence or absence of mountains, hills, and ridges, and presence or absence of objects exemplified by large cultural objects such as buildings. Because both scene properties are conceivably pertinent for the task of visual low-altitude flight, MDS appears to be an effective method for analyzing real-world scenes in this task context.

The generality of Kleiss's (1990) results is potentially limited in several ways, however. First, the small number of stimuli (nine) used in the investigation precluded consideration of dimensionalities higher than two (Kruskal and Wish, 1986). In consequence, some relevant scene properties may remain unidentified. Second, both altitude and speed varied considerably across stimulus segments. These factors could have influenced pilots' judgments. Third, videotapes were originally recorded in helicopters and were subsequently edited to produce the appearance of high-speed flight. As a result, maneuvering apparent in some segments not only was atypical of fixed-wing aircraft but was exaggerated by the increased video speed. The present experiments attempt to replicate and extend Kleiss's (1990) preliminary results using a larger stimulus set with better control over altitude and speed. Formal methods for interpreting spatial configurations will also be explored in these experiments.

EXPERIMENT 1

The videotape presentation format used by Kleiss (1990) is labor intensive with respect to both stimulus preparation and presentation time. A more economical alternative would be still photographs, but results of a study by De-Maio et al. (1983) suggest that important information might be lost with this presentation format. To address this issue, both dynamic and static presentation formats were used. Differences in presentation format were reflected in differences in the number, type, or relative importance of dimensions.

Method

Subjects and design. Subjects were 14 Arizona Air National Guard instructor pilots in the A-7 or F-16 aircraft assigned to the 162d Tactical Fighter Group, Tucson, Arizona; one U.S. Air Force (USAF) instructor pilot in the F-5 aircraft assigned to the 425th Tactical Fighter Training Squadron, Williams Air Force Base, Arizona; and one USAF pilot who had previously flown the F-111 aircraft and was currently assigned to the Armstrong Laboratory, Aircrew Training Research Division, Williams Air Force Base. Routine missions for all pilots required visual low-altitude flight.

Subjects were randomly assigned to either the *dynamic presentation* condition, in which stimuli were videotape segments, or the *static presentation* condition, in which stimuli were still photographs. To reduce the amount of time required for stimulus presentation in the dynamic presentation condition, an incomplete data design was used (Schiffman, Reynolds, and Young, 1981). To compensate for fewer observations per subject in this condition, 10 subjects were assigned to the dynamic presentation condition (mean total flying time was 4170 h) and only 6 were assigned to the static presentation condition (mean total flying time was 3642 h). Mean total flying time did not differ significantly between groups, $t(14) < 1$.

Stimuli and materials. A list of terrain types was compiled that encompassed as wide a variety of scene properties as possible. Care was taken in compiling this list to include exemplars identified by pilots in informal interviews to be of potential relevance. Some constraints were imposed by the geographic region in which filming occurred. Flights originated in Mojave, California. There was, for example, no snow-covered terrain, nor was there a dense canopy of trees unbroken by clearings. However, subject

matter experts were in general agreement that most relevant scene properties were represented.

Scenes were photographed from a T-33 jet aircraft using a 16-mm motion picture camera mounted in the nose section and a 35-mm still camera mounted in a wing pod. The motion picture camera ran at 30 frames/s and was equipped with a 12.5-mm wide-angle lens. The still camera was equipped with a standard 50-mm lens. Both cameras were canted down slightly so that the horizon appeared approximately one-quarter of the way down from the top of the frame. All scenes were filmed in color.

A radar altimeter was mounted in a second wing pod and was used to monitor altitude during filming. A straight-and-level pass was made over each terrain type while maintaining altitude at 38 m (125 ft) above ground level and airspeed at 350 knots as closely as conditions allowed. When hills were present, altitude was relative to the tops of hills. Still photographs were shot simultaneously with the motion picture film so that identical scene properties were represented in each condition.

Motion picture film was transferred to videotape, and video speed was subsequently increased to produce the appearance of high-speed flight. A miscalculation in this process resulted in a higher-than-anticipated final speed of approximately 630 knots (kn). Many modern jet fighters attain speeds equal to or exceeding this value during combat missions, so the high speed was not deemed to be inappropriate. However, speeds are more typically in the range of 400 to 500 kn. Because initial filming occurred at a relatively high speed, the increase in videotape speed produced little apparent exaggeration of motion attributable to wind buffeting and minor positional adjustments.

Stimuli in the dynamic presentation condition were 17 videotape segments, each 5 s in duration, depicting a wide variety of scene properties. Kruskal and Wish (1986) suggested that 17 stimuli are sufficient to reveal up to four dimensions if that level of structure is present in the data. A frame from each of the 17 segments is shown in Kleiss (1992). Scenes included the following:

1. Airport: flat terrain with hangars, runway, parked aircraft, and mountains occluding the horizon.
2. Desert: flat terrain with small, dense bushes and mountains occluding the horizon.
3. Dry lake: flat terrain with no vegetation, mountains occluding the horizon.
4. Ridges: multiple ridges perpendicular to the flight path, little vegetation.
5. Trees/pasture: flat terrain with groups of large trees and mountains occluding the horizon.
6. Dense trees: flat terrain with large, closely spaced trees, intermittent clearings, and mountains occluding the horizon.
7. Valley: river valley with surrounding mountains, trees and bushes, ridges, and large rocks.
8. Forested mountain: sharply contoured terrain with dense trees and intermittent clearings.
9. Hills with trees: hilly terrain with a high density of individual trees.
10. Barren hills: hilly terrain with sparse grassy vegetation and occasional roads.
11. Sand dunes: large sand dunes with no objects or vegetation.
12. Desert with trees: gently rolling terrain with small desert trees and bushes, mountains occluding the horizon.
13. Ocean: smooth water with no objects or land visible.
14. Coastline: flight path over water parallel to coastline.
15. Agricultural: flat terrain with small, dense vegetation; clearly delineated field boundaries; a road; a vehicle; power lines.
16. Grassland: flat terrain with grass, scattered bushes, and mountains occluding the horizon.
17. Shore approach: flight path over water approaching shore, mountains occluding the horizon.

These 17 stimuli yielded a total of 136 unique stimulus pairings. Stimulus pairs were randomly assigned to one of two subsets (68 pairs each) with the constraints that (a) individual scenes appeared approximately equally often in each subset and (b) no specific scene appeared in consecutive pairs. MacCallum (1978) provided evidence that structure can be successfully recovered from data with as many as 60% missing observations, provided that the sample size is 10 or larger and that different observations are missing across subjects.

Two additional subsets were made in a similar fashion, except that the order of scenes within each pair was reversed. A number preceded each stimulus pair on the videotape to indicate its position within the sequence (1 through 68). The two 5-s videotape segments in each pair were separated by a 1-s blank and were followed by a 3-s blank, during which responses were entered.

Stimuli in the static presentation condition were 16 6½- × 10-in. color still photographs depicting all but the shore approach scene. The smaller stimulus set and inherently faster pace of viewing still photographs allowed each subject in the static presentation condition to view all 120 stimulus pairs in a single session. Stimulus pairs were randomly arranged in ring binders with photographs in each pair appearing on facing pages. A second stimulus set contained the same pairs in a different random order with the positions of photographs within pairs reversed. The pairs were numbered sequentially 1 through 120.

Following Schiffman, Reynolds, and Young (1981), I recorded similarity judgments on 120-mm, ungraduated lines anchored at the left with the label *exact same* and at the right with *completely different*. Rating scales were arranged in a booklet with four scales per page, each numbered in sequence. An instruction page appeared at the front of the booklet and described the purpose of the experiment as well as the rating procedure.

To facilitate dimensional interpretation, each scene was also rated on eight bipolar attribute scales included at the end of the booklet. These scales consisted of 120-mm lines anchored at each end with dichotomous labels. The attributes were chosen based on a number of factors, including demonstrated importance in previous flight simulation research, pilot interviews, and USAF instructional materials. Attribute labels and the rationale for choosing them are as follows: (a) *prefer* versus *not prefer:* the degree of preference for scene properties, which was assumed to reflect the quality of information depicted in scenes; (b) *hilly/mountainous* versus *flat:* degree of terrain vertical development; (c) *objects* versus *no objects:* object density is an important factor affecting performance in flight simulators (Kleiss and Hubbard, 1993; Martin and Rinalducci, 1983); (d) *known size references* versus *no known size references:* the apparent size of familiar features is used by pilots as a cue for distance (162d Tactical Fighter Group, 1986); (e) *texture/detail* versus *no texture/detail:* apparent detail is used by pilots as a cue for distance (162d Tactical Fighter Group, 1986); (f) *complex* versus *simple:* intended as a measure of global scene complexity; (g) *regular* versus *random:* intended to reflect the orderliness or predictability in the positioning of scene elements; and (h) *high contrast* versus *low contrast:* the degree to which scene elements stand out against the background. Attributes were selected to capture a range of potentially relevant scene properties without regard for possible correlations among attributes.

Procedure. Data were collected in small groups of two to four subjects. Subjects first read the cover sheet, and then major aspects of the procedure were emphasized verbally with opportunities for questions. Dupnick (1979) provided evidence that orienting instructions are effective in focusing attention on stimulus properties pertinent for a given task. Toward this end, subjects were asked to imagine themselves flying over the terrain depicted in each stimulus segment using only out-of-the-cockpit visual cues (i.e., no instrumentation). Subjects were instructed to rate the extent to which the two scenes in each pair looked the same or different with respect to scene properties important for that task. If the two scenes looked the same, subjects were instructed to place a mark at the extreme left end of the rating scale. If the two scenes looked different, they were instructed to place a mark somewhere along the scale indicating how different. They were told that a general impression of similarity was desirable and that there was no need to identify specific scene properties that they felt were important. Subjects were encouraged to use the entire range available on the rat-

ing scales. To familiarize subjects with the range of stimuli used in the experiment, scenes were shown individually before presentation of stimulus pairs.

Videotapes were displayed on standard video monitors in squadron briefing rooms. The physical layout of squadron briefing rooms precluded precise control over viewing conditions. However, subjects were always seated 2 to 4 m in front of the monitor with a clear view of the screen. There were no apparent problems attributable to the fast pace of videotape presentation. Subjects in the static presentation condition had unlimited viewing time. Approximately equal numbers of subjects within presentation conditions viewed each set of stimuli.

On completing similarity ratings, subjects rated individual scenes on each of the eight bipolar attribute scales. Pilots are familiar with most of the attributes as a result of training received during the normal course of their duties. To avoid confusion, however, subjects were first asked to review all anchor labels so that questions regarding their meanings could be addressed. Scenes were presented individually, and subjects completed all eight bipolar scales before progressing to the next scene. Each scale was marked at a location that corresponded to the perceived amount of a given property. The entire session took approximately 1 h.

Results

Data for all analyses were distances in millimeters measured from the left end of each scale to the point at which the subject marked the scale. The maximum range of values was 0 to 120 mm. For pairwise ratings, larger values indicated greater dissimilarity.

Multidimensional scaling. Pairwise ratings were submitted to MDS analyses using ALSCAL for PCs (Young, Takane, and Lewyckyj, 1978). A weighted (individual differences), nonmetric approach was used. Not only does the weighted approach yield the most robust and reliable results, but spatial configurations are fixed (i.e., not rotatable) in relation to dimensional axes and are, therefore, directly interpretable

(Schiffman et al., 1981). The weighted approach also provides subject weights, which indicate the relative importance of each dimension for individual subjects. Ratings were assumed to be continuous. Missing stimulus pairs in the dynamic presentation condition were treated as missing values.

ALSCAL provides two measures of fit between similarity/dissimilarity data and interstimulus distances in spatial configurations of various dimensionalities: stress and squared correlation (R^2). Stress indexes the discrepancy between proximity measures derived by ALSCAL from raw similarity data and interpoint Euclidian distances among stimuli; smaller stress corresponds to better fit. R^2 is the squared correlation between proximity measures and interpoint Euclidian distances. Correct dimensionality (i.e., dimensionality with maximum structure) was identified using a method suggested by Isaac and Poor (1974), in which stress for experimental data was compared with stress for purely random data (i.e., data with 100% error). Correct dimensionality was taken to be that at which stress for experimental data was smallest in comparison to that for random data.

Figure 1 shows stress and R^2 for the dynamic and static presentation conditions, plus stress for random data, as a function of dimensionality. ALSCAL does not compute a one-dimensional solution with the weighted approach. Stress values for random data were taken from Spence and Ogilvie (1973) and are estimates derived from Monte Carlo studies using the TORSCA-9 MDS algorithm (Young, 1968) with a stimulus set size equal to 17.

Note first that stress is consistently smaller and R^2 is consistently larger in the dynamic presentation condition than in the static presentation condition. The dynamic presentation condition therefore provides a better fit of the data. Various MDS algorithms optimize different indices (Schiffman et al., 1981), precluding direct comparisons between stress values for experimental and random data. Indeed, stress values in the static presentation condition are somewhat larger than random stress at higher dimen-

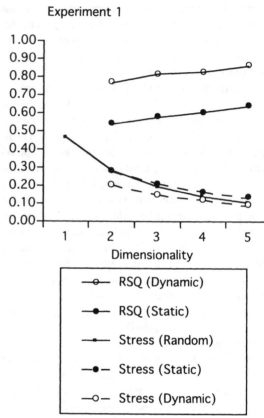

Figure 1. *Stress and R^2 for dynamic presentation and static presentation, plus stress for random data, as a function of dimensionality: Experiment 1.*

sionalities. What is important in the present case is the difference between random stress and experimental stress, which is largest at dimensionality equal to two for both presentation conditions. One-dimensionality cannot be entirely ruled out with this method because stress values for one-dimensional solutions are not available for experimental data. However, evidence for meaningful two-dimensional structure in Kleiss (1990) suggests consideration of two-dimensional solutions.

It should be noted that stress values are relatively large even in the best case of dynamic presentation. Kruskal and Wish (1986) suggested that values of 0.010 or less are to be expected at the correct dimensionality. Hence the fit of the data is not particularly good in either presentation condition. ALSCAL provides a plot of stimulus disparities (transformed dissimilarity data) versus stimulus distances in spatial configurations, which is informative regarding the distribution of error in the data. Plots for both presentation conditions reveal that many medium dissimilarities are fit to very small distances in the spatial configurations.

In the dynamic presentation condition, distances of 0.4 or less were fit to disparities as large as 2.1. In the static presentation condition, distances of 0.4 or less were fit to disparities as large as 1.8. This pattern could simply reflect noise in the data. It is also evident with three-dimensional solutions (though to a lesser degree), suggesting that the poor fit does not result simply from an underestimation of dimensionality in these data. However, it could be that some structure in the data cannot be recovered because of insufficient disparity among stimuli.

A feature of ALSCAL output provided with the individual differences approach is subject weights, which reflect the relative importance of each dimension for individual subjects. Subject weights are root mean square measures that, when squared, sum to R^2 for individual subjects. When averaged across subjects, squared subject weights provide estimates of variance explained by each dimension for the group. Because the data are ordinal and do not satisfy the metric properties that underlie usual interpretations of variance, these estimates must be taken as approximations.

Table 1 shows average squared subject weights for Dimensions 1 and 2 of the two-dimensional ALSCAL solutions for the dynamic and static presentation conditions. Average squared subject weights are larger for Dimension 1 in both presentation conditions, suggesting that the scene property captured by Dimension 1 is most important. The difference favoring Dimension 1 is larger for the dynamic presentation condition.

The extent to which an individual subject's weights are proportional to the group average is indexed by *weirdness*. A weirdness value of zero indicates that a given individual's weighting of dimensions is exactly proportional to the group average. A weirdness value approaching one indicates that a given individual has weighted one dimension disproportionately large in comparison with the group average and that the other weight(s) is correspondingly small. Data for one subject in each presentation condition indicated a weirdness value exceeding 0.500. The pattern for these two subjects was similar to the group in that both weighted Dimension 1 more heavily than Dimension 2. They were simply more extreme in that pattern than the group. In general, subjects appear to be consistent in their relative weighting of dimensions.

Bipolar attribute ratings. Bipolar attribute ratings for each scene were averaged across subjects. Table 2 shows intercorrelations among mean bipolar attribute ratings in the dynamic and static presentation conditions. Labels correspond to the leftmost (numerically smallest) end of each scale and generally denote presence of the rated attribute. Intercorrelations tend to be higher in the dynamic than in the static presentation condition. With the exception of the hilly/mountainous and regular attributes, intercorrelations tend to be high among attribute ratings, suggesting a high degree of interrelatedness among attributes.

Mean bipolar attribute ratings were also analyzed using a multiple regression approach described by Kruskal and Wish (1986). Separate analyses were performed for each bipolar attribute scale. Mean ratings were regressed on dimensional coordinates derived from the two-dimensional ALSCAL solutions. Tables 3 and 4 show regression weights and multiple correlations for each of the eight bipolar attribute scales in the dynamic and static presentation

TABLE 1

Average Squared Subject Weights for Dimensions 1 and 2 for Static and Dynamic Presentation: Experiment 1

Presentation Mode	Dimension 1	Dimension 2
Dynamic	0.524	0.242
Static	0.350	0.189

TABLE 2

Intercorrelations among Mean Bipolar Attribute Ratings in the Dynamic and Static Presentation Conditions: Experiment 1

	Prefer	Hilly	Objects	Size	Detail	Complex	Regular	Hi-Contrast
			Dynamic Presentation					
Prefer	1.00							
Hilly	0.13	1.00						
Objects	0.94	0.27	1.00					
Known size	0.93	0.19	0.97	1.00				
Texture/detail	0.94	0.28	0.98	0.96	1.00			
Complex	0.75	0.63	0.82	0.81	0.81	1.00		
Regular	−0.20	−0.60	−0.24	−0.20	−0.23	−0.55	1.00	
High contrast	0.95	0.33	0.95	0.91	0.96	0.83	−0.33	1.00
			Static Presentation					
Prefer	1.00							
Hilly	−0.16	1.00						
Objects	0.85	0.17	1.00					
Known size	0.84	0.08	0.96	1.00				
Texture/detail	0.91	0.01	0.81	0.78	1.00			
Complex	0.38	0.70	0.66	0.62	0.53	1.00		
Regular	0.27	−0.36	0.13	0.16	0.24	−0.22	1.00	
High contrast	0.82	0.04	0.75	0.75	0.95	0.52	0.22	1.00

conditions, respectively. Polarity of ALSCAL dimensional coordinates (positive vs. negative) is arbitrary. Coordinates in each presentation condition were therefore scaled in such a way that presence of attributes was associated with positive dimensional coordinates for the largest proportion of attribute scales.

Because the presence of attributes was generally denoted by small numerical values, regression weights were typically negative. Regression weights were converted to direction cosines by normalizing so that they summed to 1.00 when squared. Regression weights, therefore, describe property vectors indicating the direction in multidimensional space that best fits rated increase in various attributes. A weight of 1.00 for a given dimension indicates perfect alignment of the property vector with the dimensional axis and the closest possible relationship between attribute ratings and stimulus ordering along the axis.

Kruskal and Wish (1986) suggested that a bipolar attribute scale may provide a satisfactory interpretation of a dimension if the following conditions are met: First, the multiple

TABLE 3

Multiple Regression Analyses of Bipolar Ratings: Experiment 1, Dynamic Presentation

	Regression Weights		
Scale	Dimension 1	Dimension 2	Multiple R
1. Prefer	−0.135	−0.991	0.933*
2. Hilly/mountainous	−0.956	0.294	0.859*
3. Objects	−0.326	−0.945	0.957*
4. Known size references	−0.273	−0.962	0.930*
5. Texture/detail	−0.357	−0.934	0.923*
6. Complex	−0.775	−0.632	0.894*
7. Regular	0.999	0.040	0.451 *ns*
8. High contrast	−0.437	−0.899	0.913*

* $p < 0.001$; *ns* = nonsignificant.

TABLE 4

Multiple Regression Analyses of Bipolar Ratings: Experiment 1, Static Presentation

Scale	Regression Weights		Multiple R
	Dimension 1	Dimension 2	
1. Prefer	−0.589	−0.808	0.882*
2. Hilly/mountainous	−0.755	0.656	0.888*
3. Objects	−0.904	−0.427	0.861*
4. Known size references	−0.880	−0.476	0.804*
5. Texture/detail	−0.717	−0.697	0.939*
6. Complex	−0.956	0.292	0.944*
7. Regular	0.314	−0.949	0.447 ns
8. High contrast	−0.754	−0.657	0.860*

* $p < 0.001$; ns = nonsignificant.

correlation *(R)* for the scale must be large and statistically reliable (0.90s are good, but 0.80s and 0.70s may suffice). If the multiple *R* is large and statistically reliable, then a regression weight must also be comparatively large for one of the two dimensions. A criterion value of 0.940 or larger was selected for regression weights in the present experiment. This value corresponds to a deviation of the property vector from the dimensional axis of 20 deg or less.

Inspection of Tables 3 and 4 reveals that multiple *R*'s are large and statistically reliable for all but the regular scales in each presentation condition. However, the pattern of regression weights differs between presentation conditions. In the dynamic presentation condition (Table 3), only the hilly/mountainous scale has a significant multiple *R* and a regression weight exceeding the 0.940 criterion value for Dimension 1 coordinates. Several scales have significant multiple *R*'s and large regression weights for Dimension 2 coordinates. However, only the Prefer, Known Size references, and Objects scales have regression weights exceeding the 0.940 criterion. High intercorrelations among these attributes (Table 2) suggest that they are not independently predictive of positioning along the Dimension 2 axis.

In the static presentation condition (Table 4), only the complex scale has a significant multiple *R* with a regression weight exceeding the 0.940 criterion for Dimension 1 coordinates. None of the scales reaches these criteria for Dimension 2 coordinates. Based on these results, it would appear that different structures are being represented in the dynamic and static presentation conditions.

Spatial configurations. Because multiple regression results differed between presentation conditions, spatial configurations will be discussed separately.

Figure 2 shows the two-dimensional spatial configuration for the dynamic presentation condition. Dotted lines drawn through the origin are property vectors for attributes with largest regression weights in Table 3. Considering Dimension 1 first, five scenes clustered at the extreme right end of the dimensional axis (ridges, barren hills, hills with trees, valley, and forested mountain) are rich in hills or ridges, whereas scenes positioned at the extreme left end of the dimensional axis are flat in the vicinity of the eye point. This pattern suggests an interpretation of Dimension 1 consistent with variation in terrain vertical development, which is supported by alignment of the hilly/mountainous property vector with the Dimension 1 axis. Some scenes at the left end of Dimension 1 (e.g., shore approach, dry lake, desert, grassland, desert with trees) contain mountains occluding the horizon, whereas some scenes at the right end of Dimension 1 (e.g., hills with trees) contain no large vertical obstructions. Hence the specific property related to Dimension 1 is undulations

Experiment 1:
Dynamic Presentation

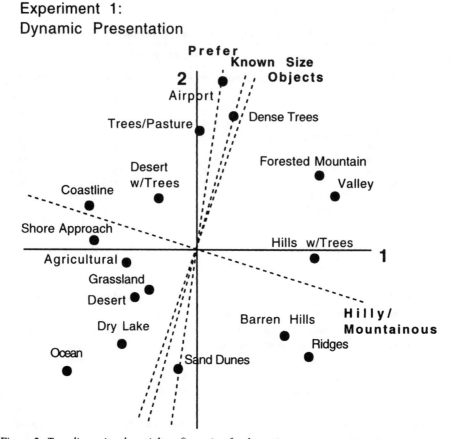

Figure 2. *Two-dimensional spatial configuration for dynamic presentation: Experiment 1.*

in the terrain surface over which the aircraft is flying, rather than presence of large vertical obstructions.

Scenes positioned at the extreme top of the Dimension 2 axis contain large buildings or trees clustered compactly into large groups, whereas scenes positioned at the extreme bottom of the dimensional axis are essentially devoid of objects or vegetation. This pattern suggests an interpretation of Dimension 2 consistent with presence or absence of objects in scenes, which is supported by alignment of the Objects property vector with the Dimension 2 axis. Scenes containing a uniform distribution of vegetation are positioned near the middle of the dimensional axis, suggesting that this spatial arrangement is not optimal. It therefore appears that when trees are clustered compactly into groups such as in the dense trees and trees/pasture scenes, a new property emerges that is similar to that exhibited by large buildings in the airport scene. Alignment of the Known Size property vector with Dimension 2 indicates that size is more readily discernible with groups of trees and large buildings than with individual objects scattered evenly on the terrain. Because Known Size information is assumed to derive from knowledge of the identity of objects, this suggests a possible influence of semantic factors in scenes.

It is of interest to note, however, that individual trees in the dense trees and trees/pasture scenes are about the same size and shape as those in the hills with trees scene. Hence familiar size in the dense trees and trees/pasture scenes appears to be associated with the occurrence of grouping rather than knowledge of the size of individual trees. It is noteworthy that known size is associated with natural objects because manufactured objects are sometimes assumed to be privileged in this regard (162d Tactical Fighter Group, 1986). Thus the common characteristic shared by large objects is presence of large bounded surfaces or, perhaps, enclosed volume. Alignment of the Prefer property vector with Dimension 2 indicates that pilots prefer large objects to hills and ridges in Dimension 1. Preference for objects contrasts with the importance of hills and ridges (Dimension 1) reflected in estimates of explained variance in similarity ratings (see Table 1). This suggests that different information underlies preference for scene elements versus importance of scene elements reflected in the perceived magnitude of differences between scenes.

Figure 3 shows still photographs depicting the trees/pasture, hills with trees, and ocean scenes representing prototypical exemplars of the dimensions discussed earlier.

Figure 4 shows the two-dimensional spatial configuration for the static presentation condition. The dotted line corresponds to the complex property vector, which was the only attribute related to the spatial configuration in Table 4. Scenes positioned at the extreme right end of Dimension 1 are rich in a variety of scene properties, including objects as well as hills or ridges. Scenes at the extreme left end of the dimension generally lack notable scene elements. Alignment of the complex property vector with Dimension 1 suggests an interpretation consistent with variation in global scene complexity. There is no obvious pattern in the ordering of scenes along the Dimension 2 axis, suggesting that data in the static presentation condition may be one-dimensional.

Discussion

Evidence in the dynamic presentation condition for one dimension related to terrain vertical development and another dimension related to

objects replicates results of Kleiss (1990) and argues for the generality of this two-dimensional structure. Lack of evidence for higher dimensionality with the present larger stimulus set suggests that two dimensions are sufficient to describe the perceptual structure of real-world scenes in the context of low-altitude flight. It is, of course, possible that the present stimulus set lacks some relevant exemplars that would reveal additional structure if included.

Care was taken during the development of this stimulus set to consult both experienced pilots and U.S. Air Force instructional materials (162d Tactical Fighter Group, 1986) for input regarding potentially relevant scene properties. Subjects were also routinely debriefed after data collection to obtain further input as to possible exclusions from the data set. No obvious omissions were noted, though pilots sometimes commented on the absence of a scene with an unbroken canopy of trees. This issue will be addressed in Experiment 2.

Pilots also occasionally cited the absence of certain problematic environments, such as snow-covered trees or dense fog. No imagery was available representing these conditions, but it was assumed that scenes exhibiting smooth water, sand dunes, or a dry lake bed would exemplify similarly problematic environments. The pattern of error in the data revealing many medium-sized disparities that are not well fit in the spatial configurations does suggest, however, the possibility that additional structure might be revealed with more disparate stimuli. Given that the present stimulus sets reflect a nearly twofold increase in stimuli over the set used by Kleiss (1990), it is not obvious what enhancements would lead to greater stimulus disparity.

The present set of videotape segments exhibited more subtle variation among stimuli than the stimulus set of Kleiss (1990), and this allowed more precise specification of relevant scene properties. For example, the importance of terrain vertical development (Dimension 1) was revealed to derive from undulations in the terrain rather than the presence of large vertical

Figure 3. *Photographs depicting the trees/pasture (top), hills with trees (middle) and ocean (bottom) scenes. Note: Photographs have been cropped.*

obstructions. Also, it was found that an uneven distribution of vegetation better exemplified Dimension 2 than a uniform distribution of individual objects, such as have been common in investigations of simulated low-altitude flight (e.g., Kleiss and Hubbard, 1993; Martin and Rinalducci, 1983).

Replication of Kleiss's (1990) results under controlled conditions of altitude and speed argues that variation in these factors in the experiment of Kleiss (1990) had little effect on similarity judgments. Nevertheless, lack of interpretable multidimensional scene structure in the static presentation condition

Experiment 1:
Static Presentation

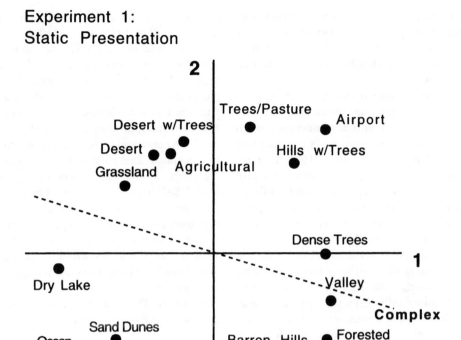

Figure 4. *Two-dimensional spatial configuration for static presentation: Experiment 1.*

indicates that some minimum amount of motion is important.

It is interesting to note that the lack of interpretable multidimensional scene structure in the static presentation condition does not result from a failure to perceive either of the exemplars defining the two dimensions in the dynamic presentation condition. Both hills and ridges, as well as large objects, are clearly represented at the right end of Dimension 1 in Figure 4. What is not perceived in still photographs is a more subtle form of information that uniquely distinguishes between these scene properties. It would appear that this information is related to motion.

EXPERIMENT 2

In both previous experiments using videotape segments for stimuli (Kleiss, 1990, and Experiment 1, dynamic presentation condition), terrain vertical development (Dimension 1) has proved to be the most important scene property, in that it explained the largest proportion of variance in similarity ratings. However, two subjects in Kleiss's (1990) investigation showed the reverse pattern, in which objects (Dimension 2) were most important. These two subjects were also unique in two other potentially important ways. First, they were the only pilots assigned to the A-10 aircraft, a relatively slow but maneuverable jet fighter. They were also the only pilots stationed in the eastern United States at the time of the investigation. All other pilots (including those in Experiment 1) were stationed in the southwestern desert, where vegetation and cultural features are relatively uncommon. It is possible that the reversal in the relative importance of dimensions for these two pilots derives from unique properties of the A-10 aircraft or of its mission, or from familiarity with environments richer in vegetation and cultural features.

This second hypothesis is consistent with recent accounts of perceptual motor skill that postulate a process of perceptual attunement to relevant information underlying skill acquisition (e.g., Flach, Lintern, and Larish, 1990; Lintern, 1991). Support for perceptual attunement also comes from pilots who report that on first flying in an unfamiliar geographic region, an adaptive period is required to become accustomed to new environmental features (Lt. Comdr. D. Shinn, U.S. Navy, personal communication, June 23, 1993). To determine whether experience flying in a geographic region rich in vegetation and cultural features influences the relative importance of dimensions, it was decided to replicate the experiment using a sample of pilots stationed in Germany, where the geography differs notably from that in the southwestern United States. The static presentation condition was retained in hopes of shedding additional light on the difference between presentation modes.

Method

Subjects and design. Subjects were 32 mission-qualified USAF pilots in Europe. Of these, 14 were F-16 pilots assigned to the 86th Tactical Fighter Wing, Ramstein Air Base, Germany; 14 were F-4 or F-16 pilots assigned to the 52d Tactical Fighter Wing, Spangdahlem Air Base, Germany; and 4 were RF-4 pilots assigned to the 26th Tactical Reconnaissance Wing, Sweibruken Air Base, Germany. Mission requirements for all pilots included low-altitude, high-speed flight. Nine pilots from the 86th Tactical Fighter Wing were assigned to the static presentation condition, and the remaining 23 pilots were assigned to the dynamic presentation condition. Mean total flying time was 896 h for the static presentation condition and 1329 h for the dynamic presentation condition. Mean fly-

ing time did not differ significantly between groups, $t(30) = 1.604$, $p = 0.119$.

Stimuli and materials. The apparent speed in videotape segments was reduced to approximately 420 knots in this experiment, a value closer to that used by pilots during routine training missions. Stimulus segments remained 5 s in duration. One additional change was made: Whereas no scene in Experiment 1 exhibited a continuous canopy of trees unbroken by clearings, a videotape segment was obtained from USAF files that depicted a dense canopy of trees on highly contoured terrain. This segment replaced the forested mountain scene in the present experiment and retained that name. This segment was recorded during a routine low-altitude training mission, though speed and altitude were unknown.

Procedure. The procedure was identical to that described in Experiment 1.

Results

Data for all analyses were distances in millimeters measured from the left end of each rating scale to the point at which the subject marked the scale. The maximum range of values was 0 to 120 mm.

Multidimensional scaling. Pairwise ratings were analyzed as described in Experiment 1. Figure 5 shows stress and R^2 for the dynamic and static presentation conditions, plus stress for random data, as a function of dimensionality. Note first the small R^2 values and large stress values for the static presentation condition, indicating poor fit of the data. Stress in the dynamic presentation condition is, once again, smallest in comparison with random stress at dimensionality equal to two. Stress in the static presentation condition is larger than random stress and differs about equally in magnitude at all dimensionalities. A two-dimensional solution will once again be considered in the static presentation condition.

Plots of stimulus disparities versus stimulus distances revealed a pattern of medium disparities fit to small distances similar to that

Experiment 2

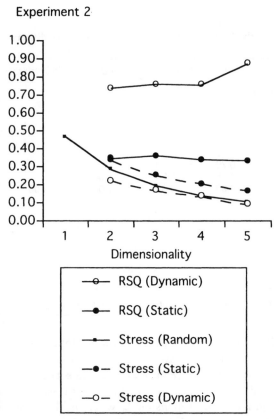

Figure 5. *Stress and* R^2 *for dynamic presentation and static presentation, plus stress for random data, as a function of dimensionality: Experiment 2.*

observed in Experiment 1. In the dynamic presentation condition, distances of 0.4 or less were sometimes fit to disparities as large as 2.0. In the static presentation condition, distances of 0.4 or less were sometimes fit to disparities as large as 1.8.

Table 5 shows average squared subject weights for Dimensions 1 and 2 in the dynamic and

TABLE 5

Average Squared Subject Weights for Dimensions 1 and 2 for Static and Dynamic Presentation: Experiment 2

Presentation Mode	Dimension 1	Dimension 2
Dynamic	0.461	0.275
Static	0.168	0.176

static presentation conditions. Subject weights in the dynamic presentation condition are notably larger than in the static presentation condition, reflecting the superior fit of the data. The difference favoring dynamic presentation is particularly large for Dimension 1. Data for one subject in each of the dynamic and static presentation conditions showed a weirdness value exceeding 0.500. In each case, Dimension 2 was weighted more heavily than Dimension 1. In general, however, subjects were consistent in their ratings.

Bipolar attribute ratings. Table 6 shows intercorrelations among mean bipolar attribute ratings in both the dynamic and static presentation conditions. Once again, with the general exceptions of the hilly/mountainous and regular attributes, intercorrelations are high among most attribute ratings.

Bipolar attribute ratings were analyzed using the multiple regression approach described in Experiment 1. Tables 7 and 8 show regression weights and multiple correlations for each of the eight bipolar attribute scales in the dynamic and static presentation conditions, respectively. Dimensional coordinates were scaled so that the presence of attributes was associated with positive coordinate values for the largest proportion of scales.

Inspection of Tables 7 and 8 shows that multiple R's are large and statistically reliable for all scales. In the dynamic presentation condition (Table 7), regression weights for Dimension 1 coordinates exceed the 0.940 criterion for both the hilly/mountainous and regular scales. For Dimension 2 coordinates, regression weights exceed the 0.940 criterion for the Prefer, Objects, Known Size references, Texture/Detail, and High Contrast scales. Results for the static presentation condition (Table 8) are more similar to those of the dynamic presentation condition than was the case in Experiment 1. Although none of the regression weights exceeds the 0.940 criterion for Dimension 1 coordinates, they are largest for the hilly/mountainous and regular scales. For Dimension 2 coordinates, regression weights exceed the 0.940 criterion for the Prefer,

TABLE 6

Intercorrelations among Mean Bipolar Attribute Ratings in the Dynamic and Static Presentation Conditions: Experiment 2

	Prefer	Hilly	Objects	Size	Detail	Complex	Regular	Hi-Contrast
			Dynamic Presentation					
Prefer	1.00							
Hilly	0.20	1.00						
Objects	0.89	0.27	1.00					
Known size	0.93	0.18	0.98	1.00				
Texture/detail	0.95	0.35	0.92	0.93	1.00			
Complex	0.62	0.73	0.78	0.70	0.76	1.00		
Regular	−0.17	−0.73	−0.12	−0.13	−0.25	−0.54	1.00	
High contrast	0.93	0.07	0.88	0.92	0.93	0.62	−0.08	1.00
			Static Presentation					
Prefer	1.00							
Hilly	0.22	1.00						
Objects	0.67	0.21	1.00					
Known size	0.73	0.27	0.95	1.00				
Texture/detail	0.77	0.42	0.92	0.89	1.00			
Complex	0.48	0.70	0.69	0.71	0.77	1.00		
Regular	−0.35	−0.86	−0.21	−0.27	−0.45	−0.74	1.00	
High contrast	0.82	0.27	0.76	0.80	0.84	0.66	−0.42	1.00

Objects, Known Size references, Texture/Detail, and High Contrast scales. Once again, high intercorrelations among attributes associated with Dimension 2 in Table 6 suggest that they are not independently predictive of Dimension 2.

Spatial configurations. Figures 6 and 7 show the two-dimensional spatial configurations for the dynamic and static presentation conditions, respectively. Dotted lines drawn through the origins are property vectors for attributes with largest regression weights in Tables 7 and 8.

Both spatial configurations appear globally similar to that in the Experiment 1 dynamic presentation condition (Figure 2). The five scenes rich in hills and ridges are positioned at the extreme right end of Dimension 1, whereas the three scenes with large objects or groups of objects are positioned at the extreme top of Dimension 2. These patterns support dimensional interpretations similar to those in the Experiment 1 dynamic presentation condition.

The hilly/mountainous property vector is aligned with the Dimension 1 axis only in the

TABLE 7

Multiple Regression Analyses of Bipolar Ratings: Experiment 2, Dynamic Presentation

	Regression Weights		
Scale	Dimension 1	Dimension 2	Multiple R
1. Prefer	0.015	−1.000	0.938**
2. Hilly/mountainous	−0.997	−0.072	0.982**
3. Objects	−0.079	−0.997	0.869**
4. Known size references	0.015	−1.000	0.877**
5. Texture/detail	−0.138	−0.990	0.911**
6. Complex	−0.703	−0.711	0.874**
7. Regular	0.984	0.179	0.732*
8. High contrast	0.163	−0.987	0.889**

* $p < 0.010$; ** $p < 0.001$.

TABLE 8

Multiple Regression Analyses of Bipolar Ratings: Experiment 2, Static Presentation

	Regression Weights		
Scale	Dimension 1	Dimension 2	Multiple R
1. Prefer	0.161	−0.987	0.796**
2. Hilly/mountainous	−0.919	−0.395	0.909**
3. Objects	−0.077	−0.997	0.830**
4. Known size references	−0.099	−0.995	0.852**
5. Texture/detail	−0.166	−0.986	0.871**
6. Complex	−0.551	−0.835	0.937**
7. Regular	0.803	0.596	0.755*
8. High contrast	0.114	−0.993	0.933**

* $p < 0.010$; ** $p < 0.001$.

dynamic presentation condition (Figure 6). Although hills and ridges were perceived as unique properties in the static presentation condition, scenes nevertheless appear to lack some information unique to the attribute of being hilly or mountainous. Alignment of the regular property vector with Dimension 1 in the dynamic presentation condition (Figure 6),

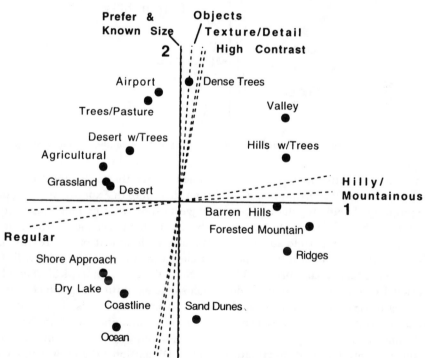

Figure 6. *Two-dimensional spatial configuration for dynamic presentation: Experiment 2.*

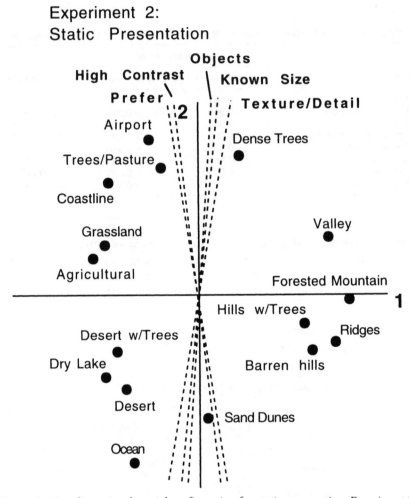

Figure 7. *Two-dimensional spatial configuration for static presentation: Experiment 2.*

pointing toward scenes with flat terrain, also suggests that motion over flat terrain is associated with the property of being regular and predictable. Flat terrain, therefore, should be conceptualized not purely with regard to absence of hills and ridges but also with regard to the positive attribute of being regular and predictable. Hence Dimension 1 can more accurately be described as relating to variation in terrain shape (flat vs. undulating) rather than to presence or absence of hills and ridges.

In the dynamic presentation condition (Figure 6), scenes with small, dense vegetation are positioned nearer the top of the Dimension 2 axis than they were in Experiment 1 (Figure 2). This suggests somewhat greater sensitivity to small objects in this experiment, which may explain the closer alignment of the Texture/Detail and High Contrast property vectors with Dimension 2. Nevertheless, prototypical exemplars of Dimension 2 exhibiting large objects are the same as in Experiment 1. It is of interest to note that the forested mountain scene in the dynamic presentation condition is positioned nearer the bottom of Dimension 2 than it was in Experiment 1 (Figure 2). Recall that in the present experiment, this scene exhibited a dense canopy of trees with no clearings. The positioning of this scene nearer the bottom of Dimension 2 provides further evidence of the importance of boundaries delineating regions of dense vegetation as a defining characteristic of large objects.

Discussion

Results in the dynamic presentation condition are similar to those of Experiment 1 (dynamic presentation) as well as those of Kleiss (1990), suggesting that familiarity with the rich European environment had little effect on pilots' judgments in this experiment. Hence the apparent difference in the relative weighting of dimensions reported for two A-10 pilots by Kleiss (1990) probably does not reflect perceptual attunement to geographic features common in nondesert environments. It is interesting to note, however, that in the three experiments using videotape stimuli (Kleiss, 1990, and Experiments 1 and 2 in this investigation), only one other pilot in the total of 48 sampled weighted Dimension 2 more heavily than Dimension 1. This is highly suggestive that the pattern is not common in the general population of fighter pilots. This fact suggests further consideration of the remaining possibility that some property unique to the A-10 aircraft or its mission accounts for the difference.

In Experiment 2 dimensional structure in the static presentation condition was more similar to that in the dynamic presentation condition than was the case in Experiment 1. However, the fit of the data remained considerably better in the dynamic presentation condition, particularly for Dimension 1 (see Table 5). This implies that the importance of terrain shape (Dimension 1) in the dynamic presentation condition can be attributed to dynamic scene characteristics.

Present evidence for two relevant scene properties, one dependent on motion and another related to objects, bears a strong resemblance to the pattern of results obtained by DeMaio et al. (1983) with a simulated altitude estimation task. The notable difference in the present case is that the scene property affected by motion was terrain shape, whereas in the investigation of De-Maio et al. it was texture on the terrain (specifically, a mosaic of geometric shapes defining the terrain surface). When moving over a flat surface, texture elements flow continuously outward in an expanding pattern centered on the heading direction. When moving over undulating terrain, optic flow is discontinuous at boundaries between foreground and background surfaces.

Optic flow discontinuities are useful for perceiving depth separation between surfaces (see Stevens, 1994, for a review of the literature pertaining to perception of depth discontinuities). USAF instructional materials (162d Tactical Fighter Group, 1986) emphasize an active strategy of searching for optic flow discontinuities as a means for detecting hills and ridges set against background terrain. A reasonable hypothesis, therefore, is that Dimension 1 in the dynamic presentation condition reflects variation in terrain shape mediated by continuous (flat terrain) versus discontinuous (undulating terrain) optic flow. Alignment of the hilly/mountainous property vector with Dimension 2 only in the dynamic presentation condition suggests that this attribute is uniquely associated with dynamic scene characteristics. Hills and ridges in the dynamic presentation were also associated with the property of being random and unpredictable. The salient difference between predictable and unpredictable terrain could account for the importance of Dimension 1 in the dynamic presentation condition.

GENERAL DISCUSSION AND CONCLUSIONS

The foregoing results provide consistent evidence for a multidimensional conceptualization of scene content in which scenes vary with respect to two different properties: terrain shape and object size or grouping. These results have both practical implications for the design of flight simulator visual scenes as well as theoretical implications for basic self-motion perception research.

The finding of a dimension related to objects

(Dimension 2) is consistent with the large body of evidence that objects provide important information for perceiving and controlling altitude in flight simulators (Buckland et al., 1981; DeMaio et al., 1983; Engle, 1980; Kleiss and Hubbard, 1993; Martin and Rinalducci, 1983; McCormick et al., 1983). A common finding is that performance in flight simulators improves with increases in object density (Engle, 1980; Kleiss and Hubbard, 1993; Martin and Rinalducci, 1983). The present results are useful in suggesting that a dense array of individual objects may not be optimal. Scenes positioned nearest the top of Dimension 2 in Experiments 1 and 2 contained either large buildings or smaller trees clustered compactly into larger groups. McCormick et al. (1983) manipulated a variety of objects in flight simulator visual scenes but found that altitude control during low-altitude flight was no better with large warehouses than with smaller trees. The best performance was obtained with a mixture of objects, suggesting that variation in object size may be an important factor. Large size appears to be an emergent property of small objects clustered compactly into groups. Because it may not always be appropriate to model large buildings in flight simulator visual scenes, an important goal for future research will be to examine the influence of natural object groupings.

Previous flight simulation research has established the importance of vertical objects that extend above the terrain surface (Martin and Rinalducci, 1983). Consistent with that finding, scenes with the tallest objects were positioned nearest the top of Dimension 2 in the present experiments. However, the most salient property of verticality was exhibited by elements of terrain shape in Dimension 1, suggesting that terrain shape defines a second unique type of verticality. The importance of terrain shape has not been widely investigated in flight simulators. Hills or mountains have been included in a few investigations of simulated low-altitude flight (e.g., Buckland et al., 1981; DeMaio et al., 1983), but they served to define large vertical obstructions to vision and navigation and had

no influence on task performance. Barfield et al. (1989) modeled smaller hills on the approach path to a simulated runway and found that estimation of impact point did improve significantly compared with a condition in which the terrain was flat. The results of Barfield et al. are consistent with the present finding that the important property of terrain shape relates to smaller-scale terrain elements rather than large vertical obstructions.

This fact has important implications for the design of flight simulator visual scenes. Terrain in flight simulators comprises polygonal surfaces joined to conform to the desired shape of the terrain. A few large polygons are sufficient to model large-scale terrain elements. However, to model undulating terrain such as that exemplified in present scenes would require a higher density of smaller polygons, which would considerably increase the load on CIG processing resources. It is not yet known what level of terrain resolution is optimal in flight simulators, and this question defines a topic for future research.

A high density of terrain polygons alone does not guarantee effective perception of terrain shape. The importance of motion for perceiving terrain shape suggests that optic flow phenomena are involved. Stevens (1994) discussed several factors affecting perception of depth discontinuities from visual motion cues. A primary requirement is a high density of texture elements on surfaces. He noted that humans exhibit hyperacuities in motion perception on the order of approximately 25 arcsec/s for a centrally fixated target (e.g., a ridgeline). This implies a need for sophisticated texturing capabilities in flight simulators as well as high display resolution capable of revealing small spatial displacements of texture elements. Because both factors affect display design and cost, another important goal for future research will be to establish minimum texturing and display resolution requirements for rendering terrain shape in flight simulators.

Attempts to establish quality criteria for flight simulator visual displays have typically focused on factors such as polygon count, texturing capabilities, luminance, contrast, resolution, and color palette (e.g., Padmos and Milders, 1992). Although display quality undoubtedly varies with these factors, they do not by themselves guarantee high-quality scenes because they do not take into account the degree to which specific scene properties are effectively rendered and perceived. The present MDS methodology suggests a criterion by which to assess the quality of simulated imagery, based on the perceived content of real-world scenes. That is, if an MDS analysis were performed using simulated scenes as stimuli, the obtained dimensional structure would reveal the degree to which scene properties perceived in real-world scenes were being perceived in simulated scenes. Differences in the number, type, and/or relative importance of dimensions compared with present results would reflect differences in display quality. An example of this approach to image quality is provided in the present investigation, which revealed poorer quality of static imagery that was traced to absence of motion information useful for perceiving terrain shape.

Available data generally indicate that training effectiveness improves with increases in flight simulator visual scene detail (Lintern and Koonce, 1992; Lintern et al., 1987; Lintern and Walker, 1991). One might assume, therefore, that maximum training effectiveness would be obtained with scenes rich in elements of both terrain shape and objects. There is evidence that argues against this assumption, however. Warren and Riccio (1985) found that training effectiveness can be improved in some cases by reducing scene content. In their experiment, a powerful cue for controlling altitude (an outline roadway providing perspective information) dominated a less powerful cue (a distribution of dots on the terrain) such that learning associated with the less powerful cue improved when the dot pattern was presented in isolation. The present multidimensional conceptualization of scene content is useful in suggesting dimensions along which scene content can be varied.

The present results are based on similarity judgments between stimuli depicting either no motion or rectilinear self-motion at constant speed and altitude. Because a primary concern in flight simulation is perception and control of changing motion, it is important to verify the relevance of these scene properties for performance-based tasks.

Previous research examining perception of change in altitude (e.g., Flach et al., 1992; Johnson et al., 1989; Wolpert, 1988; Wolpert et al., 1983) and change in speed (e.g., Larish and Flach, 1990; Owen, Warren, Jensen, Mangold, and Hettinger, 1981; Owen, Wolpert, and Warren, 1984) has typically used stimuli consisting of simple grids or dot patterns on horizontal, flat surfaces. Despite their apparent simplicity, these stimuli are rich in a variety of information useful for perceiving change in altitude and speed.

Examples of information useful for perceiving change in altitude are change in the lateral displacement of environmental elements relative to the heading direction (perspective angle) and change in the vertical displacement of environmental elements relative to the horizon (depression angle). Examples of information useful for perceiving change in speed are change in global optic flow rate and change in optical edge rate. Present scenes bearing the closest resemblance to such stimuli (e.g., agricultural, desert, and grassland, exhibiting a uniform distribution of elements on flat terrain) were positioned farthest from prototypical exemplars of Dimension 1 and midway along the Dimension 2 axis. Hence stimuli used in previous investigations are representative of a comparatively limited range of scene properties found to be relevant in present real-world scenes.

Using the terminology of Gibson (1979), prototypical exemplars of dimensions in present experiments exhibit considerable *clutter*, defined as "objects or surfaces that occlude parts of the ground and divide the habitat into semi-enclosures" (p. 307). The present results suggest that elements of clutter are perhaps the more important sources of information for perceiving self-motion events. Evidence for multidimensional scene structure implies a classification of clutter along two separate dimensions. The foregoing results therefore suggest extending analyses of stimulus structure useful for perceiving self-motion events to include that related to (a) occluding terrain surfaces and (b) large objects with vertical extent.

ACKNOWLEDGMENTS

This work was conducted under contract F33615-90-C-0005 from the Armstrong Laboratory, Aircrew Training Research Division. The task monitor was Elizabeth Martin, who provided many helpful comments on this research. The author thanks Tom Nygren and Don Polzella for helpful comments on the use of ALSCAL, Deke Joralmon for assistance with videotape editing, Celeste Howard and Julie Lindholm for helpful comments on an earlier draft of this manuscript, Fred Previc for providing the forested mountain videotape segment used in Experiment 2, and Marge Keslin for editorial assistance. I also thank two anonymous reviewers for their insightful and helpful comments.

REFERENCES

Barfield, W., Rosenberg, C., and Kraft, C. (1989). The effects of visual cues to realism and perceived impact point during final approach. In *Proceedings of the Human Factors Society 33rd Annual Meeting* (pp. 115–119). Santa Monica, CA: Human Factors and Ergonomics Society.

Biederman, I. (1981). Do background depth gradients facilitate object identification? *Perception, 10,* 573–578.

Biederman, I., Mezzanotte, R. J., and Rabinowitz, J. C. (1982). Scene perception: Detecting and judging objects undergoing relational violations. *Cognitive Psychology, 14,* 143–177.

Biederman, I., Rabinowitz, J. C., Glass, A. L., and Stacy, E. W. (1974). On the information extracted from a glance at a scene. *Journal of Experimental Psychology, 103,* 597–600.

Buckland, G. H., Edwards, B. J., and Stevens, C. W. (1981). Flight simulator visual and instructional features for terrain flight simulation. In *Proceedings of the Image Generation/Display Conference II* (pp. 351–362). Phoenix, AZ: IMAGE Society.

Buckland, G. H., Monroe, E., and Mehrer, K. (1980). *Flight simulator runway visual textural cues for landing* (AFHRL-TR-79-81, AD A089434). Williams Air Force Base, AZ: Operations Training Division, Human Resources Laboratory.

DeMaio, J., Rinalducci, E. J., Brooks, R. and Brunderman, J. (1983). Visual cueing effectiveness: Comparison of perception and flying performance. In *Proceedings of the 5th Annual Interservice/Industry Training Equipment Conference* (pp. 92–96). Arlington, VA: American Defense Preparedness Association.

Dupnick, E. G. (1979). *The effects of context on multidimensional spatial cognitive models.* Unpublished doctoral dissertation, University of Arizona, Tucson.

Engle, R. L. (1980). *Use of low altitude visual cues for flight simulation.* Unpublished master's thesis, Arizona State University, Tempe.

Fender, E. G. (1982). Analysis of non-metric data. In W. Richards and K. Dismukes (Eds.), *Vision research for flight simulation* (AFHRL-TR-82-6, AD-A118 721, pp. 31–37). Williams Air Force Base, AZ: Operations Training Division, Human Resources Laboratory.

Flach, J. M., Hagen, B. A., and Larish, J. F. (1992). Active regulation of altitude as a function of optical texture. *Perception and Psychophysics, 51,* 557–568.

Flach, J. M., Lintern, G., and Larish, J. F. (1990). Perceptual motor skill: A theoretical framework. In R. W. Warren and A. H. Wertheim (Eds.), *Perception and control of self-motion* (pp. 327–355). Hillsdale, NJ: Erlbaum.

Gibson, J. J. (1979). *The ecological approach to visual perception.* Boston: Houghton Mifflin.

Isaac, P. D., and Poor, D. D. S. (1974). On the determination of appropriate dimensionality in data with error. *Psychometrika, 39,* 91–109.

Johnson, W. W., Tsang, P. S., Bennett, C. T., and Phatak, A. V. (1989). The visually guided control of simulated altitude. *Aviation, Space, and Environmental Medicine, 60,* 152–156.

Kleiss, J. A. (1990). *Terrain visual cue analysis for simulating low-level flight: A multidimensional scaling approach* (AFHRL-TR-90-20). Williams Air Force Base, AZ: Operations Training Division, Human Resources Laboratory.

Kleiss, J. A. (1992). *Perceptual dimensions of visual scenes relevant for simulating low-altitude flight* (AL-TR-1992-0011). Williams Air Force Base, AZ: Aircrew Training Research Division, Armstrong Laboratory.

Kleiss, J. A., and Hubbard, D. C. (1993). Effect of three types of flight simulator visual scene detail on detection of altitude change. *Human Factors, 35,* 653–671.

Kruskal, J. B., and Wish, M. (1986). *Multidimensional scaling* (Sage University Paper Series on Quantitative Applications in the Social Sciences, 07-011). Newbury Park, CA: Sage.

Larish, J. F., and Flach, J. M. (1990). Sources of optical information useful for perception of speed of rectilinear self-motion. *Journal of Experimental Psychology: Human Perception and Performance, 16,* 295–302.

Lintern, G. (1991). An informational perspective on skill transfer in human-machine systems. *Human Factors, 33,* 251–266.

Lintern, G., and Koonce, J. M. (1992). Visual augmentation and scene detail effects in flight training. *International Journal of Aviation Psychology, 2,* 281–301.

Lintern, G., Thomley-Yates, K. E., Nelson, B. E., and Roscoe, S. N. (1987). Content, variety, and augmentation of simulated visual scenes for teaching air-to-ground attack. *Human Factors, 29,* 45–59.

Lintern, G., and Walker, M. B. (1991). Scene content and runway breadth effects on simulated landing approaches. *International Journal of Aviation Psychology, 1,* 117–132.

MacCallum, R. C. (1978). Recovery of structure in incomplete data by ALSCAL. *Psychometrika, 44,* 69–74.

Marr, D. (1982). *Vision.* San Francisco: Freeman.

Martin, E. L., and Rinalducci, E. J. (1983). *Low-level flight simulation: Vertical cues* (AFHRL-TR-83-17, AD-A133612). Williams Air Force Base, AZ: Operations Training Division, Human Resources Laboratory.

McCormick, D., Smith, T., Lewandowski, F., Preskar, W., and Martin, E. (1983). Low-altitude database development evaluation and research (LADDER). In *Proceedings of the 5th Interservice/Industry Training Equipment Conference* (pp. 150–155). Arlington, VA: American Defense Preparedness Association.

162d Tactical Fighter Group. (1986). *Academic text: Low altitude training.* Tucson: Arizona Air National Guard.

Owen, D. H., Warren, R., Jensen, R., Mangold, S. J., and Hettinger, L. (1981). Optical information for detecting loss in one's own forward speed. *Acta Psychologica, 48,* 203–213.

Owen, D. H., Wolpert, L., and Warren, R. (1984). Effects of optical flow acceleration, edge acceleration, and viewing time on the perception of egospeed acceleration. In D. H. Owen (Ed.), *Optical flow and texture variables useful in detecting decelerating and accelerating self-motion* (Interim technical report for Contract No. F33615-83-K-0038, Task 2313-T3, pp. 79–133). Columbus: Ohio State University,

Department of Psychology, Aviation Psychology Laboratory.

Padmos, P., and Milders, M. V. (1992). Quality criteria for simulator images: A literature review. *Human Factors, 34,* 727–748.

Schiffman, S. S., Reynolds, M. L., and Young, F. W. (1981). *Introduction to multidimensional scaling: Theory, methods, and applications.* New York: Academic.

Spence, I., and Ogilvie, J. C. (1973). A table of expected stress values for random rankings in nonmetric multidimensional scaling. *Multivariate Behavioral Research, 8,* 511–517.

Stevens, K. A. (1994). *The detection of depth discontinuities from visual motion cues.* Manuscript submitted for publication.

Warren, R. (1990). Preliminary questions for the study of egomotion. In R. Warren and A. H. Wertheim (Eds.), *Perception and control of self-motion* (pp. 3–32). Hillsdale, NJ: Erlbaum.

Warren, R., and Riccio, G. E. (1985). Visual cue dominance hierarchies: Implications for simulator design. In *Transactions of the SAE* (pp. 931–937). Warrendale, PA: Society of Automotive Engineers.

Wolpert, L. (1988). The active control of altitude over differing texture. In *Proceedings of the Human Factors Society 32nd Annual Meeting* (pp. 15–19). Santa Monica, CA: Human Factors and Ergonomics Society.

Wolpert, L., Owen, D. H., and Warren, R. (1983). The isolation of optical information and its metrics for the detection of descent. In D. H. Owen (Ed.), *Optical flow and texture variables useful in simulating self motion (II)* (Final Tech. Report for Grant AFOSR-81-0078, pp. 98–121). Columbus: Ohio State University, Department of Psychology, Aviation Psychology Laboratory.

Young, F. W. (1968). *A FORTRAN IV program for nonmetric multidimensional scaling* (L. L. Thurstone Psychometric Laboratory Report No. 56). Chapel Hill: University of North Carolina.

Young, F. W., Takane, Y., and Lewyckyj, R. (1978). ALSCAL: A nonmetric multidimensional scaling program with several differences options. *Behavior Research Methods and Instrumentation, 10,* 451–453.

Date received: January 5, 1994
Date accepted: November 4, 1994

Training Maintenance Technicians for Troubleshooting: Two Experiments with Computer Simulations

WILLIAM B. JOHNSON[1] *and* WILLIAM B. ROUSE, *University of Illinois, Urbana, Illinois*

Aviation maintenance trainees participated in two experiments designed to assess the relative effectiveness of traditional instruction versus two types of computer simulation in the context of aircraft power-plant troubleshooting. Simulations ranged in nature from abstract, context-free problems to those involving specific aircraft power plants. Traditional instruction included reading assignments, television programs tailored to aircraft power-plant troubleshooting, and on-line quizzes. The first experiment compared the three training methods, and the second considered a mixture of the two computer simulations versus traditional instruction. The primary conclusion was that an appropriate combination of low- and moderate-fidelity computer simulations can provide sufficient problem-solving experience to be competitive with the more traditional lecture/demonstration form of instruction.

INTRODUCTION

In pursuit of such goals as improved performance, energy efficiency, and safety, many technical systems (e.g., aircraft, ships, and process plants) are becoming increasingly complex. This trend has resulted in many benefits during normal operation of systems; however, considerable difficulties can arise when such complex systems fail. Therefore, the human troubleshooter is forced to deal with increasingly complicated situations that, due to increased reliability, seldom occur (Rasmussen and Rouse, 1981).

The advent of fault-tolerant systems and automatic test equipment has helped to an extent. However, the addition of such features can, in themselves, increase system complexity. Occasionally, a problem occurs that cannot be handled by redundancy and automatic

test equipment. In those instances, the human may be faced with an extremely complex troubleshooting task. Thus, troubleshooting, which has always been one of the maintenance technician's most demanding tasks, is becoming even more complex.

This paper addresses the question of training maintenance technicians for troubleshooting tasks. Specifically, two training methods involving computer simulations are compared to more traditional methods of instruction. The evaluation involved the study of maintenance trainees in the process of troubleshooting actual aircraft power plants.

BACKGROUND

Fault diagnosis training research is certainly not a new topic (Johnson, Rouse, and Rouse, 1980). A report published in the mid-1950s (Standlee, Popham, and Fattu, 1956) stated that psychological journals had discussed problem solving for "a long time";

---

[1] Requests for reprints should be sent to Dr. William B. Johnson, Aviation Research Lab., Willard Airport, University of Illinois, Savoy, IL 61874.

however, the literature of troubleshooting was "comparatively new." Whereas much research on troubleshooting training was done in the '50s using instruments like the Tab Item Test (Glaser, Damrin, and Gardner, 1954), current computational power has changed the way information is presented (e.g., CRT displays as opposed to cardboard mock-ups). Many of the same questions remain, but a new technology is being used to derive the answers.

Examples of significant computer-based fault diagnosis simulators that have emerged in the past half decade are SOPHIE (Brown, Burton, and Bell, 1975), ACTS (Crooks, Kuppin, and Freedy, 1977), and GMTS (Rigney, Towne, King, and Moran, 1978). SOPHIE (SOPHisticated Instructional Environment) is an artificial intelligence (AI), natural-language-based electronics training simulator designed for the advanced trainee. ACTS (Adaptive Computerized Training System) is another AI-based project that provides the trainee with expert, logic-based advice on an electronic circuit. GMTS (Generalized Maintenance Training Simulator) is different from SOPHIE and ACTS in that it increases its physical fidelity by the use of projected pictures of the real equipment in various operational states. Each of the three systems allows the trainee to interact with the CRT via a keyboard or touchpen.

An important issue in the design of a computer-based simulation—or of any simulation—is the degree of fidelity necessary. Fidelity may be important in at least two ways. First, and probably foremost, fidelity should be high enough to assure transfer of skills from simulator to real equipment. Second, simulator fidelity is of interest in the design of simulators for use in selection or evaluation rather than in training. This paper examines transfer of training and the degree of fidelity of training methods for maintenance trainees in the process of troubleshooting aircraft power plants.

EXPERIMENT 1

Tasks

Thirty-six advanced aviation maintenance trainees, enrolled in the final course prior to FAA certification and averaging 21 years of age, were each trained with one of the following three methods:

(1) *Task.* This method involved three sessions of 20 problems each, using increasingly complex variations of Rouse's two context-free fault diagnosis tasks (Rouse, 1979a, 1979b). Complexity was varied by changing problem size, frequency of feedback loops, level of redundancy, and availability of computer aiding.
(2) *Fault.* This method involved three sessions, totaling 35 problems, using Hunt's context-specific simulation package (Hunt, 1979; Hunt and Rouse, 1981). Simulations of an automobile power plant and two aircraft power plants were used. Two of these power plants were similar to the live equipment that the trainees would use during the transfer-of-training evaluation.
(3) *Video.* This method included three sessions of traditional instruction involving reading assignments, videotaped lectures, and on-line quizzes. Lectures included explicit demonstrations of solutions to three of the five real equipment problems that subjects would encounter in the transfer-of-training evaluation.

Although one might at first choose to use as the control treatment traditional instruction in the fullest sense of the phrase (i.e., classroom lectures and homework), it quickly became apparent that such a choice was unacceptable. There were logistical problems in employing a completely traditional approach. More importantly, however, Task and Fault provide individualized, computer-based instruction. Previous experiments have shown that trainees find Task and Fault to be very interesting and motivating (Rouse, 1979a, 1979b; Hunt and Rouse, 1981). If the control treatment did not also provide such an environment, it would have been quite possible for an artifact to result simply because subjects trained with traditional instruction would feel that they were not receiving as much attention as did the trainees in the other two groups.

To avoid this problem, we decided that instructional television would be the most suitable control treatment. The Video programs were locally produced to include the information normally presented in traditional troubleshooting training. These programs presented a combination of fault diagnosis situations, including actual demonstrations with live engines. The Video programs were supplemented with troubleshooting reading assignments and on-line quizzes.

It is important to emphasize the essential differences among Task, Fault, and Video. Task provides considerable problem-solving experience but no context-specific information. Fault provides a mixture of problem-solving experience and some context-specific information. Video provides no problem-solving experience, but does give considerable context-specific information. To determine the value of the type of context-specific information provided by Video, procedures were presented within the Video programs that provided a nearly optimal solution (as judged by expert technicians) to three of the five problems to be encountered by subjects when troubleshooting the live equipment. This enabled an evaluation of the effects of level of specificity of procedural information presented during training.

Experimental Design

The design of the experiment was a between-subjects design involving 12 subjects in each of three groups. There were three training sessions, requiring approximately six hours. Training was followed by a set of five problems on live aircraft engines. The operational power plants consisted of a four- and a six-cylinder engine found in modern general aviation aircraft. The live system performance was recorded on a standardized checklist that was developed for the experiment. The evaluator was unaware of the training method received by individual subjects.

The five real problems chosen represented the following four engine subsystems: electrical, ignition, lubrication, and fuel. The electrical problem was an "open" lead between the starter solenoid and the starter motor. When the ignition/start switch was engaged there was a loud click generated by the closing solenoid; however, the engine would not crank. The two ignition system malfunctions showed excessive rpm reductions during single magneto operation. These problems were induced by the use of a fouled spark plug and a defective secondary lead, respectively. An oil-pressure reading of zero was induced by obstructing a fitting between the engine and the gauge. The zero oil pressure indication necessitated an immediate engine shutdown. The final problem was a fuel exhaustion symptom caused by the closure of a small in-line valve. With the closed valve, the engine ran for about one minute and then died.

Performance Measures

The transfer-of-training performance measures (i.e., performance on the real equipment) were the following: average performance index (a measure reflecting the average quality of the sequence of actions taken by the subject), adjusted time (the real time to solution adjusted to the manufacturer's labor time schedule), and evaluator's rating (an overall performance score established by the evaluator immediately upon completion of each problem).

The average performance index and the evaluator's rating both used five-point scales, with five indicating superior performance and one indicating poor performance. The average performance index was calculated on the basis of the sequence of actions recorded on standardized observation forms (Johnson, 1981). First, a rating (between five and one) was assigned to each action. Then the total number of points for all actions was divided by the total number of actions (typically from

TABLE 1

Average Performance Index for Each Problem
in Experiment 1

Problem	Task	Fault	Video
Spark Plug	3.62	3.93	4.29*
Secondary Wire	3.76	3.74	4.52*
Fuel Obstruction	4.37	4.32	4.42
Starter Lead	4.07	4.12	4.29
Oil Pressure	3.99	3.77	4.32*

* Problems explicitly shown in Video program.

10 to 20) to obtain the average performance
index for the sequence of actions.

Results

The Video trainees had a significantly
higher average performance index than those
trained with either of the computer simulations, $F(2,33) = 6.27$, $p < 0.01$. The performance index was also significantly affected
by the different problems, $F(4,132) = 3.57$,
$p < 0.01$. Although the interaction between
training methods and problems was not significant, $F(8,132) = 1.44$, a post hoc analysis
of each problem showed that Video had a significantly higher performance index than did
Task or Fault on Problem 1, $F(2,33) = 4.80$,
$p < 0.025$; on Problem 2, $F(2,33) = 8.32$, $p <$
0.005; and, to a lesser degree, on Problem 5,
$F(2,33) = 2.51$, $p < 0.10$. As shown in Table 1,
these problems, in the ignition and lubrication systems, were those that were explicitly
demonstrated during the instruction programs. There was no significant performance
difference among the three groups on the two
problems (i.e., Problems 3 and 4) that were
not explicitly demonstrated in the Video programs, $F(2,33) = 0.09$ and $F(2,33) = 0.44$, respectively.

The Video trainees were not significantly
faster than the Task and Fault trainees; however, the Video trainees did have a significantly higher evaluator's rating, $F(2,33) =$
4.93, $p < 0.025$, averaging 3.21, 2.90, and 3.75
for Task, Fault, and Video, respectively.

EXPERIMENT 2

Tasks

Twenty-two advanced aviation maintenance trainees participated in the second experiment. These subjects were drawn from a
population of trainees at the same point in
the same training program as were the
trainees used for the first experiment. Task,
Fault, and Video were again the training
methods. Rather than concentrating on the
differences between Task and Fault, however,
an attempt was made to merge the two computer simulations into a reasonably complete
alternative to traditional instruction. This
required a substantial modification of Fault.

As noted earlier, Task provided the trainee
with a logical approach to problem solving
but offered no context-specific knowledge. On
the other hand, Fault provided context-
specific information concerning component
names, functional relationships, and symptoms of failures; however, Fault did not tell
the trainee how to utilize test equipment and
interpret test results. Thus, Task and Fault
both provided problem-solving experience,
and Fault provided some context-specific information, but neither Task nor Fault provided sufficient information for actually
making tests and interpreting results.

In an effort to provide the trainee with
complete information, Fault was modified for
the second experiment. The modified version
of Fault provided an explanation after each
action, if appropriate, regarding equipment
hook-up and interpretation for each simulated test. In addition, in an effort to enhance
Fault for problem-solving training, Fault was
extended to provide feedback to trainees regarding any actions that were unnecessary,
redundant, or premature. Thus, both the
context-specific and context-free aspects of
Fault were enhanced.

Experimental Design

Again, the experiment employed a between-subjects design involving 11 subjects

in each of the two groups. Training time was increased from three to five sessions (i.e., from 6 h to 10 h) for both groups. This permitted the computer trainees (i.e., Task/Fault) to solve a total of 40 Task problems and 45 Fault problems in the five visits. The Video group participants viewed two of the programs twice in their five visits. For Experiment 2, all trainees were given a brief reading assignment on troubleshooting. In addition, all trainees took an on-line quiz at the end of each treatment.

The real system failures remained essentially unchanged for Experiment 2. The evaluation form was the same but the evaluators were changed. Evaluation forms were analyzed by a group of three power-plant experts in order to establish the performance index ratings.

Performance Measures

The performance measures used in Experiment 2 were identical to those of Experiment 1.

Results

There were no significant differences between training methods for any of the three performance measures: average performance index, $F(1,20) = 0.89$; average time, $F(1,20) = 1.31$; and average evaluator's rating, $F(1,20) = 0.09$. Although there was a significant effect of problems for average performance index, $F(4,80) = 6.37, p < 0.001$, and average time, $F(4,80) = 8.12, p < 0.001$, the lack of a significant main effect of training methods or a significant interaction provided no motivation for a problem analysis.

Thus, the significant difference for the average performance index found in Experiment 1 (Table 1) did not appear in Experiment 2 (Table 2). A comparison of the overall means of the performance index showed that Experiment 1 yielded 3.97 for Task/Fault and 4.37 for Video, while Experiment 2 resulted in 3.93 for Task/Fault and 4.15 for Video. The

TABLE 2

Average Performance Index for Each Problem in Experiment 2

Problem	Task/Fault	Video
Spark Plug	4.20	4.62*
Secondary Wire	3.83	4.26*
Fuel Obstruction	4.11	4.10
Starter Lead	3.78	3.99
Oil Pressure	3.74	3.86*

* Problem explicitly shown in Video program.

overall decrease in performance index averaged across all training methods can perhaps be attributed to differences in the two groups of trainees or differences in the evaluators used in the two experiments. Nevertheless, Task/Fault did improve relative to Video in Experiment 2.

CONCLUSIONS

By contrasting the two experiments discussed in this paper, several insights into the issue of fidelity and transfer of training can be gained. In the first experiment, Task and Fault placed considerable emphasis on problem-solving strategy and also provided some context-specific information. Performances with these simulations were good enough to equal those with Video, as long as explicit solution sequences were not presented in Video. For those problems for which Video explicitly presented the solution, Video was superior to Task and Fault.

The results of Experiment 1 led to a detailed exploration of the data (Johnson and Rouse, 1982). From these post hoc analyses emerged the conjecture that Fault was inadequate in providing information about how to make and interpret tests. For Experiment 2, Fault was modified to include this information. Fault was further enhanced to include feedback regarding unnecessary, redundant, and premature actions. Finally, Task and

Fault were combined to provide an integrated computer-based alternative to traditional instruction as embodied in Video. The results of the second experiment showed that the Task/Fault combination was equal to Video for all problems, even those for which Video provided explicit solution sequences.

These results lead to the conclusion that an appropriate combination of low- and moderate-fidelity computer simulations can provide sufficient problem-solving experience to be competitive with the more traditional lecture/demonstration form of instruction. Trainees do not have to be given explicit procedures for dealing with each problem that they might encounter. Beyond the implications of this conclusion for training effectiveness, the computer-based methods described herein also offer possibilities for lowering the cost of training.

ACKNOWLEDGMENTS

This research was supported by the U.S. Army Research Institute for the Behavioral and Social Sciences under Contract No. MDA 903-790C-0421. The authors are indebted to their colleague, Ruston Hunt, for his many significant contributions to this research.

REFERENCES

Brown, J. S., Burton, R. R., and Bell, A. G. SOPHIE: A step toward creating a reactive learning environment. *International Journal of Man-Machine Studies*, 1975, 7, 675-696.

Crooks, W. H., Kuppin, M. A., and Freedy, A. Application of adaptive decision aiding systems to computer-assisted instruction: Adaptive computerized training system. Arlington, VA: US Army Research Institute for the Behavioral and Social Sciences, Technical Report No. PATR-1028-77-1, January, 1977.

Glaser, R., Damrin, D., and Gardner, F. M. The Tab Item: A technique for the measurement of proficiency in diagnostic problem solving tasks. *Educational and Psychological Measurement*, 1954, *14*, 283-293.

Hunt, R. M. A study of transfer of problem solving skills from context-free to context-specific fault diagnosis tasks. Urbana, IL: University of Illinois, Coordinated Science Laboratory, Report No. T-82, July, 1979.

Hunt, R. M., and Rouse, W. B. Problem-solving skills of maintenance trainees in diagnosing faults in simulated powerplants. *Human Factors*, 1981, *23*, 317-328.

Johnson, W. B. Computer simulations in fault diagnosis training: An empirical study of learning transfer from simulation to live system performance. *Dissertation Abstracts International*, 1981, *41* (11), 4625-A. University Microfilms No. 8108555.

Johnson, W. B., Rouse, S. H., and Rouse, W. B. An annotated selective bibliography on human performance in fault diagnosis tasks. Arlington, VA: U.S. Army Research Institute for the Behavioral and Social Sciences, Report No. TR-435, January, 1980. ERIC Document Reproduction Service No. ED 192 736.

Johnson, W. B. and Rouse, W. B. Analysis and classification of human errors in troubleshooting live aircraft powerplants. *IEEE Transactions on Systems, Man, and Cybernetics*, 1982, in press.

Rasmussen, J., and Rouse, W. B. (Eds.) *Human detection and diagnosis of system failures*. New York: Plenum Press, 1981.

Rigney, J. W., Towne, D. M., King, C. A., and Moran, P. J. Field evaluation of generalized maintenance trainer-simulator: I. Fleet communications system. Los Angeles, CA: University of Southern California, Behavioral Technology Laboratories, Report No. 89, 1978.

Rouse, W. B. Problem solving performance of maintenance trainees in a fault diagnosis task. *Human Factors*, 1979, *21*, 195-203. (a)

Rouse, W. B. Problem solving performance of first semester maintenance trainees in two fault diagnosis tasks. *Human Factors*, 1979, *21*, 611-618. (b)

Standlee, L. S., Popham, W. J., and Fattu, N. A. A review of troubleshooting research. Bloomington, IN: Indiana University, Research Report No. 3, December, 1956.

Maintenance Training Simulator Fidelity and Individual Differences in Transfer of Training

JOHN A. ALLEN,[1] *George Mason University, Fairfax, Virginia*, ROBERT T. HAYS, *U.S. Army Research Institute for the Behavioral and Social Sciences, Alexandria, Virginia*, and LOUIS C. BUFFARDI, *George Mason University, Fairfax, Virginia*

This study was undertaken to investigate the relationship between simulator fidelity and training effectiveness. Two aspects of simulator fidelity were manipulated, namely, the degree to which a training simulator "looked like" actual equipment (physical fidelity), and the extent to which it "acted like" real equipment (functional fidelity). A transfer of training design was used to assess learning. Performance on an electromechanical troubleshooting task was correlated with a number of individual difference variables. Results indicated that physical and functional fidelity were interdependent and that temporal measures were most sensitive to fidelity manipulations. Low functional fidelity was associated with longer problem solution and inter-response times. Persons with high analytic abilities took longer to solve problems, but required fewer troubleshooting tests and made fewer incorrect solutions.

INTRODUCTION

A key issue in the design of training devices and training programs based around training devices is the degree to which the training device must duplicate the actual equipment for which training is required. This degree of similarity to the actual equipment is called simulator fidelity (Hays, 1980). Despite their proven effectiveness, simulators and training devices with too high a level of fidelity may not always be cost-effective. Unfortunately, at present there is a conspicuous lack of empirical data that training developers may use to specify minimal fidelity values for training simulators. The present study was undertaken as part of a larger program of research aimed at identifying such values in the area of maintenance training. A major objective of this ongoing work is to provide the U.S. Army with the empirical data upon which engineering guidelines for maintenance training simulators might be based.

Previous research has, for the most part, been concerned with evaluating the training effectiveness of full-fidelity devices rather than systematically investigating the effect of various degrees of similarity (Ayers, Hays, Singer, and Heinicke, 1984). Notable exceptions are the research efforts of Wheaton and Mirabella (1972), Mirabella and Wheaton (1974), Johnson and Rouse (1982), Baum, Riedel, Hays, and Mirabella (1982), and Johnson and Fath (1983). Although each study systematically manipulated the char-

[1] Requests for reprints should be sent to John A. Allen, Department of Psychology, George Mason University, 4400 University Drive, Fairfax, VA 22030.

acteristics of the devices used for training on the criterion tasks, only Baum et al. (1982) varied the physical and functional characteristics of the training simulators.

In the Baum et al. (1982) study, two aspects of fidelity were manipulated: the degree to which a training simulator "looked like" the actual equipment it was simulating (physical fidelity), and the degree to which it "acted like" real equipment (functional fidelity). Training simulators of varying degrees of physical and functional fidelity were used for training people to perform a simple mechanical adjustment task. Although the authors noted a significant effect for physical fidelity, no main effect for functional fidelity or for interaction effects between the two fidelity dimensions were found. Based on these results, they concluded that, at least for simple adjustment tasks, high physical similarity may be very important from the standpoint of maximum learning transfer. They proposed additional research using different tasks to further specify the effects of physical and functional fidelity. In part, the present study may be viewed as a response to this suggestion, since one of the major aims was to investigate the effects of simulator fidelity during training on a relatively simple troubleshooting task. An additional aim was to collect data on a number of individual difference variables to help determine how such variables might interact with fidelity during training.

Literature reporting individual difference predictors of troubleshooting performance is quite sparse. By taking advantage of test information already available on subjects participating in a troubleshooting experiment, Henneman and Rouse (1984) investigated the relationships between performance and general academic ability (ACT scores and grade-point average), mechanical aptitude (Survey of Mechanical Insight), and cognitive style (reflective/impulsive, field dependent/inde-

pendent). They found evidence of a statistically significant relationship between troubleshooting performance and both cognitive style and academic ability. Rouse and Rouse (1982) confirmed the finding for cognitive style, in that reflective and field-independent individuals were found to be generally more effective at fault diagnosis.

The present study took a broader and more systematic approach than did previous research, by investigating a wide array of individual differences that might be related not only to troubleshooting performance but also to different modes of training people on such tasks. Thus, measures were chosen to reflect the logical capacity, analytic abilities, and general interests apparently required by the task. Also, a specially constructed test-bed device for high-fidelity simulation training and transfer-of-training testing was used. Degraded simulations of this device produced the various levels of physical and functional fidelity used to investigate the effects of fidelity on transfer of training. Physical fidelity manipulations involved variations in the way components and their spatial relationships were represented in the simulator. Functional fidelity, on the other hand, was defined in terms of the degree of information feedback (Bilodeau, 1966) available to the subject, or what might be called the informational aspect of equipment function. This view of functional fidelity may be contrasted with its "stimulus/response options" aspect, which refers to opportunities that the equipment provides to the subject to receive stimuli (e.g., a dial moves) and/or give responses (e.g., the subject can turn a knob) (Hays, 1980).

A wide range of performance measures was chosen to investigate fidelity effects. This strategy was selected because of the exploratory nature of the research and because previous studies have shown that no one variable fully describes how subjects solve troubleshooting problems (Finch, 1971; Glass, 1967).

METHOD

Subjects

One hundred college undergraduates (40 males and 60 females) ranging in age from 17 to 55 (M = 23.05, S.D. = 9.20) were drawn from introductory psychology courses at George Mason University and served as paid subjects ($5.00/hour). Subjects also received course credit for participation.

Materials

Individual differences measures. All subjects completed a two-hour test battery that included the following measures: (1) Group Embedded Figures Test (GEFT), (2) Bennett Mechanical Comprehension Test (BMC), (3) Graduate Record Examination—Analytic (GRE-A), (4) Rotter's Locus of Control Scale (LOC), and (5) Holland's Vocational Preference Inventory (VPI). Six of the VPI scales were used: (1) Realistic (VPI-R), (2) Intellectual (VPI-I), (3) Social (VPI-S), (4) Conventional (VPI-C), (5) Enterprising (VPI-E), and (6) Masculinity (VPI-M).

Computer. All experimental activities were controlled and monitored by a MINC-11/23 (Digital Equipment Corp.) computer. In addition to controlling the reference system and simulators, it also recorded and stored all subject actions during the training and testing phases of the experiment.

Reference System. The reference system, which served as both the "actual equipment" and the high physical fidelity trainer, may be described as follows. Twenty-eight electromechanical relays and five solid-state pullup panels were interconnected to eight output devices (fan, water pump, solenoid valve assembly, three lights, TV monitor, and sound generator and speaker). When operating properly (i.e., when no faults had been intro-

duced into the system) all of the output devices worked. However, when a relay or pullup panel was faulted by the experimenter, associated relays/pullup panels and one or more of the output devices did not work. A subject's job was to discover which relay or pullup-panel assembly was at fault. To do this, the subject had to detect which output devices were not working, and then, by testing individual relays and pullup panels, discover which component was at fault.

Hand-held tester. Relays were tested by means of a specially designed hand-held tester. To test a relay, a trainee dialed in the number of the relay, placed the probe on the relay checkpoint, and then pressed a test button located on top of the unit. If the component under consideration was operating properly, a green light illuminated on the face of the tester. If the relay was not working, the green light did not go on. A cable connecting the hand-held tester to the MINC-11 computer was used to signal when a test had been requested. Recording of the number and type of response and of the time between responses was done automatically.

Faulter panels. Individual relays and pullup panels were faulted by means of two faulter panels (one for the reference system and one for the medium and low physical fidelity simulators) located in the experimenter's room. These panels consisted of a number of lights and switches, each of which corresponded to a particular relay or pullup-panel assembly. To fault a component, the experimenter merely threw the switch appropriate to that component. This caused the component to stop working. All dependent relays, pullup panels, and output devices also ceased to function. Lamps on the face of the faulter panels provided the experimenter with instant information as to which component had been faulted.

Simulators. During the training phase,

three major training simulators were used, each differing along the physical fidelity dimension (i.e., high, medium, and low). These simulators, when coupled with one of three levels of functional fidelity (i.e., high, medium, or low), provided nine simulators for use during training:

Levels of physical fidelity.

(1) *High physical fidelity simulator.* This simulator was the reference system, as described previously.

(2) *Medium physical fidelity simulator.* The medium fidelity simulator was quite similar to the reference system in terms of its size and general appearance. Half of its components (relays, output devices, etc.) actually duplicated those found on the reference system whereas half were wooden or pictorial mockups.

(3) *Low physical fidelity simulator.* The low physical fidelity simulator was different in appearance from either the medium or high physical fidelity simulators. Essentially, it was a symbolic representation of the reference system. Relays, pullup panels, and output devices were depicted by rectangles, and lines between rectangles indicated wiring connections.

Levels of functional fidelity.

(1) *High functional conditions.* Subjects in high functional conditions were able to obtain status information about both components and output devices. That is, the hand-held unit was used to test components, whereas status information about outputs could be obtained by activating push buttons associated with either actual or mockup output devices.

(2) *Medium functional conditions.* In these conditions, subjects were provided only with status information about components (via hand-held tester). The status of output devices could not be checked.

(3) *Low functional conditions.* Subjects experiencing low levels of functional fidelity received no feedback about components and outputs. That is, testing of components and output devices was not possible.

Design and procedure. The experiment was conducted in two sessions. During the first session, subjects were tested in groups of 5 to 10 on the battery of individual difference measures previously described. Testing lasted approximately two hours.

One week later, subjects were randomly assigned to one of nine training groups, formed by combining one of three levels of physical fidelity with one of three levels of functional fidelity, or to a no-training control group. Ten subjects were trained in each condition. Note that separate one-way analyses of variance indicated no significant age or sex differences in assigning subjects to the 10 training groups, $F(9,99) = 0.47, p > 0.89$, and $F(9,99) = 0.59, p > 0.80$, respectively.

At the start of training, a subject listened to one of nine versions of taped instructions that explained the use of the apparatus and the general intent of training. Subjects in the no-training control group also heard one of the nine instructional sets (assigned randomly), and then read from a set of unrelated psychology readings for a period of time equal to that taken by subjects in the simulator training conditions. During the taped instructions, the experimenter remained in the room to demonstrate various equipment operations and answer questions. When instructions were completed, the experimenter left the room.

During training, subjects received eight training problems (or "faults"). Three sets of eight problems each were used. Sets were matched in terms of problem difficulty based on earlier pilot data. Each subject was trained on one randomly assigned set.

When a training problem was begun (through the initiation of a fault by the experimenter), subjects were signaled by a start tone. The subject's task was then to determine which of the 28 possible components was at fault. Depending on the simulator in use, this either involved testing relays and output devices (high functional fidelity), testing relays only (medium functional fidelity), or, as was the case in the low functional fidelity conditions, merely examining and studying components and their relationships. An interval of approximately 30 seconds elapsed between each problem. After training, the subject received a five-minute rest period.

After the break, the transfer portion of the session was begun. First, subjects were introduced to the reference system via a second set of recorded instructions that were identical to those heard by subjects in the high physical/high functional training condition. The subjects then were asked to solve six new problems on the reference system. All subjects solved the same set of six problems, but with the presentation order randomized. Degree of transfer was assessed using the following dependent variables: time to first solution, inter-response time (which refers to the time between any two actions, either tests or solutions), intertest time, time to correct solution, number of tests, number of tests repeated, number of attempted solutions, and number of solutions repeated.

The training and transfer phase lasted between 90 minutes and two hours, and this time was evenly divided between training and transfer.

RESULTS AND DISCUSSION

Comparisons of Control and Training Group Data

Because it was important to know whether performance on the criterion task was affected by the various training procedures, performance data from each of the nine simulator conditions was compared with that of the no-training control group. The results of these comparisons are shown in Table 1.

As can be seen, subjects across training conditions generally solved problems more quickly, attempted solutions earlier, repeated fewer tests, and took less time between tests and solutions than did their no-training counterparts. However, the fact that many of the performance differences were not statistically significant suggests that the

criterion task may not have been sufficiently difficult and/or that training was not long enough. Nevertheless, the effects of the training manipulations were powerful enough to significantly affect a number of performance measures during transfer.

ANOVAs on Dependent Variables

Three × three factorial analyses of variance comparing the transfer performances of the groups trained in each fidelity condition were performed on each of the eight dependent variables.

A significant main effect for both physical fidelity, $F(2,81) = 3.24$, $p < 0.01$, and functional fidelity, $F(2,81) = 9.89$, $p < 0.01$, on the total-time-to-solutions variable indicated that persons trained on devices with high physical and high functional fidelity reached correct solutions more quickly than did persons trained in lower fidelity conditions. (See Figures 1 and 2 below.) In addition, main effects for functional fidelity manipulations were found to be significant for time to first solution, $F(2,81) = 3.15$, $p < 0.05$; number of repeated tests, $F(2,81) = 5.48, p <$

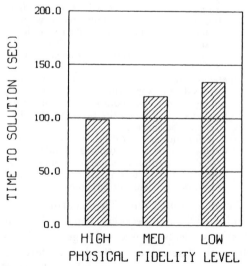

Figure 1. *Mean times to solution for high, medium, and low physical fidelity levels.*

TABLE 1

Results of Dunnett's Test Comparing the Performance of Training Groups with That of a No-Training Group on Eight Dependent Variables

			Physical Fidelity		
			High	Medium	Low
	High	Tests	—	—	$p < 0.05$
		Solutions	—	—	—
		Time to solution	$p < 0.01$	—	—
		Time to first solution	—	—	—
		Intertest time	—	—	—
		Inter-response time	—	$p < 0.05$	—
		Test repeats	$p < 0.01$	$p < 0.01$	$p < 0.01$
		Solution repeats	—	—	—
Functional fidelity	Medium	Tests	$p < 0.01$	—	—
		Solutions	—	—	—
		Time to solution	$p < 0.01$	—	—
		Time to first solution	$p < 0.05$	—	—
		Intertest time	—	—	—
		Inter-response time	$p < 0.01$	$p < 0.01$	$p < 0.05$
		Test repeats	$p < 0.01$	—	—
		Solution repeats	—	—	—
	Low	Tests	—	—	$p < 0.05$
		Solutions	—	—	—
		Time to solution	—	—	—
		Time to first solution	—	—	—
		Intertest time	—	$p < 0.05$	—
		Inter-response time	—	—	—
		Test repeats	—	—	—
		Solution repeats	—	—	—

0.01; intertest time, $F(2,81) = 8.44, p < 0.01$; and inter-response time, $F(2,81) = 9.68, p < 0.01$. As can be seen in Figures 3 and 4, for the time-to-first-solution and repeated-tests measures, higher functional fidelity levels were associated with fewer test repeats and slightly earlier solution attempts during transfer. Also, subjects experiencing medium levels of functional fidelity tended to spend less time between tests and solution attempts, whereas subjects in low functional fidelity training conditions took relatively more time in testing. (see Figures 5 and 6.)

As for interactions, only one between physical and functional fidelity on the number of repeated tests was noted, $F(4,81) = 3.00, p < 0.05$. That is, fewer repeated tests were made

by persons trained with high-physical/high-functional and high-physical/medium-functional simulators than by persons trained in other groups. On the other hand, the largest number of repeated tests was made by subjects trained in the medium-functional/low-physical fidelity condition, closely followed by subjects experiencing low levels of functional fidelity at high and medium levels of physical fidelity, respectively. No completely satisfactory explanation of this pattern of results is available.

As can be seen in Figures 2 through 6, the level of functional fidelity had a strong effect on performance, as measured by a number of dependent variables. In general, decreasing levels of functional fidelity were associated

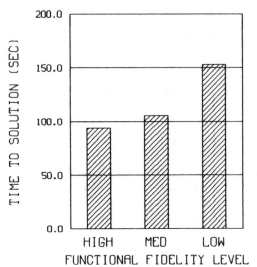

Figure 2. *Mean times to solution for high, medium and low functional fidelity levels.*

with longer solution times and fewer repeated tests. This, of course, makes sense if one considers that with more information during training, problem solving and learning may have been easier and more effi-

cient, resulting in more effective transfer during reference system testing.

The reason why lower levels of physical fidelity were also associated with longer solution times is less clear, however. One possibility may be that subjects trained on the lower fidelity devices required more time to orient themselves to the reference system during transfer. This seems quite likely when one considers how different from the reference system the medium physical fidelity simulators, and particularly the low physical fidelity simulators, appeared.

The fact that temporal measures were especially sensitive to the fidelity manipulations was not surprising and confirms similar findings in numerous field experiments involving training manipulations of various kinds (Bresnard and Briggs, 1956; Rigney, Towne, Moran, and Mishler, 1978; Unger, Swezey, Hays, and Mirabella, 1984).

One important finding, however, was the significant interaction between fidelity levels

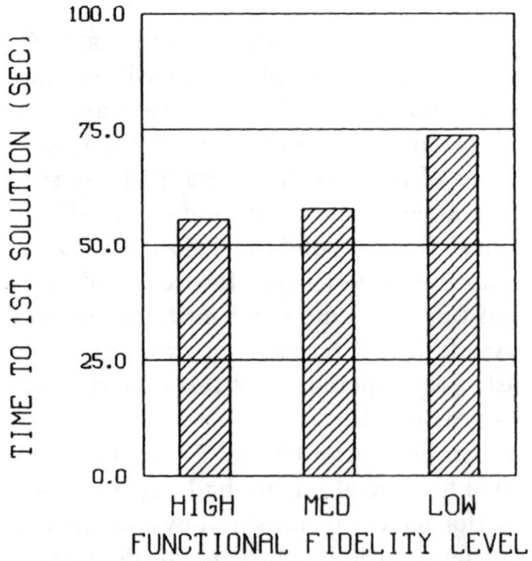

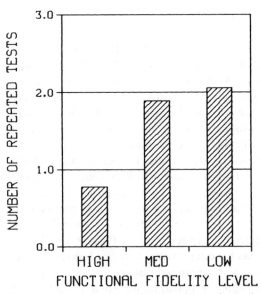

Figure 3. *Mean times to first solution for high, medium, and low functional fidelity levels.*

Figure 4. *Mean number of repeated tests for high, medium, and low functional fidelity levels.*

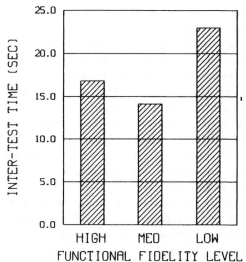

Figure 5. *Mean intertest times for high, medium, and low functional fidelity levels.*

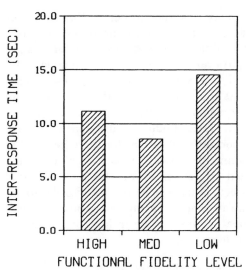

Figure 6. *Mean inter-response times for high, medium, and low functional fidelity levels.*

with respect to the number of repeated tests. The interaction emphasizes the fact that physical and functional features in equipment are all fundamentally related and should not be dealt with in isolation. As Figure 7 shows, the effect of decreasing physical similarity produced opposite effects in the medium and low physical fidelity groups. That is, for subjects trained in low functional fidelity conditions, there was a tendency to repeat fewer tests during transfer as physical similarity decreased. Subjects trained with medium functional fidelity, on the other hand, tended to repeat more and more tests with increasing dissimilarity between the training and transfer devices.

In summary, two general conclusions seem warranted by the results of the ANOVAs. First, it is clear that temporal performance measures were most sensitive to fidelity manipulations, as statistical significance was found on three of the four temporal measures. Especially sensitive was the total-time-to-solution variable, which was found to be significantly affected by variations in

both functional and physical fidelity. Number of tests, number of solution attempts, and number of repeated solutions, on the other hand, seemed to operate quite independently of fidelity manipulations. Second,

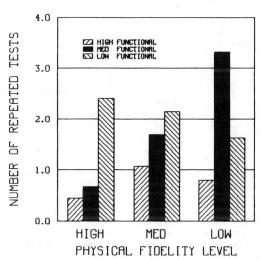

Figure 7. *Mean number of repeated tests for high, medium, and low levels of physical and functional fidelity.*

evidence was found to suggest that simulator fidelity is not a single, uniform concept, but a multidimensional one consisting, at least, of a physical and a functional component. Indeed, it was noted that each of these components often produced different effects depending on the dependent measure.

This latter finding would appear to have important implications for those involved in training-device design, as it emphasizes the need to determine carefully the critical aspects of tasks to be trained in order to ensure an appropriate mix of physical and functional features in the training device under consideration.

Individual Difference Correlates of Performance

In order to reduce the number of dependent variables for the individual difference analyses, the eight dependent variables were factor analyzed (principal factoring with iteration, Varimax rotation). The rotated factor matrix in Table 2 shows the emergence of three factors (Time, Tests, and Attempted Solutions), which parallel those found by

Henneman and Rouse (1984) in terms of time, inefficiency, and errors.

Separate one-way analyses of variance on each of the 10 individual difference measures indicated significant gender differences for three of the VPI scales. Males scored higher on realistic, $F(1,88) = 11.48$, $p < 0.05$, and masculine interests, $F(1,88) = 29.00$, $p < 0.001$. Females scored higher on social interests, $F(1,88) = 11.18$, $p < 0.01$. None of the other individual differences showed significant gender effects.

Pearson correlations. Pearson correlations were computed between each of the individual difference measures and each of the three performance factors. Significant correlations appear in Table 3. As can be seen, positive relationships were noted between time and GRE-analytic, VPI-intellectual, and VPI-masculinity scores, indicating that persons with high intellectual or masculine interests, or high analytic ability, generally took longer to solve the criterion problems. With respect to the Tests factor, persons high on either

TABLE 2

Varimax-Rotated Factor Matrix

	Factor 1— Time	Factor 2— Tests	Factor 3— Attempted Solutions
Time to solution	0.740*	0.509	0.046
Time to first solution	0.674*	0.281	−0.402
Intertest time	0.831*	−0.312	−0.015
Inter-response time	0.941*	−0.103	−0.006
Tests	−0.226	0.854*	0.269
Test repeats	−0.003	0.818*	0.131
Solutions	−0.026	0.116	0.918*
Solution repeats	0.068	0.275	0.789*

* loadings on each factor

TABLE 3

Correlations Between Individual Difference Measures and the Three Performance Factors

	Performance Factors		
	Time	Tests	Attempted Solutions
Individual Differences			
GEFT			−0.24*
BMC		−0.30**	
GRE-A	0.23*	−0.22*	−0.24*
LOC			
VPI-realistic			
VPI-intellectual	0.23*		
VPI-social			
VPI-conventional			
VPI-enterprising		0.26*	
VPI-masculinity	0.19*		
Gender			
Age			

* $p < 0.05$
** $p < 0.01$

mechanical or analytic ability, or with low enterprising interests, tended to make fewer tests. Field-independent persons with high analytic abilities made fewer incorrect solutions.

Although the pattern of results for analytic ability (i.e., negative correlations with tests and solution factors, positive correlations with time) appears paradoxical, an explanation may be found in subjects' troubleshooting strategies. That is, analytic individuals may have taken longer to reach solutions because they may have approached problems more thoughtfully than did non-analytic subjects. Indeed, non-analytic subjects may have relied mainly on multiple testing because of a lack of understanding of the system. Given the modest level of correlation between individual differences and performance found here, one may speculate that individual differences may play an important role in determining troubleshooting strategy, but that strategy may be a more direct predictor of performance (Buffardi, Allen, and Hays, 1985). Further research focusing on troubleshooting strategies appears warranted.

Interaction of Individual Differences and Training Conditions

To determine the predictive strength of physical and functional fidelity, the individual difference measures, and individual difference and fidelity interactions, 12 multiple regressions were computed for each of the three performance factors. Due to sample size limitations, however, it was not feasible to include the large number of individual difference measures and their interactions with fidelity in a single equation. Thus, in accordance with standard procedure for moderated regression equations (Ghiselli, Campbell, and Zedeck, 1981), physical fidelity level, functional fidelity level, and a given in-

dividual difference measure were forced into the equation prior to their respective interactions. This procedure allowed tests of interaction effects over and above the contributions made by the main effects of individual differences and fidelity. The results of the multiple regression analyses appear in Table 4.

It is clear that the Time factor was the most sensitive to fidelity manipulations. The subject's functional and physical fidelity training condition predicted a significant amount of variance in each regression for this factor. This, it should be noted, is consistent with the results of the ANOVAs reported previously.

No individual difference variables significantly predicted the Time factor, however, and only two interactions involving individual difference measures were significant. The VPI-intellectual scale interacted with functional fidelity in such a way that individuals with high intellectual interests took more time under the low functional fidelity conditions than did their counterparts with low intellectual interest. Also, a three-way interaction between age, functional fidelity, and physical fidelity indicated that older individuals took significantly longer to reach a solution when trained under conditions of low functional and low physical fidelity.

As for the Tests factor, only three significant predictors were found. Individuals with either high scores on mechanical comprehension or low enterprising interests attempted fewer tests. In addition, a significant interaction between VPI-social and functional fidelity indicated that individuals high in social interests attempted more tests under conditions of low and medium functional fidelity but not under conditions of high functional fidelity.

With respect to the Attempted Solutions factor, two main effects and four interactions were noted. First, individuals high on field

TABLE 4

Significant Predictors of Performance on Three Performance Factors

| | Performance Factors | | |
	Time	Test	Attempted Solutions
Individual Differences			
GEFT	Func**, Phys*		GEFT*
BMC	Func**, Phys*	BMC*	
GRE	Func**, Phys*		GRE*, GRE × F × P*
LOC	Func**, Phys*		
VPI-realistic	Func**, Phys*		
VPI-intellectual	Func**, Phys* VPI-I × F*		
VPI-social	Func**, Phys*	VPI-S × F*	VPI-S × P*
VPI-conventional	Func**, Phys*		
VPI-enterprising	Func**, Phys*	VPI-E*	
VPI-masculinity	Func**, Phys*		VPI-M × P*
Age	Func**, Phys* Age × F × P*		
Gender	Func**, Phys*		Gender × F*

* $p < 0.05$
** $p < 0.01$

independence or analytic ability generally attempted fewer solutions than did their more field-dependent or less analytic counterparts. Second, a three-way interaction between GRE-A scores, physical fidelity, and functional fidelity revealed that individuals high in analytic ability attempted fewer solutions after training under conditions of high physical and high functional fidelity than did less analytic subjects.

Third, a significant interaction between gender and functional fidelity indicated that male subjects attempted fewer solutions after training in low functional fidelity conditions, whereas female subjects attempted fewer solutions after training in high functional fidelity conditions.

Finally, two-way interactions between VPI-social and physical fidelity and VPI-masculinity and physical fidelity showed similar patterns. Individuals with low social or high masculine interests tried fewer solutions under low physical fidelity training condi-

tions, whereas individuals with high social or low masculine interests also attempted fewer solutions, but under medium and high physical fidelity training conditions. These parallel findings may be viewed as mirroring the negative correlation found between social and masculine interest scores on the VPI, $r = -0.57$. This pattern of results may be interpreted as corresponding to that reported in work by Snow and Lohman (1984), who indicate that highly able learners thrive on abstract instruction whereas less able learners may be best trained by highly structured, more concrete demonstrations. Although this latter instructional treatment improves the performance of the less able, it does so at the expense of diminishing the performance of their more able counterparts. The present results suggest that, in addition to ability, a subject's level and type of interest may also make one type of instruction preferable to others. Viewed from such a perspective, the low physical fidelity conditions in the present study, which essentially provided subjects with an abstract functional diagram, may have best suited those with high masculine interests, whereas those with low masculine interests may have benefited from the more direct and concrete high physical fidelity simulations.

Explanations of the other interactions involving individual differences are less apparent. However, the number and pattern of significant interactions would seem to support the concept of an adaptive training model; that is, optimal training may require different modes of instruction for individuals with different abilities and aptitudes. However, more definitive conclusions must await cross-validation.

SUMMARY AND CONCLUSIONS

The results of the present study may be summarized as follows:

(1) In the present task, the physical and functional aspects of fidelity were separately manipulated and yielded differential effects. This is important, as previous research has not systematically investigated these two aspects of fidelity.

(2) Functional fidelity level was a very potent determinant of performance. Decreasing levels of functional fidelity were generally asociated with longer solution times and longer inter-response times.

(3) The manipulation of the physical aspect of fidelity also produced a significant performance effect, as measured by time to solution. That is, persons trained on simulators with lower physical fidelity took longer to reach correct solutions.

(4) An interaction between physical and functional fidelity was noted only for the number of repeated tests. The effect of decreasing the amount of physical fidelity during training produced different performance levels, depending on the level of functional fidelity. Subjects in the medium functional conditions repeated more tests as physical fidelity decreased, whereas subjects in the low functional conditions repeated fewer tests as physical fidelity decreased.

(5) Although the performance of subjects trained on simulators was better than that of subjects in the no-training control group, the data suggest that the criterion problems should have been more difficult. Ongoing experiments using more difficult problem sets are currently under way.

(6) Factor analysis of the eight dependent variables yielded three performance factors: Time, Tests, and Attempted Solutions. These factors are similar to those found by Henneman and Rouse (1984) in their studies of human fault detection.

(7) Persons with high analytic abilities (GRE-A) took longer to solve criterion problems but required fewer tests and attempted fewer incorrect solutions. This may reflect a more thoughtful approach as opposed to a reliance on simple multiple testing.

(8) The Time factor was the most sensitive to fidelity manipulations. Subjects' training condition was the most potent predictor of performance on this factor.

(9) Mechanical comprehension level was the strongest predictor of performance on the Tests factor. Those subjects with high mechanical comprehension required fewer tests to reach a solution. This probably indicates that these subjects had a more complete understanding of the system than did other trainees.

(10) Several interactions predicted performance on the Attempted Solutions factor. Both the subjects' masculine and social interests were found to interact with physical fidelity. One implication is that training designers may be able to use low physical fidelity simulators (at, presumably, lower cost) with trainees with high masculine interests, but may need to use high physical fidelity simulators with trainees who are low in masculine interests.

In conclusion, the data indicate that simulator design decisions must account for both physical and functional aspects of fidelity in order to maximize training effectiveness, and that fidelity effects must be interpreted in the context of both task and trainee characteristics. Criterion performance measures should also be considered carefully, as the present data revealed differential fidelity effects depending on the performance criterion under study. Certainly, further research is needed not only to specify more clearly the effects of simulator design for training tasks like the present one, but also to determine the nature of those effects in other areas. In particular, additional research on functional fidelity is required to specify more fully the dimensions of functional fidelity and their training effects. In addition, studies aimed at further delineating the relationships between trainee abilities/interests and simulator trainer characteristics are needed. Indeed, only when a comprehensive database is accumulated will training system designers be able to make informed decisions on simulator characteristics and use.

ACKNOWLEDGMENTS

Special thanks are extended to Bob Holt, Evans Mandes, Bob Pasnak, Jim Sanford, Zita Tyer, Margaret Baechtold, Beth Drum, Rick Hall, Rob Huggins, Lou Matzel, and Ray Moffett for their valuable assistance in this work.

REFERENCES

Ayers, A., Hays, R. T., Singer, M. J., and Heinicke, M. (1984). *An annotated bibliography of abstracts on the use of simulators in technical training* (ARI Research Product 84-21). Alexandria, VA: U.S. Army Research Institute.

Baum, D. R., Riedel, S., Hays, R. T., and Mirabella, A. (1982). *Training effectiveness as a function of training device fidelity* (ARI Technical Report 593). Alexandria, VA: U.S. Army Research Institute.

Bilodeau, I. M. (1966). Informational feedback. In E. A. Bilodeau (Ed.), *Acquisition of skill* (pp. 255-289). New York: Academic Press.

Bresnard, G. G., and Briggs, L. J. (1956). *Comparison of performance upon the E-4 fire control system simulator and upon operational equipment* (Development Report: AFPTRC-TN-56-47). Lowery Air Force Base, CO: Armament Systems Personnel Research Laboratory, AFPTRC.

Buffardi, L. C., Allen, J. A., and Hays, R. T. (1985). Predicting troubleshooting performance: Cognitive strategy, individual differences, and simulator fidelity. Paper presented at the meeting of the American Psychological Association, Los Angeles, CA.

Finch, C. R. (1971). *Troubleshooting instruction in voca-*

tional technical education via dynamic simulation (Report 10064, 19-1024). University Park, PA: Pennsylvania State University.

Ghiselli, E. E., Campbell, J. P., and Zedeck, S. (1981). *Measurement and theory for the behavioral sciences.* San Francisco: Freeman.

Glass, A. A. (1967). Problem solving techniques and troubleshooting simulations in training electronic repairmen (Doctoral dissertation, Columbia University). *Dissertation Abstracts, 28,* 1717B.

Hays, R. T. (1980). *Simulator fidelity: A concept paper* (ARI Technical Report 490). Alexandria, VA: U.S. Army Research Institute.

Henneman, R. L., and Rouse, W. B. (1984). Measures of human problem solving performance in fault diagnosis tasks. *IEEE Transactions on Systems, Man, and Cybernetics, SMC-14*(1), 99-112.

Johnson, W. B., and Fath, J. L. (1983). *Implementation of a mixed-fidelity approach to maintenance training* (First-year report Contract No. MDA 903-82-CC-0354). Norcross, GA: Search Technology.

Johnson, W. B., and Rouse, W. B. (1982). Training maintenance technicians for troubleshooting: Two experiments with computer simulations. *Human Factors, 24,* 271-276.

Mirabella, A., and Wheaton, G. (1974). *Effects of task index variations on transfer of training criteria* (Technical Report NAVTRAEQUIPCEN 72-C-0126-1). Orlando, FL: Navy Training Equipment Center.

Rigney, J. W., Towne, D. M., Moran, P. J., and Mishler, R. A. (1978). *Field evaluation of the generalized maintenance trainer-simulator: II—AN/SPA-66 radar repeater* (Tech. Report 90). Arlington, VA: Personnel and Training Research Programs, Office of Naval Research.

Rouse, S. H., and Rouse, W. B. (1982). Cognitive style as a correlate of human problem solving performance in fault diagnosis tasks. *IEEE Transactions on Systems, Man, and Cybernetics, SMC-12*(5), 649-652.

Snow, R. E., and Lohman, D. F. (1984). Toward a theory of cognitive aptitude for learning from instruction. *Journal of Educational Psychology, 76,* 347-376.

Unger, K. W., Swezey, R. W., Hays, R. T., and Mirabella, A. (1984). *Army Maintenance Training and Evaluation Simulation System (AMTESS) device evaluation: Volume II, Transfer-of-training assessment of two prototype devices* (ARI Technical Report 643). Alexandria, VA: U.S. Army Research Institute.

Wheaton, G., and Mirabella, A. (1972). *Effects of task index variations on training effectiveness criteria* (Technical Report NAVTRAEQUIPCEN 71-C-0059-1). Orlando, FL: Naval Training Equipment Center.

Training High-Performance Skills: Fallacies and Guidelines

WALTER SCHNEIDER,[1] *Learning Research and Development Center, University of Pittsburgh, Pittsburgh, Pennsylvania*

A high-performance skill is defined as one for which (1) more than 100 hours of training are required, (2) substantial numbers of individuals fail to develop proficiency, and (3) the performance of the expert is qualitatively different from that of the novice. Training programs for developing high-performance skills are often based on assumptions that may be appropriate for simple skills. These assumptions can be fallacious when extended to high-performance skills. Six fallacies of training are described. Empirical characteristics of high-performance skill acquisition are reviewed. These include long acquisition periods, heterogeneity of component learning, development of inappropriate strategies, and training of time-sharing skills. A tentative set of working guidelines for the acquisition of high-performance skills is described.

INTRODUCTION

This article examines special considerations and problems associated with high-performance skill acquisition. Much of skill-learning experience and most skill-learning research relate to learning simple skills (e.g., lever positioning). Generalizations based on improvements over short training periods can produce fallacious training assumptions. These assumptions are frequently implicitly assumed in training programs. This paper explicitly identifies some of the more prevalent assumptions. The section on fallacies is written from a devil's advocate position. It is intended to cause the training program designer to question frequently held implicit assumptions. The next section describes empir-

[1] Requests for reprints should be sent to Walter Schneider, LRDC Building, 3939 O'Hara St., Pittsburgh, PA 15260.

ical results illustrating special considerations for training highly skilled performance.

The recent increased use of microprocessor-based training emphasizes the need to develop explicit guidelines for skill training. Microcomputers can provide feedback, graphic illustrations, and drill on many components of critical tasks. In the past, training amounted to a combination of classroom instruction and laboratory or on-the-job experience in the final work environment. Now microcomputer programs can be easily modified to train individual component skills, graphically represent the problem, provide augmented cues, sequence the training, and so on. If training-program developers blindly make computers perform the same type of simulation activities that were previously done with simulators, there is no reason to expect training efficiency to improve (with the exception of possibly decreasing the number of trainers). For example, an Advanced Controller Exerciser (McCauley, Root, and Muckler, 1982) was developed to replace the traditional multiperson simulation system for training air intercept control. The microprocessor-based system resulted in poorer trainee performance than the original system. Greater awareness of the special considerations of high-performance skill acquisition may enable better use of the flexibility provided by microprocessors.

For the purposes of this paper, high-performance skills will be defined as having three characteristics. First, the trainee must expend considerable time and effort to acquire a high-performance level (i.e., greater than 100 h). Second, the training programs that produce such skill levels will characteristically experience substantial failure rates even among individuals motivated to acquire the skill (i.e., greater than 20%). Third, there will be substantial qualitative differences in performance between a novice and an expert.

Military air traffic and air weapons control provide examples of high-performance skills. To develop proficiency requires from one to two years of training. Washout rates for training programs vary from 25% to 70%, with 50% being typical. Novices and experts show very different performance characteristics. For example, when performing a two-aircraft live intercept, novices have difficulty estimating the turn radii of the aircraft. The novice (13 weeks of training) continues to watch the display for minutes to determine whether the specified turn maneuvers produce the desired effect. In contrast, an expert watching the intercept could specify after only 20 s (two scope sweeps) that the one aircraft is coming in too hot and would pass in front of the aircraft that it was supposed to come in behind. For the novice, decisions are slow and uncertain, and the trainee appears very overtaxed. In contrast, the expert makes decisions quickly and with little effort, and can simultaneously perform other duties.

The training of a fighter pilot provides another example of a high-performance skill. This training typically requires 350 flight hours over two years, and washout rates range in excess of 30% (Griffin and Mosko, 1977).

A more mundane example of a high-performance skill is that shown in professional-level typing. Typical training time necessary to develop a 50 word-per-minute typing speed is in excess of 200 h (Deighton, 1971). Most of the people who try to develop typing skill never obtain that level. A highly skilled typist independently moves his or her fingers to different keys simultaneously, whereas the novice makes individual movements to keys sequentially (Rumelhart and Norman, 1982).

It is difficult to generalize research for the training of highly skilled performance. First, there are few parametric empirical studies. The studies that do exist often confound effects of training procedure, trainers, and subject criterion differences (Eberts, Smith, Dray, and Vestewig, 1982). Also, because performance changes qualitatively over time, training techniques that may be quite useful for initial acquisition may be very ineffective for later skill development. Finally, our theoretical understanding of the nature of performance change with practice is very limited. There are some theoretical perspectives that predict qualitative changes in performance (e.g., Pew, 1966, 1974; Schneider and Fisk, 1983; Shiffrin and Schneider, 1977). However, the theoretical development does not yet specify which training techniques would be best at different stages of skill acquisition.

TRAINING FALLACIES

Many training programs are based on implicitly assumed fallacies. These are fallacies in the sense that they are misleading and are based on unsound generalizations. The next section provides empirical evidence for the unsoundness of the assumptions. All of these fallacies have some truth. However, when taken to extremes, they often produce inefficient training programs. Examples are provided from the training of military air traffic controllers. Examples could easily be drawn from tasks such as typing, pilot training, or reading. The reader is encouraged to assess whether these generalizations can be seen in training programs familiar to the reader. These generalizations are described from a devil's advocate position to encourage the reader to critically consider some commonly held training assumptions.

Fallacy 1—Practice Makes Perfect

"Practice makes perfect" assumes that if individuals continue to perform a task, their performance will improve, reaching near-optimal levels. For learning simple tasks such as memorizing a phone number, practice makes perfect. However, this assumption does not prove to be a valid generalization for high-performance training. For example, in air traffic control training, a large portion of the training time is occupied with the student simply practicing the task. However, many students show only very slow acquisition rates by practicing the task and do not obtain acceptable performance levels by the end of the training program (recall the 50% washout rate).

The statement that practice makes perfect is an overgeneralization. Not only does practice often fail to make perfect, it sometimes produces no improvement in performance at all. For example, if subjects practice a digit-span task for weeks, subjects who do not consistently group the digits show little improvement in their ability to maintain information in memory (see Chase and Ericsson, 1981). Practice on consistent component tasks does improve component skills (see below). Consistent components are those elements of the task where the subject can make the same response to the stimulus whenever it occurs. When given explicit training on using a strategy to consistently encode the incoming stream, subjects' digit span can increase substantially (Chase and Ericsson, 1981).

Fallacy 2—Training of the Total Skill

The second fallacy is that it is best to train a skill in a form similar to the final execution of the skill. Total task training is necessary because the final performance is in the target task. However, belief in this fallacy tends to shift most of the training into a target-task format. A belief in this fallacy seduces one to maximize fidelity even when it yields little training benefit (e.g., see Hopkins, 1975). Training an air traffic controller to perceive turn points at which an aircraft should start a turn illustrates the inefficiency of total-task training. A normal aircraft requires four minutes to sweep out a complete turn on the radar screen. It is difficult to learn to perceive turn radii from such observations. First, because of perceptual decay, it is difficult for the controller to integrate more than about 15 s of the display. Hence, the observer must learn to perceive a circular pattern without ever having seen more than about 20 deg of the circle on the display. Second, the trainee receives very little training at this component skill even after long periods of experience. The trainee may experience only eight 90-deg turns in an hour. In contrast, a training module designed to teach this component could expose the trainee to hundreds of accelerated observations of a turn radius in an hour (Vidulich, Yeh, and Schneider, 1983). Third, the trainee receives poor feedback as to the quality of his or her judgment of the turn. The primary feedback indicates how closely the aircraft missed its final destination. There are a number of potential causes that could produce the same final error (e.g., misjudging the wind velocity, misjudging the rollout of the turn, midjudging the initial heading, too sharp a turn, etc.). When trained in the target task, the trainee will have difficulty in determining the source of an error.

Training in the target task can be inefficient. The real situation does not sequence events optimally, results in resource overload, and often produces frustration and panic. Those who support training primarily in the final situation should examine the assumptions that are implied by this "fallacy." If one believes the best way to train is in the target task one believes the following:

(1) The real world optimally presents consistent elements of the task. One is assuming that the world is fortuitously organized such that the typical execution of the skill best illustrates the consistent components of that task for optimal learning.
(2) One assumes that the real world optimally orders the sequence of events for training. Again one is assuming that the world is fortuitously constructed such that the spacing of practice at consistent task elements is optimal for learning.
(3) It is best to train a task when attentional capacities are overloaded. If one wants to learn to drive and converse at the same time, one should begin practicing doing both tasks together.
(4) It is acceptable to be confused about how errors influence performance. Whenever one performs a complex task, it is often difficult to tell which errors caused poor performance or whether those errors were caused due to lapses of attention or an inability to perform the given task.
(5) One assumes that frustration due to errors and poor performance does not reduce student effort.
(6) One assumes that there is little transfer from component task training to total performance.

If this last assumption is true, training the total task is the only way to substantially improve performance. However, in many situations there is substantial transfer of component training. For example, training with cardboard models of the cockpit can produce substantial savings in performing the task in the aircraft (Caro, 1973).

Fallacy 3—Skill Learning Is
Intrinsically Enjoyable

The third fallacy is that skill training is intrinsically motivating and thus, adding ex-

trinsic motivators is inappropriate. One example is air traffic control training. Air traffic controllers are professionals. Their futures depend on how well they do in the training program. Hence, one would expect little benefit from providing extrinsic motivators. However, being in a darkened room controlling simulated aircraft for 8 h can get boring. As the training day wears on, it can become difficult to concentrate on one's work. The problem with the belief in this fallacy is that it can justify a training program designer's lack of concern about motivating the learner. The problem of motivating the learner is left to the training personnel.

In my laboratory (Human Attention Research Laboratory, University of Illinois), about 3000 subject-training hours are executed per year. Probably the most cost-effective piece of equipment in my laboratory is a noise synthesis chip. This $15 chip can be programmed to emit interesting noises when important events occur. In an air traffic control task, for example, whenever a subject identifies the correct turn point, the aircraft flies the appropriate trajectory with an interesting frequency sweep auditory shot. Before the addition of extrinsic motivators, about 30% of the subjects failed to develop sufficient accuracy in our skill-acquisition experiments. After adding extrinsic motivators (e.g., interesting sound effects, interesting visual display patterns, providing criterion-based feedback), failure rates were reduced to less than 5%.

In many training programs the most important determinant of performance is how long a learner actively practices the task. When designing a training program, one must include motivational events to maintain active participation.

Fallacy 4—Train for Accurate Performance

The fourth fallacy is that the primary goal of training a skill is to produce highly accurate performance. In air traffic control, controllers are trained to maintain optimal separation between the aircraft. Training for maximal performance accuracy can be counterproductive. In many skill-training programs the goal should be to obtain acceptable accuracy on a component skill while allowing attention to be allocated to other components of the task. In air traffic control, an operator who can maintain optimal separation of only two aircraft would not be an acceptable controller. What is desired is an operator who can maintain safe separation among 10 aircraft.

Training programs following this fallacy tend to produce operators who can perform individual component skills well but who cannot operate well in high-workload situations. Specialized training may be necessary to develop skills that will operate well under high workload (see below). Also, in order to achieve reliable performance under high workload, substantial overtraining may be necessary. For example, LaBerge (1976) has shown that training subjects to compare symbols when not directly attending to those symbols requires about six times more training than does training them to compare symbols while attending to the task.

Fallacy 5—Initial Performance Is a Good Predictor of Trainee and Training Program Success

Belief in this fallacy suggests that if we measure a learner's performance in the first few hours of training, one can predict performance after hundreds of hours of training. In reality, initial performance of complex skills is very unstable and often provides a poor prediction of final performance. For example, the correlation between the first and fifteenth hour of performing a simple grammatical reasoning task was only 0.31 (Kennedy, Jones, and Harbeson, 1980). As the skill becomes more complex, more novel, and re-

quires longer training times, correlations between initial and final performance decrease. Note also that performance may not be a good measure of learning. For example, augmented training may greatly facilitate performance but may slow learning.

This fallacy presents a particular problem in evaluating training programs. Certain techniques may work very poorly in a short training program but be very beneficial in a long training program. For example, in a six-month air traffic control training program, it might be very beneficial to have a six-hour training module on identifying the heading angles of aircraft. However, in a pilot project, the researcher may have to demonstrate the benefit of a particular module with a simulated training program that is only six hours in length. It is likely that whole-task training would be the most effective training in a six-hour time scale even though a combination of part- and whole-task training would be better for a six-month training program.

Fallacy 6—Once the Learner Has a Conceptual Understanding of the System, Proficiency Will Develop in the Operational Setting

This fallacy leads to training programs that present technical information in a classroom setting and provide minimal instruction on how to use this information in performing the skill. For example, in air traffic control, the classroom teaching describes the aircraft performance characteristics. However, the student may not be shown explicitly what those performance characteristics look like on the radar scope. Often operators need a great deal of experience with the system even after they have learned to conceptualize it accurately. For example, it is relatively easy to visualize a mental model for a manual transmission. However, many hours have to be spent in a car before gear shifting becomes proficient.

Fallacy Summary

Most training programs regard one or more of the preceding fallacies as true principles. If one rejects these fallacies, training design becomes much more difficult, but potentially more successful. First, instead of assuming that practice is sufficient to produce high skill levels (Fallacy 1), one might emphasize practicing consistent components of the task. Second, as opposed to training the learner in the total task (Fallacy 2), one should break down the task and re-represent the task to maximize the learning rate on each component. Then one should sequence the various components to maximize the integration of the task. Third, one cannot assume that the material itself is sufficiently motivating for the learner (Fallacy 3). One may need to design extrinsic motivators into the task. One should find extrinsic methods for motivating the trainee without interfering with the actual teaching process. Fourth, one cannot train for perfect performance (Fallacy 4); one should train to what would be considered an acceptable performance level, but also train so that the learner can perform the task with little or no attention allocated to consistent components of the task. Fifth, one should be very cautious when extrapolating results from short training periods to predicting either successful training procedures or successful trainees (Fallacy 5). Sixth, one should recognize that providing the learner with a theoretical understanding is often only the first stage in developing a high-performance skill (Fallacy 6).

Empirical Characteristics

To develop effective training programs, it is useful to know the prominent features of high-performance skill acquisition. The characteristics of developing a high-performance skill are quite different than those of learning declarative information. Teaching declarative information generally includes presenting the information once in a classroom-type setting. In contrast, developing a skill entails presenting comparatively fewer "facts" per unit of time but requires the learner to devote a great deal of effort in developing and practicing component skills. The training program designer must be cautious not to use the academic training model as a model for high-performance skill acquisition. The training program designer should be aware of the nature of skill acquisition functions, the heterogeneity of component tasks, the need to discourage poor strategies, and the need to train time-sharing skills.

Extended practice function. The first class of training problems relates to acquisition functions. High-performance skill acquisition is characterized by log-log acquisition functions, improvement over long periods of training, initial instability, and false asymptotes. Practice curves are fit with a variety of functions. Reaction time data are most commonly fit with power functions and exponential functions (see Newell and Rosenbloom, 1981). Performance rating scale data are generally fit by logistic and exponential functions (see Spears, 1982). These various curve-fitting procedures show very high correlations (see Newell and Rosenbloom, 1981). The following discussion uses the power function for illustration, but none of the arguments would change if any of the other curves were utilized, since all show fast initial acquisition rates with gradual approach to an asymptote. The speed of responding is well fit as a power function of the number of trials (see Newell and Rosenbloom, 1981). The power law predicts that the log of the time to complete a response will be a linear function of the log of the number of executions of that particular response (see Figure 1). The power law is stated in the form of:

$$T = B N^{-\alpha} \qquad (1)$$

$$\log(T) = \log(B) - \alpha\log(N) \qquad (2)$$

T is the time to respond, N is the number of trials, and B and α are constants (with $\alpha < 1$). The power law predicts that if reaction time to perform a response decreased from 10 s to 5 s over the first 100 trials of training, at 440 trials response time will be 4 s, at 3978 trials response time will be 3 s, and a total of 10 000 trials of training would be necessary to reduce reaction time to 2.5 s. This law predicts that performance will improve rapidly for the initial trials but will continue to improve with a decreasing rate with more and more trials.

Newell and Rosenbloom (1981) refer to the power law as the "ubiquitous law of practice." The power law holds for a wide range of response-time tasks. Predictions from the law fit data including: operating a cigar-rolling machine (Crossman, 1959), adding digits (Crossman, 1959), editing text (Moran, 1980), playing card games (Newell and Rosenbloom, 1981), learning a choice reaction-time task (Seibel, 1963), detecting letter targets (Neisser, Novick, and Lazar, 1963), and performing geometry proofs (Neves and Anderson, 1981). The training-system designer can use the power law to predict performance improvement of component skills as a function of practice (for an excellent example of this type of prediction, see Card, Moran, and Newell, 1983).

A major feature of high-performance skills is that they show improvement over extended periods of time. For example, Crossman's (1959, see Figure 1) subjects showed improvement in operating a cigar-rolling machine over 3 million trials and two years.

The continued improvement after extended practice has two implications. First, high performance requires a great deal of practice even after the trainee understands the nature of the task and can perform the task accurately. Second, it may be beneficial to design training procedures that will allow the trainee many trials of performing critical component tasks.

The third characteristic of acquisition functions is that initial performance (e.g., the first 100 trials) is likely to be an unstable predictor of later performance. For example, Kennedy et al. (1980) found that the correlation between Day 1 and Day 15 of performing a grammatical reasoning task was 0.31. The subjects showed different initial performance levels, acquisition rates, and final asymptotes.

This instability of early acquisition stems from at least four sources. First, the rate of improvement during the first few hundred trials is very rapid, causing a large within-subject variation. Second, subjects with differential experience with related tasks start out at different performance levels. For example, assume one wanted to assess an individual's ability to perform an air traffic control task by running a simulated control session. Assume also that subjects who had played 30 h of a particular video game would start at this task with the equivalent of 2 h of training. In a 1-h test of performance these

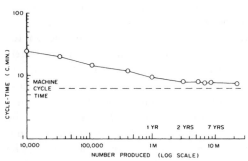

Figure 1. *Time taken to make cigars as a function of practice (Crossman, 1959).*

video-game-wise subjects are likely to perform substantially better than would subjects who had not played video games. However, since the actual training program would require hundreds of hours of training, an individual with a faster learning rate will surpass someone with a 2-h head start in training.

A third source of initial instability is that individuals vary in their rate of acquisition of skills (e.g., see Kennedy et al., 1980). It typically requires hundreds of trials to reliably assess learning rate, and hence it is difficult to get a quick estimate of this parameter. A fourth source of initial instability is that different abilities appear to limit performance at different stages of practice. Early in the training program, general cognitive abilities are critical for following instructions. Later, the basic psychomotor abilities may become the limiting factor (Fleishman and Rich, 1963).

A fourth characteristic of acquisition functions is that the willingness of the learner to continue to practice the task strongly influences final performance level. Most individuals can be motivated to perform even very boring tasks for a few hours. However, many individuals cannot maintain motivated performance when practicing a skill for tens or hundreds of hours. Most of the people who purchase a musical instrument never practice long enough to achieve even basic levels of proficiency with the instrument. In dual-task studies (e.g., Schneider and Fisk, 1984), substantial numbers of college student subjects fail to continue to put their full effort into learning the task after about six hours of training (even when they risk loss of bonus pay). This often does not result in a decrement in performance but rather a plateau or lack of improvement in performance. Bryan and Harter (1899) commented about the difficulties of overcoming plateaus in the development of skill. A training-program designer

must help the trainee continue to expend the effort to improve performance over very long practice periods. Note that, with proper motivation, performance plateaus seem less likely (see Keller, 1958).

Heterogeneity of component improvement rates. The second class of problems in skill development relates to the heterogeneity of component improvement rates. Performance on most complex tasks is determined by a variety of component skills. For example, in air traffic control, performance is determined by perceptual skills for identifying the locations and trajectories of the aircraft, cognitive skills to predict events and schedule traffic, and output skills relating to handling communications and operating equipment. Improvement rates for these different task components may vary widely. For example, the keying time may decrease by 5% over thousands of trials of training, whereas the time to decide how to schedule traffic may be reduced by 90% in the same number of trials.

Consistent task components show large improvements with practice whereas varied components do not. A consistent component is defined as one in which the subject can make the same response to a particular stimulus situation every time the stimulus situation occurs. For example, in a consistent letter-search task the subject would push a specific button every time the letter *E* appears on the display. In contrast, a varied component is one in which the mapping between the stimulus and response varies across trials. For example, in a varied letter-search task, a subject might search for and respond to *E*s on one trial, but on the next trial, the subject might search for *T*s and not respond to the letter *E*. Figure 2 shows the data in a letter-search experiment manipulating consistent and varied practice at the task. In the consistently mapped condition subjects responded to target stimuli letters

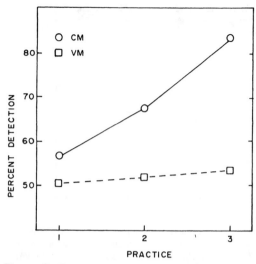

Figure 2. *Detection accuracy as a function of training trials in a letter-search task (from Schneider and Fisk, 1982b). CM refers to a consistently mapped letter search, VM to variably mapped.*

whenever they occurred. Detection accuracy improved substantially over some 840 training trials. In contrast, in the variably mapped condition, in which subjects responded to different stimuli on different trials, there was no improvement over trials (Schneider and Fisk, 1982b).

Similar to letter search, motor responding tasks show improvement primarily on consistent sequences. In a motor output sequential responding task subjects reproduced sequences of eight button-pushes. After the first 10 trials, execution of varied response sequences showed no improvement in speed or accuracy. The execution with consistent sequences improved in accuracy (30%), speed (22%), and response variability (50% relative to the varied sequences) during 50 training trials (see Schneider and Fisk, 1983). Chase and Ericsson (1981) found that subjects who varied in their grouping of digits showed little improvement with practice in a digit-

span task. However, when they consistently grouped the digits and associated them with salient classes of events, there was substantial improvement with practice. With a year of practice, one subject was able to increase his digit span to 80 digits.

Consistent task components show substantial improvement in processing speed with practice, whereas varied components do not. In a category-search experiment, Fisk and Schneider (1983) determined the increase in reaction time as a function of the number of category judgments made. The slope relating comparison time to number of comparisons was 200 ms per comparison in the varied search condition and only 2 ms per comparison in the consistent search condition. In this case there was an increase of two orders of magnitude in processing speed for the consistent component relative to the varied component.

Consistent task components can be performed with little or no attention, whereas varied task components are resource sensitive even after extended practice. Figure 3 shows the data from a dual-task experiment (Schneider and Fisk, 1984). The primary task required comparing two digits in memory to two digits on the display that changed every 400 ms. The secondary task required detecting words from a given semantic category. If the category task was consistently mapped, subjects could perform the category task equally well whether they performed the digit task or not. In contrast, if the category task was variably mapped, extended training produced no improvement in the ability of the subjects to time-share the category and digit search tasks. Kennedy and Bittner (1980) provide another example of lack of improvement in time-sharing ability in a task requiring varied responding. They found that when subjects had to count tones on two channels, there was no performance improve-

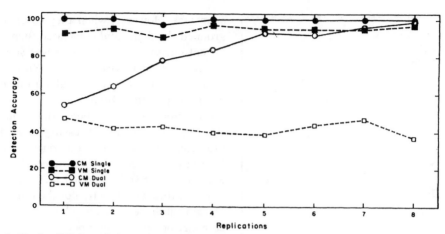

Figure 3. *Single- (filled symbols) and dual- (open symbols) task category-search detection accuracy as a function of sessions of practice (Schneider and Fisk, 1984). Note the elimination of the dual-task deficit in the CM condition (solid lines for consistent category search), and stability of the dual-task deficit for the VM condition (dashed line for varied category search).*

ment over 15 days of practice. Most time-sharing studies have consistent components and show substantial improvement with practice (e.g., Damos and Wickens, 1980; Schneider and Fisk, 1982a).

Heterogeneity of component skills complicates the problem of assessing trainee performance and providing knowledge of results. For example, in air intercept control of a stern attack, the final performance measure is how accurately the fighter is placed behind the enemy aircraft. This single score is determined by perceptual errors in assessing the heading, speed, and turn point of the aircraft; cognitive errors in planning the strategy or deciding the turn point; or motor errors in controlling the equipment and giving commands to the pilot. A single performance score provides the learner with little detail about how to improve performance. In heterogeneous tasks it is important to provide the learner with knowledge of results about the performance of individual component skills (see Newell, 1976). By developing component skill tests, an operator's weaknesses

in various phases of a complex task can be illustrated.

Heterogeneity of component skill learning rates may require differential training of components. In the air traffic control training modules, improvement rates range from a 29% improvement during the first 100 trials for identifying the turn point of an aircraft to only a 4% improvement over the first 100 trials at identifying heading angles. After the first 1000 trials at training to identify turn point, inaccuracy in this component no longer limited total performance. In contrast, heading identification would improve for thousands of trials. To optimize final trainee performance, training time should be allocated so as to maximize total performance improvement per unit of training time.

Eliminating poor strategies. A third class of problems in high-performance skill training is the need for special training to eliminate poor strategies. In many skills, a variety of strategies may be used to perform a particular task component. However, one strategy usually allows better development of more

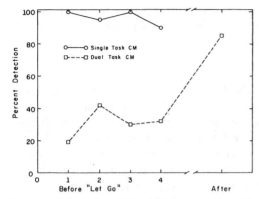

Figure 4. *Subjects' single- and dual-task consistently mapped category-search detection accuracy. After Session 4, subjects were given alternative training to facilitate "letting go" of attentionally processing the words (from Schneider and Fisk, 1983).*

advanced skills. For example, in driving an automobile one can drive by lining up the hood ornament to the road stripe or by estimating the angle of the curve and turning the steering wheel appropriately for that angle. The former strategy is easier to learn but requires many more decisions and results in higher workload.

Trainees may be resistant to discontinuing the use of a strategy that is easy to learn but results in high workload. In laboratory search tasks, some subjects require specialized training in order to enable a low-workload strategy. Most subjects can perform a consistent category search task in combination with a high-workload digit task without deficit (Figure 3, also Schneider and Fisk, 1983). However, occasionally a subject seems unable to perform a category search task under high workload. Figure 4 illustrates data from two such subjects. In dual-task conditions, the subjects' category detection rate was only about 30% of what it was in the single-task conditions. During 4 h of testing there was no evidence that these subjects were improving. This lack of improvement in consistent category search contrasted to the data of most of

the subjects (see Figure 3). At the point of Session 4, these subjects would be considered washout subjects. They could not adequately perform this task under high workload. These subjects were then trained to perform an easier semantic search and digit detection task. When the subjects were successful at learning easier categories, they returned to the original category condition in which they were having difficulty. The subjects' dual-task performance increased from the previous 30% to 84% accuracy even though they had had no training on the specific category detection task. The subjects reported that during the interim training they had learned to "let go" and to respond to the words without thinking about them. Once these subjects had learned to "let go," they could perform the category task with high accuracy even while performing a high-workload task (for details, see Schneider and Fisk, 1983).

The acquisition of reading skill provides interesting illustrations of the need to break bad habits. LaBerge and Samuels (1974) report that some readers become overly concerned about word-encoding accuracy. These readers focus too much effort at doing the word-encoding skill and have few attentional resources available for semantic integration of the task. Frederiksen (Frederiksen, Weaver, Warren, Gillotte, Rosebery, Freeman, and Goodman, 1983) reports that some poor readers develop a strategy of looking at the first and last letters of the word and guessing what the word would be. These readers can continue to use this strategy for years without substantial improvement in reading. One method that seems to successfully break this habit is to have the reader identify word units within computer-presented words ("push the button any time the letter pattern *min* appears"). Such training enables the learner to focus on elements of the language that do produce accurate performance and hence show substantial practice effects.

Developing time-sharing skills. High-performance tasks often require the development of specialized time-sharing skills. In order to perform two tasks simultaneously, it is critical that the two tasks be practiced together (e.g., Damos and Wickens, 1980). Training with different task priorities may be necessary to help subjects determine the optimal allocation of attention for maximal skill performance (Gopher and North, 1977). It is difficult to train subjects to respond to multiple channels simultaneously (see Duncan, 1980). Schneider and Fisk (unpublished data) have found that practice in conditions requiring a high frequency of simultaneous responding can greatly improve the operator's ability to deal with occasional situations requiring simultaneous responding. It may also be necessary to train subjects so that they can determine how to trade off speed and accuracy. In reading, for example, with full attention allocated to word encoding, word-encoding accuracy may be 98%; however, only a small amount of attentional resources are available for semantic processing. In contrast, if word encoding is done with minimal attention to the word encoding, encoding accuracy may be 90%, but the majority of attentional resources will be available for semantic processing. The latter strategy may result in far better comprehension. Finally, in very high workload situations, the operator must often triage through the list of priorities deciding which tasks must be left undone. Munro, Cody, and Towne (1982) have shown that as workload increases in a simulated air-traffic control task, the operators intentionally ignore information in order to cope with the critical aspects of the situation.

This review illustrates salient features for training in high-performance environments. This review is not intended to be exhaustive (for general reviews on skill acquisition, see Anderson, 1981; Welford, 1976). The emphasis here has been on identifying those features of the training program that enable the novice to become a high-performance practitioner.

Working Guidelines

The training program designer implicitly or explicitly develops the training program in accordance with certain rules. The six fallacies discussed in this paper indicate the assumptions that the designer should not make. In the last five years at the Human Attention Research Laboratory, subjects have been trained for over 10 000 h in a variety of skill-acquisition experiments. Current laboratory training programs include air traffic control and electronic troubleshooting tasks. The following is a list of rules used in the design effort. The rules developed out of basic research on developing visual search skills (see Schneider, 1982). These rules should be treated as an initial set of working guidelines and are provided to help focus discussion and research in order to develop an explicit set of training rules. The laboratory is engaged in a long-term research project to evaluate the effectiveness of these guidelines in applied training programs. (For a more detailed discussion of these guidelines, see Schneider, 1982.)

These rules are based on the proposition that human performance results from the interaction of two qualitatively different forms of processing (James, 1890; LaBerge, 1976; Norman, 1976; Posner and Snyder, 1975; Schneider and Fisk, 1983; Shiffrin and Schneider, 1977). These two forms are referred to as *automatic* and *controlled* processing. Automatic processing is a fast, parallel, fairly effortless process that is not limited by short-term memory capacity, is not under direct subject control, and performs well-developed, skilled behaviors. Automatic processing typically develops when subjects deal with the stimulus consistently over many trials. Controlled processing is characterized as a slow, generally serial, effortful, capacity-limited, subject-controlled processing mode that must be used to deal with novel or inconsistent information (see Schneider and Shiffrin, 1977; Shiffrin and Schneider, 1977). Controlled processing is expected when the subject's response to the stimulus *varies* from trial to trial. From the automatic/controlled processing perspective, training should develop automatic component skills to perform consistent task components and develop strategies to allocate limited controlled processing resources to inconsistent or poorly developed task components (see Schneider and Fisk, 1983).

Rule 1—Present information to promote consistent processing by the operator. In order to develop fast, low-workload processing, the operator must perceive and deal with situations consistently. Developing consistent processing can be done in a variety of ways, including the use of analogies, the provision of specialized representations of the problem, and adaptive training. In teaching electronic circuits it may be beneficial to teach the operator an analogy of a water-flow process for a particular logical element. As the operators visualize the analogy, they consistently make the same response to a given situation, developing automatic component processes. After several hundred trials, the operator can specify the output with little effort and without the use of the analogy. In the air traffic control task for inflight rendezvous (for mid-air refueling) we graphically illustrated all of the possible points at which the two aircraft could rendezvous (see Schneider, Vidulich, and Yeh, 1982). Trainees observe how the total space of rendezvous varies as a function of the intercept angle and displacement of the aircraft. In this way the operators learn to see the consistent relationships among patterns of rendezvous.

Rule 2—Design the task to allow many trials of critical skills. Training modules should be designed to provide the learner with many trials of experience in a short period of time. In the air traffic control task, to rapidly train visualization of flight patterns, we compress simulated time by a factor of 100; for example making a judgment of where an aircraft should turn (a maneuver that normally takes about 5 minutes) would take about 0.5 s. By compressing time in this way, we can provide the trainee with more trials at executing this particular component in a single day of training than he or she could get in a year of training with conventional methods. In order to offer extended training experience it may be necessary to compromise on simulation fidelity to increase the number of simulator hours.

Rule 3—Do not overload temporary memory and do minimize memory decay. In training the air traffic control task, during initial acquisition the flight path is drawn on the display so that operators need not retain that information in memory. This facilitates the maintenance of a consistent memory representation and speeds the development of automatic component processes to identify turn radii. The number of new tasks to be performed concurrently should be limited to minimize attentional overload.

Rule 4—Vary aspects of the task that vary in the operational situation. When developing automatic components, the components must generalize to the entire class of situations to which they are appropriate. For example, when training the operator to identify the turn point for an aircraft, the test intercepts occur at all possible locations on the screen. If all turn identification occurs with the aircraft in the center of the screen, the skill may not generalize well to other positions.

Rule 5—Maintain active participation throughout training. Active participation is enhanced if subjects need to respond every few seconds. For example, to train subjects to visualize solution spaces, subjects observe the solution space going through a range of intercept angles and then are presented a test vector. The subject must identify whether that test vector is appropriate for that intercept angle. Without these frequent tests, subjects' observation becomes passive, and there is little improvement with practice.

Rule 6—Maintain high motivation throughout the training period. Provide the trainee with extrinsic motivation to maintain high levels of effort. When subjects respond incorrectly, a simulated crash can occur. Adaptive training can sequence subjects to ever more difficult training conditions but still allow them to experience a high degree of success throughout. In order not to significantly increase training time, the motivational feedback should be limited to a small portion of the training period (e.g., less than 5%).

Rule 7—Present the information in a context that illustrates more than the to-be-learned task. For example, when subjects identify proper intercept points, feedback shows how the planes would fly to that intercept point, thus illustrating the trajectories of the flight path while the operator tries to perceive the final rendezvous point. Caro (1973) recommends training flight skills within a functional mission context. The trainer must, however, be careful to not overload the subject (Rule 3) and efficiently train component skills (Rule 2).

Rule 8—Intermix component training. Intermix the training on various component skills rather than training each component individually before proceeding to the next component. This intermix training distributes the practice and facilitates perception of the interrelationships of the components. The proportion of component and total-task training time should be allocated so as to maximize final total-task performance.

Rule 9—Train under mild speed stress. Au-

tomatic components are fast processes, probably occurring in less than half a second. Speed stressing subjects improves the development rate. When not speed stressed, subjects tend to use slow, controlled processes that may not be acceptable in the operational environment. For example, in the turn-point identification task, air traffic control operators are expected to make a response in less than 2 s.

Rule 10—Train strategies that minimize operator workload. In many tasks there are multiple strategies that involve differential workload. In air traffic control, the operator is allowed only one decision to get the airplanes together in a rendezvous. If operators develop a strategy of many small corrections during training, the workload imposed by that strategy makes it difficult to handle sets of five aircraft.

Rule 11—Train time-sharing skills for dealing with high-workload environments. Train operators in situations that require using different speed and accuracy trade-offs, frequent simultaneous responding, and triaging through task priorities. Structure the training so that the expert can perform reliably even during those rare occasions when his or her skills are pushed beyond reasonable limits.

CONCLUSION

There are special problems associated with training high-performance skills. It is difficult to get an appropriate perspective of the changes that occur during months and years of training. Certain assumptions that work well in short-term training programs may be fallacious when extended to long-term training programs. The training-program designer needs to understand the assumptions underlying each given training procedure. The trainer should be aware of the special problems of acquiring high-performance skills. The research community must work to develop and test an adequate set of guidelines for high-performance skill acquisition. With appropriate perspective, research, and guidelines, the current computer revolutions can flower into a training revolution.

ACKNOWLEDGMENT

This research was sponsored by Personnel and Training Research Programs, Psychological Sciences Division, Office of Naval Research, under Contract No. N000-14-81-K-0034, Contract Authority Identification No. NR 154-460.

REFERENCES

Anderson, J. R. (Ed.). (1981). *Cognitive skills and their acquisition.* Hillsdale, NJ: Erlbaum.

Bryan, W. L., and Harter, N. (1899). Studies on the telegraphic language: The acquisition of a hierarchy of habits. *Psychological Review, 6,* 345-375.

Card, S. K., Moran, T. P., and Newell, A. (1983). *The psychology of human-computer interaction.* Hillsdale, NJ: Erlbaum.

Caro, P. W. (1973). Aircraft simulators and pilot training. *Human Factors, 14,* 502-509.

Chase, W. G., and Ericsson, K. G. (1981). Skilled memory. In J. R. Anderson (Ed.), *Cognitive skills and their acquisition* (pp. 141-189). Hillsdale, NJ: Erlbaum.

Crossman, E. R. F. W. (1959). A theory of the acquisition of speed-skill. *Ergonomics, 2,* 153-166.

Damos, D. L., and Wickens, C. D. (1980). The identification and transfer of timesharing skills. *Acta Psychologica, 46,* 15-39.

Deighton, L. (Ed.). (1971). *Encyclopedia of education.* New York: Macmillan.

Duncan, J. (1980). The locus of interference on the perception of simultaneous stimuli. *Psychological Review, 87,* 272-300.

Eberts, R., Smith, D., Dray, S., and Vestewig, R. (1982). *A practical guide to measuring transfer from training devices to weapon systems* (Final Report 82-SRC-13). Minneapolis, MN: Honeywell Systems and Research Center.

Fisk, A. D., and Schneider, W. (1983). Category and word search; Generalizing search principles to complex processing. *Journal of Experimental Psychology: Learning, Memory, and Cognition, 9,* 177-195.

Fleishman, E. A., and Rich, S. (1963). Role of kinesthetic and spatial-visual abilities in perceptual-motor learning. *Journal of Experimental Psychology, 66,* 6-11.

Frederiksen, J. R., Weaver, P. A., Warren, B. M., Gillotte, H. P., Rosebery, A. S., Freeman, B., and Goodman, L. (1983). *A componential approach to training reading skills.* (Final Report 5295). Cambridge, MA: Bolt Beranek and Newman.

Gopher, D., and North, R. A. (1977). Manipulating the conditions of training in time-sharing performance. *Human Factors, 19,* 583-593.

Griffin, G. R., and Mosko, J. D. (1977). *Naval aviation attrition 1950-1976: Implications for the development of future research and evaluation* (Report NAMRL-1237). Pensacola, FL: Naval Aerospace Medical Research Laboratory.

Hopkins, C. O. (1975). How much should you pay for that box? *Human Factors, 17,* 533-541.

James, W. (1890). *Principles of psychology* (Vol. 1). New York: Holt.

Keller, F. S. (1958). The phantom plateau. *Journal of the Experimental Analysis of Behavior, 1,* 1-13.

Kennedy, R. S., Jones, M. B., and Harbeson, M. M. (1980). Assessing productivity and well-being in Navy workplaces. In *Proceedings of the 13th Annual Meeting of the Human Factors Association of Canada,* (pp. 8-13). Ottawa: Human Factors Association of Canada.

Kennedy, R. S., and Bittner, A. C. (1980). Development of performance evaluation tests for environmental research (PETER): Complex counting test. *Aviation, Space, and Environmental Medicine, 51,* 142-144.

LaBerge, D. (1976). Perceptual learning and attention. In W. K. Estes (Ed.) *Handbook of learning and cognitive processes* (Vol. 4) (pp. 237-273). Hillsdale, NJ: Erlbaum.

LaBerge, D., and Samuels, S. J. (1974). Toward a theory of automatic information processing in reading. *Cognitive Psychology, 6,* 293-323.

McCauley, M. E., Root, R. W., and Muckler, F. A. (1982). *Training evaluation of an automated training system for air intercept controllers* (Final Report NAVTRAEQUIPCEN 81-C-0055-1). Westlake Village, CA: Canyon Research Group.

Moran, T. P. (1980). *Compiling cognitive skill* (AIP memo). Palo Alto, CA: Xerox PARC.

Munro, A., Cody, J. A., and Towne, D. M. (1982). *Instruction mode and instruction intrusiveness in dynamic skill training* (Report ONR-99). Los Angeles: Behavioral Technology Laboratories, University of Southern California.

Neisser, U., Novick, R., and Lazar, R. (1963). Searching for ten targets simultaneously. *Perceptual and Motor Skills, 17,* 955-961.

Neves, D. M., and Anderson, J. R. (1981). Knowledge compilation: Mechanisms for the automatization of cognitive skills. In J. R. Anderson (Ed.), *Cognitive skills and their acquisition* (pp. 57-84). Hillsdale, NJ: Erlbaum.

Newell, A., and Rosenbloom, P. S. (1981). Mechanisms of skill acquisition and the law of practice. In J. R. Anderson (Ed.), *Cognitive skills and their acquisition* (pp. 1-55). Hillsdale, NJ: Erlbaum.

Newell, K. M. (1976). Knowledge of results and motor learning. *Exercise and Sport Science Reviews, 4,* 195-228.

Norman, D. A. (1976). *Memory and attention: An introduction to human information processing.* New York: Wiley.

Pew, R. W. (1966). Acquisition of hierarchical control over the temporal organization of a skill. *Journal of Experimental Psychology, 71,* 764-771.

Pew, R. W. (1974). Human perceptual-motor performance. In B. H. Kantowitz (Ed.), *Human information processing: Tutorials in performance and cognition* (pp. 1-39). Hillsdale, NJ: Erlbaum.

Posner, M. I., and Snyder, C. R. (1975). Attention and cognitive control. In R. L. Solso (Ed.), *Information processing and cognition: The Loyola Symposium* (pp. 55-85). Hillsdale, NJ: Erlbaum.

Rumelhart, D. E., and Norman, D. A. (1982). Simulating a skilled typist: A study of skilled cognitive-motor performance. *Cognitive Science, 6,* 1-36.

Schneider, W. (1982). *Automatic/control processing concepts and their implications for the training of skills.* (Tech. Report HARL-ONR-8101). Champaign, IL: University of Illinois, Human Attention Research Laboratory.

Schneider, W., and Fisk, A. D. (1982a). Concurrent automatic and controlled visual search: Can processing occur without resource cost? *Journal of Experimental Psychology: Learning, Memory, and Cognition, 8*, 261-278.

Schneider, W., and Fisk, A. D. (1982b). Degree of consistent training: Improvements in search performance and automatic process development. *Perception and Psychophysics, 31*, 160-168.

Schneider, W., and Fisk, A. D. (1983). Attention theory and mechanisms for skilled performance. In R. A. Magill (Ed.), *Memory and control of action* (pp. 119-143). New York: North-Holland.

Schneider, W., and Fisk, A. D. (1984). Automatic category search and its transfer. *Journal of Experimental Psychology: Learning, Memory, and Cognition, 10*, 1-15.

Schneider, W., and Shiffrin, R. M. (1977). Controlled and automatic human information processing: I. Detection, search, and attention. *Psychological Review, 84*, 1-66.

Schneider, W., Vidulich, M., and Yeh, Y. (1982). Training spatial skills for air-traffic control. In *Proceedings of the Human Factors Society 26th Annual Meeting* (pp. 10-14). Santa Monica, CA: Human Factors Society.

Seibel, R. (1963). Discrimination reaction time for a 1,023 alternative task. *Journal of Experimental Psychology, 66*, 215-226.

Shiffrin, R. M., and Schneider, W. (1977). Controlled and automatic human information processing: II. Perceptual learning, automatic attending, and a general theory. *Psychological Review, 84*, 127-190.

Spears, W. D. (1982). *Processes of skill performance: A foundation for the design of training equipment* (Report NAVTRAEQUIPCEN 78-C-0113-4). Orlando, FL: Naval Training Equipment Center.

Vidulich, M., Yeh, Y., and Schneider, W. (1983). Time compressed components for air intercept control skills. In *Proceedings of the Human Factors Society 27th Annual Meeting* (pp. 161-164). Santa Monica, CA: Human Factors Society.

Welford, A. T. (1976). *Skilled performance.* Glenview, IL: Scott Foresman.

Automaticity, Resources, and Memory: Theoretical Controversies and Practical Implications

GORDON D. LOGAN,[1] *University of Illinois, Champaign, Illinois*

This article describes a theoretical controversy over the nature of automaticity and suggests implications of the controversy for the design of training programs. One side of the controversy describes automaticity in terms of processing resources, in that automatic processes require little or no resources. The other side describes automaticity as a memory phenomenon dependent on direct-access, single-step retrieval from memory. The two sides differ in their ability to account for four basic questions about automaticity: (1) why automatic processing has the properties it does; (2) how automaticity is learned; (3) how the properties of automaticity emerge with practice, and (4) why consistency of practice is so important to the development of automaticity. The memory view provides better answers than the resource view, particularly for questions about training. Training is an important practical issue, and implications of the memory view for training are spelled out in some detail.

INTRODUCTION

There is a battle raging in the ivory tower over the concept of automaticity. One faction represents the "old guard," the modal view in the field, and construes automaticity as a way to overcome resource limitations. The other faction is revolutionary (or sees itself as such) and construes automaticity as a memory phenomenon reflecting the consequences of running a large data base through an efficient retrieval process. The battle may turn out to be a tempest in a teapot, affecting no more than academic promotion and tenure. Or it could be the turning point in an intellectual revolution that reaches far beyond the

ivory tower to the most practical nooks and crannies in the land.

The purpose of this article is to describe the battle as it has developed so far, sketching each position and speculating on the outcome, and to suggest what the world would be like if the revolutionary position were even partly true. The idea that automaticity is a memory phenomenon has many implications that may prove important in practice even if the old guard maintains its hold on the ivory tower. Automaticity is a major factor in skill acquisition, so new perspectives on automaticity may shed light on practical issues in training.

FACTS ABOUT AUTOMATICITY

Empirically, automaticity is reasonably well understood. There are a number of basic

[1] Requests for reprints should be sent to Gordon D. Logan, Department of Psychology, University of Illinois, 603 E. Daniel Street, Champaign, IL 61820.

facts, well documented and replicable, upon which both sides agree. Disagreement arises over the interpretation of the facts. What follows is a brief review describing the major phenomena that theories of automaticity must address. For more extensive reviews, see Kahneman and Treisman (1984), LaBerge (1981), Logan (1985a), and Schneider, Dumais, and Shiffrin (1984).

There is considerable evidence that automatic processing differs from nonautomatic processing in several respects. Automatic processing is fast (Logan, 1988; Neely, 1977; Posner and Snyder, 1975); effortless (Logan, 1978, 1979; Schneider and Shriffin, 1977); autonomous or obligatory (Logan, 1980; Posner and Snyder, 1975; Shiffrin and Schneider, 1977; Zbrodoff and Logan, 1986); consistent or stereotypic (Logan, 1988; McLeod, McLaughlin, and Nimmo-Smith, 1985; Naveh-Benjamin and Jonides, 1984); and unavailable to conscious awareness (Carr, McCauley, Sperber, and Parmalee, 1982; Marcel, 1983).

There is also abundant evidence that automaticity is learned, which is what makes it relevant to skill acquisition and training. Some have argued that automaticity is a limiting factor in skill acquisition, that the rate at which components are automatized limits the rate at which skill is acquired (Bryan and Harter, 1899; LaBerge and Samuels, 1974). Practice in consistent environments seems to be a necessary condition in order for learning to occur (Fisk, Oransky, and Skedsvold, 1988; Logan, 1979; Schneider and Fisk, 1982; Schneider and Shiffrin, 1977; Shiffrin and Schneider, 1977; but see also Durso, Cooke, Breen, and Schvaneveldt, 1987). Most of the properties of automaticity emerge through practice in consistent environments (Logan, 1978, 1979; Shiffrin and Schneider, 1977).

The major questions to be answered by theories of automaticity are as follows: (1) Why do automatic processes have the pre-

viously mentioned properties? (2) How is automaticity learned? (3) How do the properties of automaticity emerge with practice? and (4) Why is consistency important? Theories differ markedly in the answers they provide and even in their ability to provide answers.

THEORETICAL APPROACHES TO AUTOMATICITY

Thesis: Automaticity and Resources

According to the modal view, automatic processing is processing without attention, and the development of automaticity represents the gradual withdrawal of attention (Hasher and Zacks, 1979; Logan, 1979, 1980; Posner and Snyder, 1975; Shiffrin and Schneider, 1977). The idea is usually expressed in the context of a single-capacity theory of attention such as Kahneman's (1973) or Posner and Boies' (1971). In such theories attentional capacity is thought to *energize* performance, and the amount of capacity allocated to a process determines the amount or rate of processing. Processes can vary in the amount of capacity they require, and automatic processes are assumed to require no capacity. Somehow capacity demands diminish with practice, though none of these theories provides an explicit learning mechanism.

The modal view provides a clear answer to the first question about why automatic processes have the properties they do. Automatic processes are fast because they are not limited by the amount of available capacity. They are effortless because effort is proportional to the amount of capacity allocated to the process. Automatic processes require no capacity; therefore, they require no effort. Automatic processing is autonomous or obligatory because attentional control is exerted by allocating capacity, and a process that does not require capacity cannot be controlled by allocating capacity. It will be triggered whenever the appropriate stimuli appear, whether or not capacity is allocated to it. Automatic processing is also unconscious because attention is the mechanism of consciousness, and we are conscious only of those things to which we pay attention (allocate capacity; see Posner and Boies, 1971; Posner and Klein, 1973).

The modal view *describes* but does not *explain* the development of automaticity. Many authors argue that the demand for attention diminishes with practice: for example, attention is initially required to support performance by establishing temporary connections between stimuli and responses or by controlling the order in which mental operations are sequenced. Practice somehow strengthens the connections or allows sequential processes to be executed in parallel, so that attention progressively becomes less necessary to support performance and finally is not required at all.

This description fits well with intuition, and many experiments show that attentional effects diminish with practice. Two important examples are the reduction of information load effects in visual- and memory-search tasks (Schneider and Shiffrin, 1977) and the reduction of interference between concurrent tasks (Logan, 1979). But description is not explanation. A serious weakness of the modal view is that no theory specifies how the reduction in demand comes about.

The modal view provides a descriptive—not explanatory—answer to the third question of how the properties of automatic processes develop with practice. The properties develop because they are characteristic of processes that require little or no capacity, and capacity demands diminish with practice.

The modal view also has difficulty dealing with the importance of consistency, as it assumes that automatization makes the process underlying performance faster and more efficient. Such *process-based* learning depends on the number of times the process is exercised, regardless of the stimuli exercised with. For example, in a flat city like Champaign, Illinois, the effect of jogging on cardiovascular fitness depends on the distance jogged and not the route chosen. Thus process-based learning predicts no effect (or weak effects) of consistency, contrary to the facts.

By the late 1970s, the modal view had a stronghold on the ivory tower. It seemed able to accommodate the facts of automaticity, and its underlying single-capacity theory seemed able to account for most of the facts about attention (see, for example, Kahneman, 1973). But then *multiple-resource* theory arose to challenge the single-capacity view of divided attention and dual-task performance arguing that more than one capacity or resource limited performance (Navon and Gopher, 1979; Wickens, 1980, 1984). There was abundant evidence that interference depended on the modality of input, output, and central processing rather than on the total amount of information to be processed (see Wickens, 1980), so multiple-resource theory won a quick victory. Adherents of the single-capacity view admitted the existence of multiple resources but argued that one resource —an executive resource associated with attentional capacity—dominated the rest (e.g., Logan, 1979).

Multiple-resource theories were directed primarily at dual-task and divided-attention situations and never dealt seriously with automaticity. Those that broached the issue often reiterated the modal view—that automatic processes use fewer resources (e.g., Navon and Gopher, 1979; Wickens, 1984). But the meaning of this claim is not as clear in multiple-resource theories as it was in the single-capacity theories that inspired the modal view. In multiple-resource theories it is not clear *which* resources an automatic process should use fewer of. One possibility is that automatic processes use less of some particular single resource. Some theorists have argued for an executive resource that is used in all tasks (e.g., Logan, 1979); automatic processes would use less of that resource. But it is not clear whether any single resource acts as an executive or whether the executive uses only one resource. Logan, Zbrodoff, and Fostey (1983), for example, argued that executive processing uses several resources.

Another possibility is that automatic processes use fewer of whatever resources the task used to begin with. Performance becomes more efficient—that is, cheaper—with practice. But there is no easy way to differentiate this alternative from the possibility that practice changes the resource composition of a task by shifting from one set of resources to another (e.g., West, 1967). Performance will appear more efficient with respect to the resources relied on initially but *less* efficient with respect to the resources relied on after practice. Automaticity will have *increased* the amount of those resources required, and this is difficult to reconcile with the modal view that automatic processing is efficient.

One could salvage the modal view by arguing that the total amount of resources used was less after practice than before, but this creates a serious measurement problem that no theory is yet equipped to solve: How can the different resources be made commensurate? How many units of one resource equal a unit of another? Without an answer to these questions, one cannot compare the amount of resources used before and after a shift in resource composition or evaluate the claim that automatic processes use fewer resources. The bottom line is that multiple-resource theories raise difficult problems for resource-based theories of automaticity.

To make matters worse, serious attacks have been made on the idea of resources and the economic metaphor on which it is based (e.g., Allport, 1980; Navon, 1984; Neisser, 1976). Some researchers challenge the idea that performance is limited by entities or energies that can be used by only one process, and that entities or energies released by one process can be used by another (Navon, 1984). Others argue against a resource interpretation of the modality effects that defeated the single-capacity view, suggesting other mechanisms of interference such as noise and crosstalk (Navon and Miller, 1987).

These issues represent internal conflicts in the resource theory camp that remain to be resolved. They may be the death knell for resource theories or may only be difficult stages in a healthy development. Only time will tell. But they do represent serious problems for the modal view of automaticity: if resource theory dies, the modal view loses the core assumption—single capacity—that provided theoretical coherence to the concept of automaticity. If resource theory is to survive, the modal view must grapple with the problems of multiple resources. Clearly the dominant force in the ivory tower is under duress, ripe for a revolution.

Antithesis: Automaticity and Memory

The revolutionary view is that automaticity is a memory phenomenon. Performance is considered automatic when it depends on single-step, direct-access retrieval of solutions from memory rather than on some sort of algorithmic computation. For example, a person familiar with the spatial layout of a town navigates automatically, in that the goal and the current features of the landscape retrieve a travel plan or a direction from memory. A person unfamiliar with the town could not navigate automatically, as he or she would have nothing to retrieve from memory and so would have to depend on general map reading skills, and close attention to road signs to reach his or her destination.

This idea is both old and new. It was at the core of Bryan and Harter's (1897, 1899) analysis of the acquisition of telegraphy skill and it is the essence of several current theories of skill acquisition and automatization (e.g., Logan, 1988; Newell and Rosenbloom, 1981; Rosenbloom and Newell, 1986; Schneider, 1985). It appears in various guises in the literature. Though theories differ in their assumptions about internal processes and representations and the role of resources, they all assume (implicitly or explicitly) that automatization reflects the build-up of information in memory.

The challenge to the resource view lies in the possibility, expressed in the more radical theories (Logan, 1988; Newell and Rosenbloom, 1981), that resources play no role in automatization. These theories propose that novice performance is limited by a lack of knowledge rather than by a lack of resources. Through practice with specific problems novices learn specific solutions, which they can apply when faced again with the same problem or generalize when faced with a similar problem. At some point they will have learned enough to be able to retrieve solutions for all or most of the problems encountered in the domain: they will have achieved the automaticity associated with expertise.

The automaticity-as-memory view provides clear answers to the four general questions raised at the beginning of the paper. First, automatic processes have certain properties because those properties are characteristic of memory retrieval. Automatic processing is fast, effortless, and unconscious because the conditions that prevail in studies of automaticity (i.e., extensive practice) are good for memory retrieval. The memory traces that support automatic processing are "strong" in some sense, which allows them to be retrieved rapidly (hence the speed) and reliably (hence the effortlessness) in a single

step. Single-step retrieval would seem unconscious because there are no intervening steps or stages upon which to introspect. Automatic processing is obligatory because memory retrieval is obligatory; attention to an object is sufficient to cause retrieval of whatever information has been associated with it in the past (Keele, 1973; Logan, 1988; also see Kahneman and Treisman, 1984).

Second, the automaticity-as-memory view provides several mechanisms by which automaticity can be acquired. The most common mechanism is *strengthening*, in which a connection between a stimulus and a response (or the like) becomes progressively stronger with practice (e.g., LaBerge and Samuels, 1974; Schneider, 1985; Shiffrin and Schneider, 1977). Another common mechanism is *table look-up*, in which specific responses to specific input patterns are learned and held in a table that is consulted whenever another input appears. If the input matches a pattern in the table, the associated response is executed; if not, some other procedure is carried out to compute the response (e.g., Newell and Rosenbloom, 1981; Rosenbloom and Newell, 1986). A third mechanism is *instance learning*, in which each encounter with a stimulus is represented separately in memory, and multiple exposures to the same stimulus result in multiple representations of that stimulus in memory. These representations are retrieved when the stimulus (or one like it) is encountered again; the more practice, the more representations, and thus the greater the impact on retrieval (i.e., the faster and more reliable the retrieval; see Landauer, 1975; Logan, 1988).

Third, the memory view accounts for the emergence of the properties of automaticity with practice. The general principle is that the properties of memory retrieval may be very different from the properties of the algorithms on which novices rely. Early in practice, performance will reflect the properties of the algorithm, whereas later it will reflect the properties of memory retrieval. The properties of memory retrieval will "emerge" from the properties of the algorithm as subjects switch from one to the other. Because no theory as yet makes specific predictions about when specific properties emerge, the account at present is more descriptive than explanatory.

Fourth, consistency is very important in the automaticity-as-memory view. It assumes that subjects learn specific responses to specific stimuli, and that with such *item-based* learning, transfer to new stimuli should be poor. Changing the mapping rules should also impair performance because the old responses retrieved from memory are no longer relevant. Thus automaticity-as-memory accounts easily for consistency effects.

The automaticity-as-memory interpretation appears to provide a better account of the basic phenomena of automaticity than does the modal resource view. But there are other questions to be answered. The most difficult problem is why novice performance is so poor. The stock answer is that it is unskilled: skilled performance is good, so unskilled performance must be poor. But this is not an explanation. The difficulties experienced by novices reflect deep principles of mental functioning that must be elucidated. Why do novices find certain things difficult and others things easy? Resource theories provide a ready answer, but the automaticity-as-memory view has little to say (but see Anderson, 1982; Schneider, 1985).

Synthesis: Points of Compatibility and Compromise

At this point we may speculate on the outcome of the battle and consider the consequences. It seems certain that the automaticity-as-memory view will capture important ground in the academic arena. The question is whether it will drive out resource theories

or coalesce with them to form a composite theory of learning and performance.

It is hard to say whether automaticity-as-memory will entirely replace the modal view; it depends in part on the resource theorists, their interest in the phenomena of automaticity, and their cleverness and tenacity in solving the problems raised previously. However, there is nothing in the automaticity-as-memory view that *necessitates* the idea of resources. Memory theories typically do not invoke the concept of resources, and one can envision a relatively complete theory of performance without it, as evidenced by Anderson's (1982, 1983) work. So the automaticity-as-memory view is not likely to be challenged by attacks on resource theory but will remain strong and ready to replace the modal view if resource theory dies.

But if resource theory survives, can it be integrated with memory theory? Possibly. The strengths and weaknesses of resource and memory theories seem complementary: resource theory is strong on explaining novice performance and weak on learning, whereas automaticity-as-memory is the opposite. There appears to be no basic incompatibility between resource theory and memory theory. Even though there have been few points of contact in the literature, there is no reason in principle why performance could not be described both in terms of resources and memory. A new theory would have to be developed, but the building blocks have been identified.

In the new theory, reduction of resource demands could be a cause or a consequence of automatization. For example, one could adapt Crossman's (1959) learning theory to make resource reduction a causal factor. Crossman argued for a kind of "natural selection" process in which methods were selected randomly; if they were faster than average, the probability of their being selected again was increased. Over practice, the fastest method comes to dominate. The same sort of process could work with resource demands rather than speed as the criterion: methods could be selected randomly, and if they demanded less capacity than average, their selection probability would increase. Eventually, the least demanding method would dominate. In the meantime, resource demands would decrease gradually with practice.

Alternatively, in Logan's (1988) memory-instance theory, resource reduction could be seen as a consequence of automatization rather than a cause. The single-step, direct-access retrieval of a solution from memory that Logan characterizes as automaticity could involve few resources compared with the demands of the multistep algorithms that govern novice or intermediate performance. Automatic performance is simpler than novice performance and so would demand fewer resources. However, automatization is a consequence of experience, not a consequence of a need to reduce resource demands. Logan assumes that encoding into memory and retrieval from memory are obligatory consequences of attention. Attending to an object causes it to be encoded into memory and at the same time makes available whatever was associated with that object in the past. Practice forces a person to attend to aspects of the task, which are obligatorily encoded into memory. Repeated exposure adds more to memory, and the more there is in memory, the faster and more reliable the retrieval. Performance becomes fast and effortless. Resource requirements (if any) would diminish as single-step retrieval comes to dominate performance. However, the reduction in demand would be a consequence of automatization, not a cause or a motivation for it.

Whether resource reduction is a cause or a consequence of automatization, memory will be important in the new theory. Even if resource theory dies and the modal view of automaticity dies with it, memory and the theoretical and empirical principles that govern

it will have important implications for automaticity. In the remaining sections I will outline some of these implications.

IMPLICATIONS OF AUTOMATICITY-AS-MEMORY

Somewhere, beyond the shadow of the ivory tower, it may not seem to matter much in a practical sense whether automaticity is a resource phenomenon or a memory phenomenon. Many battles have raged in the ivory tower with no effect in the practical world, and this may be another. I would argue the contrary, for the issue is fundamental. The different approaches provide vastly different perspectives on automaticity, and what was difficult to see from one perspective can become obvious from the other. The new perspective may be illuminating for theorist and practitioner alike. This paper so far has demonstrated that the memory view has gained significant ground and can no longer be ignored. Whether or not we abandon the old (resource) perspective, we must give serious attention to the new one and see what insights it provides.

The disparity of perspective is most apparent on the issue of what is learned. The modal resource view argues that people learn about the processes underlying their behavior, becoming faster and more efficient with practice (e.g., Anderson, 1982; Kolers, 1975). The memory view argues that people learn about the environment in which they perform, remembering behaviors appropriate to the different states of the environment (e.g., Logan, 1988; Newell and Rosenbloom, 1981). These views raise different questions about training. The modal view leads one to ask about the resource demands of the situation and ways to reduce resource demands. The memory view leads to questions about what must be learned to perform the task and ways to improve that learning.

In general, the two perspectives predict different gradients of transfer of automaticity. The modal view predicts broader transfer than the memory view because learning is process-based and the specific stimuli do not matter. The memory view predicts narrower transfer because only specific responses to specific stimuli are available in memory. These predictions have important practical implications for the design of training programs: anticipating narrow transfer, the memory view would emphasize fidelity in training programs and simulators so that situations encountered in training were as similar as possible to situations encountered in the field when training is applied. For the same reason, the memory view would emphasize breadth of training, so that the situations encountered in training were representative of those in the field. Neither fidelity nor representativeness would be particularly important in the modal view, provided that resource demands do not change much from training to application.

There is a an extensive literature on the psychology of memory that can provide answers to questions about automaticity. Its relevance, hidden or disguised from the modal perspective, is immediately recognizable from the memory perspective wherein some factors take on new meaning and other factors become apparent for the first time. Levels of processing is an example of the former. It was viewed as a difficulty manipulation in the resource view, with deeper levels demanding more resources (e.g., Johnston and Heinz, 1978). But from the memory perspective, levels of processing affect memorability, with better memory for deeper processing (e.g., Craik and Tulving, 1975). The better memory may also facilitate automatization.

Other factors, not readily apparent in the modal view, may have powerful effects on memory and thus on automatization. Context and the match between conditions of encoding and retrieval are important examples

(e.g., Eich, 1980; Tulving and Thompson, 1973) with particular relevance to automatization. These factors will be explored in the remainder of the paper.

Context Effects

Context can have powerful effects on memory. One of the most dramatic demonstrations of its power was Godden and Baddeley's (1975) experiment with deep-sea divers. Once under water, divers apparently forget instructions they had been given on the surface and Godden and Baddeley discovered why. They had divers learn materials on the surface and tested them either underwater or on the surface. They found better memory on the surface than underwater, replicating the anecdotal reports. However, divers who learned the same materials underwater showed better memory underwater than on the surface, suggesting that the match between context at encoding and context at retrieval was the crucial variable.

Similar context effects have been produced by less drastic manipulations (for reviews, see Eich, 1980; Tulving and Thompson, 1973), though the effects are usually smaller in magnitude. Some context manipulations, such as the room in which the items were encoded and retrieved, have very little effect (Eich, 1985). Others, such as the location in which an item appears, can have strong effects (Winograd and Church, 1988). In general these effects are explained by assuming that the context at encoding is associated with the item in the memory trace and that context at retrieval is an important retrieval cue. The difficulty of retrieval depends on the match between the retrieval cues and the memory trace, and the match will be better when encoding and retrieval contexts are the same than when they differ. Thus similar contexts produce better memory performance than dissimilar ones (Metcalfe, 1985; Tulving and Thompson, 1973).

If automaticity is a memory phenomenon, governed by the theoretical and empirical principles that govern memory, then context effects should be important to automaticity and automatization. There is no published research on context effects in automaticity, so the question remains open. However, one can make some tentative predictions.

In most learning mechanisms, common features become progressively stronger with experience and unique features are buried more deeply in noise. In Logan's (1988) theory, for example, experiences are stored separately and amalgamated at retrieval. When there is only a single experience in memory, its common and unique features should be (roughly) equally retrievable. As other experiences are added, unique features become harder to retrieve. The unique features of one experience act as noise with respect to the unique features of other experiences, and the more experiences there are, the greater the amount of noise, hence the more difficult the retrieval. By definition, common features are present in nearly every experience. The common features of one experience reinforce the common features of other experiences, so they remain strong relative to the noise and easily retrievable. Other theories make similar predictions (e.g., Metcalfe, 1985; Schneider, 1985).

This analysis suggests that context effects will be more important early in practice, when unique features are easily retrieved, than later in practice, when unique features are hard to retrieve. Thus the impact of context in practical situations may depend on the amount of practice involved. At very high levels of practice context effects may be safely ignored, but at lower levels of practice or at mixtures of high and low levels some attention should be paid to context effects.

The consistency of the context is also an important consideration. Context is encoded and retrieved just as task-relevant informa-tion is encoded and retrieved, and contextual consistencies should be strengthened with practice just as task-relevant consistencies are. Contextual consistencies may come to be associated with task-relevant consistencies, locking the ability to perform the task into a specific context. Different contexts may inhibit retrieval of task-relevant information. It may be possible to immunize the learner against these effects by training in different (i.e., inconsistent) contexts so that no contextual features become uniquely associated with task-relevant features. Context would contribute noise to retrieval but would not bias it, locking out task-relevant information.

Contextual consistency can have facilitating effects if the context at application matches the context of training. In practice, application may differ from training considerably, but there may be aspects of the context that remain consistent. For example, a given procedure may always be executed in the context of a certain goal or a certain class of goals though the goals are sought in different situations. Consistency of goals may facilitate retrieval of task-relevant information even in the face of different situational contexts.

The practitioner may have little control over context at application, which is often determined by other factors such as the mission to be accomplished and the equipment to be used. However, he or she can control the context at training and should design it to take into account the breadth and nature of the context at application. Broad contexts at training may lead to better transfer but are likely to be more expensive than narrower contexts and may not be worth the extra expense if the application context is narrow. Another possibility for the practitioner is to discourage or prevent attention to context during training so that context does not become associated with task-relevant information in the memory trace. There will then be no interference from mismatching context at application (though there will also be no benefit from matching context).

Dual-Task Conditions as Context

The concurrent task in dual-task conditions may be interpreted as a context for the primary task, and some of the difficulty involved in performing and learning dual tasks may be attributed to context effects. A task may be viewed as a stream of events that become associated as the subject attends to them. Two tasks performed together present concurrent streams of events, and it is possible that events from one stream become associated with the other. But if the tasks are really separate, then the streams of events that comprise them will tend toward independence and associations between streams will be spurious. When an event occurs again in one stream, it is unlikely that the simultaneous event from the other stream that occurred with it will occur with it once again. The contextual associations will not match, and retrieval will be less effective; consequently, performance will seem less automatic than it could be.

To make the point concretely, imagine two concurrent categorization tasks, one in which subjects search for names of Mexican restaurants (e.g., Casa Lupita) and one in which they search for birthdates of Canadian psychologists (e.g., the present author). Imagine that there are 10 restaurant names and 10 psychologists. In single-task conditions learning is relatively simple: there are 20 associations to learn. But in properly balanced dual-task conditions each restaurant would be paired with each psychologist, so there would be 100 associations to learn. To make matters worse, dual-task associations would recur less frequently than single-task associations: in the 100 trials required to present

each dual-task pair once, each single-task stimulus could be presented five times. Dual-task conditions present less opportunity to show evidence of learning.

There is no direct evidence on the importance of context in dual-task performance. There is, however, evidence that dual-task conditions impair memory (Fisk and Schneider, 1984) and inhibit learning (Nissen and Bullemer, 1987), which is consistent with the idea of mismatching context. The modality effects in dual-task performance (Wickens, 1980) are also consistent with the idea of mismatching context: subjects may be more likely to associate events from the two streams if the tasks involve the same modality than if they involve different modalities. Thus context effects would be stronger within modality than between modality.

The dual-tasks-as-context hypothesis predicts more specific modality effects than the multiple-resources view. Not only does the modality matter, but the materials within the modality matter as well. The context hypothesis thus has an easier time than resource theory in dealing with Hirst and Kalmar's (1987) results. Hirst and Kalmar found more interference when subjects performed two concurrent searches for members of the same category than when they performed two concurrent searches for members of different categories. To account for these results, a resource theory would have to argue that each category was a separate resource. The context hypothesis would merely have to argue that each category was represented separately in memory, a proposition for which there is a lot of independent evidence (Medin and Smith, 1984).

The context hypothesis may also explain why training does not appear to produce immunity to dual-task interference. A number of studies show substantial interference when a dual task is introduced, even when subjects have been trained to some criterion of automaticity (Logan, 1979; Schneider and Fisk, 1984; Smith and Lerner, 1986). Single-task training may reduce dual-task interference but apparently does not eliminate it (Bahrick, Noble, and Fitts, 1954). The context hypothesis would interpret these results in terms of a shift in retrieval cues: the stimuli from the added task work together with the stimuli from the trained task as compound retrieval cues, and there is less available to the compound cue than there would be to the well-trained single-task cue. Consequently memory is slower and less reliable and performance is less automatic.

The message for the practitioner is to pay attention to the tasks that will be performed concurrently with the task one is training. One should try to incorporate into training concurrent tasks similar to those anticipated in application so that the trainees will have had experience with the appropriate context. Also, training programs should encourage trainees to treat the two tasks separately—to ignore the context provided by the other task so as not to incorporate it into memory traces relevant to the primary task. Perhaps training under a variety of dual-task conditions would encourage inattention to context and develop immunity to interference from new dual tasks at application.

Levels-of-Processing Effects

Level of processing is a potent variable in memory research. By manipulating subjects' orienting task, experimenters require them to attend to different properties of the stimulus. In general, subjects who attend to low-level physical features do not remember the materials as well as subjects who attend to higher-level semantic features (Craik and Tulving, 1975). The effect is well established and easily replicable, though not understood theoretically. Some hypotheses have been advanced (e.g., deeper processing is more distinctive or more elaborate), but as yet no one

has discovered a way to define depth independent of memory performance. This inherent circularity has led memory researchers to lose interest in the phenomenon in recent years; but at an empirical level it is robust and important.

If automaticity is a memory phenomenon, then levels of processing might have the same effect on automatization as it does on "standard" measures of memory such as recognition and recall—we might expect faster learning and better retention the deeper the level of processing. Logan (1985b) reported experiments consistent with this hypothesis. He presented pairs of words and had subjects make category or rhyme judgments about them (e.g., is *professor* a *profession?* does *sleigh* rhyme with *play?*). On the initial presentation, reaction times were the same for the two judgments. But on the second presentation, reaction times were much faster for the category judgments than for the rhyme judgments, as if subjects had learned more. It remains to be seen whether this initial advantage for deeper processing would persist throughout practice and produce a faster learning rate, but the results are suggestive.

There is some indication that levels-of-processing effects may not generalize to memory as reflected in studies of automaticity. Jacoby and Dallas (1981) exposed subjects to words under orienting tasks that varied in level of processing and then transferred the subjects to standard memory tests (recognition and recall) or to a perceptual identification task, in which the words were exposed briefly and subjects were to identify them. The standard levels-of-processing effects were replicated in recognition and recall but not in perceptual identification. Words presented during the orienting task were identified more accurately than words not presented before (indicating the presence of the former in memory), but the advantage was no greater for deep orienting tasks than for shallow ones. If the memory processes that underlie automaticity are the same as those that underlie perceptual identification, then these results suggest that levels of processing may have no impact on automatization. But if automatic memory processes are similar to those that underlie the repetition effect that Logan (1985b) studied, then there is reason to be encouraged. At present the question seems relatively open and promises to be a fertile ground for future research.

The results thus far, although speculative, may have some implications for the practitioner. They suggest that training may be more effective if it involves a deeper level of processing—that is, if trainees can become involved in the training task in a meaningful way. The more commitment and interest they show in the training program, the faster they may learn and the better they may retain what they learn in applying their knowledge once training is over.

The implications of levels of processing are less clear than the implications of context because the former variable is less well understood theoretically than the latter. Implications of context can be deduced from the various theoretical mechanisms proposed to account for context effects, whereas the implications of levels of processing must be induced from empirical observations. Deduction is better than induction, if one's premises are true, so there is more to be said about context effects than about levels of processing. This contrast illustrates once again that theoretical disputes can have important practical implications. As Barry Kantowitz once said, there is nothing so practical as a good theory.

Implicit and Explicit Memory

The contrast between Logan's (1985b) findings and those of Jacoby and Dallas (1981) points to an important distinction in the memory literature that is currently the focus of a hot debate: the distinction between *implicit* and *explicit* memory (for reviews, see Jacoby and Brooks, 1984; Schacter, 1987; Tulving, 1983). The distinction addresses the nature of the retrieval task used to probe memory. In explicit memory tasks such as recognition and recall, the subject is explicitly told to retrieve something from memory (e.g., the words that appeared on a previous list or the context associated with a currently presented item). The subject understands it as a memory task and consciously tries to use memory. In implicit memory tasks, the subject is given a task to perform of which some of the materials have been presented before. The experimenter infers memory for those materials if the subject performs better on them than on comparable materials that were not presented before. The requirement to use memory is implicit in the instruction to perform the same task on familiar materials, but there is no explicit instruction to use memory. The subject may not understand the task as a memory task and may not consciously try to use memory.

At a procedural level, the distinction amounts to a contrast between retrieval tasks: recognition and recall versus all the others. But some theorists have argued for a deeper distinction, contending that the tasks tap different memory systems (e.g., Tulving, 1983). Others agree that the distinction is deep but argue that the tasks tap the same memory system in different ways (e.g., Jacoby and Brooks, 1984). The debate is fueled by a provocative and important data base in the form of dissociations between the different measures of memory. There are several indications that factors affecting explicit memory have no effect or a different effect on implicit memory. Jacoby and Dallas's (1981) results on levels-of-processing effects in recognition versus perceptual identification are an example of the dissociation. Perhaps the most compelling dissociations come from amnesic patients, who show evidence of learning on problem-solving tasks even though they cannot recall having performed the task before (e.g., Cohen and Squire, 1980). However, dissociations are not always found. In some cases there are associations; for example, in Logan's (1985b) levels-of-processing experiments. And there is some debate over the interpretation of dissociations (see Dunn and Kirsner, 1988).

The explicit/implicit memory issue is another battle raging in the ivory tower. The debate may have strong implications for the role of memory in automaticity because automaticity seems more likely to be a phenomenon of implicit memory than of explicit memory. But it is too early to tell. The theoretical alternatives need to be worked out in more detail before they can be rigorously tested, and the relation between these memory effects and automaticity needs to be spelled out. Dissociation experiments typically use only one presentation, whereas automaticity experiments can involve thousands. The data sets may represent different points on the same acquisition curve, and effects at any one point may generalize to any other. But from some perspectives, different factors affect different parts of the learning curve. In Logan's (1988) theory, for example, performance depends on a race between memory retrieval and a general algorithm for performing the task. At low levels of practice memory retrieval is slow and unreliable, so performance is dominated by the algorithm. At high levels memory retrieval is fast and accurate and dominates performance. Thus the first exposure or two investigated in studies of implicit and explicit memory may reflect properties of the algorithm more than properties of memory. Consequently, the results may not generalize to higher levels of practice when performance is based entirely on memory retrieval.

In addition, the two kinds of memory may not be as mutually exclusive as the distinction suggests. There is no reason why someone may not try to use memory consciously or explicitly while performing an implicit memory task. Stuart Klapp and I performed some experiments that suggest that the distinction may be blurred when applied to automaticity. We trained people on an alphabet arithmetic task in which they were presented with equations to verify, such as $A + 2 = C$ or $B + 3 = E$. In essence, we were asking whether C was two letters down the alphabet from A, whether E was three letters down from B, and so on. Initially, subjects reported performing the task by counting through the alphabet for a number of steps determined by the digit addend and then comparing the counted letter with the presented one. Consistent with subjects' reports, verification times increased linearly with the magnitude of the digit addend with a slope of between 400 and 500 ms/count. With practice, however, subjects reported remembering specific problems, and the slope diminished to zero. Performance had become automatic.

We were able to reproduce this automaticity by having subjects learn the facts by rote memory, without counting. When transferred to a verification task that used the facts they had learned, subjects showed a zero slope, suggesting automaticity. But when subjects were transferred to verify unfamiliar facts, their slopes were around 400 ms/count—as were those of other untrained subjects. In subsequent experiments we compared learning by rote with learning by doing and found very little difference. Basically, the slope in the verification task depends on the number of prior exposures to the facts in question, regardless of whether the exposures occurred in a verification task (learning by doing) or in a memory task (learning by rote). We are currently trying to test the limits of this effect (or lack of effect) and to generalize the findings to other criteria for automaticity, such as immunity to dual-task interference. For now, the results are provocative in that they suggest that training under implicit and explicit memory conditions may have equivalent effects on automatization.

The moral of the story for practitioners is to keep an eye on the debate over explicit and implicit memory. It is too early to tell the outcome, but it may have profound implications for theories of automaticity and the training programs that apply them.

Procedural and Declarative Knowledge: Automaticity as Motor Skill

Automaticity often seems to be on one side of another distinction related to the explicit/implicit memory debate and important in cognition in general: the distinction between *declarative* and *procedural* knowledge. Declarative knowledge is knowledge of facts, "knowing that" something is the case; whereas procedural knowledge is knowledge of process, "knowing how" to do something. Automaticity is often characterized as procedural knowledge. A common interpretation is that skill acquisition reflects a transition from declarative knowledge to procedural knowledge (e.g., Anderson, 1982). This view is buttressed by the tendency to use perceptual-motor skills, such as riding a bicycle, as paradigm cases of automaticity. Perceptual-motor skills appear to be both automatic and procedural, so the shoe would seem to fit.

However, the fact that some perceptual-motor skills are automatic and some procedural skills are automatic does not imply that all automatic skills are procedural or perceptual-motor. There is no reason why declarative knowledge or cognitive skills cannot be automatized. My experiments with Klapp, described previously, suggest that declarative knowledge (acquired when learning by rote) can be used to support automatic performance. More generally, Logan's (1988) theory argues that automaticity results from

building up memory traces regardless of whether the traces represent procedural or declarative knowledge. There may be few studies of automaticity based on declarative knowledge in the literature, but this may reflect an absence of curiosity about the topic rather than an absence of phenomena to study.

The moral for the practitioner here is to look more broadly for automatic processes. They need not be restricted to procedural knowledge or perceptual-motor skill but may permeate the most intellectual activities in the application environment. For example, common lore and the modal view might lead one to expect that pilots should be able to automatize the perceptual and motor aspects of flying. The memory view suggests that they may also be able to automatize some of the strategic and intellective aspects of carrying out their mission.

SUMMARY AND CONCLUSIONS

There is a battle raging in the ivory tower over the concept of automaticity. The defenders of the established view—adherents of resource theories of automaticity—have suffered serious losses through both a frontal attack from proponents of the memory view of automaticity and a flank attack from enemies of resource theory in general. The memory view seems to have gained a solid foothold and consequently will influence the field for some time to come. One purpose of this paper has been to argue that practitioners should take an interest in the battle because the outcome thus far has strong implications for the design of training programs and for the understanding of expertise.

The memory view provides a new perspective on issues in automaticity and training that was not available from the modal resource view, just as one eye provides a perspective not available to the other. Ivory tower theorists may build their careers by arguing about which eye provides the better view. The wise practitioner will look with both eyes open and enjoy the benefits of binocular vision.

ACKNOWLEDGMENTS

This research was supported by Grant No. BNS 87-10436 to the author from the National Science Foundation. It was inspired by a series of workshops on multiple resources and automaticity organized by Lawrence E. Reed under the sponsorship of the Air Force Human Resources Laboratory at Wright-Patterson Air Force Base and the Air Force Office of Scientific Research. I am grateful to Dr. Reed for providing the opportunity for many very stimulating discussions and to the other members of the working group, particularly the automaticity panel (Tom Eggemeier, Dan Fisk, and Walt Schneider) for helping to shape the ideas in the paper. I am also grateful to Jane Zbrodoff, Tom Eggemeier, Art Kramer, and two anonymous reviewers for helpful comments on a previous version of the manuscript.

REFERENCES

Allport, D. A. (1980). Attention and performance. In G. Claxton (Ed.), *Cognitive psychology* (pp. 112–153). London: Routledge and Kegan Paul.

Anderson, J. R. (1982). Acquisition of cognitive skill. *Psychological Review, 89,* 369–406.

Anderson, J. R. (1983). *The architecture of cognition.* Cambridge, MA: Harvard University Press.

Bahrick, H. P., Noble, M., and Fitts, P. M. (1954). Extratask performance as a measure of learning a primary task. *Journal of Experimental Psychology, 48,* 298–302.

Bryan, W. L., and Harter, N. (1897). Studies in the physiology and psychology of telegraphic language. *Psychological Review, 4,* 27–53.

Bryan, W. L., and Harter, N. (1899). Studies of the telegraphic language. The acquisition of a hierarchy of habits. *Psychological Review, 6,* 345–375.

Carr, T. H., McCauley, C., Sperber, R. D., and Parmalee, C. M. (1982). Words, pictures, and priming: On semantic activation, conscious identification, and the automaticity of information processing. *Journal of Experimental Psychology: Human Perception and Performance, 8,* 757–777.

Cohen, N. J., and Squire, L. R. (1980). Preserved learning and retention of pattern-analyzing skill in amnesia: Dissociation of "knowing how" and "knowing that." *Science, 210,* 207–209.

Craik, F. I. M., and Tulving, E. (1975). Depth of processing and retention of words in episodic memory. *Journal of Experimental Psychology: General, 104,* 268–294.

Crossman, E. R. F. W. (1959). A theory of the acquisition of speed skill. *Ergonomics, 2,* 153–166.

Dunn, J. C., and Kirsner, K. (1988). Discovering functionally independent mental processes: The principle of reversed association. *Psychological Review, 95,* 91–101.

Durso, F. T., Cooke, N. M., Breen, T. J., and Schvaneveldt, R. W. (1987). Is consistent mapping necessary for high-speed scanning? *Journal of Experimental Psychology: Learning, Memory, and Cognition, 13,* 223–229.

Eich, J. E. (1980). The cue-dependent nature of state-dependent retrieval. *Memory and Cognition, 8,* 157–173.

Eich, E. (1985). Context, memory, and integrated item/context imagery. *Journal of Experimental Psychology: Learning, Memory, and Cognition, 11,* 764–770.

Fisk, A. D., Oransky, N., and Skedsvold, P. (1988). Examination of the role of "higher-order" consistency in skill development. *Human Factors, 30,* 567–581.

Fisk, A. D., and Schneider, W. (1984). Memory as a function of attention, level of processing, and automatization. *Journal of Experimental Psychology: Learning, Memory, and Cognition, 10,* 181–197.

Godden, D. R., and Baddeley, A. D. (1975). Context-dependent memory in two natural environments: On land and underwater. *British Journal of Psychology, 66,* 325–331.

Hasher, L., and Zacks, R. T. (1979). Automatic and effortful processes in memory. *Journal of Experimental Psychology: General, 108,* 356–388.

Hirst, W., and Kalmar, D. (1987). Characterizing attentional resources. *Journal of Experimental Psychology: General, 116,* 68–81.

Jacoby, L. L., and Brooks, L. R. (1984). Nonanalytic cognition: Memory, perception, and concept learning. In G. H. Bower (Ed.), *The psychology of learning and motivation.* New York: Academic Press.

Jacoby, L. L., and Dallas, M. (1981). On the relationship between autobiographical memory and perceptual learning. *Journal of Experimental Psychology: General, 110,* 106–340.

Johnston, W. C., and Heinz, S. P. (1978). Flexibility and capacity demands of attention. *Journal of Experimental Psychology: General, 107,* 420–435.

Kahneman, D. (1973). *Attention and effort.* Englewood Cliffs, NJ: Prentice-Hall.

Kahneman, D., and Treisman, A. M. (1984). Changing views of attention and automaticity. In R. Parasuraman and R. Davies (Eds.), *Varieties of attention* (pp. 29–61). New York: Academic Press.

Keele, S. W. (1973). *Attention and human performance.* Santa Monica, CA: Goodyear.

Kolers, P. A. (1975). Memorial consequences of automatized encoding. *Journal of Experimental Psychology: Human Learning and Memory, 1,* 689–701.

LaBerge, D. (1981). Automatic information processing: A review. In J. Long and A. D. Baddeley (Eds.), *Attention and peformance IX* (pp. 173–186). Hillsdale, NJ: Erlbaum.

LaBerge, D., and Samuels, S. J. (1974). Toward a theory of automatic information processing in reading. *Cognitive Psychology, 6,* 293–323.

Landauer, T. K. (1975). Memory without organization: Properties of a model with random storage and undirected retrieval. *Cognitive Psychology, 7,* 495–531.

Logan, G. D. (1978). Attention in character classification: Evidence for the automaticity of component stages. *Journal of Experimental Psychology: General, 107,* 32–63.

Logan, G. D. (1979). On the use of a concurrent memory load to measure attention and automaticity. *Journal of Experimental Psychology: Human Perception and Performance, 5,* 189–207.

Logan, G. D. (1980). Attention and automaticity in Stroop and priming tasks: Theory and data. *Cognitive Psychology, 12,* 523–553.

Logan, G. D. (1985a). Skill and automaticity: Relations, implications, and future directions. *Canadian Journal of Psychology, 39,* 367–386.

Logan, G. D. (1985b). On the ability to inhibit simple

thoughts and actions: II. Stop-signal studies of repetition priming. *Journal of Experimental Psychology: Learning, Memory, and Cognition, 11,* 675–691.

Logan, G. D. (1988). Toward an instance theory of automatization. *Psychological Review, 95.*

Logan, G. D., Zbrodoff, N. J., and Fostey, A. R. (1983). Costs and benefits of strategy construction in a speeded discrimination task. *Memory and Cognition, 11,* 485–493.

Marcel, A. T. (1983). Conscious and unconscious perception: An approach to the relations between phenomenal experience and perceptual processes. *Cognitive Psychology, 15,* 238–300.

McLeod, P., McLaughlin, C., and Nimmo-Smith, I. (1985). Information encapsulation and automaticity: Evidence from the visual control of finely timed actions. In M. I. Posner and O. S. Marin (Eds.), *Attention and performance XI* (pp. 391–406). Hillsdale, NJ: Erlbaum.

Medin, D. L., and Smith, E. E. (1984). Concepts and concept formation. *Annual Review of Psychology, 35,* 113–138.

Metcalfe, J. (1985). Levels of processing, encoding specificity, and CHARM. *Psychological Review, 92,* 1–38.

Naveh-Benjamin, M., and Jonides, J. (1984). Maintenance rehearsal: A two-component analysis. *Journal of Experimental Psychology: Learning, Memory, and Cognition, 10,* 369–385.

Navon, D. (1984). Resources—A theoretical soup stone? *Psychological Review, 91,* 216–234.

Navon, D., and Gopher, D. (1979). On the economy of the human processing system. *Psychological Review, 86,* 214–255.

Navon, D., and Miller, J. (1987). Role of outcome conflict in dual-task interference. *Journal of Experimental Psychology: Human Perception and Performance, 13,* 435–448.

Neely, J. H. (1977). Semantic priming and retrieval from lexical memory: Roles of inhibitionless spreading activation and limited-capacity attention. *Journal of Experimental Psychology: General, 106,* 226–254.

Neisser, U. (1976). *Cognition and reality.* San Francisco: Freeman.

Newell, A., and Rosenbloom, P. S. (1981). Mechanisms of skill acquisition and the law of practice. In J. R. Anderson (Ed.), *Cognitive skills and their acquisition* (pp. 1–55). Hillsdale, NJ: Erlbaum.

Nissen, M. J., and Bullemer, P. (1987). Attentional requirements of learning. *Cognitive Psychology, 16,* 1–32.

Posner, M. I., and Boies, S. J. (1971). Components of attention. *Psychological Review, 78,* 391–408.

Posner, M. I., and Klein, R. (1973). On the functions of consciousness. In S. Kornblum (Ed.), *Attention and performance IV* (pp. 21–35). New York: Academic Press.

Posner, M. I., and Snyder, C. R. R. (1975). Attention and cognitive control. In R. L. Solso (Ed.), *Information processing and cognition: The Loyola symposium* (pp. 55–85). Hillsdale, NJ: Erlbaum.

Rosenbloom, P. S., and Newell, A. (1986). The chunking of goal hierarchies: A generalized model of practice. In R. S. Michalski, J. G. Carbonell, and T. M. Mitchell (Eds.), *Machine learning: An artificial intelligence approach* (Vol. II, pp. 247–288). Los Altos, CA: Morgan Kaufmann.

Schacter, D. L. (1987). Implicit memory: History and current status. *Journal of Experimental Psychology: Learning, Memory, and Cognition, 13,* 501–522.

Schneider, W. (1985). Toward a model of attention and the development of automatic processing. In M. I. Posner and O. S. Marin (Eds.), *Attention and performance XI* (pp. 475–492). Hillsdale, NJ: Erlbaum.

Schneider, W., Dumais, S. T., and Shiffrin, R. M. (1984). Automatic and control processing and attention. In R. Parasuraman and R. Davies (Eds.), *Varieties of attention* (pp. 1–27). New York: Academic Press.

Schneider, W., and Fisk, A. D. (1982). Degree of consistent training: Improvements in search performance and automatic process development. *Perception and Psychophysics, 31,* 160–168.

Schneider, W., and Fisk, A. D. (1984). Automatic category search and its transfer. *Journal of Experimental Psychology: Learning, Memory, and Cognition, 10,* 1–15.

Schneider, W., and Shiffrin, R. M. (1977). Controlled and automatic human information processing: I. Detection, search and attention. *Psychological Review, 84,* 1–66.

Shiffrin, R. M., and Schneider, W. (1977). Controlled and automatic human information processing: II. Perceptual learning, automatic attending, and a general theory. *Psychological Review, 84,* 127–190.

Smith, E. R., and Lerner, M. (1986). Development of automatism of social judgments. *Journal of Personality and Social Psychology, 50,* 246–259.

Tulving, E. (1983). *Elements of episodic memory.* New York: Oxford University Press.

Tulving, E., and Thompson, D. M. (1973). Encoding specificity and retrieval processes in episodic memory. *Psychological Review, 80,* 352–373.

West, L. J. (1967). Vision and kinesthesis in the acquisition of typewriting skill. *Journal of Applied Psychology, 51,* 161–166.

Wickens, C. D. (1980). The structure of attentional processes. In R. S. Nickerson (Ed.), *Attention and performance VIII* (pp. 239–258). Hillsdale, NJ: Erlbaum.

Wickens, C. D. (1984). Processing resources in attention. In R. Parasuraman and R. Davies (Eds.), *Varieties of attention* (pp. 63–102). New York: Academic Press.

Winograd, E., and Church, V. E. (1988). Role of spatial location in learning face-name associations. *Memory and Cognition, 16,* 1–7.

Zbrodoff, N. J., and Logan, G. D. (1986). On the autonomy of mental processes: A case study of arithmetic. *Journal of Experimental Psychology: General, 115,* 118–130.

The Role of Practice in Dual-Task Performance: Toward Workload Modeling in a Connectionist/Control Architecture

WALTER SCHNEIDER[1] *and* MARK DETWEILER, *University of Pittsburgh, Pittsburgh, Pennsylvania*

The literature on practice effects and transfer from single- to dual-task performance is briefly reviewed. The review suggests that single-task training produces limited transfer to dual-task performance. Past theoretical frameworks to explain multitask performance are reviewed. A connectionist/control architecture for skill acquisition is presented. The architecture involves neural-like units at the micro level of processing, with information transmitted between modules at the macro level. Simulations within the architecture exhibit five phases of skill acquisition. Dual-task interference and performance are predicted as a function of the phase of practice a skill has reached. Seven compensatory activities occur in the architecture during dual-task training that do not appear in single-task training, including (1) shedding and delaying tasks and preloading buffers, (2) letting go of high-workload strategies, (3) utilizing noncompeting resources, (4) multiplexing over time, (5) shortening transmissions, (6) converting interference from concurrent transmissions, and (7) chunking of transmissions. Future research issues suggested by the architecture are discussed.

INTRODUCTION

Despite the vast number of studies that have attempted to chart the kinds and nature of interference that occurs when two or more tasks are performed simultaneously, little is known about the role of practice in reducing interference. The few studies that have investigated practice have demonstrated that human performance can change radically over time. But few attempts have been made to provide sufficiently detailed theoretical frameworks to address such issues as (1) why human performance becomes more accurate,

(2) why it becomes faster, (3) why interference between tasks decreases with practice, and (4) why single-task training does not always transfer to dual-task performance. This paper examines issues of practice and workload. First we summarize a number of issues concerning single- and dual-task performance and training. Next we describe a runnable simulation architecture used to investigate a variety of single- and dual-task phenomena. We then use this architecture as a framework to clarify a few central issues and problems surrounding the kinds and nature of dual-task interference that occurs and illustrate how it may change as a function of practice. Finally, we speculate on the kind of research and modeling needed to develop

[1] Requests for reprints should be sent to Walter Schneider, Learning Research and Development Center, University of Pittsburgh, Pittsburgh, PA 15260.

more predictive and theoretically based approaches to understanding dual-task training and performance.

INTERFERENCE BETWEEN HIGHLY PRACTICED TASKS

A number of studies have shown that highly practiced tasks can be performed jointly with little interference. For example, some people have developed the ability to read and dictate simultaneously (see Downey and Anderson, 1915; Solomons and Stein, 1896). Recently, Spelke, Hirst, and Neisser (1976) and Hirst, Spelke, Reaves, Caharack, and Neisser (1980) demonstrated that with substantial practice—over 50 hours—subjects could with little interference read one text while taking dictation from a second auditorily presented text. When subjects were asked to read aloud or shadow prose while taking dictation, performance dropped markedly. Following substantial dual-task training, subjects achieved reading and comprehension rates similar to single-task control rates. Shaffer (1975) documented one highly practiced typist's ability to type visually presented material while concurrently either reciting nursery rhymes or shadowing prose or random letters with only about a 10% cost in typing speed and accuracy over single-task baselines. Allport, Antonis, and Reynolds (1972) reported that piano players could shadow prose while sight reading material of varying difficulty.

Dual-task studies show that even after extensive single-task practice, however, additional dual-task training is needed to perform two tasks concurrently. Damos, Bittner, Kennedy, and Harbeson (1981) offer evidence of the need for dual-task practice and illustrate the time needed to regain single-task performance levels after the two tasks have been combined. Subjects received 45 15-minute sessions of single-task training followed by 15 sessions of two concurrent tracking tasks.

One task required use of the left hand and the other use of the right. Damos et al. found that performance on the dual task improved sharply over the first four days of practice and continued to improve throughout the testing period. After 15 sessions, dual-task performance approximated the levels obtained after three single-task practice sessions.

The ability to perform combined dual tasks varies greatly depending on the tasks and amount of practice. Bahrick and Shelly (1958) found that even after prolonged single-task training the addition of a secondary task can greatly reduce performance on the primary task. Bahrick and Shelly's subjects received 25 training sessions. A visual reaction-time task (i.e., pressing buttons under one of four lights) was performed alone in Sessions 1–2, 4–13, and 15–24, with 10 trials per session. In Sessions 3, 14, and 25, the visual task was combined with an auditory task in which subjects responded to one of five keys, depending on the numbers they heard. In dual-task Sessions 14 and 25, the drop in performance ranged from about 8% for the group with the most consistent stimuli (a repeating pattern of four lights) to about 40% for the group with the least consistent stimuli (random patterning of lights).

Schneider and Fisk (1982) emphasized the importance of task consistency in minimizing dual-task interference. Their subjects performed two visual search tasks. The search tasks were either consistent or varied. In a consistent mapping (CM) condition, each specific stimulus is responded to in the same way (e.g., one responds to the letter X whenever it appears). In a varied mapping (VM) condition, the response to a stimulus varies over trials (e.g., in trials in which one searches for the letter X one always responds to X's, but in other trials X's are not responded to as other letters are searched). Performance under the dual-task CM conditions showed a dramatic improvement with practice, benefiting the shortest display durations most. Performance under the dual-task VM condition also improved somewhat but did not reach single-task VM performance at the end of the first experiment.

A second experiment further showed that subjects were able to perform concurrent VM and CM searches when the VM task was emphasized. Although VM performance did not result in a deficit, CM performance did drop by 17%. However, when the CM task was emphasized, VM performance dropped precipitously, to near chance levels. Schneider and Fisk also demonstrated that concurrent VM searches could not be performed without substantial deficit that did not diminish with additional practice. They concluded that a nonemphasized CM task could be performed with an emphasized VM task but that dual-task VM processing produced severe decrements.

The importance of consistency was replicated in a series of category search experiments by Fisk and Schneider (1983). Subjects performed a digit-span task, a CM visual category search task, and a VM visual category search task. The dual tasks were the digit task joined with either of the two search tasks. When the digit and CM search tasks were first performed together, detection accuracy dropped by about 10% from the single-task detection baseline. After 90 additional trials of dual-task practice, however, category detection accuracy reached a level comparable to single-task accuracy. In contrast, when the digit and VM search tasks were first performed together, detection accuracy dropped by about 25% from single-task levels. Moreover, additional dual-task practice did not improve detection accuracy.

The first session of dual-task training typically produces dramatic drops in either CM or VM performance. Schneider and Fisk (1984) examined the effects of extended single-task training (1755 trials) on transfer in CM and VM category visual search tasks. Subjects first performed three single-task searches: digit search, CM category search, and VM category search. Under subsequent dual-task conditions, subjects were required to detect both digits and category exemplars concurrently. The first time each of the two semantic search tasks was combined with the digit-search task, semantic search performance dropped substantially—45% for the CM task and 49% for the VM task—relative to single-task performance levels. Subjects' performance at detecting digits also dropped substantially, by 26% for the CM task and 29% for the VM task. Subjects' performance on the consistent category search task improved with additional dual-task practice, matching single-task levels after four additional sessions and ceiling after eight. Subjects' performance on the VM category search task failed to improve with additional dual-task practice; the smallest decrement was 49% and the largest was 61%, relative to single-task performance. Finally, it should be noted that even after eight sessions of dual-task practice, performance on the digit task still incurred an 11% decrement from its comparable single-task level for the CM search task and a 17% decrement for the VM search task.

The results on CM or VM practice show that given a high degree of consistent practice, subjects could simultaneously perform dual tasks with little deficit. However, even after extensive single-task training, substantial dual-task training was needed to reach their single-task detection levels when the two tasks were combined. Further, despite high degrees of single-task practice, subjects were unable to perform the VM search and digit-span tasks without substantial deficit.

Automatic/controlled processing theory suggests that two qualitatively different forms of processing can account for these types of changes with consistent practice (see Schneider, 1985; Schneider, Dumais, and Shiffrin, 1984; Schneider and Shiffrin, 1977; Shiffrin and Schneider, 1977). Controlled processing is characterized as slow, effortful, capacity-limited, and largely under subject control, and it typically occurs in novel and inconsistent processing tasks. Automatic processing is characterized as fast, parallel, relatively effortless, and largely not under subject control and typically occurs in well-practiced consistent tasks. The processing demands of most complex tasks usually reflect a combination of automatic and controlled processing components. Therefore, the theory predicts good time-sharing in CM conditions because consistent targets gain a high priority, enabling automatic processing. Practiced CM stimuli can be processed in the absence of attention (see Schneider, 1985; Schneider and Mumme, 1988). In contrast, VM stimuli are attended to on some trials and ignored on others, producing low priority. The VM targets must be control processed to be detected.

A simple application of automatic/control processing theory to dual- or multitask training might advocate practicing consistent single-task components first, prior to having the learner perform the tasks concurrently. In single task, training components become automatic, no longer requiring attention; they could subsequently be combined to perform dual tasks. However, the data are incompatible with this simple view. In the Schneider and Fisk (1984) experiment, both the consistent and inconsistent tasks dropped nearly equally in the first dual-task block.

FRAMEWORKS FOR SINGLE-TASK TO MULTITASK TRANSFER

Although psychology can offer a variety of theoretical frameworks for multitask performance, there are no models that can predict single-task to multitask transfer. Frame-works of multitask performance generally involve attentional switching (e.g., Broadbent, 1958) or allocation of some limited resource (e.g., Kahneman, 1973). Recent models have both elaborated the nature of resource sharing (e.g., Navon and Gopher, 1980) and attempted to stratify the types of resources (e.g., Wickens, 1980, 1984a, 1984b). These frameworks generally have not been applied to practice effects. From switching theories one might assume that practice speeds up the switching rate and facilitates learning the orders and rates of switching that are most effective; from resource theories one might assume that practice facilitates learning what proportions and types of resources to allocate to a task. Wickens (1984a) comments that practice can increase the resource efficiency, so that the same task can be accomplished with progressively fewer resources. The mechanism for this practice effect has not been specified.

Recent frameworks of single-task learning suggest that single-task training should transfer to dual tasks. Pew (1974), for example, extended previous models of tracking and suggested that humans develop higher-order control mechanisms. In tracking tasks novices continuously monitor the feedback at a low level of control—for example, monitoring momentary error between the actual and intended positions of the object being controlled. With practice, attention moves to higher levels, such as monitoring the drift between the control equations for the desired and intended positions. Shiffrin and Schneider (1977) extended a long line of arguments to the effect that practicing consistent task components develops automatic components that require little if any attentional resources. J. R. Anderson (1983) suggested that practice produces compiled productions, resulting in automatic component skills. Norman and Shallice (1985) proposed a conceptual model similar to Shiffrin and

Schneider (1977), suggesting that practice reduces interference by allowing processes to function without attention. Hunt and Lansman (1986) presented a computer simulation model incorporating production systems and direct activation functions. The model accounts for some dual-task paradigms due to time-sharing a limited executive control system, and assumes that learning allows an associative retrieval function to process stimuli without the use of the executive.

None of the present frameworks directly addresses the issue of transfer from single to multiple tasks. To understand issues related to transfer, much more detailed models are required. Models are needed that can specify boundary conditions and predict the appropriateness and marginal utility of various types of practice to multitask performance. The following architecture is an initial step toward being able to make such predictions.

OVERVIEW: A CONNECTIONIST/ CONTROL ARCHITECTURE

The proposed connectionist/control architecture provides a runnable simulation model for representing a class of models of dual-task performance and workload (see Schneider and Detweiler, 1987; Schneider and Mumme, 1988). We first provide a brief overview of the connectionist control architecture. Connectionist models (see Rumelhart and McClelland, 1986; Schneider, 1987) represent a radical departure from energy-based metaphors (e.g., Kahneman, 1973). We recommend that readers unfamiliar with this type of modeling carefully work through the diagrams and examples in the text. After the overview, we will illustrate how skill is acquired within this model in both single- and dual-task processing.

This architecture incorporates a variety of processing elements that provide mechanisms for accomplishing stable information processing of real-world tasks in an architec-

ture that is neurally feasible (see Schneider and Detweiler, 1987). The model implements a five-phase account of skill acquisition in which skill development is characterized as gradual and continuous. Five assumptions underlie and constrain this architecture. First, we assume that information is processed in networks of neural-like units. These units are organized into modules, and sets of modules are organized into levels and regions. The various regions—for example, auditory, visual, and motor—are connected to and communicate with one another. Second, we assume that the modules exhibit local specialization of function, processing only a restricted class of inputs. Third, we assume that knowledge is stored in the connection weights among the neural-like units. As skill is acquired, changes in knowledge are reflected in changes in the connection strengths among units and/or by the size of their weights. Fourth, we assume that the connection weights operate under the influence of a variety of learning-rate constants. These constants determine how quickly the connections change as a result of intervening learning and the duration of the retention interval. And fifth, we assume that a control processing system modulates the transmission of information within and between processing regions. These assumptions characterize the architecture as a variation of the CAP1 (Controlled Automatic Processing Model 1) system designed to model automatic and controlled processing (see Schneider and Mumme, 1988; Schneider and Shiffrin, 1977; and Shiffrin and Schneider, 1977).

The connectionist/control architecture can be described at three levels of scale. The *microlevel structure* is the first level and represents a network of neural-like units that account for associative information processing and a range of attentional phenomena (see Figure 1). A *message vector* is an output from one module (e.g., visual) and an input to an-

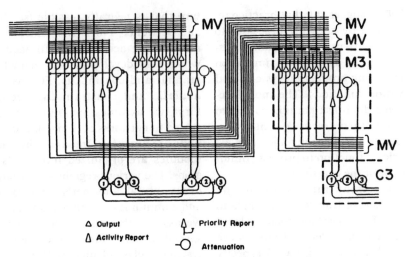

Figure 1. *Microlevel structure of the CAP1 simulation.*

other module (e.g., motor), transferring information between modules. A vector is a set of activities of the units within a module. For example, the letter *A* might be coded as a message vector 0, 1, 1, 1, 1, where the 0's and 1's represent the absence and presence of features—in this instance vertical lines, horizontal lines, backward slant, and forward slant.

Figure 1 illustrates the microlevel structure of the model. Information flowing from a module is regulated by an *attenuation unit* within each module (for details see Figure 1; Schneider and Detweiler, 1987; and Schneider and Mumme, 1988). Attention is a scalar multiplication of the activity of all of the units; for example, if partially attending to the letter *A*, the output might be 0, 0.5, 0.5, 0.5, 0.5. As depicted in Figure 1, processing is assumed to occur in networks of neural-like units. Units are organized into modules (the box labeled M3 outlines the third module) that process a particular class of inputs. Information between modules is transferred as a message vector (MV) on fibers connecting the output of one module to the input of the next. In the diagram, information flows from

left to right (e.g., the top left MV might encode visual features, the two left modules letters, and the right module words). Each module contains a vector of output units that receive input from other modules and connect autoassociatively to themselves. The recurrent connections from the bottom of each output unit going up and connecting to the other output units in the same module represent the autoassociative connections. Each of the crossing points above the output units (to message vector or autoassociative fibers) represents an associative connection that can change the strength of connection with learning.

In the rest of the diagram the reverse arrow-type connections represent excitatory influences and the flat connections represent inhibitory influences. A module's output is controlled by an attenuation unit within the module, which regulates information flow from the module. Each module's activity is regulated by a control structure (the box labeled C3 represents the control structure for the third module), and each module reports its activity to the lower-level control structure via activity report and priority report units. The lower units (labeled 1, 2, and 3) illustrate a potential control circuit. Cell 1 receives the activity reports from the module and inhibits the activity of neighboring modules. Cell 2 inhibits Cell 3, first reducing the attenuation activation, which reduces inhibition of the output units and enables an MV to transmit. Cell 2 is assumed to habituate, resulting in a burst of output and sequential switching of attention. The AP box in M3 illustrates the local circuit for automatic processing. For an automatic process, the priority report unit inhibits the attentuation unit and causes the vector to transmit from the module.

The *macrolevel structure* is the second level of scale and represents interactions among a set of modules (e.g., visual letter processing; see Figure 2B). The modules are organized as

levels and *regions* of processing. The levels correspond to successive processing stages within a region. For example, the visual module is assumed to consist of a series of stages that includes units such as features, characters, and words. The *system-level structure* is the third level of scale and represents the interactions among processing regions made up of modules (see Figure 2A). Each region consists of a series of levels of modules and their respective control structures. There are regions specializing in input (e.g., visual and auditory), output (e.g., motor and speech), and associative processing (e.g., semantic, spatial, and context). The innermost levels of each region communicate with other regions by passing vector messages.

It is assumed that all of the regions communicate with one another on an *inner loop*

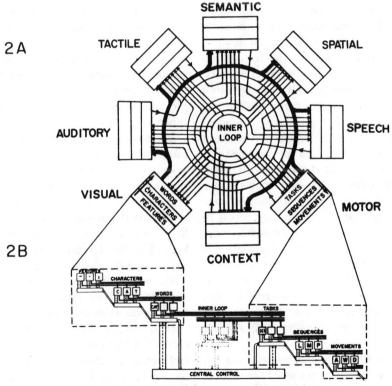

Figure 2. *System-level description of the model.*

of associative connections (see Figure 2). Each module is assumed to have associative connections to the other modules on the inner loop. (This is a simplifying assumption based on neuroanatomy; see Mishkin and Appenzeller, 1987). The associative connections allow each module to send message vectors to other modules. Having separate connections to each receiving module enables transmitted output to be transformed and received as different messages in different regions in parallel. For example, a bright flash may transmit from the visual system and associatively evoke a startle response in a motor module, a fear response in a mood module, and an attempt to retrieve related information in a semantic module. The fact that each module on the inner loop has its own connections also enables multiple modules to *transmit* messages simultaneously. In this way, for instance, the visual module may transmit to the motor module while the auditory module transmits to the speech module.

Parallel transmission on the inner loop does not necessarily imply parallel processing. All the message vectors coming into a module are summed (see Figure 1). When two messages are added, intermessage interference results. This is analogous to a cocktail party at which every speaker can speak at loud volume for extended periods with no energy limit on output. However, each listener receives the summation of all spoken messages. Parallel transmissions may result in high interference such that no message is received. In such cases more information is conveyed if individuals speak sequentially even though each has the ability to speak in parallel. Within the proposed architecture, control processing is the mechanism that moderates message transmissions on the inner loop.

In order to illustrate the model's system level, consider next the top and bottom portions of Figure 2. The top portion (2A) shows

the message vector connections between regions; the bottom portion (2B) shows the macrolevel view of some of the regions. The squares and rectangles in Figure 2B represent the modules and control structures (e.g., Figure 1, module M3, control C3). This is a top-down view of the regions of processing within the system. Each region represents a series of processing levels. The first or last level of a region (last level for input regions and first level for output regions) is assumed to input to the inner loop of connections between regions. The modules on the inner loop have separate message vectors (MV) to each of the other modules to which they connect. All the lines in Figure 2A represent message vectors (MV in Figure 1). Each module sends a message vector to all the other modules on the inner loop. The output for the visual module is highlighted to illustrate this connection pattern. Figure 2B shows the processing of a visual word to produce a button press. The first level of the visual region processes letters; the second, characters; and the third, words. The message is then transmitted to all modules on the inner loop; the dotted set of modules represents all other regions on the inner loop receiving the visual message.

The motor regions illustrate the motor output of the system. The first level of the motor system stores motor tasks (e.g., K1 represents the code for pressing the first key). The second motor level codes the sequences required to execute the motor task (e.g., L = lift finger, M = move to position of key, and P = press key). The third motor level illustrates the components of the lift sequence. The darker horizontal lines represent message vectors and the thin lines represent control signals. Each module (square) sends an ACTIVITY report to its controller (rectangle below it) and receives a FEEDBACK and TRANSMIT control signal (the three lines from the squares to the rectangles). Control modules exchange LOAD and NEXT signals (diagonal lines between rectangles) to control sequential processing between modules. For modules on the inner loop, the Central Control exchanges control signals for ACTIVITY, RESET, TRANSMIT, NEXT, and LOAD to modulate inner loop message traffic (see Schneider and Detweiler, 1987).

There are two categories of information flow in the system: message and control information flow. *Message flow* involves the transmission of a vector representing a code from one module to another, such as the visual module sending a vector coding the features of the letter *A* to the semantic module. *Control flow* involves exchange of control information between the modules and a control structure (see Figure 2B). Control information codes the importance of messages waiting to transmit and the transmission state of any of the modules—for example, signals indicating how active a module is, how important a message to be sent is, or how much to attenuate the message. At the macro level and system level, control structures receive control information to moderate information flow (e.g., sequencing transmissions to reduce interference). At the system level, the control flow processing involves a *central control structure* (see Figure 2) that receives activity reports from all regions. These reports are a single number from each module, indicating the importance of the message in the module awaiting transmission. This is analogous to each module raising its hand—the more critical the message, the higher the hand. The control structure then ranks the requests for transmission and allows the module with the highest-priority request to transmit. A three-neuron-per-module circuit (shown at the bottom of Figure 1) is sufficient to accomplish this control function.

Control processing is critical for performing novel tasks. The connectionist/control architecture can perform a novel task following verbal instruction (see Schneider and Mumme, 1988). This is accomplished by loading vectors into modules, comparing input vectors with vectors held in working memory, and releasing output vectors (see following discussion). This processing is slow, serial, and effortful, requiring many shifts among those modules that are allowed to transmit. The control information necessary to limit message interference also enables verbal rules to be executed (Oliver and Schneider, 1988). With each execution of a verbal rule, associative connections between the modules change such that the input will evoke the output without moderation by control processing. This transition is the mechanism through which processes become automatic.

In this architecture messages need not all pass through a central executive; rather, regions can communicate directly with other regions. However, the transmission of concurrent messages may cause interference (e.g., if the auditory module transmits to the motor module, a message is also sent to the speech module, perhaps disrupting the reception of other messages). The central control structure moderates transmission among regions to limit such interference. Sequencing messages in this way has the potential of causing delays or omissions. The model predicts many of the item and order loss phenomena witnessed under conditions of high workload.

A *context storage* mechanism enables the system to model the stability of human processing and to mimic a variety of learning phenomena such as episodic memory and remindings (see Schneider and Detweiler, 1987). This context mechanism acts to associate the contents of the messages on the inner loop to the temporal context in which they occur. The context vector has separate connections to all the other modules on the inner loop (see Figure 2A). One context vector can evoke a different message vector in every module on the inner loop. When information vectors within modules decay or are displaced, modules can be reloaded by retransmitting the context vector. The ability to reload information via transmitting the context vector reduces the chances of catastrophic processing failure. Further, when the system is confronted with high levels of workload, context can temporarily store associations to vectors and enable the reloading of a low-priority task delayed during the processing of high-priority tasks.

The connectionist/control architecture has been implemented as a computer simulation, CAP1 (see Schneider and Detweiler, 1987; Schneider and Mumme, 1988). The simulation can perform single and dual tasks comparable to typical laboratory dual tasks. We will describe the model and the phases it exhibits as it learns to perform single and dual tasks. It is important that the reader try to conceptualize the operations of the model. Most of the implications follow from the model's conceptual architecture. The details of the simulation provide an existence proof that this type of architecture will produce the phenomena discussed.

The model is a connectionist simulation model; details, equations, and rationale for the architecture can be found in Schneider and Detweiler (1987) or Schneider and Mumme (1988). The model simulates the behavior of populations of neural-like units. It incorporates the associative and autoassociative models of J. A. Anderson's (1983) brain-state-in-a-box model. The major components are illustrated in Figure 1. Each module is made up of a 200-element vector of output units. Output units are neuron-like units in which the pattern of active units codes the response. Information is coded as a vector of active and inactive units. Each output unit sums its input linearly with a decayed value of the activity of the previous iteration. The output of each unit is a logistic function of the input. Each output unit decays to some proportion (typically 0.9) of its activity on the previous cycle. Each output unit connects autoassociatively to half the other output units of the module. Autoassociative feedback varies from 0 for receiving a new vector to 0.6 to latch a received vector. Each output unit also connects to half of the units of other modules to which it is connected. The autoassociative and associative connection matrices are each made up of 20 000 connections per module. All associations are initialized at zero strength.

The model incorporates two types of learning. Learning involves changing connection weights via a modified Hebb-type learning rule, the delta or Widrow-Hoff learning rule (see J. A. Anderson, 1983). The first type, *associative learning*, modifies connections *between* modules. This involves changing the connection weights between the input and output units such that the input comes to elicit the output. The second type, *priority learning*, occurs *within* a module. This involves changing the associative weights between the input units and the priority units (see Figure 1). Stimuli that are consistently attended activate the priority unit strongly, which results in the vector being transmitted from the module. Stimuli that are not attended activate the priority unit weakly and inhibit the vector from being transmitted unless control processing enables the vector to be transmitted. *Attention* is implemented by the strength of the attenuation unit (see Figure 1). The attenuation unit multiplies the output unit between 0 (unattended output) and 1 (fully attended output). The multiplied output determines the activity that reaches other modules.

In the present model *controlled processing* involves modulating attenuation and monitoring the activity of the modules. *Automatic processing* involves activating the priority unit sufficiently to reliably transmit the vector to the next stage of processing in the absence of controlled processing input.

PHASES OF SKILL ACQUISITION IN A SINGLE TASK

The CAP1 simulation has been used to model single-task skill acquisition. Details of the modeling can be found in Schneider and Mumme (1988). In this paper we present only an overview of single-task learning and focus on dual-task performance.

CAP1 acquires skills through five phases. The movement between these phases is a gradual, continuous transition. The rate of movement between stages depends on the task to be learned. We will illustrate the transitions using numbers based on subjective impressions from work in search paradigms (Schneider and Fisk, 1984) and learning logic gates from electronic troubleshooting (Carlson and Schneider, 1988). The phases are as follows:

(1) *Controlled comparison* from buffered memory (Trials 1–4)
(2) *Context-maintained* controlled comparison (Trials 5–20)
(3) *Goal-state-maintained* controlled comparison (Trials 21–100)
(4) *Controlled assist* of automatic processing (Trials 101–200)
(5) *Automatic processing* (after Trial 200 per component task)

The first three phases involve extensive attentional processing. In Phases 1–3, the subject serially compares stimulus information with information in memory. Once the skill has shifted to Phase 4, there is a substantial reduction in attentional processing and a qualitative change in the processing.

Phases 4 and 5 involve direct associative retrieval of output patterns from input patterns at a series of processing levels. We will illustrate single-task learning using a category search task patterned after Fisk and Schneider (1983). In this task the subject must remember one or more categories and respond to a visually presented word by indicating whether the word is a member of the remembered categories. Subjects' initial performance is slow (e.g., 200 ms per comparison), serial (e.g., the *no* responses are twice as slow as the *yes* responses), and effortful (e.g., subjects perform poorly under a secondary-task load). This processing characterizes controlled processing. By comparison, as we discussed previously, in a CM search task subjects make the same response to the same stimulus over trials. Practice in CM search results in processing that is fast (e.g., 2 ms per comparison), parallel (e.g., equal *yes/no* slopes), and low in effort (e.g., subjects can perform a secondary task; see Fisk and Schneider, 1983). This processing characterizes automatic processing.

Controlled Comparison from Buffered Memory

The first phase, *controlled comparison from buffered memory*, involves loading and maintaining memory vectors in modules, comparing comparison vectors with the input vectors, and releasing the appropriate response vector. CAP1 performs a controlled comparison by adding two vectors to determine their similarity. Vector addition entails that each output unit (Figure 1) add the inputs it receives. The report cell receives the square or the sum of the absolute value of all of the output units. The equation for the activity report following the addition of two vectors is $SQRT(X^2 + Y^2 + 2r_{xy}XY)$, where X and Y are the vectors and r_{xy} is the correlation of the two vectors. To perform a comparison, two vectors are added together. If the two vectors are similar—for example, *cat* and *animal*—the added vector is nearly twice as long, as both vectors are pointing in the same direction. In contrast, if the two vectors are dissimilar, such as *car* and *animal*, the vector elements add orthogonally, pointing at right angles. This produces a shorter vector; for example, of length 1.41 for vectors with correlation of 0. The difference in length provides a criterion for match; for example, a length greater than 1.85 (correlation = 0.5) is defined as a match. Processing is serial because adding triplets of vectors substantially reduces the accuracy (e.g., reduced *d'*) relative to doing paired comparisons (see Schneider and Mumme, 1988).

Phase 1 processing is very effortful in the sense that it requires many shifts of attention and monitoring of the received activity of vectors (see Figure 3). For example, to perform a two-category search task with a display size of two, the subject must actively maintain the two category vectors and two response vectors (*yes* and *no* responses), shift attention four times (two visual stimuli × two categories), compare the length of the added vectors to the criterion length four times, shift attention to the output module, and release a response vector. Phase 1 performance is very error-prone. If the subject is interrupted, the vectors in the buffers will decay, causing errors. If the attention-switch-

ing operations are disturbed—for instance, by a secondary task—the comparisons cannot be made, resulting in response delays or omissions.

Context-Maintained Controlled Comparison

Phase 2, *context-maintained controlled comparison*, is similar to Phase 1 except that information is maintained in fast learning weights that associate vectors stored in modules to the context. Activating the context module can refresh information in modules (see Schneider and Detweiler, 1987). This context learning mechanism is important for robust processing and enables the model to account for a wide range of working memory phenomena including episodic and semantic memory, retroactive and proactive interference, elaborative rehearsal, and release from proactive interference effects. The mechanism is consistent with physiological data suggesting the existence of separate learning systems that learn at different rates (see Mishkin, Malamut, and Bachevalier, 1984).

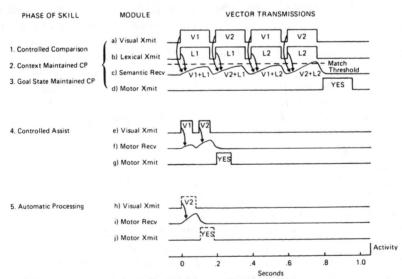

Figure 3. *Timing diagrams of the system as a function of phase of skill acquisition in a category search task. The bottom line provides the time scale in seconds after visual encoding is complete. The elevated lines show transmission or reception of messages. (For details, see text summary of phases, p. 552.)*

The connections between the context module and the modules on the inner loop are assumed to have fast learning weights (see Figure 2A). These connections can quickly associate the context to the current contents of modules on the inner loop. After a small number of trials (e.g., four repetitions), the context can evoke the category and response vectors. This context storage mechanism allows the system to reactivate the vectors if they decay or are displaced. After a few trials of searching for the same categories, subjects can perform a distractor task (e.g., counting backward by threes between trials) without disrupting primary task performance. Klapp, Marshburn, and Lester (1983) have demonstrated that if subjects rehearsed the memory set for five seconds in a memory scanning experiment, performing a digit-span task between the presentation of the memory set and the probe display did not disrupt performance.

Phase 2 processing is effortful and requires attention, but it will not be as seriously disrupted by interpolated tasks. Performing a category search task requires as many shifts of attention and comparisons as in Phase 1. However, the context can now be used to activate the vectors in the buffers. Note that the context storage works only for runs of the same task and is of little value if what the subject is searching for changes from trial to trial. Fast connection learning weights show very serious proactive interference effects (see Schneider and Detweiler, 1987) and provide little information about previous associations if new associations are made to the same context. During Phase 2, processing is reliable for runs of the same comparison set but reverts back to Phase 1 when a different comparison set is searched each time.

Goal-State-Maintained Controlled Comparison

The third phase, *goal-state-maintained controlled comparison*, is similar to Phase 2 except that the goal state can reload the modules in addition to the context-based reloading. For example, assume that the subject has to learn the three rules A, B, and C. In Phase 2, if the subject performs a series of A trials, the A vectors become associated with the context and can be reloaded if the information decays from the buffer. However, on the first trial with the B rule, the subject must be told what vectors to load in the buffers. This loading associates the B vector to the context, making the A vectors less available. In Phase 3, the subject learns to associate appropriate vectors to multiple goal states A, B, and C instead of to a single context vector. When the task changes, the subject needs a short time to remind himself or herself about the rule to reload the buffers. The subject then performs the same attentional operations as in Phase 1.

Phase 3 processing requires attention and involves the slow, serial, effortful form of processing characterized in Phases 1 and 2. It is somewhat more reliable, in that performing an interpolated task in the same modules will not disrupt performance. The difficulty that subjects encounter in going from practicing a single rule to practicing multiple rules provides evidence for the presence of Phase 3 processing. Carlson and Schneider (1988) found that subjects could perform single digital logic gate tasks (e.g., judging that the output of an AND gate is high if all inputs are high) in 0.7 s if the rules were massed after about five trials. However, more than 1000 trials per gate were needed before subjects could perform trials of mixed gate types (e.g., AND, OR, and NOR) at that speed.

Controlled Assist of Automatic Processing

Phase 4, *controlled assist of automatic processing*, produces a dramatic reduction in processing time and effort relative to the previous three phases (see Figure 3). During Phases 1–3, the subject repeatedly compared the input with the vectors in the buffers. The match processes involved four comparisons. A match was followed by a *yes* response and a nonmatch by a *no* response. In Phase 4, associative learning alters the connections between the input and output modules such that the input evokes the output (see Schneider and Mumme, 1988); for example, if the stimulus *cat* is transmitted, the learned associations evoke the "index finger" response. The resulting associative process eliminates the need for controlled comparison. If a target vector is transmitted from the input, then it will associatively evoke the appropriate response in the output. Attention is still required to transmit the vector from the input module to the output module. However, attentional switching of the comparison vector and monitoring are no longer necessary. Processing is parallel in memory in the sense that the input will evoke its output, independent of the number of input/output pairs learned by the connection matrix. Phase 4 requires a small amount of attention to transmit the input vectors on the inner loop to the output modules.

Automatic Processing

The fifth and final phase, *automatic processing*, occurs when automatic processing substitutes for attentional processing. During Phases 1–4, various vectors were attended in order to transmit them to later modules. A message that was transmitted prior to a positive event—for example, the visual input *CAT* associated with a *yes* response—would be associated *within* the module with a high-priority tag. In contrast, messages transmitted without follow-on events—for example, the word *CAR*, which is never responded to in the experiment—would have a low-priority tag. Automatic processing occurs when a message associated with a high-priority event is transmitted in the absence of attentive input. This takes place when the local circuit of the priority tag inhibits the attenuation units transmitting the message (see Fig-

ure 1). Automatic processing can occur through a series of stages. Each stage involves associating the output of the previous vector, which evokes a new vector within the module. Then that vector is categorized via autoassociative interactions and evokes the priority tag. If the priority tag is high enough, the vector is transmitted out of the current module to the next module. This process then cascades through a series of stages.

Phase 5 processing requires no attention; however, transmitting messages automatically does require some transmission time on the inner loop. In a search paradigm, priority learning eliminates the need for attentional switching. Assume that the words *CAT* and *CAR* are presented, with *CAT* being a previous target and *CAR* a previous distractor. The *CAT* message will evoke a high-priority tag and be transmitted, whereas the *CAR* message will evoke a low-priority tag and not be transmitted. Transmitting the *CAT* vector will evoke the positive response in the motor module. The motor response will evoke a high-priority tag, transmit its message, and cause the *yes* response to be made. The priority tag-based filtering in Phase 5 eliminates the need to serially transmit visual vectors as in Phase 4.

The qualitative changes that occur through the five phases of skill acquisition are depicted in Figure 3. Notice first that the bottom line provides the time scale in seconds after visual encoding is completed. Phases 1–3 involve multiple transmissions and monitoring of messages. The elevated bars in Line *a* show the transmission of the vectors from Visual Module 1 (V1) or 2 (V2); for example, the transmission of *CAT* as in Figure 2B. The two visual modules would alternate between transmitting two words. Line *b* shows the transmission of the lexical vectors (L1 and L2) for the words stored in the lexical buffers. Line *c* shows the power or activity report of the cells in the semantic module. The activity increases during transmission due to summation and feedback effects within the module. The first wave is for the sum of the V1 and L1 vectors; for example, the sum of the semantic codes for *TOP + FRUIT*. If the activity is below the threshold specifying a match, then the next pair is transmitted—for instance, *CAT + FRUIT*. When the visual vectors have all been sent, attention is switched to the next lexical module (e.g., *ANIMAL*), the visual module pointer is reset, and the V1 + L2 comparison occurs. This continues until either there is a match or the last comparison is completed. The last wave in Line *c* illustrates a match, *CAT + ANIMAL*—that is, a high activity report. After a match, the *YES* response is transmitted from the motor module (Line *d*) and initiates a motor response.

Phase 4 occurs after associations are built up between the visual input and motor response, such that the target *CAT* associatively evokes the *YES* response and bypasses the semantic match process found in Phases 1–3. Two transmissions are still needed (Line *e*) from the visual to the motor region to prevent the V1 and V2 messages from interfering. Line *f* shows the received motor activity from the visual transmissions. Line *g* shows the transmission of the motor response after it was received. Phase 5, automatic processing, occurs after priority learning occurs. Vectors with high-priority tags are transmitted automatically (dashed-pulse Line *h*) without the need for attention. This transmission evokes the response (Line *i*) that pro-

duces an automatic response transmission (Line *h*).

Table 1 contrasts the resource demands of performing the search task as a function of phase of processing. The time estimates are based on observed reaction-time slopes (Fisk and Schneider, 1983) for subjects comparing words with semantic categories. Note the dramatic drops in attentional, switching, and inner-loop time between Phases 3, 4, and 5. Phase 5 requires only 13% of the inner-loop time and none of the attentional resources required in Phases 1–3.

PROBLEMS OF MULTITASK PERFORMANCE

In many applied contexts, humans must perform multiple tasks concurrently. Performance typically deteriorates sharply when subjects are asked to perform concurrent tasks. The connectionist/control processing architecture provides an interpretation of the difficulties of dual-task performance and the improvements that occur with practice. Having a subject perform multiple tasks will have different consequences depending on what phase of practice the subject has reached before dual-task processing occurs.

Having a subject perform multiple tasks during Phase 1 will show severe disruption in performance. For example, in a memory scanning experiment, if the subject briefly sees a memory set (too briefly to allow rehearsal) and then performs a distractor task,

TABLE 1

Resource Demands as a Function of Phase of Skill Acquisition

Phase	Attention Time	Attention Switches	Comparisons	Inner-Loop Time
1. Controlled processing (CP)	1.0 s	5	4	0.8 s
2. Context-maintained CP	1.0 s	5	4	0.8 s
3. Goal-state-maintained CP	1.0 s	5	4	0.8 s
4. Controlled-assist AP	0.3 s	2	0	0.2 s
5. Automatic processing (AP)	0.0 s	0	0	0.1 s

performance is predicted to be very error-prone. The errors result because the buffer codes cannot be maintained and the comparison time requires considerable controlled processing and inner-loop resources (see Table 1). Performing a secondary task must be multiplexed with the attentional processing required by the primary task; for example, 1.0 s for the four-comparison category search task. This attentional processing for the comparison is required during Phases 1–3.

Phase 2 processing can be performed after an interruption from an irrelevant secondary task. In a search task, after the subject rehearses the words several times or performs several trials with the same memory set, the vectors representing the memory set are associated with the current context. If an irrelevant task is performed that requires attention, the memory-set vectors will decay. By reevoking the context, the memory-set vectors can be reloaded and the memory comparison process restarted. However, if the subject performs the search task with a new memory set, the new set will be associated with the current context and make the previous memory set unavailable. Therefore, the subject will not be able to time share a process that requires context storage of information in the same buffers.

Phase 3 processing can be performed after an interruption from a relevant secondary task. Once the goal state can evoke the vectors for comparison, context storage is no longer necessary. For example, if the subject is looking for words from Categories A, B, or C in Phase 3, presenting the category labels should be sufficient to allow the appropriate vectors to be evoked. However, the actual comparison would still require substantial processing (e.g., 1.0 s for the category search).

Phase 4 processing requires little attentional processing. It can be time-shared with other tasks. The other tasks must not require the same modules as the Phase 4 task, and

short periods of time must be available so that attention can be allocated to the Phase 4 task. In the category search example, the Phase 4 process required 0.3 s of attentional processing compared with 1.0 s for Phases 1–3.

Phase 5 processing requires only a short period of inner-loop transmission time and the modules necessary to process the input and output. Phase 5 processing can occur concurrently with other tasks, yet the other tasks must not completely monopolize the inner-loop transmission capacity. In the category search example, the Phase 5 process requires 0.1 s of inner-loop transmission time. Assuming one set of stimuli is presented at a rate of one per second, the category task would require time for the visual modules to process the word, time on the inner loop (0.1 s) to transmit the visual vector to the motor module, and time for the motor module to output the word. A secondary task could utilize all other modules, all of the attentional control processing, and 90% (0.9 s of 1 s) of the inner-loop transmission time. In this way two tasks utilizing different input/output modules with modest inner-loop transmissions—for example, reading while taking dictation—could be processed concurrently at single-task performance levels.

Phase 5 processing is still limited. If two modules transmit on the inner loop at the same time, there may be a loss of information. In the model, when multiple messages are transmitted, the received message is the addition of the transmitted messages. For example, consider a task in which a subject responds to a word by pressing a button and responds to a tone by speaking a word. The motor module might receive both the tone and the visual features of a word; if the messages are of equal strength and evoke incompatible outputs in the receiving modules, then no interpretable message will be received (see following discussion).

COMPENSATORY ACTIVITIES AND TRAINING IN DUAL TASK

Given the aforementioned changes in performance, *why does single-task training show such limited transfer to dual-task performance?* As we noted, a simple view of automaticity would predict that once a task is automatic, it should be possible to combine it with other tasks without deficit. If one accepts this view, one should train all single tasks and spend relatively little time in dual-task training. However, our literature review shows this prescription to be wrong; for example, Schneider and Fisk (1984) found nearly novice-level dual-task performance after eight hours of single-task training.

This section provides a model-based analysis of what might happen differently in dual-task training than in single-task training and describes how the system compensates in the dual task to facilitate performance. In general, performance in a dual-task situation puts a premium on the use of attention and inner-loop transmissions that are not critical in the single-task situation. For example, in single-task category search, a Phase 4 task requires 0.3 seconds of attention and 0.2 seconds of inner-loop transmission time. Performance is fast, accurate, and involves little effort, and there is little need to change strategy.

In the dual-task situation an entirely new set of behaviors must be executed that are *not* required in the single-task situation and may make the Phase 4 attention load unacceptable. To illustrate this, consider a category/tone dual task. The category task requires responding to visually presented animal words by pressing a button with the right hand. The tone task requires saying "target" whenever a high tone is presented. Let us assume that the category and tone tasks are at Phase 4 levels of skill. In single-task category search the subject need only switch attention between the input channels,

monitor the motor channel, and release the response. In the dual task the subject must additionally switch between the visual and auditory regions. If the regions transmit at different time scales, transmission durations must be changed. The system must switch what is being monitored; that is, during visual transmission the motor system is monitored and during auditory transmission the speech module is monitored. If different criteria are needed for received messages—if the motor association, for example, is stronger than the speech association—these must also be switched between tasks. If a module is loaded inappropriately, the inappropriate message must be cleared before the appropriate message can be received. The system must also block the interference effect of the second message transmission; for example, the motor module must not be cleared by the tone message transmission, and the motor message must output to other levels in the motor region while the tone message is being transmitted.

The types of compensatory activities that are effective in multitask situations depend on what phase of skill the subject is in on each of the tasks. If two tasks are in Phase 1, dual-task performance is highly error-prone except for very simple tasks. Errors occur because the subject is unable to maintain information about the second task while performing the first. If the subject is in Phase 2 or 3, the two tasks can be done by performing each task separately with little overlap. The subject can use context and goal-state information to reactivate the vectors of the second task after performing the first task. In Phase 4 the subject must switch attentional processing between the tasks, but the input vectors will evoke the appropriate outputs without having to reload or compare vectors. In Phase 5 the subject must modify the inner-loop transmissions to eliminate interference among messages.

There are seven behaviors that the current architecture manifests in dual-task situations that either do not occur in or are not as critical for single-task situations. These all involve decreasing the load on limited attentional resources and inner-loop transmission.

1. Shedding, Delaying, and Preloading Tasks

When two tasks must be performed concurrently, the easiest compensatory activities are either (1) to not perform one of the tasks or (2) to delay the lower-priority task. Not performing tasks eliminates the cognitive load of those tasks. Training in a dual task enables the learner to realize which tasks can be deleted and how much cognitive capacity will be made available to other competing tasks. Experience in performing multiple tasks enables one to anticipate and monitor the consequences of delaying or eliminating a task. For example, under conditions of high workload, experience guides the performer in knowing when to delay a task or to delete it altogether in order to attend to a higher-priority task component.

To manage workload, it would be best to even out potential workload peaks, either by delaying or preexecuting procedures. Delay involves buffering the input from one task while performing the more critical task. In the present connectionist/control architecture every module is a buffer that can maintain information for short periods (see Schneider and Detweiler, 1987). If stimuli from two tasks are presented simultaneously to two buffers, the lower-priority stimulus can be buffered while the higher-priority task is completed; then the lower stimulus can be processed.

Preloading involves preprocessing information prior to the onset of the critical workload segment. If vectors can be preloaded and maintained in modules, these activities need not be executed while the stimuli are being processed. For example, the context storage

system can activate vectors in many modules simultaneously. If an operator reviews the various tasks before the critical task segment begins, attending to the context can evoke vectors in multiple modules simultaneously. Techniques used to train combat helicopter pilots illustrate this technique. Pilots are encouraged to verbally rehearse possible actions before a "pop-up" maneuver exposing themselves to enemy fire—a high-workload situation. In the present architecture this rehearsal associates task-relevant vectors to the current context. The context can then maintain the vectors and eliminate the need to recall each procedure once the task has begun.

2. Letting Go of Unnecessary, High-Workload Strategies

Subjects' strategies in single-task conditions may entail a greater workload than necessary. The five phases of skill acquisition described previously represent CAP1's learning stages. Certain levels of practice are required before later phases become reliable. However, a subject's strategy may influence how long he or she continues a high-workload strategy after a lower-workload strategy is available. In our letter search experiments (see Schneider, 1985) a proportion of subjects continued to exhibit serial search (i.e., reaction times that increase linearly with the number of comparisons) long after the majority of subjects exhibited parallel search (i.e., nonlinear or flat-slope functions). When these subjects were pressured to respond faster, we often saw a dramatic break in the slope function. This break suggests that they shifted from a Phase 3 to a Phase 4 strategy. In dual-task training the learner cannot use control processing for both tasks, and this encourages the use of automatic processing.

For the CAP1 model, experience during later phases of skill acquisition results in better transfer to high-workload performance than during earlier phases. For example, practice during Phases 1–3 produced associations between the added vectors such as the stimulus and memory vectors. During Phases 4 and 5, the stimulus vectors are directly transmitted to the response modules. Transfer from Phase 3 to Phase 4 will depend on the *degree of overlap* between the combined and individual vectors (see Schneider and Mumme, 1988). *Once Phase 4 is reliable*, practice using a Phase 4 strategy will produce better transfer to Phase 5 than practice using a Phase 3 strategy.

3. Utilizing Noncompeting Resources

During multitask training, the subject can learn to allocate task components to minimize resource competition. Single-task practice facilitates developing a strategy that is fast and accurate. Multitask training facilitates developing the optimal *combined strategies* for performing all of the tasks. For example, in an air intercept control task one can determine the trajectories of two aircraft in order for them to intersect by using either quantitative or spatial procedures. In single-task training subjects will execute the strategy that is stressed by the instructor and that produces fast and accurate responding. However, if the learner practices the intercept task while performing a concurrent task requiring spatial operations—for example, navigation or such quantitative operations as calculating flying time—the optimal strategy depends on the context.

Within the proposed connectionist/control architecture multiple resources can be invoked to accomplish tasks (see Schneider and Detweiler, 1987). For example, Baddeley, Grant, Wight, and Thomson (1974) showed that when subjects could store digits verbally or spatially, verbal coding was better when the subjects had to perform a concurrent digit task. Wickens (1984b) has proposed that human factors designers should allocate

tasks to modalities to minimize resource competition. Early statements of this view (e.g., Wickens, 1980) suggested that modalities had different resource pools, and hence putting two tasks into different modalities would be advantageous. Recent reviews have shown that dividing tasks between modalities sometimes improves and sometimes degrades performance (Wickens and Liu, 1988). In the CAP1 architecture no simple prescription of dividing tasks among modalities is applicable over a wide range of tasks or practice levels.

In the present architecture, placing tasks in different regions (e.g., modalities) is sometimes advantageous and sometimes disadvantageous. Placing tasks in different regions has the benefit that there is no competition between multiple vectors in the same modules—for example, letters can be stored in a visual module and tones in an auditory module with little interference. However, it may be more difficult to switch attention between modalities than within a modality. Remember that Phase 1–3-level skills require a great deal of attention switching, whereas Phase 4 requires little and Phase 5 none. This suggests that dividing tasks between modalities may produce inferior performance for modestly practiced tasks (e.g., under 200 trials per stimulus in a search task) while at the same time producing superior performance for consistent, well-practiced tasks that do not require attention switching.

4. Multiplexing Transmissions Over Time

Training in a multitask situation provides opportunities to learn how to time-multiplex transmissions on the inner loop to accomplish the combined tasks. If two modules transmit messages at the same time, the potential of intermessage interference exists. For example, in the dual category/tone task, if both the visual and tone vectors transmit at the same time, the receiving modules will receive the summed vector—for example, the sum of the semantic vector of *CAT* and *HIGH TONE*. These vectors are likely to be unrelated, and interference will result in both messages being uninterpretable by the receiver (see Figure 4A). If the messages are multiplexed—transmitting *CAT* first and *HIGH TONE* second (see Figure 4B)—then both messages can be received accurately after a small delay for the second message.

Multitask training provides the learner with opportunities to exercise different multiplexing schemes. Given that two tasks, A and B, must be accomplished, there are many multiplexing schemes. Should A be completed before B is started? Should all the inputs be multiplexed before the outputs are multiplexed? Should A be multiplexed at a higher rate (e.g., process Task A twice as often as Task B)? Senders (1983) and Moray (1984, 1986) have shown that after extended training, human operators learn to sample instrument gauges at the optimal rate based on the relative information rate of each channel. The allocation of internal control processing may be tuned through experience in a manner comparable to the way the operators allocate attention among gauges.

5. Shortening Transmissions

In multitask situations inner-loop transmission time is at a premium. Learning to transmit messages with shorter transmissions would enable more transmissions per unit time and hence result in better multitask performance. In a single-task condition there is no benefit for shorter transmissions. When the visual module transmits to the motor module, a transmission of 100 ms or one of 500 ms may have effectively the same result. When the motor module receives the message in 100 ms, the response is begun. No damage is done if the visual system transmits for an additional 400 ms. Long transmission times have the potential benefit of increasing the reliability of the transmission in the event

that some messages require more than 100 ms to complete. Single-task training is likely to lengthen the transmission time of messages transmitting on the inner loop.

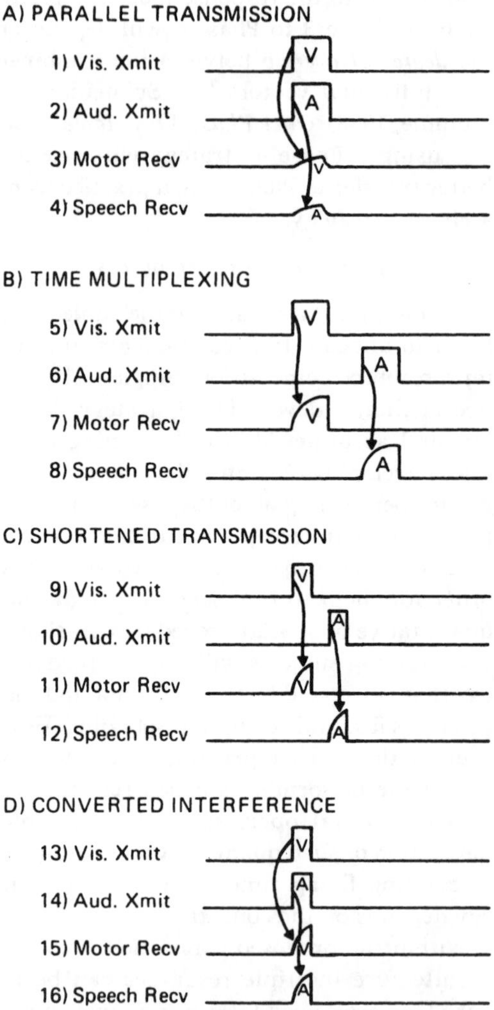

Figure 4. *Compensatory activities developed during dual-task training. The rectangular pulses indicate transmitted vectors; the waves indicate received information. Initially transmitting the visual (Line 1) and auditory (Line 2) vectors in parallel results in little information reception in the motor (Line 3) or speech modules (Line 4). Time-multiplexing the signals (Lines 5 and 6) results in greater reception with some delay of the second message (Lines 7 and 8). Dual-task training can shorten transmissions (Lines 9–12) and convert interference (Lines 13–16).*

Multitask training encourages the learner to shorten transmissions to the minimum length sufficient to transmit the information (see Figure 4C). In the simulation model multitask training enables the system to determine the minimal transmission time and to tune the receivers to reliably receive transmissions of short duration. The algorithm for varying transmission time is simple: if the transmission was successful, reduce transmission time on the next trial; conversely, if the transmission was unsuccessful, increase transmission time. The net result is that the system finds the minimum transmission time necessary to transmit the message reliably. A secondary benefit of practicing with shorter transmission times is that short transmissions can tune the receivers to categorize noisy transmissions into the appropriate messages.

6. Converting Interference from Concurrent Transmissions

In addition to avoiding interference by procedures such as multiplexing, associations can be modified to effectively tune out specific interfering messages. Such a tuning effect can be illustrated in a dual category/tone task in which a subject responds to animal names by pushing Button A and to non-animal names by pushing Button B. The tone task requires the subject to say "target" to a high tone. To illustrate interference effects, assume that the subject had learned to make a motor response (push Button D) to the high tone before the experiment. During initial dual-task training, the conflict between the visually evoked response (A or B) and the tone-evoked response (D) would produce message interference in the motor module. The motor module would receive the combined message (A + D), which is ambiguous. This interference necessitates multiplexing the messages. To eliminate interference, the tone input would have to either elicit no response in the motor system or elicit a re-sponse that did not interfere with the A or B responses.

Multitask training can cause an input with an initially incompatible response to be associated with the relevant response such that the irrelevant input does not alter the response evoked by a relevant input. The process of converting interference from concurrent transmissions occurs through changes in the association matrices between modules. This process is one of *orthogonalizing* vectors —that is, altering each connection matrix so that one task's message transmissions do not bias the receiver modules of the other task. In the dual category/tone task, multiplexing can associate the input tone with the motor responses for the visual input. In a dual-task trial, the visual stimulus would be transmitted and evoke the appropriate motor response (e.g., the word CAT evoking Response A. Shortly thereafter the tone would be transmitted and evoke the appropriate response in the speech region (e.g., if high tone, say "target"). In a short period the motor response would therefore receive two input messages (word and tone) and make one response (Response A). Associative learning would cause the input to be associated with the output (*CAT*–Response A; tone–Response A).

Thus the tone association with the previous motor response (tone–Response D) would be weakened, and the association with the current response (tone–Response A) would be strengthened. Similarly, the speech response between the word and the speech output would be weakened while the tone-to-speech (tone–"target") association would be strengthened. In other trials the tone would be associated with the motor response of the responses of alternative task stimuli (tone–Response B). After dual-task training, the motor module would receive the summed code of A for an animal name and the average of A + B for the tone. The net input (1.5A + 0.5B) would be highly correlated to the A response and would generally be categorized as the A response. At this point the tone stimulus could be transmitted simultaneously with the word without deficit, thus allowing both stimuli to be transmitted in parallel with little loss (see Figure 4D).

Note that interference conversion is specific to the messages trained, and all possible combinations of the messages must be trained. Each input stimulus must be associated with each of the responses of the *other* task. For example, to learn to associate 10 visual stimuli with 3 motor responses and 5 auditory stimuli with 2 speech responses requires learning 20 visual-to-speech patterns (10 visual stimuli to 2 speech responses) and 15 auditory-to-motor responses in order to convert the interference. Each visual stimulus would evoke its own motor response in the motor module and both speech responses in the speech module. Thus each input transmission would evoke an appropriate output in its output module and a neutral vector in the other output module. This enables concurrent transmission of both messages. A stimulus that does not evoke an interfering response—for example, the word *blue* normally evoking a keypress—would require little training to counteract message interference. However, if it evoked a strong interference—the visual word *blue* evoking the speech output of the word as in a Stroop (1935) task—extensive training would be required to convert the interference. Interference conversion is specific to particular messages and does not disconnect regions. For example, practice in the category/tone task may orthogonalize the motor output from a high tone message. However, such practice would minimally affect the motor association of dissimilar auditory inputs (e.g., responding to the auditorily presented word "jump").

The specificity of interference effects predicts the asymmetric interference effects often observed in attention and multitask situations. For example, in Stroop (1935) interference the word interferes with naming the color but not vice versa. Shaffer's (1975) report on dual-task transcription also illustrates this. He found that concurrent transcription of visually presented material and auditory shadowing were done with little cost, whereas transcription of auditorily presented material and visual shadowing produced severe interference.

7. Chunking Transmissions

If the system can transmit chunks or compact codes, more transmission time can be available for other tasks. Consider the behaviors of a copy typist. The letter pattern *THE* might be transmitted to the fingers as the *T*, *H*, and *E*, requiring three transmissions. In contrast, if the visual system can transmit a combined code of *THE*, only a single transmission is needed. Transmitting syllable chunks would reduce inner-loop transmissions as a direct function of chunk size (e.g., resulting in a 66% reduction for a three-item chunk).

Practice in a single-task condition will show little benefit from chunking unless the output is limited by the transmission time. For example, until a person types faster than 10 characters per second, there is no speed advantage in developing chunk codes for a single task. By contrast, in a multitask situation any reduction in attention or transmission time on one task provides more attention and inner-loop transmission time for other tasks.

Developing chunk codes can occur when modules transmit in parallel from input to output. In the simulation model (see Schneider and Mumme, 1988), the system modifies the connections between the input and output modules so that the input comes to evoke the output. To develop proper associations, the correct output must be in the output module *before* the to-be-learned input vector is transmitted. This can occur if all the input modules first sequentially transmit each message individually, then all the input modules transmit simultaneously before the output is associated with the input. For example, consider transmitting the word *THE*. Initially, if all three messages are transmitted simultaneously, no letter is received because of intermessage interference. However, if the letters are transmitted sequentially, the *T*, *H*, and *E* can be evoked in the output region, translated to the appropriate response, and buffered (see Schneider and Detweiler, 1987). If all the input modules then transmit simultaneously, followed by the transmission of the output, the combined input code will come to evoke the combined output code.

There are three modes of transmitting a chunk predicted by the CAP1 model (see Schneider and Detweiler, 1988). The first mode, *sequential output*, involves serially transmitting each element of the chunk (e.g., in typing *THE*, transmit from the visual to motor system the *T*, *H*, and *E* separately). The second mode, *parallel transmission*, involves transmitting all the elements of the chunk in parallel after the converting interference from concurrent transmissions (e.g., transmit *T*, *H*, and *E* simultaneously across the inner loop). The third mode, *chunk transmission and decoding*, involves the input region coding the components as a single chunk code, transmitting the chunk code from a single visual module to a single motor module, and then decoding the motor chunk into its elements (e.g., visually encode *T*, *H*, and *E* into a single visual word module; transmit the word code on the inner loop; and then decode the word code into the finger movements for *T*, *H*, and *E*).

The three modes of transmitting chunks produce dramatically different output rates and inner-loop loads. To use a typing example, if we postulate a 50-word-per-minute (wpm) typist, six characters per word, 0.2 s

transmission times, 1–5 letter syllable chunks (average three letters), then the sequential output mode would average a 100% inner-loop processing load and the parallel or chunk transmission a 33% load (one transmission every three letters). If we assumed that the encoding and decoding occurred at one character per 0.1 seconds, the typist would have an average typing rate of 50 wpm for sequential transmission (0.2 s per letter transmission), 60 wpm for parallel transmission (0.2 s per transmission and 0.3 s delay for the output of three characters from the buffer), and 100 wpm for chunked output (0.3 s to concurrently input, transmit, and output a three-letter chunk).

RESEARCH AGENDA

The present theoretical framework provides an agenda for the study of single-task to multitask transfer. The literature review at the beginning of this article illustrates that there is very little research on how single-task training transfers to multitask situations in which practice levels are sufficient to develop skilled performance. Learning how to optimize this transfer is important for applied questions such as optimizing simulator time in skill acquisition and for addressing basic issues such as how cognitive processing changes with practice.

Marginal Utility of Single-task to Multitask Transfer

The present model predicts that there is a declining benefit of single-task training for performance in multitask situations. However, there *is* a benefit for single-task training. In order for associative learning and priority learning to occur, the learner must be able to maintain the necessary vectors in memory and to perform the task accurately. Starting a learner in a multitask situation is likely to overload processing during Phases 1–3. Hence it is beneficial to instruct the learner on each component task individually. The learner should then perform the task in the single-task mode until performance is fast and accurate. To ensure that at least a goal-state-maintained level (Phase 3) is developed, it is important that subjects be able to perform well even when required to intermix the behaviors in random order. Failure to provide this single-task training can hinder progress because the component associations never become reliable.

Single-task training has a reducing marginal utility. After a certain level of skill is reached, continued single-task training can be inefficient compared with multitask training. First, multitask training provides the potential to develop the seven compensatory activities described previously. In addition, multitask training typically provides more practice on both tasks for the same total practice time, allows components to become integrated, and is generally more motivating to the learner.

Systematic research is needed to identify parameters predicting the marginal utility of single-task training and the optimal sequencing of component- and total-task training to maximize training effectiveness. Task complexity variables need to be identified to predict how many trials are needed for a skill to make transitions through various phases. More studies comparing one part-task training scheme with one whole-task scheme will be of little benefit.

To develop guidelines to improve training, criterion variables must be identified to predict when to shift from part training to aggregate training. The current model suggests that one should move from part- to whole-task training once the individual's component skills have reached at least the controlled assist phase (Phase 4) of proficiency. The phase of skill development can be independently verified using secondary task tests to see how well the skill can be performed

under workload (see Schneider and Detweiler, 1986).

Examination of Learning Multitask Compensatory Activities

The present architecture suggests seven compensatory activities that can develop in multitask situations that have little significance in single-task situations. For the most part, these compensatory activities have not been studied. For example, how long can one interrupt one task and then resume it without restarting the task? In multiplexing, how many channels can the central control structure keep track of before some requests are lost? How sensitive is transmission reliability to the duration of transmissions? How long does it take to build a higher-level chunk code, and does it require explicit practice aimed at building chunk transmission codes? How hard is it to switch between modules within a region and between regions? How fast and how fully can dual-task training orthogonalize transmissions to reduce inter-message interference? Can individuals sample information from different regions at different rates? Do they naturally develop different sampling rates? Do individuals discover, on their own, what tasks should optimally be shed in high-workload situations? Real-world multitask performance requires the development of compensatory activities, and basic research and training guidelines for understanding compensatory activities are severely needed.

Role of Part-Task Trainers for Multitask Skills

It is important to identify how part-task training can be modified to increase transfer to multitask performance. In many complex training systems, part-task training can be much cheaper than full-task training. For example, training a pilot to operate radio gear in a part-task desktop computer simulation might cost only a small fraction as much as training the skill in a full-motion/visual scene flight simulator. However, if the single-task training does not transfer to the multitask situation, then the part-task training is of little benefit. Given that most of the compensatory activities for dealing with high workload are not present in single-task situations, training a single component may have limited utility.

Training single tasks under high workload may have substantial benefits over single-task training for transfer to multitask situations. Of the seven compensatory activities described previously, all except converting intermessage interference develop under most high-workload situations and do not require the exact messages to be sent in order for compensatory activities to develop. For example, any dual-task situation will encourage the learner to delay tasks, adopt low-workload strategies, develop time multiplexing skills, and shorten and chunk transmissions.

The current modeling suggests that part-task trainers should be multitask trainers. We speculate that most of the training time required to develop high skill levels involves practice moving the skill from the goal-state-maintained phase (Phase 3) to the automatic processing phase (Phase 5). To accomplish this, secondary-task loading is critical. High workload can be produced either via presenting a calibrated workload task or by concurrently practicing other high-workload tasks. A calibrated workload task might be a VM auditory search task that requires considerable attention and inner-loop transmission but does not improve with practice (see e.g., Fisk, Derrick, and Schneider, 1986–87). Let us refer to this as Task X. The training simulators would train A, B, C, D, AX, BX, CX, DX, and then ABCD. Such part-task training simulators would cost about the same as the single-task trainers and might produce sub-

stantially more transfer. It would perhaps be more efficient to build multitask trainers so that the learner would use the practice time to develop skill on task-relevant procedures. The training simulators would train A, B, C, D, AB, CD, and then ABCD, with the bulk of the training time being in the AB and CD dual-task training. Research is required to determine the effectiveness of such trainers and to develop guidelines for task analysis and the division of tasks across training devices.

Quantitative Modeling of Skill Acquisition

There is a critical need for developing and testing quantitative models of skill acquisition with emphasis on multitask performance. Most previous modeling of high-workload performance has been at too coarse a level of analysis to have had a strong impact on the training process. We feel that general resource theories (e.g., Kahneman, 1973; Navon and Gopher, 1980; Wickens, 1984a) have neither dealt explicitly with practice nor differentiated the resources to a level of detail to suggest guidelines for training. An analogy to economics illustrates our concern. Macroeconomic theory at the level of predicting gross national product has had very limited success in predicting economic shifts or providing business managers with data to make production decisions. By contrast, linear programming techniques that predict production costs as a function of specific resources (e.g., the cost of ice cream as a function of the cost of sugar, milk, and chocolate) allow managers to evaluate alternative configurations. Models detailing the use of specific cognitive resources and functions of practice will enable better management of training time and training devices.

Models need to specify (1) the types and quantities of existing resources, (2) how those resources are utilized to accomplish specific tasks, and (3) how resource utilization changes with practice. The connectionist/control architecture illustrates the beginning of such a model. The resources involve the number and kinds of modules, attention switching, transmission time on the inner loop, and number and strength of connections. The computer simulation can perform specific tasks such as visual search and acquisition of simple digital troubleshooting skill. Resource utilization changes dramatically as skill is acquired. Initially (Phases 1–3), performance is slow, serial, and effortful. With practice, automaticity develops and performance becomes fast, parallel, and requiring little effort.

Modeling effort should emphasize cognitive architectures rather than single models. A cognitive architecture identifies a space of models rather than an individual model (see J. R. Anderson, 1983; Laird, Rosenbloom, and Newell, 1986). The present connectionist/control architecture defines such a space of models; within this architecture there may be many possible individual models (e.g., postulating different connection patterns among modules on the inner loop provides a family of related models).

Model predictions should be compared with human data to tune the modeling effort. The model's predictions of practice data should be compared with human skill acquisition data. The models should be able to predict the entire practice function.

SUMMARY

Review of multitask training and part-whole task literature shows that performance on consistent tasks changes dramatically with practice. Single-task training can transfer to multitask performance; however, that transfer can be very limited, and dual-task training produces substantial performance improvement even after extended single-task training. Existing theoretical frameworks for multitask performance generally do not predict the observed limited single- to dual-task transfer effects.

A connectionist/control architecture for skill acquisition and multitask training effects provides an interpretation of the limited nature of single- to dual-task transfer. The model predicts that as a skill is acquired, performance progresses through five phases. A qualitative change in processing occurs that enables performance of multiple tasks. Seven compensatory activities occur in the model during multitask training. The development of these compensatory activities provides an interpretation of the large practice effects observed in dual-task situations even after extensive single-task training. The connectionist/control architecture model of skill acquisition and multitask performance suggests an extensive research agenda of basic and applied issues relating to skill acquisition for high-workload tasks. It emphasizes that models of dual-task performance must deal with issues of practice. Researchers should *not* ask simply whether single-task training transfers to dual-task performance or whether part- as opposed to whole-task training is better. Rather, research should map out quantifiable performance variables to assess the marginal utility of practice and to predict optimal points to shift from single-task to multitask performance. The present connectionist/control architecture is beginning to deal with the complexity of practice effects during skill acquisition.

ACKNOWLEDGMENTS

This work was sponsored by the Army Research Institute, under contract No. MDA903-86-C-0149 and by the Personnel and Training Research Programs Psychological Sciences Division, Office of Naval Research, under Contract Nos. N-0014-86-K-0107 and N-00014-86-K-0678.

REFERENCES

Allport, D. A., Antonis, B., and Reynolds, P. (1972). On the division of attention: A disproof of the single channel hypothesis. *Quarterly Journal of Experimental Psychology, 24*, 225–235.

Anderson, J. A. (1983). Cognitive and psychological computation with neural models. *IEEE Transactions on Systems, Man and Cybernetics, SMC-13,* 799–815.

Anderson, J. R. (1983). *The architecture of cognition.* Cambridge, MA: Harvard University Press.

Baddeley, A. D., Grant, S., Wight, E., and Thomson, N. (1975). Imagery and visual working memory. In P. M. A. Rabbitt and S. Dornic (Eds.), *Attention and performance V* (pp. 205–217). New York: Academic Press.

Bahrick, H. P., and Shelly, C. (1958). Time sharing as an index of automatization. *Journal of Experimental Psychology, 56,* 288–293.

Broadbent, D. E. (1958). *Perception and communication.* Elmsford, NY: Pergamon Press.

Carlson, R. A., and Schneider, W. (1988). *Learning and using causal rules.* Unpublished manuscript.

Damos, D. L., Bittner, A. C., Kennedy, R. S., and Harbeson, M. M. (1981). Effects of extended practice on dual-task tracking performance. *Human Factors, 23,* 627–631.

Downey, J. E., and Anderson, J. E. (1915). Automatic writing. *The American Journal of Psychology, 26,* 161–195.

Fisk, A. D., Derrick, W. L., and Schneider, W. (1986–87, Winter). A methodological assessment and evaluation of dual-task paradigms. *Current Psychological Research and Reviews, 5,* 315–327.

Fisk, A. D., and Schneider, W. (1983). Category and word search: Generalizing search principles to complex processing. *Journal of Experimental Psychology: Learning, Memory, and Cognition, 9,* 177–195.

Hirst, W., Spelke, E. S., Reaves, C. C., Caharack, G., and Neisser, U. (1980). Dividing attention without alternation or automaticity. *Journal of Experimental Psychology: General, 109,* 98–117.

Hunt, E., and Lansman, M. (1986). Unified model of attention and problem solving. *Psychological Review, 93,* 446–461.

Kahneman, D. (1973). *Attention and effort.* Englewood Cliffs, NJ: Prentice-Hall.

Klapp, S. T., Marshburn, E. A., and Lester, P. T. (1983). Short-term memory does not involve the "working memory" of information processing: The demise of a common assumption. *Journal of Experimental Psychology: General, 112,* 240–264.

Laird, J., Rosenbloom, P., and Newell, A. (1986). *Universal subgoaling and chunking: The automatic generation and learning of goal hierarchies.* Boston, MA: Kluwer.

Mishkin, M., and Appenzeller, T. (1987). The anatomy of memory. *Scientific American, 256* (6) 80–89.

Mishkin, M., Malamut, B., and Bachevalier, J. (1984). Memories and habits: Two neural systems. In G. Lynch, L. McGaugh, and N. M. Weinberger (Eds.), *Neurobiology of learning and memory* (pp. 65–77). New York: Guilford Press.

Moray, N. (1984). Attention to dynamic visual displays in man-machine systems. In R. Parasuraman and D. R. Davies (Eds.), *Varieties of attention* (pp. 485–513). Orlando, FL: Academic Press.

Moray, N. (1986). Monitoring behavior and supervisory control. In K. R. Boff, L. Kaufman, and J. P. Thomas (Eds.), *Handbook of perception and human performance* (Vol. II, pp. 40-1–40-51). New York: Wiley.

Navon, D., and Gopher, D. (1980). Task difficulty, resources, and dual-task performance. In R. S. Nickerson (Ed.), *Attention and performance VIII* (pp. 297–315). Hillsdale, NJ: Erlbaum.

Norman, D. A., and Shallice, T. (1985). Attention to action: Willed and automatic control of behavior. In R. J. Davidson, G. E. Schwartz, and D. Shapiro (Eds.), *Consciousness and self-regulation: Advances in research* (Vol. IV). New York: Plenum.

Oliver, W. L., and Schneider, W. (1988). Using rules and task division to augment connectionist learning. In *Proceedings of the 10th Annual Conference of the Cognitive Science Society* (pp. 55–61).

Pew, R. W. (1974). Human perceptual motor performance. In B. H. Kantowitz (Ed.), *Human information processing: Tutorials in performance and cognition* (pp. 1–39). Hillsdale, NJ: Erlbaum.

Rumelhart, D. E., and McClelland, J. L. (Eds.). (1986). *Parallel distributed processing: Explorations in the microstructure of cognition. Volume 1: Foundations.* Cambridge, MA: MIT Press.

Schneider, W. (1985). Training high-performance skills. *Human Factors, 27,* 285–300.

Schneider, W. (1987). Connectionism: Is it a paradigm shift for psychology? *Behavior Research, Methods, Instruments and Computers, 19,* 73–83.

Schneider, W., and Detweiler, M. (1986). Changes in performance in workload with training. In *Proceedings of the Human Factors Society 30th Annual Meeting* (pp. 1128–1132). Santa Monica, CA: Human Factors Society.

Schneider, W., and Detweiler, M. (1987). A connectionist/control architecture for working memory. In G. H. Bower (Ed.), *The psychology of learning and motivation* (Vol. 21, pp. 53–119). Orlando, FL: Academic Press.

Schneider, W., and Detweiler, M. (1988). *The role of practice in dual-task performance: Toward workload modeling in a connectionist/control architecture* (expanded report). Unpublished manuscript.

Schneider, W., Dumais, S. T., and Shiffrin, R. M. (1984). Automatic processing and attention. In R. Parasuraman and D. R. Davies (Eds.), *Varieties of attention* (pp. 1–27). Orlando, FL: Academic Press.

Schneider, W., and Fisk, A. D. (1982). Concurrent automatic and controlled visual search: Can processing occur without resource cost? *Journal of Experimental Psychology: Learning, Memory, and Cognition, 8,* 261–278.

Schneider, W., and Fisk, A. D. (1984). Automatic category search and its transfer. *Journal of Experimental Psychology: Learning, Memory, and Cognition, 10,* 1–15.

Schneider, W., and Mumme, D. (1988). *A connectionist control architecture for attention, automaticity and the capturing of knowledge.* Unpublished manuscript.

Schneider, W., and Shiffrin, R. M. (1977). Controlled and automatic human information processing: I. Detection, search, and attention. *Psychological Review, 84,* 1–66.

Senders, J. (1983). *Visual scanning processes.* The Netherlands: University of Tilburg Press.

Shaffer, L. H. (1975). Multiple attention in continuous verbal tasks. In P. M. A. Rabbitt and S. Dornic (Eds.), *Attention and performance V* (pp. 157–167). New York: Academic Press.

Shiffrin, R. M., and Schneider, W. (1977). Controlled and automatic human information processing: II. Perceptual learning, automatic attending and a general theory. *Psychological Review, 84,* 127–190.

Solomons, L., and Stein, G. (1896). Normal motor automatism. *Psychological Review, 3,* 492–512.

Spelke, E., Hirst, W., and Neisser, U. (1976). Skills of divided attention. *Cognition, 4,* 215–230.

Stroop, J. R. (1935). Studies of interference in serial verbal reactions. *Journal of Experimental Psychology, 18,* 643–662.

Wickens, C. (1980). The structure of attentional processes. In R. Nickerson (Ed.), *Attention and performance VIII* (pp. 239–257). Hillsdale, NJ: Erlbaum.

Wickens, C. D. (1984a). Processing resources in attention. In R. Parasuraman and D. R. Davies (Eds.), *Varieties of attention* (pp. 63–102). New York: Academic Press.

Wickens, C. D. (1984b). The multiple resources model of human performance: Implications for display design. In *AGARD Conference Proceedings No. 371. Human factors considerations in high performance aircraft* (pp. 17-1–17-6). Williamsburg, VA: NATO.

Wickens, C. D., and Liu, Y. (1988). Codes and modalities in multiple resources: A success and a qualification. *Human Factors, 30,* 599–616.

Adaptive Team Coordination

Elliot E. Entin, ALPHATECH, Inc., Burlington, Massachusetts, and **Daniel Serfaty,** Aptima, Inc., Woburn, Massachusetts

It is hypothesized that highly effective teams adapt to stressful situations by using effective coordination strategies. Such teams draw on shared mental models of the situation and the task environment as well as mutual mental models of interacting team members' tasks and abilities to shift to modes of implicit coordination, and thereby reduce coordination overhead. To test this hypothesis, we developed and implemented a team-training procedure designed to train teams to adapt by shifting from explicit to implicit modes of coordination and choosing strategies that are appropriate during periods of high stress and workload conditions. Results showed that the adaptation training significantly improved performance from pre- to posttraining and when compared with a control group. Results also showed that several underlying team process measures exhibited patterns indicating that adaptive training improved various team processes, including efficient use of mental models, which in turn improved performance. The implication of these findings for team adaptive training is discussed. This research spawned the adaptive architectures for a command and control project investigating adaptive models that focus on changes in the structural and process architecture of large organizations. The research also produced a cadre of integrated performance assessment tools that have been used in training and diagnostic settings, and new components for a team training package focused on effective coordination in high-performance teams.

INTRODUCTION

A ubiquitous factor that teams have to confront and adapt to is stress induced by uncertainty, ambiguity, and time pressure. Stress reduces an individual's or team's flexibility and causes errors (e.g., the *Vincennes* incident). Janis and Mann (1977) stated that under stress, particularly high stress, team members may experience such an inordinate amount of cognitive constriction and perseveration that their thought processes are disrupted. Not all teams, however, appear to be equally affected. Serfaty, Entin, and Deckert (1993) found that an increase in the level of stress did not necessarily result in a decrease in the team's outcome performance. For example, increasing target uncertainty did not have a direct effect on the identification error-rate of the team; team

members simply increased their information-seeking rate. In other words, team members coped by altering their information-seeking strategy, a characteristic that LaPorte and Consolini (1988) attributed to highly reliable teams. The most striking evidence of team adaptation to stress in the experiment conducted by Serfaty, Entin, and Deckert (1993), however, came from the observation that teams were able to maintain the same level of performance with one-third of the time available to make decisions.

We maintain, as did Serfaty, Entin, and Deckert (1993), that the primary adaptation mechanism that allowed these teams and teams in general to maintain and improve their performance under a high level of time pressure was a switch from explicit to implicit coordination, a special mode of coordination (Kleinman &

Address correspondence to Elliot E. Entin, ALPHATECH, Inc., 50 Mall Rd., Burlington, MA 01803; elliot@alphatech.com. **HUMAN FACTORS,** Vol. 41, No. 2, June 1999, pp. 312–325. Copyright © 1999, Human Factors and Ergonomics Society. All rights reserved.

Serfaty, 1989; Orasanu, 1990; Wang, Luh, Serfaty, & Kleinman, 1991).

Cannon-Bowers and Salas (1990), Orasanu (1990), and others have effectively argued that implicit coordination involves the use of shared or mutual mental models among team members. It is hypothesized that shared mental models are made up of two principal components: a common or consistent model of the tactical situation among team members and a set of mutual mental models about the other team-member functions. It is the mutual mental models that team members have of one another that allow one team member to preempt the actions and needs of another so that actions can be coordinated and needs met in the absence of explicit communication. In this way implicit coordination reduces communication and coordination overhead. For example, in a study to observe how teams cope and adapt to high levels of stress, Entin, Serfaty, Entin, and Deckert (1993) reported that there was a strong correlation (.79) between the use of implicit coordination and the performance levels of the six teams that participated in the experiment. Moreover, the use of implicit coordination was adaptive to time-pressure-induced stress. The three higher-performing teams in the experiment increased their use of implicit coordination as time pressure increased, whereas the three lower-performing teams did not. Presumably some teams are able to employ their mutual mental models and shift to implicit communication modes under high stress, thereby reducing their communication and coordination load so that more attention can be focused on their mission tasks.

Orasanu's (1990) study of communication strategies used by airline pilots to cope with emergencies suggests that effective shared mental models are developed during periods of low workload and implemented during periods of high workload. To the extent that team members have accurate mental models, implicit coordination allows the team to maintain an effective level of performance under stress. To the extent that the mental models are lacking or inaccurate, high stress will lead to degraded performance.

We believe there is ample evidence that understanding superior team performance and coordination in terms of shared mental models is a promising approach to team training (Cannon-Bowers, Salas, & Converse, 1991; Kleinman & Serfaty, 1989; MacIntyre, Morgan, Salas, & Glickman, 1988). Moreover, researchers (e.g., Morris, Rouse, & Zee, 1987) hypothesize that training fostering the development of accurate mental models of a system will improve the performance of operators controlling those systems. This viewpoint can be extended to the team environment, where the development of a system model can be complemented by the development of mutual mental models of each team member's tasks and abilities. The importance of implicit coordination strategies used by effective teams in high-stress situations suggests that shared mental models are useful constructs to explain the anticipatory behavior of team members in the absence or scarcity of communications.

Decision makers have other means of reducing high cognitive loads (e.g., stress) confronting them. Payne, Bettman, and Johnson (1986) have argued that individual decision makers faced with increasing time pressure or task complexity will turn to various heuristics to lower the cognitive effort required to perform the task. Payne et al. noted that although some heuristics may be detrimental (Tversky, 1969), many prove to be adaptive to the increases in task demands. It is their contention that people's selection among strategies is adaptive; a decision maker will select strategies that are relatively efficient in terms of effort and accuracy as task demands escalate.

Decision makers operating as a team have a richer repertoire of alternative adaptive strategies available to them than does a single decision maker working alone. Extending the Payne et al. (1986) findings to the team environment, Kleinman and his colleagues (Kleinman, Luh, Pattipati, & Serfaty, 1992; Kleinman & Serfaty, 1989; Serfaty & Kleinman, 1985) hypothesized that high-performing teams will employ adaptive strategies in conditions of high workload/task complexity. Moreover, they will do so in a highly efficient manner that minimizes team errors and maintains performance. Indeed, experimental results reported by Kleinman and Serfaty (1989) and Wang et al. (1991) show that in conditions of high workload, higher performing teams adopt communication and coordination strategies that reduce the effort needed to meet task demands while maintaining performance levels.

Clearly, teams that perform well in high-stress conditions employ different strategies than do teams that perform poorly under stress (Entin et al., 1993; Entin & Serfaty, 1990; LaPorte & Consolini, 1988; Serfaty, Entin, & Deckert, 1993). LaPorte and Consolini identified three characteristics of highly reliable teams: The team structure is *adaptive* to changes in the task environment; the team maintains *open and flexible communication* lines; and team members are extremely *sensitive to other members' workload and performance* in high-tempo situations. High-performing teams possess the ability to adapt not only their decision-making and coordination strategies, but also their structure in order to maintain their performance in the presence of escalating workload and stress (LaPorte & Consolini, 1988; see also Pfeiffer, 1989, and Reason, 1990a, 1990b). We maintain that an important mechanism that highly effective teams use in the adaptation process is to develop a shared mental model of the task environment and the task itself, as well as a mutual mental model of interacting team members' tasks and abilities. These models are used to generate expectations about how other team members will behave (Cannon-Bowers & Salas, 1990; MacIntyre et al., 1988; Orasanu, 1990; Serfaty, Entin, & Volpe, 1993).

Research has shown that high-performing teams make use of such models (particularly when timely, error-free, and unambiguous information is at a premium) to anticipate both the developments of the situation and the needs of the other team members (Entin et al., 1993; Serfaty, Entin, & Deckert, 1993). This research evidence also shows that it is this *team coordination strategy* (anticipating changes in the situation and in the needs of other team members) that contributes significantly to the teams' high performance under stress. It is also the reason these teams perform consistently better under a wide range of tactical conditions.

ADAPTATION MODEL AND HYPOTHESES

Based on the adaptation model presented in Figure 1, we assume that high-performing teams adapt their (a) decision-making strategy, (b) coordination strategy, and (c) behavior and organizational structure to the demands of the situation in order to either maintain team performance or to minimize perceived stress. The experiment we shall discuss focuses on the middle loop of the model depicted in Figure 1.

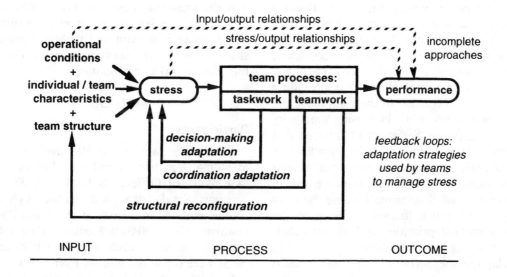

Figure 1. Theoretical framework for team adaptation.

To test the hypothesis that high-performing teams sense changes in the situational stress level and adjust their coordination strategies accordingly, we exposed ad hoc teams to the team adaptation and coordination training (TACT) procedure. (A glossary of abbreviations/acronyms is at the end of this article.) We taught ad hoc teams to recognize changes in situational stress levels, a set of adaptive coordination strategies, and the most appropriate conditions to use each adaptive strategy.

We reasoned that if teams could become aware of mounting stress and had a set of adaptive strategies to use in such conditions, then at the first signs of increased stress they would adapt accordingly. The adaptive strategies would help the team reduce their coordination overhead, freeing up more time for the increasing number of anti-air warfare (AAW) tasks they must perform. Several of these adaptive coordination strategies encourage switching to implicit modes of communication when the stress level is high. We hypothesized that teams exposed to the adaptive and coordination training (i.e., that were taught to recognize changes in the situational stress level and change to more effective coordination strategies) would perform better than teams without such training.

We also implemented an information structuring strategy that follows directly from the mental model perspective. We adapted the strategy used by Serfaty, Entin, and Deckert (1993), who instructed team leaders to periodically communicate their judgments (hostile, neutral) and related confidences about incoming targets to the other team members. The experimenters reported that providing this type of information to the team made the team more resilient to increasing ambiguity and time pressure. Serfaty et al. inferred that the information contained in the team leader's periodic briefings improved the mental models of individual team members. Improvement in the mental models occurred because the team leader's information focuses the team on the current tactical priorities and updates their understanding of the situation.

Adapting the Serfaty, Entin, and Deckert (1993) strategy, we instructed the team leader to give periodic situation-assessment briefs to the team. We held that these briefings would help team members update their mental models of the situation and leader, improving implicit communication within the team. The periodic situation-assessment briefs were combined with the TACT procedure to produce a second condition, TACT+. It is important to note that our design is not intended to provide a test of the isolated effect of the leader's updates without the adaptive coordination training. This is based on two arguments. First, we argue that the periodic information provided by the leader would be useful to the team members *only* if they had been trained to communicate and coordinate in an adaptive fashion beforehand. Second, practical considerations prevented us from conducting a full-factorial experiment with the four levels of a training manipulation necessary to test this hypothesis.

Our primary hypothesis asserts that teams receiving TACT and TACT+ will exhibit higher mission performance than teams that do not receive this training. We further hypothesize that teams receiving TACT+ will be able to formulate better shared situational mental models and, therefore, will perform at a higher level than teams that receive only TACT. We expect that teams receiving the adaptive coordination training will perform at a higher level and exhibit higher scores on process variables supporting performance than control teams receiving neither adaptive coordination training nor periodic situation-assessment briefs. We also predict that the benefit of adaptive coordination training both with and without periodic situation-assessment briefs will be greatest in high stress/workload conditions.

METHOD

Participants

The participants were 30 naval officers attending Department Head School at the Surface Warfare Officers School (SWOS) in Newport, Rhode Island, and 29 officers plus 1 civilian enrolled in courses at the Naval Postgraduate School (NPS) in Monterey, California. The participants in each sample (SWOS and NPS) were organized into six teams of 5 individuals each. An officer with tactical action officer (TAO) experience was selected to play the role of TAO in each team. The remaining 4 team members apportioned themselves to the four watch stations of identification supervisor (IDS), tactical information coordinator (TIC), anti-air warfare coordinator (AAWC), and electronic warfare supervisor (EWS). All members of the sample participated in the experiment for about 8 h, which in some cases was distributed over more than 1 day.

Design

The experimental design used in this study was a pretest/posttest control group design (Design 4 in Campbell & Stanley, 1963). The three levels of experimental condition (control, TACT, and TACT+), the two levels of testing (pre- and postintervention), and the two levels of workload-induced stress (low and high) were completely crossed. This created a 3 (experimental condition) between-subjects × 2 (testing) within-subject × 2 (stress) within-subject factorial design. To control for site differences, the experimental design was replicated at both sites; two teams from each site were randomly assigned to the control condition, the TACT condition, or the TACT+ condition.

Independent Variables

Training manipulations. The training manipulations used to establish the three conditions of the primary independent variable (TACT, TACT+, and control) are outlined here; for a more detailed description, see Entin, Serfaty, and Deckert (1994). The aim of TACT is not to teach teams new task knowledge or skills. Instead, TACT teaches teams strategies that enable them to better manage the increases in coordination and communication overhead that result from increases in workload and stress.

TACT occurred in three phases. During Phase 1, teams were given expository instruction on how to identify signs and symptoms of stress in the external environment, in the team, and in individual team members (including workload-induced physical and psychological signs). In Phase 2, teams received instruction on five adaptive strategies they might use to cope with increases in workload and stress. These strategies, which were designed to

reduce communication and coordination overhead within the team, are (a) preplanning, (b) use of idle periods, (c) favoring information transmission over action/task coordination, (d) anticipation of information needs (implicit communication), and (e) dynamic redistribution of workload among team members. (A more complete description of the strategies is in Entin et al., 1994).

Also as part of Phase 2, teams viewed three pairs of specially prepared videotape vignettes in which a team of five active-duty U.S. Navy and Air Force officers demonstrated effective and ineffective strategies, as well as indications of heightened stress on team behavior. Extensive discussions preceded each vignette, and vignettes were systematically interrupted to point out specific indications of stress and examples of effective and ineffective behavior and strategy utilization.

In Phase 3, teams were given the opportunity to practice what they had learned by completing two 12-min training scenarios. During and after each training scenario, the participants received feedback and comments on their coordination behavior and performance. Phase 3 concluded with the viewing of a summary videotape.

The TACT+ intervention was identical to TACT with one notable addition. During TACT+ the TAOs were given specific instructions and practice on how to give brief (approximately 30 s) periodic situation-assessment updates (SITREPs) to the rest of the team. These SITREPs included information on the TAOs' current priorities, targets of interest, and situation perception. TAOs were prompted by one of the role players at the control station to give the SITREPs approximately once every 3 min. Team members were taught that the SITREPs contained a digested summary of the situation as perceived by the TAO based on reports supplied by team members and other external sources. Team members were further instructed that these SITREPs represented the TAO's mental model of the tactical situation and, as such, could help unify the team's tactical picture. That is, team members could use the perspective and priorities contained in the TAO's SITREPs to select what to report to the TAO and what to filter out as nonessential information.

The control condition provided control teams with an experience comparable to that experienced by the teams in the TACT and TACT+ groups. Teams in the control condition were told they were being trained to appreciate the "big picture" and how their team's performance affected other teams on their platform and in the battle group. Care was taken to ensure that the control condition training was of the same duration, was approximately the same intensity, and appeared logical. Moreover, control teams were afforded the same practice opportunities as the teams in the two training conditions. Postexperiment interviews showed that participants in the control condition viewed the experience as a legitimate training situation and that no one suspected that he or she was in a control condition.

Testing. All teams performed two data-collection scenarios prior to the training manipulation and two data-collection scenarios after the training manipulation.

Workload-induced stress. The four-data collection scenarios used during the experiment were developed by the Naval Air Warfare Center Training Systems Division (NAWCTSD). Two scenarios were developed with high workload (stress) and two with low workload (stress). In the high-workload scenarios there were 50% more targets to process than in the low-workload scenarios. Each scenario lasted between 25 and 30 min. The primary task performed in each scenario by the five-person combat information center (CIC) team was situation assessment and deconfliction (the process of discriminating friend or foe). The team's objective was to correctly infer the identity, and thus the intentions (i.e., potentially hostile or neutral), of detected air and surface contacts. The function of each watch station was to supply the TAO with information necessary (but not always sufficient) to draw conclusions regarding a contact's identity, capability, and intention. The TAO then prescribed how the contacts were to be processed.

Prior to and again after the training intervention, each team participated in one high- and one low-stress scenario. The presentation of high- and low-stress scenarios was counter-balanced over the four trials using an a-b-b-a (or b-a-a-b) ordering. A manipulation check showed that subjective workload means derived from the Task Load Index (TLX; Hart & Staveland, 1988) were significantly higher under high than under low stress, indicating that the stress manipulation was effective.

Dependent Measures

Anti-air warfare performance outcome and teamwork assessment. Performance outcome and teamwork were assessed during each trial by two active-duty naval officers at the NPS site and two retired naval officers at the SWOS site who were trained to use the Team AAW Performance Scale. The scale consists of 12 behaviorally anchored items that assess overall AAW team performance and 15 behaviorally anchored items that assess the six dimensions of teamwork (i.e., team orientation, communication behavior, monitoring behavior, feedback behavior, back-up behavior, and coordination behavior). The performance items are based on the AAW Team Performance Index and the Behavior Observation Booklet (Hall, Dwyer, Cannon-Bowers, Salas, & Volpe, 1993; Johnston, Cannon-Bowers, & Jentsch, 1995). The teamwork items were adapted from teamwork assessment instruments used by Serfaty, Entin, and Deckert (1993) and Entin et al. (1993), which in turn are based on the AAW Team Observation Measure (ATOM) developed by NAWCTSD and a model of team evolution and maturation by Glickman and Zimmer (1989) and Glickman et al. (1987).

The agreement between the two observers scoring teamwork and performance outcome, for the NPS and SWOS sites combined, was found to be high. Coefficient alphas (Cronbach, 1970) were .79 or higher. Therefore, the scores from the two observers were averaged to form a performance outcome score and a teamwork score for each team.

Communication/coordination assessment. Two psychologists, who had previous experience coding verbal behavior and who demonstrated at least 85% agreement on practice materials, coded the behavior of the teams throughout the scenario. Observers used specifically designed observation matrices to code the verbal behavior of the teams; one matrix was designed to code the TAO's verbal behavior, and the other matrix was designed to code the subordinates' verbal behavior. Both matrices are laid out in a similar manner. Down the side of the matrix are the types and contents of communication messages to be noted (e.g., request information, request action and task, transfer information, transfer action and task, acknowledgments), and across the top are the message destinations (e.g., on the TAO's form: TAO to TIC, TAO to IDS, TAO to All; and on the subordinate's form: TIC to TAO or Team, IDS to TAO or Team, Team to Out). Each tally mark a coder makes in a matrix denotes a particular message type and content and to whom that message was directed. Both coders wore earphones connected to the communication net to monitor team communication.

To derive the various communication/coordination measures from the raw communication matrices, appropriate rows and/or columns were summed. In addition, the duration (in minutes) of each data-collection trial (scenario) was recorded and used to compute communication-rate variables.

Workload measure. At the conclusion of each trial, team members completed the TLX (Hart & Staveland, 1988) developed for the National Aeronautics and Space Administration. The TLX is a self-report subjective measure of workload that requires participants to rate the workload they just experienced on six dimensions (mental demand, physical demand, temporal demand, performance, effort, and frustration). Participants respond to each dimension using a 20-point graphical scale anchored at one end by the words *very low* and at the other end by the words *very high*. The TLX has exhibited good validity and reliability in the past (Lysaght et al., 1989). Its reliability in this study, as assessed by coefficient alpha, was .85.

Procedure

The Decision-Making Evaluation Facility for Tactical Teams (DEFTT) simulation provides a relatively realistic abstraction of five AAW CIC watch stations in "air-alley" found aboard an Aegis-capable platform. The teams received training on how to play the DEFTT simulator and engaged in three practice scenarios devel-

oped by NAWCTSD. NAWCTSD personnel were available to help any team member in need during the practice scenarios and to give detailed feedback at the end of each one. The DEFTT instruction and practice phase of the experiment lasted approximately 2 h.

A two-channel communication system provided a representation of the open-net communication network aboard ship. Team members at each of the watch stations could use one channel to communicate with team members at other watch stations and the other channel to communicate with the "outside world" (i.e., role players at the control station). Duplicating the arrangement found aboard ship, team members wore headsets equipped with a microphone and earphones (one in each ear) configured to allow simultaneous monitoring of each channel.

Before each data collection scenario, the teams received a mission brief delineating their mission, goals, potential threats, and rules of engagement. The briefing was always delivered by an active or retired naval officer. Prior to each data-collection scenario, the acting TAOs were always afforded time to brief their teams if they so desired. The first scenario was then presented. At the conclusion of the scenario, participants completed the TLX. Teams were then given a break, during which they received little or no feedback with regard to their performance. This was followed by a second scenario, which followed the same format as the first. The training intervention, which was administered after completion of the second scenario, lasted approximately 2 h and was always conducted by the same individual (EEE) regardless of site. After the adaptive or sham training, data collection Scenarios 3 and 4 were administered in the same manner as Scenarios 1 and 2.

RESULTS

Team AAW Performance

The average team AAW performance (hereafter referred to as *team performance*) was derived by taking the mean of the 12 items constituting the Team AAW Performance Scale. Analysis of the average team performance scores clearly demonstrates that the adaptation training intervention was effective. As shown in Figure 2, the team performance means of the three groups – control, TACT, and TACT+ – were not different from one another prior to the adaptation training intervention, $F(2, 9) = 1.50$, *ns,* whereas after training the performance of the three groups differed significantly, $F(2, 9) = 4.91$, $p < .04$. Figure 2 also shows that teams that received the adaptation training performed at a higher level after the intervention than before – the preintervention mean was 4.13, and the postintervention mean was 4.90; $t(9) = 2.44$, $p < .05$ – whereas the teams in the control condition showed about the same level of team performance pre- and post-intervention. Although the main effect for stress was significant, $F(1, 9) = 5.88$, $p < .04$, showing higher team performance under the low- than the high-stress condition, there was no difference between the low- and high-stress scenarios prior to or after the adaptation training intervention; nor were any significant interactions found involving stress before or after the adaptation training intervention. These findings indicate that the stress induced by the high-stress scenarios was relatively subtle, appearing only when both high-stress scenarios are compared with both low-stress scenarios over the entire sample.

Campbell and Stanley (1963) recommended that when using Design 4, one should combine the computation of pretest/posttest gain scores for each group with pretest scores as the covariate in an analysis of covariance in order to obtain the most responsive analysis procedure. To implement this analysis, we subtracted the pretest average performance score from the post-test average performance score to yield a gain score for each team in the sample. Using pretest average performance scores as a covariate on the computed gain scores, a Training Treatment × Stress analysis of covariance was computed. Inspection of the adjusted performance gain score means in Figure 3 confirms that the adaptive training intervention was effective and that teams receiving TACT and TACT+ performed significantly better than teams in the control condition, $F(2, 19) = 5.73$, $p < .02$.

The covariance analysis also shows the benefit of the intervention to help teams cope with the debilitating effects of stress. Examining the adjusted means, we see that high stress caused a drop in team performance when compared with low stress for teams in the control condition. Although the performance of teams receiving TACT was worse under high stress than under low stress, TACT provided some improvement in the performance of teams under high stress when compared with teams under high stress in the control condition. Clearly, however, the TACT+ intervention yielded the greatest benefit to teams under high stress. The TACT+ intervention allowed teams to perform as well under high stress as

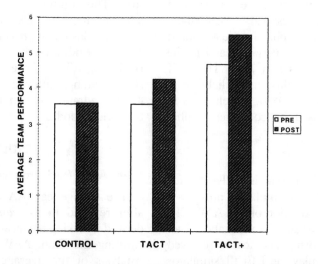

Figure 2. Average team performance by training treatments, pre- and postintervention.

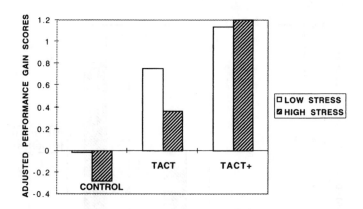

Figure 3. Adjusted performance gain scores by training treatments and stress.

they did under low stress such that their performance under high stress was significantly better than that of the control condition teams, $t(8) = 1.98$, $p < .05$, one tail.

Teamwork

To analyze the process measures of teamwork, we averaged and analyzed the six items representing the original ATOM measures (see Entin et al., 1993; Serfaty, Entin, & Deckert, 1993). Results of the ATOM teamwork-measure analysis closely parallel the performance findings. Figure 4 shows that teamwork for the three groups differed significantly after training, $F(2, 9) = 5.48$, $p < .03$, but not prior to training, $F(2, 9) = 1.78$, *ns*. The analysis also shows that the teamwork of the two groups

receiving training improved significantly after the adaptation training intervention – the preintervention mean was 4.08, and the postintervention mean was 4.85, $t(9) = 2.26$, $p < .05$ – whereas the teamwork for the control teams was about the same pre- and postintervention. We contend that adaptation training positively affected teamwork skills, which in turn gave rise to the superior performance outcomes observed for the TACT and TACT+ teams.

Similar to the performance results, ATOM teamwork was better under low stress than under high stress, $F(1, 9) = 6.76$, $p < .03$; however, none of the interactions involving stress were significant. The covariance gain score analysis performed on the ATOM teamwork

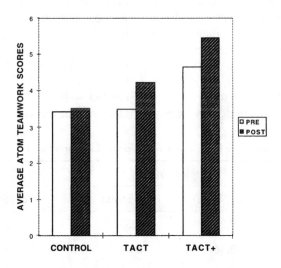

Figure 4. Average ATOM scores by training treatments, pre- and postintervention.

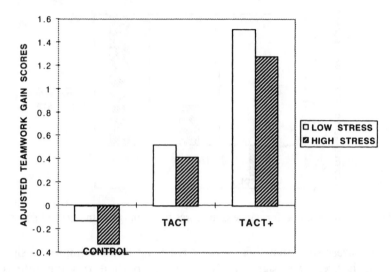

Figure 5. Adjusted teamwork gain scores by training treatments and stress.

measure is depicted in Figure 5. There is no longer a main effect attributable to stress. Teamwork increases monotonically from the control to the TACT+ condition under low stress, $F(2, 8) = 4.58$, $p < .05$, and the increase in teamwork under high stress is almost as strong. Teamwork means for the TACT+ condition are significantly higher than the control means, $t(8) = 1.87$, $p < .05$. We contend that it is more difficult to perform and exhibit good teamwork skills in a high-stress environment than in a low-stress one. That teamwork skills were significantly higher in the TACT+ condition than in the control condition, and that teamwork skills showed higher gain scores under the high-stress condition than under the low-stress one, indicates that TACT training offered a greater benefit to teams under high stress than to those under low stress. Otherwise, teamwork skills under high stress would be expected to be significantly below teamwork skills under low stress.

Communication/Coordination

Our theory predicts that TACT intervention will induce teams to shift away from explicit communication to more implicit communication. This might lead one to predict that the overall communication rate would decline as a result of this intervention; however, this is not necessarily the case because several factors come into play. The total communication rate means as a function of experimental condition, training, and stress are depicted in Table 1. A significant three-way interaction, $F(2, 9) = 11.41$, $p < .009$, shows that the communication rate did decline after the training intervention, but only for the low-stress scenarios. In the high-stress scenarios communication rates stayed about the same pre- and post-intervention. Although the TACT intervention made teams more efficient when they communicated – hence the drop in communication after training in the low-stress scenario – the

TABLE 1: Total Communication Rate Means as a Function of Experimental Condition, Training, and Stress

	Pretraining		Posttraining	
	Low Stress	High Stress	Low Stress	High Stress
TACT+	8.25	8.33	6.66	8.59
TACT	8.36	8.24	6.43	8.25
Control	7.24	6.96	6.09	7.59

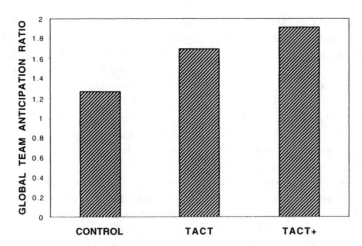

Figure 6. Global anticipation ratio by training treatments

increased demands in the high-stress scenarios required higher rates of communication among the team members. However, as we explain in the next set of results, the TACT and TACT+ interventions induced *different* communication patterns in a high-stress condition.

Anticipation ratios. Anticipation ratios are computed by dividing the number of transfers to individual X by the number of requests made by X. An anticipation ratio greater than 1.0 indicates that X's needs are being anticipated, and X receives what is needed without having to request it every time. Anticipation ratios greater than 1.0 can also be interpreted as a shifting away from explicit communication and toward implicit communication. To shift from explicit to implicit communication, team members must rely on mental models of the other team members, developed earlier, to anticipate their needs. Thus a large anticipation ratio can be taken as partial confirmation for mutual mental models.

An important anticipation ratio, because it reflects the tenor of the team as a whole, is the one based on all transfers and all requests issued in a team and is referred to as the *overall (or global) anticipation ratio.* As hypothesized, both the TACT and TACT+ interventions produced significantly higher global anticipation behavior than did the control condition (see Figure 6). Analysis revealed that the TACT intervention produced a 23% increase in the overall anticipation ratio – a shift from 1.46 in the preintervention condition to

1.79 in the postintervention condition, $F(1, 9) = 34.59, p < .001$. In the main, TACT appears to increase team members' awareness of the needs of others, thus allowing them to push needed information before it is requested.

Also as predicted, the anticipation ratio based on subordinate-to-TAO transfers contrasted with the TAO-to-subordinate requests showed a significant increase with training: a shift from 1.80 in the preintervention condition to 2.26 in the postintervention one, $F(1, 9) = 17.49, p < .003$. Moreover, the anticipation ratio based on team members' information transfers to the TAO compared with TAO information requests from team members (1.58 pre- vs. 2.31 postintervention), and TAO information transfers to team members compared with team members' information requests from the TAO (4.14 pre- vs. 6.36 postintervention), show the same pattern of results: The anticipation ratio is greater after intervention than before, $F(1, 9) = 15.70, p < .004$, and $F(1, 9) = 8.95, p < .02$, respectively. Across experimental conditions the pattern is also complementary to the effect of the training intervention. The ratio of the TAO information transfer to team members compared with team members' information requests from TAO was greater for teams receiving the training (TACT+ and TACT conditions) than in the control condition (6.69 and 3.26, respectively), $t(9) = 1.99, p < .05$, one-tail. These significant anticipation ratio results between subordinates and the TAO confirm the overall anticipation results, further indicating that TACT facilitates the development of better mutual mental models for all members of the team.

DISCUSSION

This study tested the hypothesis that highly effective teams adapt to stressful (high-workload) situations by using effective coordination strategies and that ad-hoc teams can be trained to employ many of these adaptive strategies to maintain performance under conditions of high stress and workload. We developed, implemented, and evaluated team-training procedures, TACT and TACT+, designed to teach ad-hoc CIC teams to adapt to high-stress situations by improving their coordination strategies. In the face of increasing stress, trained teams were expected to minimize their communication/coordination overhead in order to maintain performance level. The training program was predicated on the importance of a shared mental model of the situation and the task environment, as well as mutual mental models of interacting team members' tasks and abilities.

Results plainly demonstrate that the TACT and TACT+ adaptation training procedure was successful. Teams in the TACT and TACT+ conditions performed significantly better than teams assigned to a control condition. Comparisons of pre- and posttraining conditions showed that the teams' AAW performance was significantly improved, by an average of 21%, as a result of the TACT and TACT+ interventions. The four teams constituting the control group showed about the same level of performance before and after the sham training, which included the same amount of practice time as the adaptation training procedures. Teamwork for the three experimental conditions differed significantly, and the results paralleled the performance findings. We contend that the adaptation training program improves teamwork behaviors and coordination strategies, which in turn lead to better performance. Specifically, teams that received the training program reduced their coordination and communication overheads and thereby had more time and cognitive resources to devote to the task, resulting in better performance.

Another performance-related hypothesis supported by the results concerns performance levels in conditions of high stress and workload. We felt that because of its inherent design to be adaptive to stress and workload, the TACT (and TACT+) interventions would mitigate the effects of stress so that performance would reach similar levels under high stress and low stress. Findings for the adjusted gains (see Figure 3) revealed that teams receiving the TACT+ intervention demonstrated substantially improved performance for both the high- and low-stress conditions over the control and the TACT manipulations. Moreover, the TACT+ teams did as well under high stress as they did under low stress. Evidently, the adaptation training, and TACT+ in particular, made teams more resilient to stress. This resilience to stress may be attributable in part to better teamwork skills fostered by the adaptation training.

Results also support the hypothesis that TACT+ improves on TACT. We hypothesized that the SITREPs, which differentiated TACT+ from TACT, would provide all team members with more continuous and coherent situational mental models based on the TAO's current tactical priorities. As shown in Figure 2, the TACT+ procedure produced performance improvements beyond TACT alone. From these results we conclude that the addition of a SITREP component to adaptation training significantly improves the performance of teams beyond adaptive coordination training alone (TACT).

Changes in communication patterns were dominated by a stronger upward (from subordinates to leader) information push and more anticipatory behavior for the TACT and the TACT+ teams, supporting the hypothesis that the training program induces implicit team-coordination mechanisms that are essential to effective adaptation to stress. Moreover, results show that changes in team communication patterns from pre- to postintervention performance were in accord with the adaptation training objectives. Specifically, the hierarchical organizational structure of the CIC team (i.e., TAO-subordinates) and the nature of the AAW tasks are such that the lion's share of the team communications (about 50%) is concentrated around the TAO. Therefore, training manipulations focused on the "upward anticipatory behavior" (i.e., subordinates' anticipation of the TAO's needs).

We found that both the TACT and TACT+ training produced anticipatory behavior that primarily lowered the downward communication rate but also reduced the TAO's information requests. Furthermore, the TACT+ procedure had the additional effect of increasing significantly the lateral (subordinate-to-subordinate) information-exchange and action-coordination messages. Our contention is that coordination behavior is a by-product of the periodic SITREPs sent to the team by the TAO. A coherent picture of the situation, acquired through frequent SITREPS, prompts the subordinates to exchange information and resolve problems at their level first, before providing information to the TAO.

Plainly the TACT and TACT+ training procedure enhanced various types of anticipatory behaviors, which intimates that team members must have used mutual mental models of other members and situational mental models to preempt the needs of the TAO and other team members. We did not find differences in anticipatory behavior between low- and high-stress conditions. Contrary to our hypothesis, the same level of implicit coordination was present in both conditions. This lack of evidence of differential adaptation to stress might be attributable to the subtle stress differences present in the scenarios.

This study demonstrates that teams can be trained to recognize the signs of increasing workload and stress and then use adaptive coordination strategies to mitigate some of the debilitating effects of high workload and stress. The finding that appropriate training can significantly improve both teamwork skills and task performance supports the assertion that the dual concepts of shared mental models and adaptive coordination are a productive approach for understanding and developing effective teamwork. It is also noteworthy that although all of the participants in this study were military officers, the majority being Navy officers, the sample was still diverse. Of the NPS sample, 25% were drawn from services other than the Navy, and the Navy officers themselves were drawn from different programs. Despite this diversity,

the performance and process hypotheses were supported. This argues for the robustness of the experimental findings.

We hypothesize that further strengthening of the shared mental model will have a very positive effect on team processes and improve team performance. Strengthening of the shared mental model, however, employing only cross-training and interpositional training (which familiarize team members with their teammates' tasks, functions, and information needs in a dynamic fashion) will not be most effective. Such techniques are not sufficient to maintain performance under stress. Team members must also be trained to exercise their shared mental models through specific training of coordination strategies. Future team-training procedures must adopt an integrated approach to team training. Skill-based training (coordination/adaptation) supported by sound communication procedures (through information structure) – in addition to knowledge-based training (mental models) – are necessary ingredients for superior team performance.

GLOSSARY OF ABBREVIATIONS

AAW:	anti-air warfare
AAWC:	anti-air warfare coordinator
ATOM:	AAW Team Observation Measure
CIC:	combat information center
DEFTT:	Decision-Making Evaluation Facility for Tactical Teams
EWS:	electronic warfare supervisor
IDS:	identification supervisor
NAWCTSD:	Naval Air Warfare Center Training System Division
NPS:	Naval Postgraduate School
SITREP:	situation-assessment update
SWOS:	Surface Warfare Officers School
TACT:	team adaptation and coordination training
TACT+:	team adaptation and coordination training + periodic situation assessment briefs
TAO:	tactical action officer
TIC:	tactical information coordinator
TLX:	Task Load Index

ACKNOWLEDGMENTS

This research was supported by TADMUS-program contract N61339-91-C-0142 from the

Naval Air Warfare Center Training System Division. The authors are grateful to John Burns, Al Coster, and John Poirier for their assistance during data collection. We thank our contract monitor, Joan Hall Johnston, as well as Janis Cannon-Bowers and Eduardo Salas, for their help and advice throughout the project. We thank Lt. Cdr. Lonnie Green and William Kemple for bringing the particular covariance analysis technique to our attention. Finally, we are grateful to James Deckert and Katherine Masek for their editorial advice.

REFERENCES

Campbell, D. T., & Stanley, J. C. (1963). *Experimental and quasi-experimental design for research.* Chicago: Rand McNally College Publishing.

Cannon-Bowers, J. A., & Salas, E. (1990, April), *Cognitive psychology and team training: Shared mental models in complex systems.* Paper presented at the 5th annual Conference of the Society for Industrial and Organizational Psychology, Miami, FL.

Cannon-Bowers, J. A., Salas, E., & Converse, S. A. (1991). Cognitive psychology and team training: Training shared mental models of complex systems. *Human Factors Society Bulletin, 33*(12), 1–4.

Cronbach, L. (1970). *Essentials of psychological testing* (3rd ed.). New York: Harper and Row.

Entin, E. E., & Serfaty, D. (1990). *Information gathering and decision making under stress* (Report No. TR-454). Burlington, MA: ALPHATECH.

Entin, E. E., Serfaty, D., & Deckert, J. C. (1994). *Team adaptation and coordination training* (Report No. TR-648-1). Burlington, MA: ALPHATECH.

Entin, E. E., Serfaty, D., Entin, J. K., & Deckert, J. C. (1993). *CHIPS: Coordination in hierarchical information processing structures* (Report No. TR-598). Burlington, MA: ALPHATECH.

Glickman, A. S., & Zimmer, S. (1989). *Team evolution and maturation* (Final report; Contract No. N00014-86-K0732). Arlington, VA: Office of Naval Research.

Glickman, A. S., Zimmer, S., Montero, R. C., Guerette, P. J., Campbell, W. J., Morgan, B. B., Jr., & Salas, E. (1987). *The evolution of teamwork skills: An empirical assessment with implications for training* (Report No. TR-87-016). Orlando, FL: Naval Training Systems Center, Human Factors Division.

Hall, J. K., Dwyer, D. J., Cannon-Bowers, J. A., Salas, E., & Volpe, C. E. (1993). Toward assessing team tactical decision making under stress: The development of a methodology for structuring team training scenarios. In *Proceedings of the 15th Annual Interservice/Industry Training System Conference* (pp. 97–98). Washington, DC: National Security Industry Association.

Hart, S. G., & Staveland, L. (1988). Development of NASA-TLX (Task Load Index): Results of empirical and theoretical research. In P. A. Hancock & N. Meshkati (Eds.), *Human mental workload* (pp. 139–183). Amsterdam: Elsevier.

Janis, I. L., & Mann, L. (1977). *Decision making.* New York: Free Press.

Johnston, J. H., Cannon-Bowers, J. A., & Jentsch, K. A. S. (1995). Event-based performance measurement system for shipboard command teams. In *Proceedings of the 1st Internation Symposium on Command and Control Research and Technology* (pp. 274–276). Washington, DC: Center Advanced Command and Technology.

Kleinman, D. L., Luh, P. B., Pattipati, K. R., & Serfaty, D. (1992). Mathematical models of team performance: A distributed decision making approach. In R. W. Swezey & E. Salas (Eds.), *Teams: Their training and performance* (pp. 177–218). Norwood, NJ: Ablex.

Kleinman, D. L., & Serfaty, D. (1989). Team performance assessment in distributed decision-making. In R. Gibson, J. P. Kincaid, & B. Goldiez (Eds.), *Proceedings of the Interactive Networked Simulation for Training Conference* (pp. 22–27). Orlando, FL: Naval Training System Center.

LaPorte, T. R., & Consolini, P. M. (1988, January). *Working in practice but not in theory: Theoretical challenges of high reliability organizations.* Paper presented at the Annual Meeting of the American Political Science Association, Washington, DC.

Lysaght, R. J., Hill, S. G., Dick, A. O., Aamondon, B. D., Linton, P. M., Wierwille, W. W., Zaklas, A. L., Bittner, A. C., & Wherry, R. J. (1989). *Operator workload: Comprehensive review and evaluation of operator workload methodologies* (Report No. TR-851). Alexandria, VA: U.S. Army Research Institute for the Behavioral and Social Sciences.

McIntyre, R. M., Morgan, B. B., Jr., Salas, E., & Glickman, A. S. (1988). *Teamwork from team training: New evidence for the development of teamwork skills during operational training.* Paper presented at the 10th Annual Interservice/Industry Training Systems Conference, Orlando, FL.

Morris, N. M., Rouse, W. B., & Zee, T. A. (1987). *Adaptive aiding for human-computer control: An enhanced task environment for the study of human problem solving in complex situations* (Report No. TR-87-011). Norcross, GA: Search Technology.

Orasanu, J. M. (1990). *Shared mental models and crew decision making* (CSL Report No. 46). Princeton, NJ: Princeton University, Cognitive Science Laboratory.

Payne, J. W., Bettman, J. R., & Johnson, E. J. (1986). *Adaptive strategy selection in decision making* (Report No. TR 86-1). Durham, NC: Duke University.

Pfeiffer, J. (1989, July). The secret of life at the limits: Cogs become big wheels. *Smithsonian,* pp. 38–49.

Reason, J. (1990a). The contribution of latent human failures to the breakdown of complex systems. In D. E. Broadbent, J. Reason, & A. Baddeley (Eds.), *Human factors in hazardous situations* (pp. 475–484). Oxford: Clarendon.

Reason, J. (1990b). *Human error.* Cambridge, UK: Cambridge University Press.

Serfaty, D., Entin, E. E., & Deckert, J. C. (1993). *Team adaptation to stress in decision making and coordination with implications for CIC team training* (Report No. TR-564, Volumes 1 & 2). Burlington, MA: ALPHATECH.

Serfaty, D., Entin, E. E., & Volpe, C. (1993). Adaptation to stress in team decision-making and coordination. In *Proceedings of the Human Factors and Ergonomics Society 37th Annual Meeting* (pp. 1228–1232). Santa Monica, CA: Human Factors and Ergonomics Society.

Serfaty, D., & Kleinman, D. (1985, October). Distributing information and decisions in teams. In *Proceedings of the 1985 IEEE Conference on Systems, Man, and Cybernetics* (pp. 171–178). Los Alamitos, CA: IEEE.

Tversky, A. (1969). Intransitivity of preferences. *Psychological Review, 76,* 31–48.

Wang, W. P., Luh, P. B., Serfaty, D., & Kleinman, D. (1991, June). Hierarchical team coordination: Effects of team structure. In *Proceedings of the Joint Directors of Laboratories Symposium on Command and Control Research* (pp. 160–169). Washington, DC: National Defense University.

Elliot E. Entin is a senior psychologist in the Decision Systems Section of ALPHATECH, Inc. He received his Ph.D. in education and psychology from the University of Michigan in 1968.

Daniel Serfaty is president of Aptima, Inc. He received an M.S. in aeronautical engineering from the Technion, Israel Institute of Technology in 1981 and an M.B.A. in international management from the University of Connecticut in 1985.

Date received: March 19, 1997
Date accepted: October 23, 1998

Part 5: Training Evaluation

Three basic objectives exist for research on the evaluation of training systems. First, one should seek knowledge about the instructional effectiveness of the training system. Second, one should seek to increase understanding of how such training systems effect learning. Third, one should try to enhance the practice of training via improved techniques.

Although numerous studies have focused for nearly a century on one or more of these objectives, the total yield in terms of understanding instructional approaches, guiding utilization, or improving training has been disappointing. One possible cause is that much of the research in training evaluation has been concerned with demonstrating the superiority of one delivery system over another, as opposed to determining which aspects of delivery systems are effective in teaching different types of tasks. In addition, researchers have not typically been concerned with identifying appropriate instructional strategies to enhance the effectiveness of delivery systems.

A reason for this confusion may be that performance differences across training methods can interact with task complexity; performance on simple tasks shows no differences across training methods, but performance differences do occur for more complex tasks. It seems reasonable to assume that if existing experiments employ noncomparable tasks, noncomparable conditions, and/or confounded variables, they cannot be expected to yield consistent results.

This part contains seven articles on training evaluation. Adams (1979) contends that the widespread use of such techniques as transfer experiments and rating scales to evaluate training devices is flawed, and that use of known scientific principles, or "laws," in evaluating such systems is a preferred alternative. The second article, by Swezey (1979), describes a Bayesian decision-making methodology for use in training analysis and compares this technique with a separate judgment-based method. Hawley (1984) provides commentary on performance assessment issues from the unique perspective of embedding performance assessment capability within training devices. Vreuls and Obermayer (1985) also discuss a number of areas in the domain of simulation and training-oriented performance measurement and raise a variety of issues concerning problem areas and research issues with respect to these topics.

The paper by Andrews and Mohs (1987) offers information on policy capturing and Delphi techniques as means to assess the performance measurement concerns of key decision-making personnel in training evaluation situations. McDaniel and Rankin (1991) present a mathematical decision aid for use in evaluating flight training performance. Finally, the article by Dwyer, Oser, and Fowlkes (1995) comments on two methodological techniques, TARGETS and TOM, which were used to evaluate a distributed training application, again in a flight training context.

On the Evaluation of Training Devices

JACK A. ADAMS[1], *University of Illinois at Urbana-Champaign*

The theme of the paper is ways of evaluating flight simulators for aircrew training. The transfer of training experiment and the rating method are the two present-day ways of evaluating the worth of a simulator. The transfer experiment requires the trainee to practice in the simulator and then be tested in the parent aircraft to demonstrate the training value of the simulator. The rating method requires the pilot to be experienced in the parent aircraft and to rate the simulator for similarity to the aircraft. If similarity is high, the training value is assumed to be high. Arguments are presented that both of these methods are flawed. It is contended that a simulator, or any other system, need not necessarily be tested if it is based on reliable scientific laws and the success of other systems, based on the same laws, has been high. Good laws produce accurate prediction, and when outcome can be predicted it is redundant to conduct a system evaluation. With uncertainty about laws, the requirement for system testing increases. The psychological principles underlying simulators are reviewed.

Military flight training is expensive in dollars and energy, and so flying hours are being reduced by an amount expected to reach 17% by fiscal year 1981 (Orlansky and String, 1977). Planners appear to have concluded that flight training can be accomplished effectively on the ground with simulators which cost less to buy and less to operate than aircraft, and which use energy in frugal amounts. To reduce flying hours without impairing pilot proficiency, the Department of Defense in fiscal years 1975–1979 is augmenting its present arsenal of simulators with 1.5 billion dollars of new ones (Orlansky and String, 1977), and more money is certainly to follow. The simulator unquestionably has arrived. Buyers are confident that simulators will train, just as they are confident that airplanes will fly. The purpose of this paper is to examine the basis of

that confidence and analyze how flight simulators are evaluated.

A flight simulator is a training system, and so its power to train aircrew members must be demonstrated. There are two main methods for evaluating the training worth of flight simulators:

(1) The transfer of training experiment, where competence in the aircraft is required as evidence of a simulator's training value.
(2) The rating method. This method has two parts. One is an engineering evaluation of the simulator, where the hardware and software specifications are checked. The other is assessment by a pilot experienced in the aircraft, who flies the simulator and indicates its similarity to the parent aircraft.

It is a thesis of this paper that both of these methods are flawed, and that neither of them is satisfactory for evaluating the billions of dollars in simulators being acquired. To document this assertion, the transfer of training experiment and the rating method will be reviewed, and the difficulties with each will be indicated.

[1] Requests for reprints should be sent to Jack A. Adams, Department of Psychology, Psychology Building, University of Illinois, Champaign, Illinois, 61820, U.S.A.

THE TRANSFER OF TRAINING EXPERIMENT

At the outset, it should be pointed out that the transfer experiment has been criticized in articles by Mudd (1968), by Blaiwes, Puig, and Regan (1973), and by Matheny (1974; 1975), and they provide a foundation for further criticism.

Table 1 points out the essentials of the transfer of training design. The first methodologically sound transfer experiments to evaluate the training worth of flight trainers and simulators were conducted by Williams and Flexman (1949a) for the School Link Trainer and light aircraft, and for the SNJ simulator of the SNJ, or T-6, aircraft (Williams and Flexman, 1949b). There never were many transfer studies after that, but when they were done they usually were done with trainers for a class of aircraft (not simulators of particular aircraft) and light aircraft, and usually the percent transfer of training was low, of the order of 10–25% (e.g., Williams and Flexman, 1949a; Povenmire and Roscoe, 1971).

The low 10–25% level of transfer never appeared to upset anyone. Rather it was taken as assurance that aircrew training devices were worth something. Moreover, these devices had immunity from weather, could train in certain emergencies that could not be trained in the air, and had comparatively low cost. There was the general sentiment that

TABLE 1

The Basic Design of a Transfer of Training Experiment for the Evaluation of a Training Device.

Group	Practice on the Training Device	Test on the Parent System
Experimental Group	Yes	Yes
Control Group	No	Yes

corresponding studies on high fidelity simulators for advanced aircraft would have percent transfer that would be much higher. Maybe so, but left unsaid was that these transfer studies for high fidelity simulators were left mostly undone; it is hard to find a suitable transfer study of an advanced simulator. Hundreds of millions of dollars in simulators pour from production lines without the objective stamp of approval that is the transfer experiment. Why is the transfer experiment not routinely used for simulator evaluation?

The first reason is that the cost of a transfer experiment for the simulator of an advanced aircraft is high. The simulator and its parent aircraft are expensive, research staff and technicians are expensive, and the subjects, who undoubtedly would be professional pilots, are expensive. A shying from high costs is understandable.

A second reason, which is more fundamental, is that the transfer experiment is unsuited for advanced aircraft. There are problems for both the control and the experimental group where advanced aircraft are concerned. In a laboratory transfer experiment where an experimental group learns Word List A and Word List B, and a control group learns only Word List B, there must be confidence that the learning of Word List B by the control group is unaffected by prior learning. If prior learning has affected performance of the control group then it will be a biased baseline, and it will produce a biased estimate of the transfer of Word List A to Word List B. Except for routine facility with the language, for which both groups are equated, the control group must be naive with respect to Word List B. Now consider the case of the control group in an experiment to evaluate a simulator for a Boeing 747 aircraft. The control subjects will have flying experience, as the laboratory subjects have routine facility with the language, but it is insufficient for them to fly the aircraft. The control subjects can perform their task in the laboratory study of verbal transfer, but not the control subjects in the study of 747 simulator transfer. By some means, the control subjects must be raised to a minimum level of flying proficiency in the aircraft so that they fly it well enough to generate meaningful performance measures and fulfill their role. Browning, Ryan, and Scott (1973), Browning, Ryan, Scott, and Smode (1977), and Browning, Ryan, and Scott (1978), in evaluating a simulator for the U.S. Navy's P-3 patrol aircraft, gave their control groups practice on older training devices for comparison with the experimental group which practiced on a new simulator. The older devices gave the control groups competence for air tests, but obviously they are not true control groups. Such experiments tell us about the relative value of different training devices, but they lack the no-practice baseline that allows us to compute the correct percent transfer for the new simulator on which the experimental group trained. The transfer measures are attenuated by an amount proportional to the extra flying proficiency that has been given the control subjects, and so the training capabilities of the simulator are obscured.

An experimental group has problems of its own. The subjects of an experimental group must fly the aircraft well on the first flight after simulator training. If not, they would not be allowed to fly, and so there would be no measures and no experiment. The experiment cannot be done unless rather high positive transfer is an implied guarantee. It is a strange experiment where positive outcome is a precondition.

THE RATING METHOD

The rating method deserves special attention because it is the evaluation method which the simulator industry, in cooperation with pilots from the buyer, uses routinely. As mentioned earlier, the rating method includes an engineering evaluation and a pilot evaluation. The engineering evaluation, which is not central to this discussion, checks to see that the hardware items and the functions that are generated are as specified. A training evaluation is also conducted where a pilot, experienced in the parent aircraft, rates the simulator for its similarity to the parent aircraft, and this definitely does concern us. If the similarity is high, and the simulator flies like the airplane, the training worth of the simulator is assumed.

Gerlach, Bray, Covelli, Czinczenheim, Haas, Lean, and Schmidtlein (1975) recommend a rating scale for simulators as shown in Table 2. The thinking of Cooper and Harper (1969) is often cited in behalf of rating scales like this. Even if a judgment of a simulator dimension is made without the explicit use of a rating scale, the scale is implied.

The rating procedure is a straightforward adaptation of procedures that are used in aircraft design. In the case of a *design* simulator for an aircraft being built, the test pilot will judge the simulated aircraft for such properties as "handling qualities" and "controllability," and the opinion will influence the design of the aircraft. The same judgments will be made when the new aircraft is test flown. In the case of a *training* simulator, the aircraft will exist, the pilot has experience in it, and the similarity of the simulator to the aircraft is judged for the purpose of inferring about the training value of the simulator. In both cases, the same general rating procedure is used.

The acceptance of simulator ratings for inference about training value is uncritical. There are eight major problems with them:

(1) A big difficulty with the rating method is the underlying assumption that amount of transfer of training is positively related to the rated similarity between simulator and aircraft. The assumption has been explicitly

TABLE 2

A Rating Scale for Evaluating the Training Value of a Flight Simulator (Adapted from Table V in Gerlach et al., 1975)

Category	Rating	Adjective	Description
Satisfactory Representation of the Parent System	1	Excellent	Virtually no discrepancies
	2	Good	Very minor discrepancies
	3	Fair	Simulator is representative of the parent system
Unsatisfactory Representation of the Parent System	4	Fair	Simulator needs work
	5	Bad	Simulator is not representative
	6	Very Bad	Possible simulator malfunction

made by Caro (1970), Harris (1977), Gerlach, Bray, Covelli, Czinczenheim, Haas, Lean, and Schmidtlein (1975), and Davies (1975), as well as the Federal Aviation Administration (Skully, 1976) in their circular on simulator evaluation. The assumption is a half truth.

An example of the partial truth is a laboratory investigation of fidelity of simulation by Briggs, Fitts, and Bahrick (1957b). The task was two-dimensional compensatory tracking, where the subject used a control stick in attempts to keep a moving dot centered on a cathode ray tube display. The experimental variables were force required for stick displacement and stick amplitude. The experimental design required groups to train under different control stick conditions, and then all transfer to the same condition. The groups with the different control stick specifications performed differently in training, as would be expected, but unexpected was that all groups performed about the same in the transfer trials; the degree of similarity of training and transfer conditions made no difference for the amount of transfer.

There are other findings in the same vein.

Rockway and his associates (Rockway, 1955; Rockway, Eckstrand, and Morgan, 1956) found that in tracking, the control-display ratio in training left transfer to a new value uninfluenced. Briggs, Fitts, and Bahrick (1957a), in another tracking study, found that the amount of visual noise in tracking left transfer unaffected. Briggs and Rockway (1966) found that transfer from compensatory to pursuit tracking or vice versa was 100%. For cockpit procedures, Dougherty, Houston, and Nicklas (1957), and Prophet and Boyd (1970), found that training in a low fidelity photographic mockup of a cockpit produced as much transfer as a high fidelity simulator.

In all of these studies a rater would have said that the training and the transfer situations are of low similarity, and the prediction would have been low transfer. The prediction would have been wrong, and the reason is that the assumption is wrong. Instead of the simple, direct relationship between similarity and transfer, there is a family of relationships between similarity and transfer, shown in Figure 1. The lower curve shows that the direct relationship can be true, so the

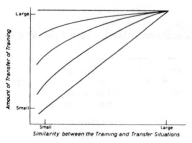

Figure 1. *Relationships between transfer of training and similarity of the training and transfer tasks that permit some decoupling of similarity and transfer. The linear function with positive slope allows for similarity and transfer to be positively related, and the functions that lie above it allow for the high transfer that has been found with low similarity.*

rating method will work sometimes. But, it will make mistakes because transfer and similarity can be decoupled, with low similarity giving high transfer, and the rating method does not allow for it.

(2) It cannot be assumed that an objective similarity of simulator-aircraft relationships can be discerned by any pilot who has flown the parent aircraft. Similarity is a psychological dimension, and the ratings that are obtained depend on the rater. There is evidence that ratings are a function of amount of experience of the pilot raters in the parent aircraft. Table 3 has data reported by Meshier and Butler (1976) on ratings of the air combat simulator by pilots of the USAF Tactical Air Command. The simulated aircraft were F-4's. The air combat simulator is a dual arrangement where two pilots, each in his/her own simulated aircraft with a visual

attachment for a simulated sky, can conduct aerial combat. The data indicate that the ratings might be more a function of the experience level of the pilots than of the adequacy of the simulation.

(3) If experience in the aircraft affects the ratings, then it is reasonable to suppose that experience in the simulator will affect them also. Hewett and Galloway (1976) say that it is necessary for the pilot to have regular refamiliarization with the aircraft during times that the simulator is being judged. In their experience, eight hours of simulator time is enough to impair a pilot's judgment and lower the validity of ratings. It is as if adaptation to the simulator makes the simulator rather than the aircraft the frame of reference. Gerlach et al. (1975) report that the pilot's rating of simulator fidelity improves with experience in the simulator—a pro-simulator bias creeps in.

(4) The dimensions of a simulator interact so that the rating of one dimension is affected by the presence of another. Cooper and Drinkwater (1971) say that the variables which determine control feel greatly affect rating of aircraft stability and control; the perception of aircraft stability and control is filtered through control system variables and affected by them. Gerlach et al. (1975) report that the rating of dynamic response is affected by the characteristics of the motion system of the simulator and by the scale factors of the visual system. Ratings are not a simple assessment of a simulator dimension,

TABLE 3

Ratings of the Air Combat Simulator as a Function of Experience Level of the Pilot Rater. Entry is percent. (From Meshier and Butler, 1976)

	Rating Category			
Experience Level	Excellent	Good	Fair	Poor
Experienced Pilots	28	60	12	0
Inexperienced Pilots	68	18	7	7

as the rating method assumes, but are influenced by other features of the flying task.

(5) Can a pilot tell where the source of poor simulation lies when flying a simulator? All that the pilot knows is that the system is not performing very well. Can human skill deficiencies be distinguished from deficiencies of the simulator? Gerlach et al. (1975) have found this to be a problem in their research on visual simulation.

(6) One would think that there would be a positive correlation between ratings and flying performance in the simulator. Not always so, apparently, Monfort (1971) reported on the task of following an ILS beam from 457 m to 61 m in a simulator. The handling qualities of the simulated aircraft were systematically degraded and, although the pilot's ratings regularly changed, tracking errors did not.

(7) The simulator is a teaching device, and it is not any better than the instructor, the instructor's station, and the training syllabus. One cannot talk about the training value of a simulator without including them. The rating method has nothing to say about these factors that are so influential in training operations. Whatever criticisms are made of the transfer experiment, it can be said that an evaluation of the instructor and the training syllabus are implicit in it.

(8) Finally, but not least, is that the rating method is based on a false premise. The rating method accepts this syllogism:

> Pilot ratings are useful for evaluating aircraft.
> A flight simulator is an earthbound aircraft.
> Therefore, pilot ratings are useful for evaluating flight simulators.

The error is the premise "A flight simulator is an earthbound aircraft." Rather, the premise should be "A flight simulator is a teaching machine." With this revised premise, the conclusion does not follow.

In the face of these problems and evidence, it is hard to believe that ratings are worth much. Yet, it is difficult to fault those who use the procedure. At worst they share the widespread misperception that a flight simulator is an earthbound flying machine and so must be tested like an airplane. Moreover, there must be an influence from the field of education that has always used expert raters for the evaluation of teaching instruments. Teachers judge textbooks, laboratory teaching apparatus, films, and the content of courses and curricula. That it would seem entirely natural to rate the training potential of a flight simulator is understandable.

A POSSIBLE SOLUTION

Do these criticisms of the transfer experiment and the rating method mean that there is no basis for believing in the validity of simulators for training? The answer is "No"—there is reason to believe in the validity of simulators. To substantiate the belief, it is necessary to list the psychological principles on which simulators depend. This may seem a digression, but it is a preface.

There are five major principles that underly the design and use of modern high fidelity flight simulators:

(1) The first and foremost principle is that human learning is dependent on knowledge of results (Adams, 1978; Annett, 1969; Newell, 1977). Knowledge of results is information given to the learner about the adequacy of behavior in meeting a criterion of accuracy; it is error information. Without learning there is no transfer, so the operations that support learning support transfer of training.

A simulator, by virtue of the aircraft it imitates, has some knowledge of results built in because the cockpit instruments allow the pilot to infer errors in the aircraft's state, but simulator designers can hardly take credit for this source of information. What they can take credit for is the design of the instructor's station with the mechanisms that it has for following the course of the trainee's behavior and providing information about it. Giving the instructor repeater instruments, or locating the station so that the instructor is always aware of a trainee's performance, is important for the quality of the knowledge of results that is provided. Printouts of performance data are sources of knowledge of results, as are devices for recording and playing back trainee performance.

(2) The second principle is perceptual learning, which is an increase in the ability to extract information from stimulus patterns as a result of experience. Flight produces a great array of dynamic stimuli, and there is an organization among them that must be learned for both normal and emergency flight situations. These stimuli can have any source in the operating environment, and they can use any of the senses. Gibson (1969), in her book on perceptual learning, engages the question of whether knowledge of results is required for perceptual learning, and so is no more than a special case of reinforced learning. She decides not, because there is too much evidence on the refinement of stimulus appreciation without knowledge of results. So, perceptual learning is a separate principle of learning.

(3) The third principle is stimulus-response learning (Honig, 1966; Honig and Staddon, 1977). If a response is to be made to a stimulus, then the stimulus and the control for response to it must be in the simulator. This is not perceptual learning, but the learning to do something in the presence of stimuli. The knowledge of results principle is combined with this principle to teach trainees how to move in response to the stimulus classes of the system. For whatever other kinds of learning there are, and for whatever criticisms cognitive psychologists make of it, this behavioristic concern with stimulus and response, with indifference to

mediating events, is a scientifically reliable way to teach a wide variety of behavior. Modern behaviorism, exemplified by B. F. Skinner, is based on it, and both its laboratory studies and its applied efforts in behavior modification show how useful a stimulus-response conception of learning can be. (Engineering psychologists have not seriously addressed the implications of mediated, cognitive learning.)

(4) The fourth principle is that transfer of training is the highest when similarity of the training and transfer situations is the highest. Osgood's transfer and retroaction surface (1949), or a successor, is often cited as the locus of this law. While it is sometimes possible to obtain high positive transfer with low similarity, as was discussed earlier, high transfer can also be obtained with high similarity, and this is the governing principle for most simulators that are built.

(5) The fifth principle is that a trainee must be motivated, and that the characteristics of the task are a source of motivation. The task must be intrinsically interesting so that the trainee will practice willingly and learn. In commercial parlance, there must be consumer acceptance.

Undoubtedly one could find lesser psychological principles that are used, but these are the main ones which guide the design and use of high fidelity flight simulators. The reason for putting forth these principles is *the assertion that a system built on sound scientific laws needs less concern with evaluation because a good scientific law produces accurate prediction, and when the outcome can be predicted it is redundant to conduct an evaluation.* A strict requirement for system evaluation is based on uncertainty about the outcome of laws and their interactions, but if uncertainty is low, should not the requirement for evaluation be low? There is an undeniable appeal in a requirement for system evaluation, because it is comforting to show that the system will do what it is supposed to do, but a ritualistic requirement can be questioned if there are good reasons for believing that the system will perform successfully. If research gives us high confidence in the laws that enter a system, and if the history of success for other versions of the system, based on the same laws, is good, then the requirement for evaluation can be relaxed. Textbooks of system engineering (e.g., Goode and Machol, 1957; Flagle, Huggins, and Roy, 1960; Shiners, 1967) have nothing to say about when systems should be rigorously tested and when the testing requirement can be relaxed.

The assertion has precedence because there are a number of visible systems that have been untested, and yet this fact upsets no one because these untested systems are successful. Nuclear weapons and space vehicles ordinarily do not have prototype versions that are repeatedly tested in the criterion environment before they are produced and used. The operational use of such systems is their first use, and there is belief in their success potential because there is confidence in the scientific laws that underlie them and because of the successes that similar systems, based on the same laws, have had. The same can be said for some flight simulators. Aircrew personnel of manned space vehicles perform competently on first flights after simulator training, and yet the simulators are never assessed for training value. Single-seat fighter aircraft have been flown successfully on the first flight before a two-seat trainer version is available and with only simulator training for the pilot. There have been single-seat fighters which have not had a two-seat trainer version produced, and so a simulator has been the only way of successfully acquiring flying skills.

DISCUSSION

If there is substance in this position, then simulators will rise or fall on evidence of the success of predecessors, and on the strength of scientific laws. Some of the laws that are now used in simulators are strong, like knowledge of results, and some seem a bit weaker, like perceptual learning. It is important that research strengthen these laws, as well as find new ones that are relevant for training.

Typically, there are two research paths. One is basic research, where uncommitted science discovers and technology applies. For all the impatience that applied scientists have with basic research, there is a tolerance of it because it makes a contribution to all technologies over the indefinite long run. The other kind of research is applied research, where there is a need which pulls scientists to solve a problem in the foreseeable future. For whatever the benefits of basic research, training problems call for research of the applied sort as a main thrust. Influential applied research is not the local engineering research efforts that are colloquially called "putting out fires." Rather, it is research that is comparatively broad in conception and which produces generalizable knowledge that spreads over a number of situations (Adams, 1972). Generalizable applied research may not always be fully appreciated. It is neither uncommitted basic research nor short-term engineering research. Perhaps it is a research encounter of the third kind.

Consider an example of applied research that might lead to generalizable training knowledge. Response-produced feedback and learning have been an interest of psychology almost since the beginning of its experimental history (Adams, 1968). A major problem area for simulators is motion subsystems, and whether they affect the learning of skills and the transfer of them to aircraft. Psychologically, motion produces change in feedback from the vestibular system, visual cues, and proprioception associated with motion and g forces. Uncommitted basic research on feedback of all sorts will be done to improve general understanding, but perhaps some could ask the more focused question of how motion affects feedback channels and the transfer of skills. Not only will the answer to the old issue of motion vs. no-motion and transfer emerge, but new directions will be suggested. Adams, Gopher, and Lintern (1977) found that vision is a powerful feedback channel that overpowers the concurrent proprioceptive feedback channel in motor learning. If vision does this for all feedback channels, and it is true for transfer of training also, then simulation of the visual accompaniments of motion without motion itself could be defended. This line of research does not exhaust issues about motion and feedback, however. Motion is one of the cues that is used in perceptual learning to interpret normal and abnormal configurations of the aircraft (Povenmire, Russell, and Schmidt, 1977). How do the various feedback channels contribute to perceptual learning and its transfer?

SUMMING UP

Flight simulators are vulnerable because billions are spent on them and yet they are not evaluated by the criteria that prevail in systems engineering today. A proposal has been made that the criteria be redefined, giving particular attention to the possibility that system testing can be set aside when the scientific laws underlying the system are known to be strong and when predecessors, based on the same laws, have been successful. It is our strength as a behavioral science that makes this suggestion possible for simulators, and that strength comes from research. With billions of dollars being spent on simulators, research programs on transfer have a payoff because they strengthen laws of learning and transfer, and because they better secure the circumstances under which training devices can be confidently used without necessarily relying on conventional system testing procedures.

ACKNOWLEDGMENT

This paper was delivered at the 1978 Toronto meeting of the American Psychological Association in acknowledgment of the Franklin V. Taylor Award for outstanding contributions to engineering psychology, given to the author by the Society of Engineering Psychologists. The paper was also presented at the symposium "Fifty Years of Flight Simulation," hosted by the Royal Aeronautical Society, London, April 23-25, 1979.

REFERENCES

Adams, J. A. Response feedback and learning. *Psychological Bulletin*, 1968, *70*, 486-504.

Adams, J. A. Research and the future of engineering psychology. *American Psychologist*, 1972, *27*, 615-622.

Adams, J. A. Theoretical issues for knowledge of results. In G. E. Stelmach (Ed.) *Information processing in motor control*. New York: Academic Press, 1978, 229-240.

Adams, J. A., Gopher, D., and Lintern G. Effects of visual and proprioceptive feedback on motor learning. *Journal of Motor Behavior*, 1977, *9*, 11-22.

Annett, J. *Feedback and human behaviour*. Baltimore: Penguin, 1969.

Blaiwes, A. S., Puig, J. A., and Regan, J. J. Transfer of training and the measurement of training effectiveness. *Human Factors*, 1973, *15*, 523-533.

Briggs, G. E., Fitts, P. M., and Bahrick, H. P. Learning and performance in a complex tracking task as a function of visual noise. *Journal of Experimental Psychology*, 1957, *53*, 379-387. (a)

Briggs, G. E., Fitts, P. M., and Bahrick, H. P. Effects of force and amplitude cues on learning and performance in a complex tracking task. *Journal of Experimental Psychology*, 1957, *54*, 262-268. (b)

Briggs, G. E., and Rockway, M. R. Learning and performance as a function of the percentage of pursuit component in a tracking display. *Journal of Experimental Psychology*, 1966, *71*, 165-169.

Browning, R. F., Ryan, L. E., and Scott, P. G. Training analysis of P-3 replacement pilot and flight engineer training. Orlando, Florida: Naval Training Equipment Center, TAEG Report No. 10, December, 1973.

Browning, R. F., Ryan, L. E., and Scott, P. G. Utilization of device 2F87F OFT to achieve flight hour reductions in P-3 fleet replacement pilot training. Orlando, Florida: Training Analysis and Evaluation Group, TAEG Report No. 54, April, 1978.

Browning, R. F., Ryan, L. E., Scott, P. G., and Smode, A. F. Training effectiveness evaluation of Device 2F87F, P-3C operational flight trainer. Orlando, Florida: Training Analysis and Evaluation Group, TAEG Report No. 42, January, 1977.

Caro, P. W. Equipment-device task commonality analysis and transfer of training. Alexandria, Virginia: Human Resources Research Organization, Technical Report 70-7, June, 1970.

Cooper, G. E. and Drinkwater, F. J., III. Pilot assessment aspects of simulation. In R. J. Wasicko and A. G. Barnes (Eds.) Simulation. Neuilly-Sur-Seine, France: NATO, Advisory Group for Aerospace Research and Development, AGARD Conference Proceedings No. 79. January, 1971.

Cooper, G. E. and Harper, R. P., Jr. The use of pilot rating in the evaluation of aircraft handling qualities. Neuilly-Sur-Seine, France: NATO, Advisory Group for

Aerospace Research and Development, AGARD Report No. 567, April, 1969.

Davies, D. P. Approval of flight simulator flying qualities. *Aeronautical Journal*, 1975, 79, 281-297.

Dougherty, D. J., Houston, R. C., and Nicklas, D. R. Transfer of training in flight procedures from selected ground training devices to the aircraft. Port Washington, New York: U.S. Naval Training Device Center, Human Engineering Technical Report NAVTRADEVCEN 71-16-16, September, 1957.

Flagle, C. D., Huggins, W. H., and Roy, R. H. *Operations research and systems engineering.* Baltimore: Johns Hopkins Press, 1960.

Gerlach, O. H., Bray, R. S., Covelli, D., Czinczenheim, J., Haas, R. L., Lean, D., and Schmidtlein, H. Approach and landing simulation. Neuilly-Sur-Seine, France: NATO, Advisory Group for Aerospace Research and Development, AGARD Report No. 632, October, 1975.

Gibson, E. J. *Principles of perceptual learning and development.* New York: Appleton-Century-Crofts, 1969.

Goode, H. H. and Machol, R. E. *System engineering.* New York: McGraw-Hill, 1957.

Harris, W. T. Acceptance testing of flying qualities and performance, cockpit motion, and visual display simulation for flight simulators. Washington, D.C.: Naval Air Systems Command, Technical Report NAVTRAEQUIPCEN IH-251, May, 1977.

Hewett, M. D. and Galloway, R. T. On improving the flight fidelity of operational flight/weapon system trainers. Neuilly-Sur-Seine, France: NATO, Advisory Group for Aerospace Research and Development, AGARD Conference Proceedings No. 198, June, 1976.

Honig, W. K. (Ed.) *Operant behavior: Areas of research and application.* Englewood Cliffs, New Jersey: Prentice-Hall, 1966.

Honig, W. K. and Staddon, J. E. R. (Eds.). *Handbook of operant behavior.* Englewood Cliffs, New Jersey: Prentice-Hall, 1977.

Matheny, W. G. Training research programs and plans: Advanced simulation in undergraduate pilot training (ASUPT). Phoenix, Arizona: Human Resources Laboratory, Flight Training Division, Report LSI-TR-74-2, November, 1974.

Matheny, W. G. Investigation of the performance equivalence method for determining training simulator and training methods requirements. American Institute of Aeronautics and Astronautics 13th Aerospace Sciences Meeting, Pasadena, California, AIAA Paper 75-108, 20-22 January, 1975.

Meshier, C. W. and Butler, G. J. Air combat maneuvering training in a simulator. In *Flight simulation/guidance systems simulation.* Neuilly-Sur-Seine, France: NATO, Advisory Group for Aerospace Research and Development, AGARD Conference Proceedings No. 198, June, 1976.

Monfort, M. Engineering Analysis. In R. J. Wasicko and A. G. Barnes (Eds.) *Simulation.* Neuilly-Sur-Seine, France: NATO, Advisory Group for Aerospace Research and Development, AGARD Conference Proceedings No. 79, January, 1971.

Mudd, S. Assessment of the fidelity of dynamic simulators. *Human Factors*, 1968, 10, 351-358.

Newell, K. M. Knowledge of results and motor learning. In J. Keogh and R. S. Hutton (Eds.) *Exercise and sport sciences reviews.* (Vol. 4) Santa Barbara, California: Journal Publishing Associates, 1977, 195-228.

Orlansky, J. and String, J. Cost-effectiveness of flight simulators for military training. Volume I: Use and effectiveness of flight simulators. Arlington, Virginia: Institute for Defense Analyses, Science and Technology Division, IDA Paper P1275, August, 1977.

Osgood, C. E. The similarity paradox in human learning: A resolution. *Psychological Review*, 1949, 56, 132-143.

Povenmire, H. K. and Roscoe, S. N. An evaluation of ground-based flight trainers in routine primary flight training. *Human Factors*, 1971, 13, 109-116.

Povenmire, H. K., Russell, P. D., and Schmidt, D. Conservation of people, planes, and petroleum through optimized helicopter simulation. Orlando, Florida: Naval Training Equipment Center, *Proceedings of the 10th/NTEC/Industry Conference*, 1977.

Prophet, W. W. and Boyd, H. A. Device-task fidelity and transfer of training: Aircraft cockpit procedures training. Alexandria, Virginia: Human Resources Research Organization, Technical Report 70-10, July, 1970.

Rockway, M. R. The effect of variations in control-display during training on transfer to a "high" ratio. Dayton, Ohio: Wright Air Development Center, WADC Technical Report 55-366, October, 1955.

Rockway, M. R., Eckstrand, G. A., and Morgan, R. L. The effect of variations in control-display ratio during training on transfer to a low ratio. Dayton, Ohio: Wright Air Development Center, WADC Technical Report 56-10, October, 1956.

Shinners, S. M. *Techniques of system engineering.* New York: McGraw-Hill, 1967.

Skully, R. P. Aircraft simulator evaluation and approval. Washington, D.C.: Department of Transportation, Federal Aviation Administration, Advisory Circular AC 121-14A, 9 February, 1976.

Williams, A. C., Jr. and Flexman, R. E. Evaluation of the School Link as an aid in primary flight instruction. Savoy, Illinois: University of Illinois Institute of Aviation, Aeronautical Bulletin No. 5, 1949. (a)

Williams, A. C., Jr. and Flexman, R. E. An evaluation of the Link SNJ Operational Trainer as an aid in contact flight training. Port Washington, New York: U.S. Naval Special Devices Center, Technical Report SDC 71-16-3, July, 1949. (b)

An Application of a Multi-Attribute Utilities Model to Training Analysis

ROBERT W. SWEZEY[1], *Science Applications, Inc., Reston, Virginia*

Edwards, Guttentag, and Snapper's, (1975) multi-attribute utilities model is applied to a decision making problem in a military training analysis situation. Multi-attribute utilities is a Bayesian-oriented decision making paradigm. Results of the application are presented and the model is compared to a second simple judgment analysis model using the same input data. Aspects of the multi-attribute utilities approach are discussed.

INTRODUCTION

One purpose of many training analytic studies is to facilitate intelligent decision making among alternative training options. Thus, models which facilitate the process of making these decisions can provide major benefit as inclusions in training-analytic methodologies. One such technique, which may be readily adapted to training evaluation research efforts, is termed the multi-attribute utilities model (Edwards, Guttentag, and Snapper, 1975). This model utilizes the general evaluation logic known as Bayesian inference.

Seldom, in traditional training analysis studies, are the effects of prior knowledge or opinion about the question at issue considered in the evaluation of data. Such information is often regarded as a member of a different domain of discourse than is information derived from inferences drawn directly from a sample. Traditional evaluation logic often tends to treat each sample as though it was the first of its sort ever considered (Hays,

1973). The multi-attribute utilities approach, by contrast, is an alternative method which is viewed as being useful to persons concerned with practical decision-making in training analysis situations. The multi-attribute utilities technique is essentially a descriptive one, which gathers and reports information as available. As additional data come in, they are processed according to the same reporting rules as previously, and existing decisions may be revised in light of the new data. The model can be exercised indefinitely as new data are acquired until a stable, asymptotic condition is reached.

It is not the purpose of this paper to describe Bayesian logic or to detail its axioms. For such a discussion see, for example, Edwards, Lindman, and Savage (1963), Hays (1973), and Schum (1970). Multi-attribute utilities analysis is a technique developed for use in applied contexts which incorporates a flexible logic and encourages quantitative combinations of evidence from various sources as well as varying lines of inquiry and investigative techniques. The multi-attribute utility technology has as an essential quality the proposition that outcomes to be evaluated are located on various dimensions of value. It is

[1] Requests for reprints should be sent to Dr. Robert W. Swezey, 11316 Links Court, Reston, Virginia 22090, U.S.A.

presumed that the outcomes may, therefore, be assessed according to a variety of techniques (which may include experimentation, judgment, or naturalistic observation). The located measures are then aggregated, using a weighted linear combination, in a fashion which weights the importance of each dimension of value, relative to all others.

The present paper has as its purpose a brief description of the multi-attribute utilities technique as well as a demonstration of its use in an applied context involving decisions about gunnery training alternatives. The technique is then briefly compared to a simple ranking of training alternatives. It is important to point out that the multi-attribute utilities method is still in an exploratory stage of development and is, in its entirety, a considerably more complex approach than the abbreviated version presented here. The version discussed here, however, is one which has been specifically designed for applied situations and which has been proceduralized by the developers for practical uses.

METHOD

Edwards, et al. (1975), have put forth a ten step, proceduralized methodology for applying the multi-attribute utilities technique. This methodology is described in an abbreviated fashion (modified from Edwards, et al. 1975). For further information, see the original article.

Step 1: Identify the person or organization whose utilities are to be maximized. This person or organization may, for example, be a training directorate or department or, for that matter, an individual job incumbent.

Step 2: Identify the issue or issues (i.e., decisions) to which the utilities needed are relevant. That is (for example): "What variables should be included in defining a multivariable training criterion?"

Step 3: Identify the entities to be evaluated. Entities are defined as the outcomes of relevant actions. Often, entities are operationally defined as a simple description of the action itself.

Step 4: Identify relevant dimensions of value. This involves discovery of what dimensions of value are important to the evaluation of the entities of interest.

Step 5: Rank the dimensions in order of importance.

Step 6: Rate the dimensions in importance, preserving ratios. This is accomplished by assigning the least important dimension an importance rating of 10, and then considering the next least important dimension in terms of how many times more important (if any) it is than the least important. This dimension is then assigned a number which reflects the ratio. One then continues up the list, checking each set of implications as each new judgment is made. Thus, if one dimension is assigned a weight of 20 while another is assigned a weight of 80, the assumption is that the dimension weighted 20 is 1/4 as important as the dimension weighted 80.

Step 7: Sum the importance weights, divide each by the sum, and multiply by 100.

Step 8: Measure the location of each entity being evaluated on each dimension by using a linear 0-100 scale in which 0 is the minimum plausible and 100 is the maximum plausible value on each dimension. (According to Edwards, et al., 1975, following the first eight rules allows for designation of Step 6 judgments in a fashion such that when the output of Step 7, or of Step 6 which differs only by a linear transformation, is multiplied by the output of Step 8, equal numerical distances between the products on different dimensions presumably correspond to equal changes in desirability.)

Step 9: Calculate utilities for the entities. The appropriate equation is: $U_i = \sum w_j u_{ij}$; where U_i is the aggregate utility for the ith entity, w_j is the normalized importance weight of the jth dimension of value, u_{ij} is the rescaled position of the ith entity in the jth dimension, and k is the number of judges. Thus w_j is the output of Step 7 and u_{ij} is the output of Step 8. It is necessary that $\sum w_j = 100$.

Step 10: Decide. If a single act is to be chosen, the rule is: maximize U_i. If a subset of i is to be chosen, then the subset for which $_iU_i$ is maximum is suggested.

Application of the Multi-Attribute Utilities Method

In the present application of the multi-attribute utilities method, the purpose was to determine what variables might be considered for inclusion in the design of improved gunnery ranges for a military anti-armor training system (Swezey, Chitwood, Easley, and Waite, 1977). Actual gunnery training ranges were felt to have only marginal relevance to existing combat situations. Typically, existing gunnery training on the system involved the use of a training simulator which allowed trainees to track movements of a target vehicle equipped with an infrared

TABLE 1

Variables Identified as Relevant for Possible Manipulation in Gunnery Training

1. Target Range	—Distance from gunner to threat
2. Target Speed	—Speed of movement of target
3. Target Number	—Number of possible targets from which the gunner must choose
4. Target Direction	—Direction of target movement
5. Terrain	—Variations in terrain features
6. Weather	—Variations in weather conditions
7. Target Angle	—Angle of movement of the target relative to the position of the gunner
8. Intervisibility	—Directness of line of sight from the gunner to the target
9. Position	—Position of the target relative to the sun
10. Illumination	—Illumination of the target
11. Stress	—Amount of psychological stress placed upon the individual gunner
12. Fatigue	—Degree of gunner fatigue
13. Noise	—Ambient noise level
14. Combat Support	—Extent to which supportive fire is directed at the same target
15. Vulnerability	—Extent to which gunner position is perceived to be susceptible to detection
16. Suppressive Fire	—Extent to which gunner is intimidated by incoming fire
17. Mode	—Mount from which gunner is firing
18. Staying Power	—Number of rounds fired before moving
19. Safety	—Extent to which gunner position is adequately camouflaged and/or concealed

source. The target vehicle moved at a constant speed on a known flat course from left to right across the visual field until it turned and moved from right to left. Typical combat conditions, thus, were either not simulated at all in the training firings or were simulated at an extremely low level of situational fidelity.

A review of existing gunnery literature on the system indicated that a large number of conditions may affect gunnery proficiency. These include such categories of variables as target characteristics, situational characteristics, individual gunner characteristics, terrain configurations, and others. In the context of designing revised gunnery training scenarios, however, such variables must be limited to those which are manipulable by training designers. Individual difference variables within the population of gunnery trainees for instance, are for training design

purposes essentially fixed (i.e., not manipulable by the training designers). The appropriate requirement was to include conditions which most adequately simulate combat parameters and which can be manipulated by the training designers.

A literature review identified a number of variables for possible inclusion in revised gunnery training situations (Swezey, et al., 1977). Table 1 shows the variables. The variables in Table 1 were identified as relevant for possible consideration in developing revised training scenarios. Using these conditions as basic entities, a multi-attribute utilities analysis was performed to identify those variables whose utilities were perceived as being most critical.

In order to decide among gunnery training alternatives, the variables shown in Table 1 were incorporated as the entities of interest in the multi-attribute utilities model (Step 3). Two dimensions of value (Step 4) were identified. These were defined as:

Ph₁—Probability of a first round hit

TS—Gunner tracking score (i.e., % tracking time on target which was available from the training simulator).

The multi-attribute utilities model was applied using the 19 entities shown in Table 1 and the two dimensions of value. Four analysts served as the judges.

It was determined that the person whose utilities were to be maximized (Step 1) was the individual gunner. The decision to which the appropriate utilities were considered relevant (Step 2) was identified as: "What should be the situational conditions included in a combat-oriented simulation for individual gunner training?"

As specified in Step 5 of the multi-attribute utilities model, the dimensions of value were ranked in importance. Then, as required by Step 6, the ranked dimensions were rated in importance, preserving the appropriate ratios. Step 7 involved conversion of the weights into probabilistic terms. As only two dimensions of value were treated in this application, a consensus among the four judges was reached concerning these decisions and the consensus estimates employed in the model. In this case, hit probability (Ph₁) was given a weight of 70 and tracking score (TS) a weight of 10. These numbers were then summed (to total 80) and probabilistic conversions calculated by dividing 70 and 10, respectively, by 80; yielding values of 87.5 for

hit probability and 12.5 for tracking score. Table 2 shows the results of Steps 5, 6, and 7.

Steps 8 and 9 in the multi-attribute utilities model require calculation of the basic cell utilities (u_{ij}) using the converted probabilistic weights, and their combination into overall utilities for each entity (U_i). These were computed using the following formula:

$$u_{ij} = 87.5 \, (Ph_1) + 12.5 \, (TS), \tag{1}$$

and

$$U_i = \sum_k w_j \, u_{ij} \tag{2}$$

where

w_j = is the normalized importance weight of the *j*th dimension of value (i.e., the output of Step 7),

u_{ij} = the rescaled position of the *i*th entity on the *j*th dimension (i.e., the output of Step 8), and

k = the number of judges.

Thus, u_{ij} was computed for each judge on each entity using Equation 1 and summed across judges (Equation 2) to obtain the U_i for each entity.

RESULTS AND DISCUSSION

Table 3 shows the cell entries (u_{ij}) and utilities (U_i) by entity.

The final step in the multi-attribute utilities model involved decisions among the entities. The appropriate decision rule is to maximize U_i (i.e., the subjective expected utility) for the situation. In the case of the gunnery training parameter application, the candidate conditions for manipulation in designing revised training situations were ranked in order by U_i. The data are shown in the righthand column of Table 3.

As is apparent, the top ranked entities were

TABLE 2

Dimension Weights

Dimension	Rank	Weight	Probabilistic Conversion
Ph₁	1	70	87.5
TS	2	10	12.5

TABLE 3

Multi-Attribute Utilities Data

Entity	U_{i1}	U_{i2}	U_{i3}	U_{i4}	U_i	Rank
1. Target Range	6750	6125	6750	9000	28625	1
2. Target Speed	7000	4125	8000	7250	26375	2
3. Target Number	3750	2625	4375	2800	13550	13
4. Target Direction	3000	1750	4375	2250	11375	18
5. Terrain	5875	4625	5250	4125	19875	6
6. Weather	7000	3000	2750	4125	16875	9
7. Target Angle	4625	1875	3500	3125	13125	15
8. Intervisibility	8000	6000	4375	3250	21625	4
9. Position	7000	2000	1750	2250	13000	17
10. Illumination	8500	2000	2000	4125	16625	10
11. Stress	8250	1875	2750	8000	20875	5
12. Fatigue	5875	3875	1875	6125	17750	8
13. Noise	3000	0	1875	2250	7125	19
14. Combat Support	6125	2625	2625	1750	13125	15
15. Vulnerability	7000	3500	4375	3125	18000	7
16. Suppressive Fire	7875	3500	4375	7875	23625	3
17. Mode	6000	2750	1875	4125	14750	12
18. Staying Power	7000	3500	2625	1750	14875	11
19. Safety	7000	2625	2625	875	13125	15

target range (1) and target speed (2). Manipulations of these conditions were, therefore, considered to be most critical in the design of gunnery training situations. Of note is the fact that these two conditions were already varied in the existing gunnery qualification requirements. Of interest, however, is that only these two gunnery variables were manipulated in the current training. Variables 3 (suppressive fire), 4 (intervisibility), 5 (stress), and 6 (terrain), for instance, were *not* systematically manipulated. Based upon such an analysis, recommendations were developed concerning how the six leading variables might be incorporated into the training situation. (See Swezey, et al., 1977.) The rationale for stopping the recommended incorporations at six, was that the seventh ranked variable, vulnerability, was not effectively manipulatable in the design of revised gunnery training ranges. Gunner vulnerability is a variable which was simply not feasible to manipulate in the training situation for reasons of trainee safety.

One might legitimately question the use of this particular multi-attribute utilities procedure, as opposed to other possible analytic techniques. What might happen, for instance, if the four analysts' recommendations were simply ranked in order of their perceived importance to training improvements and the mean employed as the basis for ranking the variables? To explore this question, the individual judges' ranks for most important dimension of value (hit probability) were obtained. These ranks are shown in Table 4.

As can be seen, the rankings changed somewhat primarily among the lower-ranked variables. The rank order correlation coefficient (Spearman's rho) between the two methods was 0.77, which is relatively high. In fact, the two methods agreed on five of the six top ranked attributes (target range, target speed, suppressive fire, stress, and terrain). This comparison of course, is merely *one* comparison of the results of a multi-attribute utilities analysis with another procedure. A variety of other possible methods exist, such

TABLE 4

Entities Ranked For Hit Probability (Ph_1)

Entity	Rating (Rank)				Mean Rank	Rank of Means
	Judge 1	Judge 2	Judge 3	Judge 4		
1. Target Range	70 (10)	60 (1.5)	70 (2)	90 (1.5)	3.75	1
2. Target Speed	70 (10)	40 (6)	80 (1)	70 (4)	5.25	3
3. Target Number	40 (17)	30 (11)	50 (6)	10 (18.5)	13.125	15
4. Target Direction	30 (18.5)	20 (15)	50 (6)	20 (15)	13.625	15
5. Terrain	60 (14)	50 (3)	60 (3)	40 (7.5)	6.875	4
6. Weather	70 (10)	30 (11)	30 (12)	40 (7.5)	10.125	9
7. Target Angle	50 (16)	20 (15)	40 (9)	30 (11)	12.75	12
8. Intervisibility	80 (5.5)	60 (1.5)	50 (6)	30 (11)	14.875	17.5
9. Position	70 (10)	10 (17.5)	20 (17)	20 (15)	14.875	17.5
10. Illumination	90 (2)	10 (17.5)	20 (17)	40 (7.5)	11	10
11. Stress	90 (2)	20 (15)	30 (12)	80 (3)	8	6
12. Fatigue	60 (4)	40 (6)	20 (17)	60 (5)	10.5	8
13. Noise	30 (18.5)	0 (19)	20 (17)	20 (15)	17.375	19
14. Combat Support	70 (10)	30 (11)	30 (12)	20 (15)	12	13
15. Vulnerability	80 (5.5)	40 (6)	50 (6)	30 (11)	7.125	5
16. Suppressive Fire	90 (2)	40 (6)	50 (6)	90 (1.5)	3.875	2
17. Mode	60 (14)	30 (11)	20 (17)	40 (7.5)	12.375	14
18. Staying Power	80 (5.5)	40 (6)	30 (12)	20 (15)	9.625	7
19. Safety	80 (5.5)	30 (11)	30 (12)	10 (18.5)	11.75	11

as, for example the Delphi method (Helmer, 1966) or the judgment analysis network (Christal, 1968).

One major aspect of the multi-attribute utilities procedure is its capability to aggregate judgments over multiple dimensions of value (in this application, hit probability and tracking score) in a probabilistically weighted fashion. A second is its iterative capability. Should additional data on the questions of interest explored in this situation be obtained, for example, the present data may be employed as the relevant prior probabilities, and the new data used to modify them, according to the specified procedures.

One frequent criticism of Bayesian inference is that it is essentially subjective. It can be argued that the inclusion of prior opinion about a phenomenon in the assessment of existing data is not appropriate, because a given parameter is constant in a given situation and cannot legitimately be treated as a random variable. On the other hand, the argument can be made that the prior opinion of expert judges is an appropriate datum to consider. Further, this prior opinion may well be (and often is) based upon empirical data. Most experts have access to a wealth of empirical data which is generally considered to be relevant. This information should not be summarily excluded from a reasonable analysis. Certainly in many training analytic situations when controlled experimentation over the range of relevant parameters is not possible for practical or for cost reasons, alternative paradigms such as the multi-attribute utilities approach might reasonably be considered.

ACKNOWLEDGMENTS

This paper was originally presented at the 39th Military Operations Research Society Symposium, United States Naval Academy, Anapolis, Maryland, June, 1977. It is based upon work supported under contract No. DAAG39-77-C-0044 for the United States Army Infantry School, Fort Benning, Georgia, and conducted while the author was associated with the Mellonics Systems Development Division of Litton Systems, Inc. The author has profited greatly from the comments of T. E. Chitwood, D. L. Easley, B. J.

Waite, J. R. Chiorini, of Litton Mellonics; of J. R. Cartner, of the U.S. Army Research Institute; and of Robert H. Sullivan, of Planning Research Corporation.

REFERENCES

Christal, R. JAN: A technique for analyzing group judgment. *The Journal of Experimental Education*, 1968, *36*(4), 24-27.

Edwards, W., Guttentag, M. and Snapper, K. A decision-theoretic approach to evaluation research. In E. Streuning, and M. Guttentag (Eds.) *Handbook of evaluation research* (vol. 1), Beverly Hills, California: Sage, 1975.

Edwards, W., Lindman, H., and Savage, L. Bayesian statistical inference for psychological research. *Psychological Review, 1963, 70,* 193-242.

Hays, W. L. *Statistics for the social sciences* (2nd ed.), New York: Holt, Rinehart and Winston, 1973.

Helmer, O. *A use of simulation for the study of future values* Santa Monica, California: The RAND Corp., 1966. (AD 640 296)

Schum, D. A. Behavioral decision theory and man-machine systems. In K. B. De Greene (Ed.) *Systems psychology.* New York: McGraw-Hill, 1970.

Swezey, R. W., Chitwood, T. E., Easley, D. L., and Waite, B. J. *Implications for TOW gunnery training development* Springfield, Virginia: Mellonics Systems, Development Division, Litton Systems, Inc., 1977.

Some Considerations in the Design and Implementation of a Training Device Performance Assessment Capability

John K. Hawley, U.S. Army Research Institute, Ft. Bliss, Texas

In an era of increasing system complexity and resource constraints, live-fire exercises on a scale necessary to insure personnel proficiency are not feasible for many weapons systems. A partial solution to the problem of insufficient performance evaluation involves the use of training devices in lieu of operational equipment in the conduct of evaluation exercises. A trainer that has been extended to include an embedded performance evaluation capability is referred to as a training device performance assessment capability (DPAC). The present paper presents a series of guidelines for the design and development of DPACs. The guidelines were prepared on the basis of background research and actual experience in developing a DPAC for the Patriot air defense missile system. Because of their association with an actual developmental effort, the guidelines have considerable practical utility for DPAC design and implementation. A number of issues important for the full realization of DPAC potential also are discussed.

INTRODUCTION

One of the artifacts of increasing sophistication on the part of the weapons systems developed and fielded by the United States is that live-fire exercises on a scale necessary to assess and maintain combat readiness often are not feasible. This problem stems from factors such as the cost and limited availability of ordnance, limited range availability, range safety considerations, and cost of logistics support items, to name several. Military commanders are thus often faced with the dilemma of knowing that if war comes their units must be able to fight immediately but of not having the training and evaluation resources necessary to provide a high expectation of success under such an eventuality.

A partial solution to the problem of limited opportunity for training and proficiency evaluation in such cases involves the use of training devices in lieu of operational equipment. In addition to their intended training applications, training devices often can provide a vehicle for individual and collective performance assessment (Crawford & Brock, 1977). Histor-

ically, the most extensive uses of training devices in performance assessment have been in the aviation community. The commercial airlines and the Federal Aviation Administration routinely use flight simulators in aircrew performance certification. Follow-up studies have indicated that performance in flight simulators is predictive of performance in actual aircraft (Weitzman, Fineberg, Gade, & Compton, 1979). In the military, the uses of training devices in performance assessment generally have mirrored civilian applications and involved aviation. There has been, however, an increasing use of training devices to assess individual and collective performance in other areas, such as antisubmarine and air defense operations (Bell & Pickering, 1979; Hawley & Dawdy, 1981).

The evaluation of aircrew members in a flight simulator or air defense missile battery personnel using an engagement simulator illustrates the concept of a training device performance assessment capability, or DPAC. The term DPAC simply means that a proficiency assessment capability is embedded within a system's training devices. Once included in the training devices, the DPAC is used to assess

Reprinted from Proceedings of the Human Factors and Ergonomics Society 28th Annual Meeting (pp. 201–205).

the job proficiency of the individuals that operate or maintain the hardware system.

A recent review of training device proficiency assessment potential indicates that the DPAC concept can be applied to the training devices for virtually any hardware system (Shelnutt, Smillie, & Bercos, 1978). Currently, actual use of training devices as DPACs is not widespread. but the potential remains. Furthermore, the potential for using training devices as DPACs will increase as more and more devices are based upon an embedded information processing capability. An embedded information processing capability greatly enhances a training device's potential for use as a DPAC (Hawley & Dawdy, 1981).

In a limited sense, DPAC capabilities are implicit in nearly any training device. Practically speaking, however, DPAC implementation may require a device's modification or extension to provide performance status information that is immediate and useful. The potential impact of such training device extensions and modifications provided the rationale for the development of the present paper. In this regard, the objective of the paper is to outline and discuss a number of practical considerations for the design and implementation of an effective DPAC. These considerations are based upon lessons learned during an effort to develop an operational DPAC for the Patriot air defense missile system using the Patriot Tactical Operations Simulator (PTOS) as a host training device (Hawley, Brett, & Chapman, 1983). The guidelines discussed in the paper were distilled from an extended experience with the Patriot system, but the lessons have broad generic applicability. Without some general developmental guidelines of the type presented, implementing a DPAC, particularly after a system is fielded, can be a lengthy and costly undertaking which in many cases will provide only marginal returns to a proponent in terms of the utility of the performance status information provided.

DPAC DEVELOPMENTAL CONSIDERATIONS

As noted in the introductory section, the objective of the paper is to present a series of

guidelines for DPAC development. To facilitate meeting this objective, the guidelines are presented in terms of a series of steps that, if followed, will significantly increase the likelihood of a successful developmental effort. Each of the steps is now addressed in turn.

Structured, Top-Down Planning

The first activity essential to successful DPAC development is structured, top-down design. A structured, top-down design approach is one in which the first action is the definition of what are termed system missions, which are derivatives of system objectives. The system missions may, in turn, consist of a number of sub-missions. All missions and sub-missions are identified. Then, since a total system usually consists of a number of distinct subsystems, or elements, all relevant subsystems are identified. The next action in the process is to characterize how the various elements serve suprasystem objectives. Once these relationships are identified and characterized, the final step in the planning process involves analyzing each element mission and sub-mission into its component functions, tasks, and task elements. At the task and task element level, the operator activities undertaken in support of the element missions and sub-missions may be described in terms of operational sequence diagrams.

At the outset of the Patriot effort, the necessity for an initial top-down approach to DPAC development was not obvious. A review of the literature pertinent to human-machine performance assessment indicated, however, that a number of attempts to develop facilities similar to that under consideration for Patriot had been made. Moreover, these previous attempts almost universally failed to live up to the expectations of their designers. An issue of obvious concern for the Patriot effort thus concerned the reasons for this apparent lack of success. A further review of previous efforts indicated that the primary problem stemmed from the fact that the performance data collected were not meaningful to decision makers. Furthermore, the apparent lack of utility to decision makers was attributable to the fact that the performance measures provided were dictated by opportunity (i.e., they were available and seemed reasonable to collect) and not through an in-depth consideration of system objectives. In other words, previous DPACs typically were designed following what is termed in the systems literature a "bottom up," or at best a "middle out," approach. Beginning with various performance measures of opportunity and apparent utility, an elaborate performance measurement scheme is developed. However, when the resulting performance structure is exercised to obtain performance data, there is an unclear relationship between the various performance indices and the system's objectives. Decision makers are left in the position of asking, "What do all of these data mean?"

Given that a bottom-up approach to DPAC design was not likely to prove useful, an obvious next question was, "What approach to operator performance assessment in human-machine systems is likely to prove useful?" In addressing this issue, it was necessary to consider two predominant characteristics of human-machine systems:

1. Human-machine systems are goal oriented.
2. Human-machine systems typically are hierarchically organized.

From these characteristics, two rules for performance assessment in human-machine systems follow:

1. To assess operator performance in human-machine systems, it is necessary to consider the total system's goal structure.
2. A human operator performance assessment scheme must be sensitive to the hierarchical structure of the total system.

The first rule merely states that the goal structure of the total system is the proper context for the development of individual and collective operator performance assessment models.[1] Operators are effective when their actions result in total system objectives being met; they are not effective when their actions result in system goals not being attained.

---

[1] A performance assessment model is defined as a procedure for characterizing performance based upon a priori conditions. A measure of performance is some physically measurable quantity that allows one to determine a specific level of performance through exercise of the model.

Rule two states that the hierarchical structure of the total system should be reflected in the performance assessment scheme. Another way of stating this rule is that the network of performance models developed within the context of a hierarchical human-machine system also should be hierarchically organized.

Although the two rules noted above may appear to a sophisticated reader to represent a statement of the obvious, most of the performance assessment schemes described in the literature appear to have been developed without regard for them. Also, going beyond the obvious, the implications of the two rules for performance assessment in human-machine systems are significant. The primary implication is that a bottom-up approach to the development of a DPAC often leads to a situation in which it is difficult to relate operator performance data to system objectives. The background research undertaken as part of the Patriot effort indicates that in order meaningfully to relate operator performance data to system objectives, a structured, top-down approach to DPAC design is required.

Screening for Utility

The second step in DPAC development concerns screening the performance structure for utility. As noted previously, the DPAC concept has potential for application in virtually any training device. An important issue, however, concerns the circumstances under which a capability should be developed. It is necessary to determine the conditions under which the cost of developing and operating a capability are offset by the value of the performance data received. Performance status information is expensive to collect. Hence it is important that only those data having actual utility for decision makers be obtained. Careful consideration also must be given to the problem of providing too much data. In many cases, decision makers can be overwhelmed with data and left with the additional task of determining which data are important and which are not. A preliminary version of the Patriot DPAC was, for example, capable of generating a virtual mountain of performance-related data, far beyond the capabilities of any operator or evaluator to interpret.

As part of a parallel effort (Hawley, Brett, & Chapman, 1982), a trade-off procedure directed at the issue of data collection costs versus information utility was developed. This trade-off procedure, termed cost and information effectiveness analysis (CIEA), is based upon a structured multiattribute utility measurement process. Application of the CIEA process will lead a design team to a decision concerning the overall utility of a DPAC in a particular situation, as well as to a prioritization of the elements of the performance structure. The priorities, or utilities, assigned to the various elements of the performance structure provide an indication of those elements that can be eliminated immediately and those that can be discarded later if the capability must be de-graded gracefully to stay within cost or other resource constraints. At a minimum, the CIEA process forces DPAC designers to think through the issues of what performance data actually are required and how the various elements are to be used. Satisfactory resolution of the information utility issue is a key aspect in the development of a successful DPAC.

Definition of Performance Assessment Models

The third step in DPAC design concerns developing performance assessment models. As noted previously, a performance assessment model is a rule for combining observable performance measures to provide an indication of operator performance for a particular mission area. The labels attached to each of the elements in the performance structure resulting from the mission analysis define in general terms what is to be assessed using the DPAC. Prior to defining an operational performance assessment scheme, however, it is necessary to state explicitly and exhaustively what is implied by each of the performance statements. This action represents an attempt to circumscribe the various performance domains of interest, thus defining what are termed performance constructs. In the case of Patriot, for example, one element sub-mission is titled, "Provide defense of sector assets." The performance construct corresponding to this sub-mission defines operator performance as the "protection of defended assets against physical penetration by non-friendly aircraft." Note that three aspects of operator performance are indicated: defended assets, physical penetration, and non-friendly aircraft. Furthermore, since defended assets are assigned priorities, asset values represent a fourth aspect of operator performance. These aspects provide a link between the performance constructs and observable performance measures within the human-machine environment.

Once satisfactory performance constructs are identified, the next action in specifying performance models concerns operationally defining each construct in terms of observables within the human-machine environment (e.g., cues, responses, reaction times, etc.). In most situations, this action will simply require a review of the design specifications for the host training device to determine where required performance measures are to be obtained. Other situations will be encountered, however, in which available performance measures do not match well with the concept underlying the performance construct. In such cases, it may not be possible to define a satisfactory performance construct in terms of measures from the performance environment. The performance assessment scheme for Patriot again provides a case in point. One of the element sub-missions is titled "Provide defense for deployed friendly forces." The corresponding performance construct is termed Airspace Defense (AD) and is defined as "timely processing of non-friendly aircraft as they enter a fire unit's sector of responsibility." Operationally defining what is implied by the term "timely processing" presented some difficulties, however. Fully treating what is implied by timely processing would have resulted in a performance model that was too complex to be of much value. Hence, a compromise was reached. Airspace Defense was defined as a function of the elapsed times that non-friendly tracks are eligible for engagement but not engaged. This definition is based upon a performance element common to all aspects of the term timely processing. Conceptually speaking, AD is not a completely satisfactory performance construct, but the definition does represent an implementable solution.

The final action in defining performance models concerns the choice of a combination rule. A combination rule, in this usage, is an expression defining how performance measures are to be combined to form an index of operator performance. In many cases, the choice of a combination rule is straightforward: The performance measure is the reported performance index. In other situations, a more complex combination rule might be desired. In the case of Patriot, for example, AD is defined in terms of the number of physical asset penetrations by non-friendly aircraft weighted by the value of the penetrated assets. The complete statement of the performance model for AD is thus given as:

$$DA = [1 - (\Sigma\Sigma AV_j d_{ij}/\Sigma\Sigma AV_j g_{ij})^b] \qquad (1)$$

In expression (1), N is the number of non-friendly tracks scripted for asset penetrations; $AV_j = (11 - ATC_j)$, where ATC is the asset-threat category (i.e., priority) assigned to asset j; $d_{ij} = (0,1)$ depending upon whether asset j is penetrated by track i; $g_{ij} = (0,1$ as a function of whether asset j is scripted for penetration by track i; and b is a constant added for generality in the shape of the operator performance penalty function.

Impact on Host Training Device

Prior to proceeding with DPAC development, it is essential that the impact of the proposed capability on the host training device be considered. This step is particularly important if the DPAC will be dependent upon an information processing capability embedded within the host device. If the host device is already fielded, the impact of DPAC-related software modifications can be significant. In the Patriot project, for example, the PTOS software modifications necessary for DPAC implementation resulted in timing problems which seriously affected the fidelity of the engagement simulation. A solution to the time fidelity problem was developed, but the cost of identifying and implementing the "fix" exceeded the cost of the original software modifications. In all likelihood, the problems associated with implementing the timing fix would have been less troublesome if more attention had been paid to potential implementation problems at the outset of the project. Retrofitting a DPAC into an existing training device is a difficult undertaking at best, but prior planning can spell the difference between success and failure.

In the case of developing a DPAC for a new device, it is essential that the capability's requirements be addressed along with other design features. Failure to consider DPAC requirements can result in many of the same problems encountered when retrofitting a capability into an existing device. An additional problem that can arise when DPAC requirements are not considered in advance is the fielding of an inadequate capability. Examples can be cited in which considerable expense was incurred to develop a DPAC only to find after fielding that information processing limitations on the part of the host device seriously restricted the realism of the scenarios that could be presented to operators. In other cases, design limitations have precluded exercises intended to exercise inter-unit command and control functions. Addressing DPAC requirements during training device planning should alleviate problems such as these and result in the deployment of a more useful operational capability.

Application Procedures

The final set of issues relevant to DPAC development concern how the capability will be used. Experience and research have shown that the two primary applications for a DPAC will be (1) readiness assessment/performance certification and (2) training management. Using the capability for readiness assessment or performance certification will require the development of formal procedures for performance evaluation and reporting. Methods for performance evaluation also must treat the issues of standards development and scenario construction. It will be necessary to differentiate reliably between acceptable and unacceptable performance. In addition, a ready supply of appropriate evaluation scenarios must be provided. Furthermore, to be useful in performance evaluation, scenarios must be organized in terms of their content and operator loading characteristics. Unfortunately, little objective guidance on proper methods of scenario construction currently is available; scenario design remains an extremely subjective process.

The second major application of current DPACs is training management. Developing an efficient and reliable training management capability to use in conjunction with a DPAC will require a means for analyzing operator performance protocols to isolate performance deficiencies. At the present time, such diagnostic analyses are performed by highly skilled evaluators, but the review process is tedious and unreliable across analysts. Two analysts reviewing the same performance trace may not identify the same performance deficiencies. Obviously, such procedures are neither efficient nor reliable.

An obvious solution to the problems of tedium and interrater reliability is to automate the protocol evaluation process. Automating the analysis process may not be straightforward, however. This statement is based upon the observation that in many human-machine systems the flow of operator responses is not completely determined. Operators often have available a number of approaches for addressing a problem situation. Some approaches may be more appropriate than others at specific times, but it is often difficult to determine whether a true error has been committed. In many cases, the consequences of an operator action cannot be assessed until later in the exercise, and even then it may not be possible to do so in an unambiguous fashion. The time lines for the performance of various operator actions often are not fixed either. Operators have available "windows" within which actions may be taken. The presentation of a system cue is a stimulus for action, but the response oftentimes need not be taken immediately. Hence, the development of absolute time standards for operator action sequences may not be appropriate.

The performance situation described above represents a dynamic process in which the development of an algorithmic protocol analysis procedure may not prove feasible or even desirable. In fact, a rigid, algorithmic diagnostic process could lead to a situation in which operators are shaped into stylized and potentially maladaptive performance patterns as a consequence of experiencing the training/evaluation program. What is required to analyze operator performance data for diagnostic purposes is a procedure capable of evaluating the situation at hand and making an "intelligent" assessment of the appropriateness of operator actions, much like the analyses currently performed by expert human evaluators. Realizing such a capability will, however, likely require a foray into the emerging field of artificial intelligence.

Realizing an effective performance diagnosis capability also will require a closer link between the scenario design and performance evaluation processes than currently exists. To realize an effective diagnostic scheme, scenarios will have to be viewed as a medium within which to embed evaluation "vignettes," much in the same fashion that individual test items are embedded within the medium of a psychological test. The vignettes must be designed and placed in the scenario action stream in order to elicit specific classes of operator behavior. The evaluation mechanism then will review operator responses across a series of vignettes and make a determination regarding specific performance deficiencies. Performance diagnostic procedures that function in such a fashion have been developed for use in a paper-and-pencil mode, and some have even been computer-aided (i.e., adaptive testing), but the procedures have not been extended to a dynamic, work sample environment like that provided by a DPAC.

DISCUSSION

The intent of the present paper is to present and discuss a series of considerations, or guidelines, for the development of a training device based performance assessment capability. The guidelines were derived from a detailed post-mortem evaluation of an effort to develop a DPAC for the Patriot air defense missile system. Given their association with an actual developmental effort, the guidelines should have considerable practical importance to anyone considering a similar undertaking. Careful attention to the points noted in the body of the text should make future DPAC developmental efforts less error-prone, thus increasing the likelihood of a successful product.

Based upon a review of the Patriot project and related efforts, it is safe to assert that

DPACs can have considerable practical utility in training and evaluation. Moreover, given the cost of operating many new systems, the DPAC concept will likely be considered for application in more and more situations. However, the full potential of the DPAC concept will be realized only if future capabilities are carefully planned. Poorly planned developmental efforts hold considerable potential for trouble – a lengthy and painful developmental process and/or operational capabilities having limited practical value. The lessons of the Patriot effort are clear: Think through the requirements and impact of the capability fully before committing to development.

REFERENCES

Bell, J.D. , & Pickering, E.J. *Use of performance measurement data from the 14A2 trainer complex in a performance assessment system.* San Diego: U.S. Navy Personnel Research and Development Center, 1979.

Crawford, A., & Brock, J. Using simulators for performance measurement. In *Proceedings of the Symposium on Productivity Enhancement: Personnel Assessment in Navy Systems.* San Diego: U.S. Navy Personnel Research and Development Center, 1977.

Hawley, J.K., Brett, B.E. , & Chapman, W.A. *Cost and information effectiveness analysis: An improved methodology (Vol. 1).* Alexandria, VA: U.S. Army Research Institute for the Behavioral and Social Sciences, 1982.

Hawley, J.K., Brett, B.E., & Chapman, W.A. *Optimizing operator performance on advanced training simulators: Volume IV – Final Technical Report.* Alexandria, VA: U.S. Army Research Institute for the Behavioral and Social Sciences, 1983.

Hawley, J.K., & Dawdy, E.D. *Training device operational readiness assessment capability (DORAC): Feasibility and utility.* Alexandria, VA: U.S. Army Research Institute for the Behavioral and Social Sciences, 1981.

Shelnutt, J.B., Smillie, R.J., & Bercos, J. *A consideration of Army training device proficiency assessment capabilities.* Alexandria, VA: U.S. Army Research Institute for the Behavioral and Social Sciences, 1978.

Weitzman, D.O., Fineberg, M.L., Gade, P.A., & Compton, G.L. Proficiency maintenance and assessment in an instrument flight simulator. *Human Factors,* 1979, *21*(6) , 701–710.

Human-System Performance Measurement in Training Simulators

DONALD VREULS[1] *and* RICHARD W. OBERMAYER, *Vreuls Research Corporation, Thousand Oaks, California*

Automated human-system performance measurement subsystems are being specified as a requirement in modern training simulators. Although hardware and software technology can support this requirement, there are many unanswered questions about the design of real-time automated measurement systems. Fundamental performance measurement problems and research issues are discussed.

INTRODUCTION

Automated human-system performance measurement subsystems are being specified as a requirement in modern training simulators. Unfortunately, most existing automated training performance measurement systems are so poorly designed that they are useless (Semple, Cotton, and Sullivan, 1981). There are many reasons for this condition, but perhaps the principal reasons are that we have oversimplified the designs and we lack sufficient design methods. In this paper, we shall examine some of the underlying problems.

Human-system performance is measured in training to acquire information for many purposes, such as (1) feedback to the instructor and student, (2) performance diagnosis, (3) student performance assessment and grading, (4) training system evaluation, quality control, and management, and (5)

prediction of job performance. If measurement information for the first three purposes is to be useful for training, it must be provided in real or near-real time, during or immediately after the completion of a training session. Also, the information must be presented in a form that is understood by the user in the context of training.

Automated measurement system technology emerged from measurement methods for research; that origin explains some of the difficulties. Often in research, the tasks and environments are constrained and tightly scripted. Frequently measurement is done by: (1) comprehensively recording everything that seems reasonable, (2) cleaning up the data by removing unwanted segments after examination (sometimes weeks after data collection), (3) processing it to convert samples of raw data into many measures, such as average error from the desired profiles, speeds, or other task performance measures, (4) performing a variety of statistical analyses, and (5) attending to those measures that show differences in experimental conditions (and ignoring the rest).

[1] Requests for reprints should be sent to Donald Vreuls, Vreuls Research Corporation, 68 Long Court, #E, Thousand Oaks, CA 91360.

All of the judgments that enter into the determination of when tasks start and end, the cleanup of the raw data, and the selection of the measures of interest (and possible weighting of several measures to develop composite performance scores) have to be part of the algorithms that control a real-time measurement system. In other words, real-time automated performance measurement systems have to be as smart about the evaluation of performance before the fact as a researcher is after months (or even years) of empirical data collection and analysis. Actually, automated performance measurement systems for training have to be even smarter, because they have to judge the quality of performance in operational terms (a grade); researchers seldom know the operational meaning of a performance change in their experiment. It is all of these judgments that constitute issues of *how* to measure, as opposed to *what* to measure.

There are accepted analytical methods to determine what to measure for fundamental tasks that are performed in accordance with scripted scenarios or "school solutions," where the performance standards are known. The typical design follows the Instructional System Development (ISD) model; a measurement analyst interacts with mission specialists and/or subject-matter experts (SMEs) to describe the tasks, performance objectives, and standards. This initial analysis primarily specifies measurement for tasks performed by single individuals where there are a limited number of acceptable solutions.

Task-analytic methods provide initial guidance on what to measure and are a necessary first step in measurement development. These methods, however, do not capture all of the behavior of interest in dynamic person-vehicle or person-process control tasks, monitoring tasks, team interactions, or decision tasks. They do not appear to handle judgmental tasks well. The ISD model fails the

performance measurement analyst when the performance standards cannot be defined quantitatively. In short, our existing analytical tools are not adequate to address all of *what* to measure, and they do not tell us much about *how* to measure.

This paper is intended to highlight the weaknesses in knowledge and methods for performance measurement design in a way that has not been expressed in prior technical reports on the subject. We will not dwell on what has been accomplished or what can be done. The existing literature has been reviewed (Mixon, 1981; Mixon and Moroney, 1982), and a comprehensive study of existing and prototype automated performance measurement systems for aircrew training has been accomplished (Semple, Cotton, and Sullivan, 1981) and need not be repeated here. Finally, to provide a manageable scope for this paper, we shall express our views based on empirical research and automated measurement system design experience and shall not attempt a comprehensive integration of the views of others.

The subsequent discussion is divided into three parts. First, the reader is reminded of four fundamental problems in training-simulator performance measurement; one should keep these general factors in mind when assessing specific measurement issues. Second, there is a discussion of specific areas of weakness in current measurement knowledge that require research; one should recognize that these issues overlap, often to a large degree, and sometimes address only a specific application. Third, a short summary is provided.

FUNDAMENTAL PROBLEMS IN PERFORMANCE MEASUREMENT

There are at least four fundamental problems that pervade all performance measurement issues in simulation: the hidden and embedded nature of performance; the lack of a general theory of human performance; determining validity of performance measures; and establishing criteria for performance. Each of these issues is addressed briefly in the following paragraphs.

Hidden Knowledge and Embedded Performance

In most simulators the purpose of performance measurement is to infer something about the knowledge, skills, or decision processes of the human participant. Performance measures, however, depend on overt actions; internal processes that produce overt actions are not directly observable. Moreover, the results of complex processes, such as decision making, may be manifest only by simple actions or no actions at all. For example, a nuclear power-plant operator may be quite busy assessing the state of the plant, but by concluding that everything is satisfactory, the operator may take no action at all. To make inferences about internal processes based on minor actions or no actions at all requires measurement of system inputs, outputs, and states, as well as a model of the internal processes of the operator.

Operator responses are embedded in the behavior of the system. Human manipulation of a system may have a large range of effects, from negligible to dramatic and from immediate to long term, depending upon the characteristics of the system. Moreover, the simulated environment and scenario affect the observable responses. That is, the features of the simulated environment importantly interact with the processes internal to the human. Thus, any particular performance measure or set of measures is indirect for most purposes, taps only a small sample of the internal processes, and must make inferences about human processes that are confounded by their interaction with the specific characteristics of the simulation. The measures selected often are specific to the task and the situation, and, conversely, are not generalizable.

Lack of Theories of Performance

The second fundamental measurement problem is the lack of unifying theories of human performance to predict actions over a wide range of circumstances. Formal theories and models tend to define what must be measured, the relative importance of each measure, and interactions with other measures under given circumstances. In the absence of theories to guide selection of performance measures, one is driven to the alternative of measuring as much as is reasonably possible. As a result, much time and effort can be wasted collecting, summarizing, and analyzing a large amount of useless data for a given task and environment.

Typically, even if desirable, measurement of all human inputs and system outputs is beyond the capacity of most simulator measurement subsystem designs, especially if the simulation involves interacting human participants, as it does in team training. Consequently, measurement specification is based either on expert opinion or on laborious empirical statistical selection, which could produce a set of measures that may be useful only in the context of the simulation in which they were developed.

Measurement Validity

Indirect measures of human performance require empirical tests to determine their validity for the intended purpose. Measures for simulator training often are derived from analytical studies and selected for use on the basis of perceived suitability; the construct, concurrent, content, and predictive validity of measures are seldom tested.

There have been isolated attempts to validate performance measures for aircrew training (Vreuls and Wooldridge, 1977; Waag and Knoop, 1977) and operations (Brictson, Burger, and Wulfeck, 1973; Lees, Kimball, Hoffman, and Stone, 1976). Typically, these efforts start with analytically derived measures; they then record performance during training or operations, generate many candidate measures from the performance data, and perform empirical data analyses. There are three common ways to validate measures: (1) use the actual performance of a representative sample of experts as criteria against which all other performance is compared, (2) use a representative sample of experts to judge performance quality, then determine which measures correlate with these judgments, and (3) correlate measures with outside criteria, such as success in training, mission performance, or peer ratings.

Although isolated measurement validation studies have been conducted, performance measurement validity is still largely an open question. For example, the validity of performance measures during skill acquisition as predictors of final task performance has been questioned. Kennedy and Bittner (1977) reported that performance on relatively simple psychomotor tasks (video games) does not approach asymptote as early as many training specialists might believe. People continued to improve with time at different rates, and more than a few slow learners eventually performed better than fast learners. If the performance data are taken too early in the learning curve, the accuracy of prediction will suffer. Because there are enormous individual differences in the rate of learning and the shape of the learning curve, the predictability of measures in a training program must be determined empirically.

The predictive validity of performance measures in training has not been established for many tasks. Where it has, there have been reasons to question the validity of commonly used measures for training. For example, one automated training study of four instrument flight tasks (Vreuls, Wooldridge, Obermayer,

Johnson, Norman, and Goldstein, 1974) found that measures that discriminated between trainees and experts were slightly different from those that are commonly used to assess flight performance. The measure set that discriminates between winners and losers in an air combat maneuvering simulation can be rationalized after it is known, but it is not totally defined by a front-end analysis (Kelly, Wooldridge, Hennessy, Vreuls, Barnebey, Cotton, and Reed, 1979; Wooldridge, Kelly, Obermayer, Vreuls, Nelson, and Norman, 1982).

Measurement validation requires substantial empirical data collection and analysis, which is costly and time consuming; as a consequence, it is seldom performed. Although this is understandable, the costs of using invalid measurement and the costs of uncertain generalizability and predictability also must be considered. Clearly, the lack of measurement validation studies is an obstacle to the advancement of the measurement state of the art.

Operational Performance Criteria

The final basic problem in simulator performance measurement is the lack of quantitative criteria for assessing the importance of performance changes. Performance changes can be measured in a simulator during training, but only infrequently can such performance changes be related to overall system or mission effectiveness. For some isolated tasks, there are criteria based on historical precedent and common practice, (as in the case of instrument flight performance limits and nuclear power-plant technical specification limits). Common rules of thumb and criteria are adjusted by instructors and inspectors to suit the environmental and operational complexities of the problem at the time of test.

Instructional System Design (ISD) analysis efforts are hampered by the lack of actual performance data and criteria for many tasks. It is rare that one finds criteria that differentiate between the performance of journeymen and experts on any basis other than the amount of time on the job or exposure to various situations, such as the number of aircraft licensed during the year by an Airframe Inspector, the number of hours flown or night landings performed by a pilot, or the number of plant evolutions performed by a power-plant operator. These metrics are useful to describe experience, but they are not performance criteria. In general, the scale of performance quality in operationally meaningful terms is unknown for many tasks.

PERFORMANCE MEASUREMENT RESEARCH ISSUES

This section discusses specific topics for which research is necessary to improve performance measurement knowledge and practice.

Operational Figures of Merit

One of the more urgent needs is for overall measures of system effectiveness for representative tasks and operating environments. Quantitative criteria for acceptable and unacceptable job performance are largely unknown, except at very general levels. Quantitative data which show how various skills are acquired and maintained would assist the training system design process as well. Until the necessary commitment is made to acquire operational performance criteria, more generally useful measurement and principles for measurement are not likely to emerge.

Automated Performance Measurement

There are many reasons for automating performance measurement. Measurement can be based on a greater number of factors than is possible through direct observation, and precision and reliability can be improved. Data can be collected, summarized and analyzed in a short period of time. Adaptive feedback and automated training can be

implemented. Personnel requirements can be reduced. However, there are many obstacles to implementing acceptable automated performance-measurement systems. One of the most important of these is the difficulty of assessing complex human performance with anything approaching the perspicacity of expert judgment. To do so will require intelligent measurement systems (i.e., large computing resources and an extensive database of knowledge).

An automated measurement system must know what tasks are to be performed. This is relatively easy for fixed profiles or procedures, but can require a great deal of intelligence if the task changes as a function of the simulated environment or if there is no exact profile or procedure, as in emergent environments. The system must know how to recognize the start of a task and the end of it. This is not a trivial issue, because all real-world operator and maintainer tasks do not start or end with easily recognized events, such as alarms sounding or ships contacting piers.

Many tasks are performed in parallel; some are executed only once, and others may continue for long periods of time. Often, there is a flow from one, or many, tasks to the next task or tasks, where the transition may be chosen arbitrarily by the operator or may be due to an unpredictable variety of scenario or environmental events. Human instructors or observers usually can take these changes into account; automated systems can handle task changes only if they are explicitly programmed to do so. This demands a great deal of built-in knowledge and processing capability. Research is needed to develop and validate algorithms that can recognize transitions in tasks, as well as to capture the essential elements of performance for concurrently performed and irregularly sequenced tasks.

The field of artificial intelligence (AI) has borne fruit and now has a number of practical successes. The techniques of AI provide promising methods to include human knowledge and capabilities in the design of automated performance measurement systems. Although AI methods provide a convenient technique for handling complex rules, these methods, programmed in languages such as Lisp and Prolog, consume large amounts of memory and are slow. They do not offer real-time measurement solutions in their native form. Research is needed to determine how to best use these methods to address the performance measurement requirements in training simulators.

Near-Real-Time Measurement

Often, to be useful, performance information must be provided at the time a task is completed, or very soon thereafter. For training, performance information loses a major part of its instructional value if summarized data, in a format that is understandable to the student and instructor, is not available at the time of student debriefing. For research, near-real-time performance data are needed for experiment quality control especially when "holistic" experimental designs are used. These designs do not gather all data at one time. Instead, the designs use a series of data-collection efforts in which the combinations of independent variables for the next set of experimental trials are determined by the results of the previous trials (Simon, 1978). Consequently, quick measurement is needed for rapid experimental design decisions.

The design of real-time or near-real-time measurement systems presents challenging issues. Working with subject-matter experts, the analyst or designer must determine the measurements, the conditions under which the measures will be sampled, the desired value for each measure, the functions that may combine several system states into one measure, weighting coefficients for each measure to derive composite figures of merit.

When multiple scenarios are used, each variation of the task and scenario may require unique measures. Also, developing a simple-to-use interface for the experimenter or instructor are important problems that have not been addressed in any general way. In short, the design of a real-time measurement system is extensive and complex.

Performance Diagnosis

The development of measurement for performance diagnosis is another major challenge. Indices of performance based on composite summary information are not likely to provide diagnostic information for analyzing training problems. After-the-fact examination of the time histories of all system states by expert personnel may lead to diagnosis, but diagnosis performed totally by machine requires a major development effort.

When a major blunder is made, diagnosis is not difficult because its cause is usually apparent. The difficulty increases at a level below the obvious blunder, where minor deviations from expected performance compound into an error. Here, measurement must be able to spot patterns over time, assess probabilities, and determine probable causes. Often, there are limited resources that can be provided for measurement development, so the value of the information must be evaluated against the cost of providing it.

Apart from transfer-of-training studies, there are no general methods to determine the training value of various levels of diagnostic information, but it is unlikely that traditional transfer-of-training methods would be practical for this purpose. The architecture for storage, processing, and retrieval of diagnostic data may be very different from that of the basic simulation. Guidelines for the collection, summarization, and presentation of information for instructional use and diagnosis for real- or near-real-time measurement systems in simulators have never been studied or formally developed.

Operator Control Policies and Strategies

It is common practice to measure error from known, desired conditions, but operators of complex systems do not always try to maintain a minimum amount of error from a desired condition. Frequently, certain performance objectives are more important than others at a given time. For example, an automobile driver may sacrifice passenger comfort momentarily to avoid a collision; a nuclear power-plant operator may exceed the desired rate of change of reactivity in order to prevent a reactor trip. In these cases, a high-level concern may cause a temporary excursion from desired states of the system; under normal operations, such excursions might be outside the limits of good practice, but they are excusable or even highly desired under certain circumstances.

There are control policies, strategies, and performance error trade-offs that may be more subtle than in impending emergency or catastrophic situations. For example, certain actions (or no action at all) may be taken to minimize workload because the shift is long, and the operator may need to preserve energy and alertness for an upcoming situation. An experienced pilot will not take action to return to a low-level course centerline if the aircraft is not outside of the designated corridor and will be close enough to the next checkpoint for visual observation of it. Under these conditions, the experienced operator may take no action at all to null system error that may be annoying, but not critical.

There is little documentation of the control strategies and policies that operators employ in the performance of tasks in the real world. In training, simplified policies are given in the solution recommended by the school, but these policies may not be appropriate as the basis for performance assessment of skilled operators. This issue becomes important as simulators are used to provide recurrent training for experienced operators, or readiness training for military combat teams. Research is needed to derive measurement algorithms that are sensitive to the presence or absence of various control policies and strategies.

Measurement of Expert Performance

Closely related to the foregoing is the issue of deliberately training people to become masters of their craft (e.g., well beyond journeyman status). We tend to train to a minimum acceptable level of operational performance for practical reasons, leaving the further development of skill to on-the-job learning. For the military, however, the inherent danger of this philosophy has been recognized, and there is increased emphasis on operational exercises to develop combat skills, and the use of simulators for this purpose.

Little is known about the differences between acceptable performance and the performance of experts. Most current performance objectives and criteria have focused on initial training. Research is needed to determine the composition of expert performance, and develop measurement algorithms for simulators to provide feedback for deliberate training of expert performance.

Measurement of Transition-State Performance

Common measures of operator control describe operator-system regulator behavior during steady states (e.g., while holding a vehicle on course or on a turn, or holding a power plant at a given power level). Transitions from one steady state to the next are just as important as the steady states, but they frequently are not measured because the algorithms may be complex and there is little guidance in the literature. Research is needed to develop useful algorithms that describe operator behavior during transitions.

Measurement of Team Performance

Most complex systems are composed of teams of individuals who have designated functions. The contribution of each person to a team effort often is difficult to define and measure (except for highly procedural tasks). It is inappropriate for training purposes to use overall team or system performance as the only metric for performance feedback because an individual may have performed properly but the team failed, or the converse.

Where verbal communications represent the interactions between team members, adequate measurement may demand speech-understanding systems that are beyond the present state of the art. Even with such systems, some communication may be non-verbal, having to do more with an individual's knowledge and expectations of the other team members' behavior than what is happening or not happening at the moment.

There is much that needs to be learned about team training and performance measurement of teams if current and projected team training simulators are to be used for their maximum effectiveness. Research in this area is needed.

Measurement of Decision Performance

The assessment of decision performance requires different kinds of measures and algorithms than operator-system control. Here, a decision maker must assess information and make decisions that will influence a plan of action or strategy. If precise decision rules are known, performance measurement is tractable but may require significant amounts of processing capacity.

Some decisions, however, may not be amenable to direct and immediate measurement, and may have to be inferred from events that follow the time when the decision should

have been made. In these cases, contingent probabilities and pattern-matching techniques might be required to identify the decision that has occurred. Also, some decisions are neither right nor wrong at the time they are made, but may become better or worse as a consequence of subsequent decisions or events that were unknown at the time.

Human decision making has been the subject of research, and various researchers have expressed decision behavior in terms of their approach or theories. A common framework for assessment of decision performance is needed. Research in this area should address the range from operator decisions to elect a particular course of action to decisions made by controllers of complex systems.

Use of Expert Opinion

In system design and training-system design, subject-matter experts represent a valuable source of information. In training-system design, there are many issues that cannot be treated with methodological rigor; the judgments of subject matter experts can provide boundaries for problems; can eliminate unsuitable training strategies, courseware, firmware and hardware; and can describe good and bad performance. In simulator training programs there is a need for instructor judgments and performance assessment, because the instructor brings to the training situation more knowledge than can be implemented in current simulator automated measurement systems. Guidelines are needed on how to best use expert judgment for measurement system development, and to decide what measurement and assessment functions should be allocated among machines and humans.

Observational Methods for Performance Measurement

Automated instrumentation and performance measurement systems are not possible in all simulations, or for certain tasks in simulators that provide some automated measurement. Military field exercises and civilian disaster control exercises are examples of simulations that may not be amenable to automated measurement. Communications is an example of a task that can be measured only in limited ways by machines (at the present time). Furthermore, there are behaviors that can be measured by observers that are very difficult to automate, such as appropriate visual scan patterns of displays, lookout doctrine in automobile and aircraft simulators, the physical position and movements of operators in a nuclear power plant, or subtle cues that indicate an operator or controller is getting prepared for an oncoming event or considering alternative problem solutions. In these situations, observers frequently are required to capture the performance of interest as objectively as possible.

There are no commonly accepted methods to determine person-machine allocation of function for performance assessment in simulators; what behaviors should be measured automatically and manually, what are the trade-offs between performance data accuracy, validity, and costs to acquire the information? It has been argued that all measurement is subjective at some stage of development (Muckler, 1977), so we ought to forget about pseudo-issues such as subjective versus objective measurement and just refer to observational and automatic methods. Research on observational performance assessment is needed and should not be forgotten in the pursuit of automated measurement methods for simulators.

SUMMARY

The use of simulators for training provides a tantalizing environment for performance measurement: quantities that would be difficult to sense in operational environments are readily available, and the built-in computing capability permits a wide range of measurement algorithms to be implemented. Given this opportunity, one may be surprised that the design of effective performance measurement systems is not simple. It is hoped that this paper has provided some information to impatient designers and users as to why needed performance measurement doesn't exist. It is hoped also that this paper has provided some information about measurement issues in need of investigation and development.

The problem defies simple summarization; however, several issues stand out: (1) Front-end analyses fall somewhat short of defining the molecular behavioral requirements and the obscure cognitive activity of human performance in complex systems. They also fall short in specifying expected levels of performance representative of expert behavior in the operational environment. (2) There are a number of tough human tasks that are not understood well enough to enable comprehensive training measurement. These include team tasks, decision making, and complex system control. (3) Although we may not be fully satisfied with observational measurement by knowledgeable personnel, we go on to attempt to measure automatically, and pursue automation that reflects the knowledge of expert training personnel. Of course, the pursuit of automated systems may ultimately lead to better observational or hybrid systems. (4) We also are limited in our ability to implement performance measurement because of the complexity of the training task, the difficulty of dividing performance into distinct measurement segments, or the speed of required measurement system response.

We must make advancements in the "how" as well as the "what" of measurement. In particular, there is much promise in the approach of integrating artificial-intelligence techniques into performance measurement systems for training. Performance measurement systems must be made "smarter" to function automatically in real time and to have capabilities that compare to those of the expert observer.

Clearly there is much to be gained through better measurement; measurement should provide the information needed by the instructor, the operational command, the training manager, the designer, and the student. A number of issues have been discussed in this paper, and, of course, the relative importance of these issues depends on the specific training application. It is likely, however, that research in any one area will have payoffs in several other of the areas; we urge any research which addresses the issues of measurement.

REFERENCES

Brictson, C. A., Burger, W. J., and Wulfeck, J. (1973). *Validation and application of a carrier landing performance score: The LPS*. La Jolla, CA: Dunlap and Associates.

Kelly, M. J., Wooldridge, A. L., Hennessy, R. T., Vreuls, D., Barnebey, S. F., Cotton, J. C., and Reed, J. C. (1979). *Air combat maneuvering performance measurement* (Tech. Report AFHRL-79-3). Brooks AFB, TX: Air Force Human Resources Laboratory.

Kennedy, R. S., and Bittner, A. C. (1977). Development of a Navy performance evaluation test for environmental research (PETER). In L. T. Pope and D. Meister (Eds.), *Symposium proceedings: Productivity enhancement: Personnel performance assessment in Navy systems* (pp. 393-408). San Diego, CA: Navy Personnel Research and Development Center.

Lees, M. A., Kimball, K. A., Hoffman, M. A., and Stone, L. W. (1976). *Aviator performance during day and night terrain flight* (Tech. Report USAARL-77-3). Fort Rucker, AL: Army Aeromedical Research Laboratory.

Mixon, T. R. (1981). *A model to measure bombardier/navigator performance during radar navigation in Device 2F114, A-6E Weapon System Trainer*. Unpublished master's thesis, Naval Postgraduate School, Monterey, CA.

Mixon, T. R., and Moroney, W. F. (1982). *An annotated bibliography of objective pilot performance measures* (Tech. Report NAVTRAEQUIPCEN IH-330). Orlando, FL: Naval Training Equipment Center.

Muckler, F. A. (1977). Selecting performance measures: "Objective" versus "subjective" measurement. In L. T. Pope and D. Meister (Eds.), *Symposium proceedings: Productivity enhancement: Personnel performance assessment in Navy systems* (pp. 169-178). San Diego, CA: Navy Personnel Research and Development Center.

Semple, C. A., Cotton, J. C., and Sullivan, D. J. (1981). *Aircrew training devices: Instructional support features* (Tech. Report AFHRL-TR-80-58). Brooks AFB, TX: Air Force Human Resources Laboratory.

Simon, C. W. (1978). *Applications of advanced experimental methodologies to AWAVS training research* (Tech. Report NAVTRAEQUIPCEN 77-C-0065-1). Orlando, FL: Naval Training Equipment Center.

Vreuls, D., and Wooldridge, A. L. (1977). Aircrew performance measurement. In L. T. Pope and D. Meister (Eds.), *Symposium proceedings: Productivity enhancement: Personnel performance assessment in Navy systems* (pp. 5-31). San Diego, CA: Navy Personnel Research and Development Center.

Vreuls, D., Wooldridge, A. L., Obermayer, R. W., Johnson, R. M., Norman, D. A., and Goldstein, I. (1974). *Development and evaluation of trainee performance measurement in an automated instrument flight maneuvers trainer* (Tech. Report NAVTRAEQUIPCEN 74-C-0063-1). Orlando, FL: Naval Training Equipment Center.

Waag, W. L., and Knoop, P. A. (1977). Planning for aircrew performance measurement R&D. In L. T. Pope and D. Meister (Eds.), *Symposium proceedings: Productivity enhancement: Personnel performance assessment in Navy systems* (pp. 381-392). San Diego, CA: Navy Personnel Research and Development Center.

Wooldridge, A. L., Kelly, M. J., Obermayer, R. W., Vreuls, D., Nelson, W. N., and Norman, D. A. (1982). *Air combat maneuvering performance measurement state space analysis* (Tech. Report AFHRL-TR-82-15). Brooks AFB, TX: Air Force Human Resources Laboratory.

Key Army Decision Maker Concerns About Training Performance Measurement and Assessment

Dee H. Andrews, Army Research Institute, and **Betty Mohs**, HAY Systems, Inc., Orlando, Florida

This study explored an area of Army training performance measurement and assessment (PMA) which has apparently not been examined. It provides an understanding about Army training PMA requirements and uses, and reveals a number of PMA issues which should be more closely examined in the future. The methodology adapted for the study combined elements of Policy Capturing Analysis with elements of Policy Implications Analysis and the Delphi Technique.

The 1985 Army Science Board stated that, "Measurement of training is necessary to provide: feedback to trainers and training designers, and Return on Investment information to senior managers to guide expenditure of Army training resources" (p. 29). These 'trainers' and 'senior managers' are those identified as Army Key Decision Makers (KDMs) in this paper. In order to keep the study described here focused as much as possible on training, the KDMs surveyed were those who manage the Directorates of Training Developments at Army schools.

A broad state of the art survey echoed the Army Science Board's 1985 Summer Study's criticisms on the way the Army measures training performance. The survey did not reveal any literature which addressed KDM performance measurement and assessment concerns. This void makes it difficult to draw conclusions about the effectiveness of the Army's current system. How can we know if the Performance Measurement and Assessment (PMA) systems meet Army requirements if we don't know what those requirements are?

To fill the information void about KDMs, a methodology was devised for determining their PMA requirements. Key questions addressed were: (1) Who are the KDMs in the Army with regard to training PMA? (2) What are their key information requirements? (3) Are they currently receiving the information necessary to satisfy their requirements? (4) Is the information in qualitative or quantitative form (a major concern of the 1985 Army Science Board). (5) Which information is most and least useful for the decisions the KDMs have to make?

METHODOLOGY

The methodology adapted for the survey combined elements of Policy Capturing Analysis (Christal, 1963; Madden, 1963) with elements of Policy Implications Analysis (Madey and Stenner, 1981) and the Delphi Technique (Dalkey, 1969). These techniques are used by management analysts, industrial-organizational psychologists, and evaluation specialists to determine critical aspects of policy development and information use by KDMs.

An abbreviated form of the methodologies described above was adapted for the KDM survey for two reasons. First, those methods described above are primarily intended to aid analysts in predicting future actions of managers and experts in policy making questions. This study was conducted for descriptive rather than predictive purposes. The intent was to capture current concerns and interests of KDMs, while identifying training PMA issues for further study. The methodologies were also modified to reduce the time demands on respondents. Initial contacts with some of the respondents indicated that their schedules were extremely

Reprinted from Proceedings of the Human Factors and Ergonomics Society 31st Annual Meeting (pp. 1251–1255).

tight. The researchers determined that administration of the more formal methodologies may not have resulted in an adequate response rate had the respondents felt too much time was required.

Two questionnaires were developed. The first was intended to query KDMs about their training PMA information needs. The questionnaire was sent out to 23 schools within the Army, of which nine responded. In the second questionnaire, consensus responses were developed for the key PMA items explored in the first questionnaire. Again, to obtain the highest response rate possible, and in the interest of the respondents' time, the instrument was kept brief. The instructions asked that either the Commanding Officer (CO) or the Executive Officer (XO) complete the questionnaire. The goal of the consensus instrument was to have individuals respond who are truly KDMs. The results of the first inquiry had shown that the CO and XO were considered, by all schools re-

sponding, to be the KDMs. A copy of the first questionnaire was included in the mailing of the second so that any new respondents could understand the context of the study. Eleven second round responses were received.

RESULTS/FINDINGS

The key data obtained from the two questionnaires, along with the questions themselves, is presented in this section. Other data of less interest to this paper were obtained but are not reported here. Due to the descriptive purpose and nature of the study no attempt has been made to perform additional analyses. Our hope is that this descriptive data will serve as the foundation for more detailed analyses of the topic.

Tables 1 and 2 display the results of the first questionnaire. The intent of the questions, which are repeated in the tables, was to determine: how KDMs effect the PMA process, what PMA

TABLE 1: How often do the KDM's decisions affect the following?

		Never	Seldom	Sometimes	Generally	Always
a.	What is measured?			3	6	1
b.	When to measure?		1	1	7	1
c.	How to measure?		2	2	4	2
d.	What training devices are used to measure?	1		5	2	1

Note: The responses, in some cases, exceed the number of questionnaires returned, because some respondents gave more than one answer.

TABLE 2: Rank ordered importance of PMA data in the PMA decision process

Selection	Average Rank	Standard Deviation
End of Cycle Test Results	1.7	.8
Feedback from Active Units	2.2	1.9
Department of Evaluation and Standardization of Data	2.3	1.0
Test Validity/Reliability Data	3.2	1.3
Feedback from Instructors	3.8	2.2
Training Device Effectiveness Data	4.0	1.4
Skill and Qualification Test Results	4.4	2.1

Note: 1 = Most Important, 10 = Least Important

data are required by KDMs, and the overall importance of the PMA data that are collected.

Table 3 presents the results of the second questionnaire. This questionnaire presented consensus statements derived from the information gathered in the first questionnaire. The respondents in all cases were either the CO or XO of the Army School. Where the respondents disagreed with the statements, we asked for reasons for disagreement. Due to space limitations those reasons are not presented in this paper.

DISCUSSION

Although this study did not derive data from every Key Decision Maker (KDM) in the Army schools, it did access information from a good percentage (39% for the first questionnaire and 48% for the second). It explored an area of training PMA which has apparently not been previously examined. It allowed the researchers to gain some understanding about

PMA information requirements and uses. Perhaps most importantly it revealed a number of PMA issues which should be more closely examined in future research.

In general, the results of the second questionnaire validated the first. There are two possible exceptions to that statement. The first questionnaire revealed that the KDMs did not view information about training device effectiveness as being particularly important. However, certain responses on the second instrument gave indication that a number of the KDMs are beginning to value the PMA information which can be provided by the training device.

One of the problems with past and present training devices has been their confusing and complicated PMA systems, for those devices which have such systems. The instructors are usually inundated with PMA data since digitally based devices can record every possible trainee and instructor action. This "ocean" of data usually overwhelms the instructor and is often not used for that reason. As this problem

TABLE 3: Agreement/Disagreement of KDMs with PMA consensus statements from first questionnaire

Question	Consensus Derived From First Questionnaire	Do You Agree With Consensus?	
		Yes	No
1. How often do your decisions affect WHAT performance is measured?	Generally	9	2
2. How often do your decisions affect WHEN to measure?	Generally	10	1
3. How often do your decisions affect what TRAINING DEVICES are used to measure performance?	Sometimes	7	4
4. What kind of information do you need to make your decisions?	Feedback from Units	8	3
	Feedback from Students	10	1
	Training Effectiveness Data	10	1
5. Is the information you need qualitative or quantitative?	Qualitative	7	2
6. What type of data is most important when making decisions on training and assessment?	Feedback from Active Units and End-of-Cycle Tests	8	3
7. What type of data is least important when making decisions on training and assessment?	Training Effectiveness Data	8	3

becomes better understood, PMA systems are beginning to be designed more often with the user and KDMs in mind.

The second questionnaire results were less unanimous about the importance of feedback from active units as a PMA data source. There were indications that some KDMs felt the units were too narrowly focused on the units' specific concerns to provide valid feedback data. In other words, such data may be so biased that it is of little use to a KDM who must train individuals for all of the Army. It is possible that the data might be more useful if the units received interviews/questionnaires that specifically addressed the feedback instruction about the PMA needs of the schools and were given structured needs of the school.

The survey revealed that the KDMs need information about cost-effectiveness data and trainee learning rates. However, it is clear that the KDMs do not receive this information on a regular basis. Cost-effectiveness data can help KDMs to determine whether the Army is getting the best training outcome for its training resource investment. It can also aid in obtaining additional resources, if required, by quantitatively showing potential sponsors areas of strength and weakness in the training program.

However, such cost-effectiveness data is obviously difficult to obtain. It requires a firm, quantitative understanding of both the resource input and the learning outcome of a training program. Quantifying resource input is difficult in the military because many costs are indirectly charged to the training program. For example, while we can calculate the cost of training an instructor for school duty, it is nearly impossible to determine the lost cost benefit of not having that instructor in a combat unit.

In like manner, it is often not possible to quantitatively determine the learning outcome of a training program. Our measurement techniques are still developing and it is clear, as evidenced by the responses to the questionnaires, that KDMs are not receiving enough of this information. The end result is a general quandary about the cost effectiveness of training programs. Managers would like this PMA data but it is not forthcoming.

Data about the learning rates of trainees can help KDMs answer such questions as: (1) Is this group of trainees learning as quickly as previous trainees? (2) Based upon their learning rate can we cut the length of a course? Do we need to lengthen the course? (3) Should remediation sessions be made available to trainees based upon their slower than average learning rates?

To the degree that quantitative learning data is not provided to KDMs they will have to rely mainly on instructor feedback. Certainly such feedback is valuable, but it is always prone to bias and thus becomes less valuable than if it were paired with quantitative learning rate data.

The difficulty in measuring training outcomes, as mentioned above, is largely responsible for the seemingly contradictory finding about the need for training effectiveness data. When asked about their need for such data, the KDMs unanimously responded that it was needed. However, when they were asked to rank order the importance of various PMA sources, training effectiveness data ranked last. This discrepancy is likely due, again, to the difficulty in obtaining valid training effectiveness data. For example, the survey showed that 79% of the respondents received this type of data and yet they still ranked it as least important of all the PMA sources. Such a finding can be interpreted as a non-vote of confidence for the efficacy of the training effectiveness data which is presently being generated.

Another seeming anomalous finding, concerns the importance of feedback from instructors as a PMA data source. The KDMs unanimously stated that such feedback is needed. Yet, such data were only ranked fifth out of ten when compared to other PMA sources. This may be yet another indication that KDMs value instructor feedback but in absence of quantitative training effectiveness data as a cross-check, the KDMs tend to often look elsewhere for valid PMA information.

One final area of discussion is that the 1985 Army Science Board (ASB) made a strong point about the need for quantitative PMS data. The concept is that significant increases in training effectiveness can only come about as we are better able to measure where the training systems have come from, where they are now, and where they are going. The findings of this survey tend to support the ASB's recommendation that more quantitative data is required. The ASB's contention that quantitative data is generally unavailable is not supported however by one finding. The KDMs indicated that the PMA data they have access to are at least as much quantitative in nature as it is qualitative. The one exception is in the area of feedback from instructors, which was discussed above as being a problematic data source.

It is entirely possible that the KDMs and the ASB are defining qualitative and quantitative data differently. It is not clear in the ASB study how the Board defines the terms. In this study, we did not ask the respondents for their definitions, although such additional information would be helpful. Despite these definitional issues, it appears that the true amount and type of quantitative PMA data which exists in the Army's training systems should be a continuing source of study.

RECOMMENDATIONS

This brief survey has shown that KDMs in Army schools do not, in many cases, receive the type of PMA data they feel they require. Future research should continue to define the types of data required. Effective methods should be developed for gathering such data. This survey topic should also be examined with other KDMs not directly associated with the daily operation of the schools (e.g., Training Doctrine Command (TRADOC), Deputy Chief of Staff for Operations and Plans (DCSOPS), Deputy Chief of Staff for

Personnel (DCSPER), US Forces General Command (FORSCOM), US Army Europe (7th Army) (USAREUR), Department of Army (DA), Department of Defense (DoD), and Congress. Each of these agencies and organizations has their own PMA information needs, but the literature this study has examined has not yet revealed any information about those needs. Detailed techniques such as Policy Capturing and Policy Implications Analysis could help those interested in improving Army training to describe the present PMA information needs and uses of and by school KDMs. They also could help predict what data would be of most use. In turn, those techniques applied to non-school organizations would predict their future PMA information needs. As our technologies and capability for gathering PMA data increase, such information about KDM needs is vital.

REFERENCES

Christal, R. (1968). JAN: A technique for analyzing group judgement. *Journal of Experimental Education, 4.*

Dalkey, N. C. (1969). *The Delphi Method: An experimental study of group opinion.* Rand Corporation, Santa Monica, CA.

Final Report of the 1985 Summer Study on Training and Training Training Technology Applications for Airland Battle and Future Concepts. (Dec. 1985). Prepared for Department of the Army Assistant Secretary of the Army Research Development and Acquisition. Washington, DC 20310-0103.

Madden, J. M. (May 1963). *An application to job evaluation of a policy capturing model for analyzing individual and group judgments.* Report PRL-TDR-63-15, Personnel Research Laboratory, Aerospace Medical Division, Lackland AFB, TX.

Madey, D. L. & Stenner, J. (1981). Policy implications analysis: A Method for improving policy research and evaluation, pp. 23–39, in C. B. Aslanian (Ed.). *Improving educational evaluation methods: Impact on policy.* Sage Publications, Beverly Hills, CA.

Determining Flight Task Proficiency of Students: A Mathematical Decision Aid

WILLIAM C. McDANIEL,[1] *Navy Personnel Research and Development Center, San Diego, California, and* WILLIAM C. RANKIN, *Naval Training Systems Center, Orlando, Florida*

Accurate appraisal of student performance during and after training is important for the proper functioning of the training system and realization of training goals. Training systems—and particularly flight training programs—rely heavily on expert assessors' determination of student proficiency. Research is needed that will lead to improvements in the reliability and accuracy of these assessments. Recent research in decision making suggests that errors are frequently introduced because of the limited capabilities of people to integrate information and reach accurate conclusions. Mathematical decision aids appear to be helpful in reducing these errors. A decision aid using Wald's binomial probability ratio test and the sequential examination of student task performances was adapted to a training application. The decision aid required significantly less task trial information and predicted subsequent task performance more accurately than did expert assessors using the current assessment method. When students performed inconsistently and below the required flight task standards, instructors were more willing to declare proficiency than was the decision aid. This finding was especially apparent on the more difficult flight tasks.

INTRODUCTION

A major problem in flight training is determining the amount of in-flight training that should be given to students to meet the objectives of the flight training program. The difficulty centers on determining student performance and assessing skill levels appropriate for the graduate. This inability to achieve precise determination of student performance at given times hampers the achievement of training goals (Goldstein, 1987), creates dysfunction within the training system

(Flexman and Stark, 1987), and confounds training effectiveness evaluations and transfer of training studies (Caro, Shelnutt, and Spears, 1981).

Flight instructors, acting in the role of expert assessors, have traditionally and commonly determined the level of student proficiency demonstrated on flight tasks. The flight instructor represents the greatest single source of variability in proficiency assessment during flight training. Roscoe, Jensen, and Gawron (1980) observed that flight instructors often maintain their personal identities by developing personalized styles of preparing students to perform flying duties.

[1] Requests for reprints should be sent to William C. McDaniel, Route 2, Box 95, East Bernstadt, KY 40729.

Caro (1988) noted that evaluations of student pilot knowledge and performance are more a function of the instructors conducting the assessments than of the performance of the students being assessed. Considering the importance of a reliable and valid proficiency assessment method within a training system and the recognition of problems using the current method of flight instructor assessments, a research need exists to develop more effective ways of eliciting expert assessments and determining the validity of these assessments (Meister, 1985).

This paper describes the development and evaluation of a proficiency assessment method in a Navy flight training program that is more reliable and valid than the current method.

Current Proficiency Assessment Method

Preparatory to entering aircraft flight training, the student has completed extensive academic and ground-based simulator training. Generally, in Navy aircraft flight training one can observe three distinct phases: training, testing, and flight evaluation. During the flight training phase students perform a subset of tasks from the training syllabus during a training flight. The student performs each flight task a varying number of times at the discretion of the instructor pilot. The instructor pilot grades the student on the tasks performed using a standard scale. However, the instructor also employs his or her own personal criteria; that is, the instructor attempts to grade in terms of the average student performance at this stage of training. If performance meets these personal criteria, the student is advanced to the next training flight; if it does not meet the criteria, the student repeats the flight. It is usual for the student to be exposed to several instructor pilots during the training phase.

After a specified minimum number of flights and a recommendation by an instruc-

tor pilot, the student is scheduled for testing in a check flight. Performance on selected tasks is graded by an instructor pilot acting in the independent role of check pilot. The check pilot employs the same standard scale used during the training phase; however, rather than using personal criteria, the check pilot assesses student performance against an established standard. If the student performance does not meet the established standard, the student repeats the check flight until performance meets or exceeds the standard.

Upon successful completion of the check flight, the student is scheduled for a flight evaluation. Flight evaluations are conducted by highly experienced, specially trained pilots who ensure that performance conforms to the established standards. Successful completion of the evaluation flight qualifies the student for aircraft operation in an assigned on-the-job position.

The current method of proficiency assessment does not provide a clear qualitative or quantitative record of the student's flight task performance. The grades recorded on the standard scale represent a score based on one or more observations of each flight task performance. Further, student proficiency on specific tasks cannot be assessed against the established standard during flight training. The inadequacy of the current method of proficiency assessment may result in insufficient or excessive flight task practice by the student.

Proficiency Grading System

In a series of training effectiveness evaluations of an operational flight trainer, the inadequacies of the current proficiency assessment method were recognized (Browning, Ryan, and Scott, 1978; Browning, Ryan, Scott, and Smode, 1977). A proficiency grading system (PGS) was devised to overcome these inadequacies. The purpose of the PGS was to increase performance measurement sensitivity during training effectiveness evaluations and for this study of a mathematical decision aid. Although possessing desirable metric attributes for training, the PGS is not used in current practice.

The PGS provides a clearer picture of the student's flight task performance in aircraft training but still requires a subjective assessment by the instructor and check pilots. However, the instructor grades task performance against a more precise standard—that is, the required terminal level of performance.

Actual grading of performance was accomplished using a dichotomous scale. Task performance that met or exceeded the flight evaluation standard was recorded as "P"; task performance that did not meet the standard was recorded as "1." The PGS includes a further requirement. Performance was graded each time the task was performed, and the graded trails for each flight task were recorded in the order performed by the student. The PGS resulted in a series of graded practice trials and provided a qualitative and quantitative record of student performance for each flight task.

The PGS was expected to increase the reliability of performance assessment by focusing on specific tasks and the recording of performance contiguous to the behavior (Smith, 1976). Although the PGS may have reduced error variance in performance assessment attributable to flight instructor error, it appeared to increase the salience of two sources of variability in student performance: learning rate and task difficulty. During flight training, students are transitioning from a low proficiency level to a level required by the established standards. This transition reflects different learning rates by the individual student; these rates are highly variable within and between individuals (Sidman, 1960). Each student also exhibited variability in flight task performance as a function of task difficulty. The performance variability on simple tasks was considerably less than the variability evidenced on more difficult tasks. This variability of skilled task performance as a function of task complexity and difficulty has been well documented (Fitts and Posner, 1968).

Table 1 illustrates the variability found in the PGS trial sequences over a series of flights. In comparisons of the trial sequence of a student with that of a trained pilot, the student's learning rate is evidenced by initial unsatisfactory performance. Learning rates among tasks differ as shown by comparing Task A (a difficult task) with Task B (an easy task). During later flights the trial sequences are not readily distinguishable from those of the trained pilot. Examination of the trial sequences for Task A and Task B suggests that both the student and the trained pilot perform the more difficult tasks inconsistently. The essence of the problem lies in determining, with a minimum number of trial observations and with a specified degree of confidence, the point at which the student's trial sequence is equivalent to the sequence of the trained pilot.

Determination of proficiency levels by assessing the trial sequences shown in Table 1 is difficult; however, this assessment is simplified because the full trial sequence including both the initial and final task performances is provided. The problem can be better exemplified by considering the following sequence of trial performances:

11P1PP11PPP1P111PPP1PPPPP1PPPP

By covering the sequence and exposing one trial at a time, the problem more closely parallels the situation an expert assessor faces in determining flight task proficiency. After exposing each trial, an assessor makes a decision, based on demonstrated proficiency, either to terminate training or to continue to collect trial performance information. The

TABLE 1

Comparison of Task Trial Sequences for Two Different Tasks and Two Levels of
Pilot Proficiency

Task/Pilot	Trial Sequence during Six Flights					
	One	Two	Three	Four	Five	Six
Task A (difficult)						
Student	111	P11	1P1	1PP	P1P	PP
Trained pilot	PPP1	PPP	1PP	P	P1	PPP
Task B (easy)						
Student	11	1P	P	PPP	PP	PP
Trained pilot	PP	PPP	P1	PP	P	PP

P = task performance that met or exceeded the flight evaluation standard; 1 = task performance that did
not meet the standard.

human assessor encounters difficulty in integrating the information needed to reach a proper conclusion. Hogarth (1987) attributed this difficulty for people to reach proper conclusions to their limited information processing.

Kahneman and Tversky (1973) provided an excellent illustration of vagaries in judgments of student flight performance and the putative factors leading to the performance. Experienced flight instructors observed that student pilots who receive praise for making an exceptionally smooth landing typically make a poorer landing on the next try and that students who receive harsh criticism after a rough landing usually improve on the next try. The instructors concluded that verbal rewards are detrimental to learning and that verbal punishments are beneficial, which is contrary to accepted psychological doctrine. The instructors failed to consider the well-known statistical phenomenon of regression toward the mean.

As in other cases of repeated examinations, performance will usually improve after a poor performance and deteriorate after an outstanding performance, even if the instructor does not comment on or respond to the student's achievement on the first attempt. Because the instructors had praised their trainees after good landings and admonished them after poor ones, they reached the erroneous and potentially harmful conclusion that punishment is more effective than reward. This example demonstrates the fallibility of human judgment and a failure to integrate the information needed to reach a proper conclusion.

Practical solutions are available to aid the human assessor in the integration of information. Wisudha (1985) described a number of decision aids that were developed to overcome information integration problems. Research has shown that decision aids that integrate information by mechanical or mathematical methods are more consistent and accurate than are decisions based on human information integration (Dawes and Corrigan, 1974; Meehl, 1954; Sawyer, 1966). Human decisions require more data than do mathematical aids because of poorly defined parameters and biases in the processing of information for decisions (Slovic, 1976; Tversky and Kahneman, 1974). Dawes (1979) concluded that mathematical methods are especially efficient at integrating information.

The mainstream approach to decision making emphasizes heuristics and biases that affect judgment under uncertainty (Kahneman, Slovic, and Tversky, 1982). These heuristics and biases occasionally lead to suboptimal decisions. Optimal decisions may be reached through rigorous analysis of structuring the problem, assessing the likelihoods, assessing the importance of the potential outcomes, and integrating all the information about probabilities and values (Slovic, 1989). Klein (1989) questioned the applicability of this approach in naturalistic conditions and time-pressured tasks. Dreyfus and Dreyfus (1986) critically reviewed previous decision-making research and concluded that observed biases in human judgment are often the result of contrived experimental situations and inexperienced or poorly motivated human subjects. Alternatively, they suggested that optimal decision making is the product of skill acquisition and experience rather than a rigorous analysis to overcome potential biases. The human ability to give meaning to information is common to both approaches. Dawes (1979) observed that humans have expertise in perceiving and sorting information that cannot be matched by a computer or mathematical model.

Given that people excel in perceiving and sorting information and that mathematical decision aids are especially good at combining or aggregating information, it appears that a combination of these methods would enhance the decision-making process. It follows that a combination of experts assessing trial performance and a decision aid determining the integration and quantity of these assessments should substantially increase the validity and reliability of training proficiency decisions.

The preferred approach would be to examine the graded task trials one at a time and accumulate this information into a decision aid. Using this approach one would expect to be in a better position to make proficiency assessments than if no attempt were made to look at the trial data until a sample of a fixed number of trials had been observed. Available methods using sequential sampling techniques and a statistical procedure operate on this accumulation of information and require fewer observations than the fixed-size sample methods. The statistical procedure is limited to two choices in decision making (three choices if one considers deferring a decision as a decision). This limitation is not troublesome when applied to proficiency assessment. The two decisions of primary concern are (1) Is the student proficient? and (2) Is the student not proficient? Sequential sampling techniques have the advantage of reaching a decision with minimum observations at an established level of confidence.

Wald's Binomial Probability Ratio Test

Wald (1947) used a sequential sampling technique and formulated the binomial probability ratio test for use in an industrial quality control setting. The procedure was used to determine if a lot (collection of manufactured items) should be rejected because of a high proportion of defective items or accepted because the proportion of defectives was below a criterion. The procedure also provided for the postponement of a decision until sufficient evidence was accumulated to make the decision at required confidence levels. This deferred decision is based on prescribed values of alpha (α) and beta (β). The value of α limits errors of declaring a product lot to be acceptable when it is unacceptable, which is referred to as a Type I statistical error. The value of β limits errors of declaring a product lot unacceptable when it is acceptable, which is called a Type II statistical error.

In an industrial quality control setting, the inspector uses a chart similar to Figure 1 to perform a sequential test to determine whether the manufacturing process has produced a lot with too many defective items or if the proportion of defects is acceptable. Two

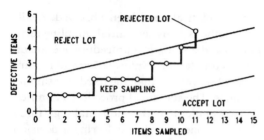

Figure 1. *A quality control sequential sampling chart developed from Wald's binomial probability ratio test.*

parallel lines divide the chart into three regions that correspond to appropriate decisions. The region above the upper parallel line represents a *reject lot decision*, the region below the lower parallel line represents an *accept lot decision*, and the area between the two parallel lines forms a *postpone decision* region requiring that additional items be sampled. The parallel lines are constructed from the binomial probability ratio test calculations using four preselected parameters. These parameters are (1) the proportion of defective items at which the lot is considered to be acceptable, (2) the proportion of defective items that renders the lot unacceptable, (3) the value of α, and (4) the value of β.

As each item is observed the inspector plots a point on the chart one unit to the right if the observed item is not defective or one unit to the right and one unit up if the item is defective. If the plotted line crosses the upper parallel line, the inspector will reject the production lot. If the plotted line crosses the lower parallel line, the lot will be accepted. If the plotted line remains between the two parallel lines of the sequential decision chart, another sample item will be drawn and observed/ tested.

This sequential sampling procedure has been applied in educational and training settings (Ferguson, 1970; Kalisch, 1980). Both applications employed the procedure to improve testing. Results from both applications

indicate that this procedure has greater efficiency than tests composed of fixed numbers of items and that it substantially reduced testing time.

Adapting Wald's Procedure to Assess Flight Task Proficiency

Wald's (1947) procedure was altered in three ways to better accommodate the proficiency assessment application. First, the decision regions were inverted to place the focus on proficiency rather than unacceptable performance. Second, in the industrial quality control setting, sampling occurs *after* the manufacturing process. In the educational and training applications cited (Ferguson, 1970; Kalisch, 1980), sequential sampling occurred *after* the learning period. In this proficiency assessment application, the sequential sampling occurs *during* the learning period and eventually terminates it. Third, if the proficiency assessment results in a *not proficient* decision, remedial training is the most reasonable action to be taken. This remedial training is assumed to occur on the trial following the not proficient decision. This trial is considered the first of another trial sequence and is entered into the decision aid. This modification precludes the requirement for the remaining trial sequence to traverse from the not proficient area, through the postpone decision area, to cross the *proficient* boundary.

The following equations are based on Wald's binomial probability ratio test and are used to calculate the decision regions of the sequential sampling decision aid.

$$d_n \le \frac{ln\dfrac{\beta}{1-\alpha}}{ln\dfrac{P_2}{P_1} + ln\dfrac{1-P_1}{1-P_2}} + n\frac{ln\dfrac{1-P_1}{1-P_2}}{ln\dfrac{P_2}{P_1} + ln\dfrac{1-P_1}{1-P_2}}$$

(1)

and

$$d_n \geq \frac{ln\dfrac{1-\beta}{\alpha}}{ln\dfrac{P_2}{P_1} + ln\dfrac{1-P_1}{1-P_2}} + n\frac{ln\dfrac{1-P_1}{1-P_2}}{ln\dfrac{P_2}{P_1} + ln\dfrac{1-P_1}{1-P_2}},$$

$$(2)$$

where d_n = number of proficient trials in the sequence; n = total number of trials in the sequence; ln = natural logarithm; P_1 = proportion of proficient trials for a beginning flight student to the total number of trials observed; P_2 = proportion of proficient trials for a trained pilot to the total number of trials observed; α = probability of a Type I error; and β = probability of a Type II error.

The first term of Equations 1 and 2 determines the intercepts of the two linear functions. The width between these intercepts is determined primarily by values selected for α and β. The width between the intercepts translates into a region of uncertainty; thus as lower values of α and β are selected, this region of uncertainty increases.

The second term of Equations 1 and 2 determines the slopes of the two linear functions. Because the second term is the same for both equations, the result will be parallel lines with equal slopes. Values of P_1 and P_2 as well as differences between P_1 and P_2 affect the slope of the lines. The concept of differences in P_1 and P_2 is analogous to the concept of effect size in statistically testing the difference between the means of two groups. In such statistical testing when α and β remain constant, the number of observations required to detect a significant difference is reduced as the anticipated effect size increases.

Characteristics of this mathematical aid complement the requirements for determining appropriate levels of proficiency. The parameters, P_1 and P_2, reflect initial and final performances for specific training tasks. Typically the more difficult tasks exhibit lower rates of skill acquisition with greater vari-

ability and lower proportions of acceptable trials on completion of the training program. Tasks involving simple procedures are generally discernible by large differences in initial and final performances; skills are acquired quickly, and final performance is characterized by relatively consistent adherence to an established standard. The slope of the decision boundaries and the postponement region between the boundaries reflect differing skill acquisition rates as well as differences in initial and final performances specific to each flight task.

The sequential sampling of observed trials results in an efficient sampling technique. Hoel (1971) estimates that sequential methods may be 40% to 50% more efficient in reaching a decision than are fixed sample methods. Effect size and confidence levels for both Type I and Type II errors are predetermined; thus sample size becomes a random variable. Decision alternatives are evaluated after each observation, so no further sampling is necessary if a proficient decision is reached. This characteristic is especially important to skill training. The capability to discern proper performance levels with minimum observation can preclude costly additional trials or interject remedial or additional training when required.

Prescribing appropriate confidence levels to reduce Type I and Type II errors is made operational through the setting of α and β. Type I and Type II errors engender different risks. The capability to specify these risks quantitatively is particularly desirable in a training program.

Finally, once parameters are established and the decision regions are plotted using the mathematical equations, further computations and complex mathematical manipulations are not required. This feature simplifies use of the decision aid in the training system application.

Field Evaluation of the Decision Aid

The mathematical algorithm used in the decision aid closely parallels the current method of proficiency assessment. Both the decision aid and the expert assessor base decisions on varying numbers of task trials rather than on requiring a fixed number of trial performances. Consistency of student task performance is also considered in determining the appropriate number of trials. Both the decision aid and the expert assessor consider the risks involved in making an inappropriate decision. An evaluation was conducted to examine the practicality of the decision aid in an actual training situation.

The practical utility of the decision aid depends on several outcomes. First, decisions reached by the decision aid should be consistent; that is, once proficiency is declared, decision reversals based on subsequent trial information should not occur beyond established confidence levels. Second, the number of observations required by the decision aid to reach a decision should be equal to or less than the number of observations required by the flight instructors. Third, the decision aid should integrate trial information in a reasonable or logical manner to determine proficiency. Consistent trial performances typical of easy flight tasks should require less information (i.e., fewer trial observations) than the less consistent trial performance found in more difficult tasks. Finally, the decision aid should be useful in predicting future performance on the flight evaluation. Proficient assessments by the decision aid should be followed by a proficient grade on the flight evaluation.

METHOD

Students

The student sample consisted of 29 newly designated naval aviators undergoing fleet replacement pilot training in the SH-3 aircraft at Helicopter Antisubmarine Squadron One (HS-1), Naval Air Station, Jacksonville, Florida. The students were recent graduates of undergraduate pilot training at the Naval Air Station at Pensacola, Florida; had no prior flight experience in the SH-3 aircraft; and had completed academic ground training in SH-3 aircraft systems, cockpit procedures training, and instruction in the 2F64C, a high-fidelity operational flight trainer for the SH-3 aircraft.

Instructors

Flight task training was provided by the 28 regular HS-1 flight instructors. All instructors had completed at least one tour in an operational assignment and the training course for flight instructors at HS-1. The instructors were given training in the current grading procedures and the PGS. Eight of the flight instructors were designated as flight evaluation check pilots and were qualified to perform annual flight evaluations in addition to their regular duties.

Flight Tasks and Task Parameters

Each student was required to master approximately 190 flight tasks to become qualified in the SH-3 aircraft. From the task inventory of 190 tasks, 18 were selected as representative of the range of difficulty in the inventory. Flight instructors acting as subject matter experts sorted the 18 tasks into categories of difficulty (i.e., easy, medium, and difficult) and then rank ordered the tasks in each category. The three tasks ranked easiest were selected from the six tasks in the *easy* category, the three intermediate-ranked tasks were selected from the six tasks constituting the *medium* category, and three tasks ranked as the most difficult were selected from the *difficult* category. This procedure provided tasks in clear categories of task difficulty and are shown in Table 2.

The values for the preset parameters are

TABLE 2

Flight Tasks, Task Difficulty, and Parameters Used in the Decision Aid

Tasks and Difficulty Level	Parameters			
	P_1	P_2	α	β
Easy				
Normal start	0.12	0.65	0.10	0.10
Shutdown checklist	0.29	0.91	0.10	0.10
Normal landing	0.18	0.78	0.10	0.10
Medium				
Search and rescue manual approach	0.25	0.80	0.10	0.10
Alternate approach pilot procedures	0.18	0.69	0.10	0.10
Single engine malfunction takeoff abort	0.41	0.75	0.10	0.10
Difficult				
Windline search and rescue pilot procedures	0.38	0.80	0.10	0.10
Automatic stabilization system off landing	0.30	0.86	0.10	0.10
Freestream recovery	0.12	0.55	0.10	0.10

crucial to the functioning of the decision aid. The parameters shown in Table 2 were determined using the following procedure. The P_1 parameter values for each of the 18 tasks were determined from an examination of first-trial performance data, in the aircraft, of a separate group of 17 students who had previously undergone flight training. These students had flying experience in other types of aircraft, had received cockpit procedures training, and had completed each of the flight tasks in a high-fidelity simulator. The proportion of flight task performances judged proficient on the first trial to the total number of first-trial performances was used to determine P_1.

The P_2 parameter values were determined from the performance of 50 experienced naval aviators on the Naval Air Training and Operating Procedures Standardization (NATOPS) Program Flight Evaluation Worksheet. Each naval aviator is required to undergo a NATOPS flight evaluation annually to maintain currency and qualification in the appropriate type/model/series aircraft. The evaluation worksheet provides specific grad-

ing information about performance on the selected tasks; that is, task performance by the aviator is graded as *qualified, conditionally qualified, or unqualified.* The proportion of the number of qualified grades to the total number of assessments for each flight task was used to set an initial value for P_2. Because experienced naval aviators performed these tasks, the standard of performance might have been too ambitious for graduating students. The standard deviation was calculated for each proportion, and the final value for P_2 was set at one half the standard deviation units below the initial value.

The experimenters established α and β values at 0.10. A confidence level of 90% in decisions made by the decision aid appeared reasonable; however, as rigorous field testing of the decision aid was conducted, these parameters would be modified as indicated by empirical evidence and training policy.

Materials and Equipment

Standard training materials and equipment were used by the students and instructors at HS-1. The primary data collection instruments were the standard syllabus grade card with a provision for grading and recording the sequence of trial performances on each task as required by the PGS. Flight evaluation grades were recorded on the NATOPS Flight Evaluation Worksheet.

Procedure

As students proceeded through the flight training syllabus, instructors recorded student performance on the syllabus grade card using the current procedure and the PGS. The experimenter collected the syllabus grade cards after each training flight. The training trial sequence for each of the selected tasks was evaluated using the decision aid; however, the proficiency assessments made by the decision aid were withheld from students, instructors, and the training staff.

If the flight training phase and check flights were completed and the decision aid required additional trial information to reach a proficiency assessment, an estimate for the number of trials needed to reach a decision was calculated and added to the number of trials already performed. This estimate used the previously established parameter values and is based on the following equation from Hoel (1971):

Additional estimated trials to P_2 =

$$\frac{(\beta)ln\dfrac{\beta}{(1-\alpha)} + (1-\beta)ln\dfrac{(1-\beta)}{\alpha}}{(1-P_2)ln\dfrac{(1-P_2)}{(1-P_1)} + (P_2)ln\dfrac{P_2}{P_1}} \quad (3)$$

Using this approach to estimate trials, the information required for the decision aid to reach a decision could be less than, equal to, or more than the amount of information needed by the current proficiency assessment method to reach a proficient decision.

On completion of the training syllabus and satisfactory completion of the check flight, and at the discretion of the training staff, each student was scheduled for a final NATOPS flight evaluation by one of the eight designated instructor pilots. The flight evaluator graded and recorded the student's performance on the NATOPS Flight Evaluation Worksheet. The experimenter collected the worksheets, reviewed grades and performance discrepancies, and determined the evaluation grade for each student's performance on each of the selected tasks.

Dependent Measures

Two dependent measures were extracted from the data collected: the number of task trial performances required to declare the student as proficient, and the number of proficiency assessments made for each student and task.

Information requirements for the current assessment method were the total number of task trials that the student performed during the training phase and check flight. The decision aid information requirements were either the number of trials required to reach a proficient decision or the sum of the number of trials performed and the estimated trials required to reach proficiency.

The number of proficiency assessments made by the current assessment method was the number of tasks graded as qualified on the check flight. The number of decision aid proficiency assessments was the number of proficient decisions based on the evaluation of the student's task trial sequence for each task. The predictive accuracy of the current proficiency assessment method and the decision aid was the percentage of proficient decisions by each method which resulted in qualified task performance on the flight evaluation.

RESULTS

Information Requirements

Data bearing on the number of task trials needed to make a proficiency assessment were analyzed using a 2 (decision method: current procedure vs. decision aid) × 3 (task difficulty: easy, medium, difficult) Treatment × Subjects ANOVA. The main effect of decision method, $F(1,28) = 109.81, p < 0.05$, and the interaction of Decision Method × Task Difficulty, $F(2,56) = 43.22, p < 0.05$, were significant.

The decision aid required less information than the current method to reach a decision across all levels of task difficulty. The mean task trial performances required by the decision aid was 6.4 trials. The current method needed a mean of 11.8 trials on each task before a decision was reached.

Figure 2 shows the relationship between task difficulty and mean number of task trials to determine proficiency for both decision

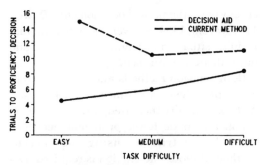

Figure 2. *Number of training trials to reach a proficient assessment by decision method for three categories of task difficulty.*

methods. Reliable differences within this interaction were determined using Tukey's Wholly Significant Difference (WSD) test; any differences in the means greater than 2.2 may be considered significant at the 0.05 level. The number of trials required to reach a proficiency assessment by the decision aid increased with task difficulty. Reliable differences were found between information requirements for easy (mean = 4.8 trials) and difficult tasks (mean = 8.2 trials). The reverse occurs for the current method; that is, more trial information to reach a decision was collected on the easy tasks (mean = 14.6 trials) than on the medium (mean = 10.2 trials) or difficult tasks (mean = 10.5 trials).

Decision Consistency

The decision aid assessed 261 trial sequences (9 tasks × 29 students); for 189 of those sequences, additional trials were recorded after a proficiency decision had been declared by the decision aid. The remaining trial sequences for each of these 189 tasks were assessed using the decision aid. No reversals in decisions were reached by the decision aid; that is, once the student was declared proficient on a task by the decision aid, additional trials only confirmed the initial decision.

Predictive Accuracy

Across all categories of task difficulty, the decision aid resulted in 192 proficient assessments. Qualified grades on the subsequent NATOPS evaluation revealed that 172 of these assessments were accurate in predicting NATOPS flight evaluation performance. The current procedure resulted in 216 proficient assessments; 182 of these assessments predicted NATOPS flight evaluation performance accurately. A test of chi-square revealed a significant association between decision method and decision accuracy (χ^2 = 6.81, $p < 0.01$). For each category of task difficulty, 87 decisions (3 tasks × 29 students) could be made by the current procedure and the decision aid. For flight tasks categorized as easy, both the decision aid and the current procedure declared that students attained proficiency on 86 of the trial sequences. Both decision methods predicted task performance on the flight evaluation with the same percentage of accuracy (99%).

The decision aid made 53 proficient assessments on flight tasks of medium difficulty. The current procedure resulted in a greater number of proficient assessments (68 of the possible 87). However, the decision aid predicted qualified task performance on the NATOPS evaluation flight more accurately than did the current assessment method (83% vs. 78% accuracy). This difference was not significant (χ^2 = 0.65, $p > 0.05$).

A similar result was found for difficult tasks. The decision aid reached a proficient assessment in 53 of the 87 trial sequences evaluated; 81% of these 53 proficiency decisions were later assessed as qualified on the NATOPS evaluation flight. The current procedure made 62 proficient assessments. The predictive accuracy of the proficient assessments was lower than the decision aid; the current method resulted in only 71% of the proficient decisions being followed by qualified scores on the NATOPS evaluation flight. However, this difference was not significant ($\chi^2 = 1.60, p > 0.05$).

Considering this trend toward better predictive accuracy for the decision aid, the entire sample of 18 tasks was evaluated for proficiency assessments by the decision aid and the current method. These proficiency assessments were then compared with the student's performance on the NATOPS evaluation flight. Results indicated that for 12 of the 18 tasks, the decision aid predicted task performance on the NATOPS evaluation flight more accurately than did the current method. The decision aid and the current method demonstrated equal predictive accuracy on 2 of the 18 tasks. The decision aid predicted task performance on the NATOPS evaluation flight less accurately on only 4 of the 18 tasks. A sign test revealed that the decision aid reliably predicted task performance on the NATOPS evaluation flight better than did the instructors using the current proficiency assessment method ($p < 0.05$).

DISCUSSION

Results of this study indicated that the decision aid, using the parameter values established in this study, required less information to make a decision than did the current method based solely on expert human judgment. Decisional consistency was maintained by the decision aid; that is, once a proficient assessment was made, additional task trial performances only confirmed that assessment. The decision aid reflected better predictive accuracy for subsequent task performance on the NATOPS flight evaluation. The finding that the decision aid required less trial information to reach a decision of greater precision strongly supports the aid's superior efficiency when compared with the current method of proficiency assessment. Results of this study extend previous research results suggesting greater consistency and accuracy of mathematical decision aids (Dawes, 1979; Meehl, 1954; Sawyer, 1966) to an actual training situation in a more unstructured and naturalistic environment.

Both decision methods made a large number of proficiency assessments for easy tasks. As task difficulty increased, however, the decision aid made fewer proficient decisions than did instructors using the current method. If more information is required by a decision method to reach final proficiency assessment, the method exhibits *conservatism* in reaching the final decision. Likewise, methods needing less information have often been characterized as demonstrating *riskiness*. The conservatism or riskiness of the decision aid is established through the parameter values, specifically α and β. Because these parameter values were held constant across all tasks and levels of task difficulty in the decision aid, the instructors seemed willing to take more risks in their proficiency decisions for the medium and difficult tasks. This willingness to take greater risks may have resulted in the lowered predictive accuracy of proficient decisions made by the instructors. Results of this study are similar to a study of human decision-making behavior in a sequential testing situation reported by Becker (1958). Becker's subjects responded in a manner more similar to Wald's sequential sampling method when the problem was difficult than when the problem was easy. Typically subjects required relatively more samples of information on easy problems and relatively less information on the difficult ones, as if they set α and β lower for the easy problems.

Why instructors continued to collect task trial information well beyond the decision aid requirements to assess proficiency on easy tasks remains unclear. This may have been because easy tasks were introduced earlier in the training program, allowing more time to practice, or because the easy tasks were prerequisite to the performance of the more difficult tasks. Instructors may have allowed students to perform easy tasks so that successful performance would motivate each student to perform better on the more difficult tasks. It is possible that the student's demonstrated high level of performance on easy tasks may have reinforced the instructor, thereby increasing the probability that the instructor would request the student to perform the tasks again. Finally, instructors obtaining information about student performance on easy tasks do so at a lower cost. Easy tasks do not require the degree of actual physical risk to the instructor, student, and aircraft that the more difficult tasks present.

Whether this effect is attributable to single or multiple causes, easy tasks are overlearned because significantly more practice is allowed. This overlearning probably resulted in consistent task performance by the student and led to high agreement among the current proficiency assessment method, the decision aid, and the NATOPS flight evaluation. Considering the increased performance consistency and the high agreement between decision methods and evaluation procedures, overlearning appears to be desirable.

Undertraining is a far more critical issue than overtraining. Neither the decision aid nor instructors using the current method rendered proficiency assessments in 20% to 40% of the decisions concerning the medium and difficult tasks. Further, the results appear paradoxical. The decision aid was more conservative (made fewer proficient assessments because of insufficient trial information) than did instructors using the current method. However, results suggest that the decision aid required fewer trials than did the instructors to make a proficiency assessment. Task trial sequences were examined to resolve this apparent paradox. This examination revealed that students demonstrating consistent, proficient task performance were allowed to per-

form task trials well beyond the decision aid's proficiency determination. Conversely, students who performed inconsistently with greater numbers of unsatisfactory trials were not afforded the opportunity to continue practice on the task until a sufficient number of trials were accumulated for the decision aid to reach a proficient decision.

It seems reasonable that instructors would require more trial observations to make a proficiency assessment if the student's performance were inconsistent. These results appear to be counterintuitive because more observations were collected on the more consistent performers. Despite being counterintuitive, these findings are similar to those of a previous study, which concluded that human raters do not acquire a greater sampling of behavioral information for inconsistent performers than for consistent ones (Padgett and Ilgen, 1989).

Results confirmed that the selected parameter values were reasonable for each of the flight tasks; however, these values may be modified as indicated by future empirical evidence or dictated by changes in training policy. Kalisch (1980) outlined three methods for selecting P_1/P_2 values. First, external criteria directly related to the instructional objectives can be used. These criteria can be in terms of demonstrated levels of proficiency either on the job or in a training environment. Second, experts in the subject area who understand the relation of the training objectives to on-the-job performance may select the P_1/P_2 values to reflect their estimation of the necessary levels of performance. Specifying two values creates an indecision zone— neither proficient nor nonproficient. Third, the scores of prior students, who demonstrate the entire range from extremely poor to exemplary performance on objectives, are used to estimate P_1/P_2 values. The proportion correct for the entire sample is used to obtain an initial cutting score. Scores are separated

Readings in Training and Simulation

into two categories: those scores greater than or equal to the initial cutting score and those less than the initial cutting score. The mean proportion correct for the first category equals P_2; the mean for the second category equals P_1.

Selection of α and β should be based on the criticality of accurate proficiency decisions. Smaller values of α and β require more task trials to make decisions with greater confidence. Risk factors that are important in selecting values of α are (1) safety—potential harm to the student or to others because of actual nonmastery of the task; (2) prerequisite in instruction—potential problems in future instruction, especially if the task is prerequisite to other tasks; (3) time/cost—potential loss or destruction of equipment either in training or the job assignment; and (4) student's view of training—potential negative view by the student when he or she is classified as proficient but lacks confidence in that decision. In addition, the student may have a negative view of the training program if he or she is not sufficiently prepared for the final job assignment.

Risk factors that need to be considered in selecting values of β are (1) instruction—requirement for additional training resources for unnecessary training in case of misclassification as not proficient; (2) student attitude—the attitude of students when tasks have been mastered but training continues, student frustration, and the corresponding effect on performance in the remainder of the training program and job assignment; and (3) cost/time—the additional cost and time required of students to complete additional training that is not really needed.

Constraints of the field test setting placed major restrictions on assessing the practical use of the decision aid in a training system. Students proceeded through the training program at the discretion of the training staff

using the current proficiency assessment method. Although the decision aid may have reached a proficiency decision, training often continued. Alternatively, if the decision aid required further observations to reach a proficient decision, further training was deemed inappropriate by the training staff using the current proficiency assessment method. Despite these constraints, these results provide evidence that flight task training is poorly distributed to meet student requirements, as indicated by the overtraining and undertraining of specific flight tasks.

Training and Training Research Implications

A major difficulty in flight training and flight transfer research centers on variability in performance assessment. Lintern, Roscoe, Koonce, and Segal (1990) found a significant correlation between experimental and control subjects paired by instructors indicating that instructors had a small but real effect on performance measures. It is unclear whether this variance is evidence of differential subjective criteria or that some instructors provide better training than others. This mathematical decision aid offers a potential tool both to reduce variance in the performance assessments by instructors and to investigate instructor effects.

The issue of so-called standardized instructor-based proficiency assessment goes beyond instructor training aimed at using consistent criteria. Because flight instructors must pass judgment on a student's achievement with some degree of confidence (uncertainty), the question arises as to whether instructors themselves can be trained to be sensitive to the parameters of the Wald decision algorithm. If optimal decision making is a matter of experience and skill acquisition (Dreyfus and Dreyfus, 1986), this decision aid may provide a learning tool for information integration. Interinstructor variability may be investigated using a decision aid of the type described in this study. Such an approach would take proficiency assessment out of the realm of one-on-one (student with one instructor judging proficiency) to one-on-several (student with two or more instructors), acting as a team or as individuals judging proficiency. Helmreich, Wilhelm, Gregorich, and Chidester (1990) found striking and unexpected degrees of variation in cockpit resource management performance ratings among crews flying different aircraft within the same organization. Specific behaviors that triggered observer ratings of above- or below-average performance differed markedly between organizations. This mathematical decision aid provides a method to evaluate characteristics of experts' ratings against quantitative standards established a priori through performance and risk parameters.

Student proficiency in training simulators is equal in importance to that in the aircraft. Here is an excellent opportunity to apply the decision aid with parameter manipulation—both performance and risk parameters—in a far more controlled setting. Equivalent flight tasks performed in a simulator and in the aircraft represent different degrees of risk and stress to the aircrew and instructors/evaluators. The decision aid would be useful in comparing information integration of instructors under differing degrees of risk or stress.

Regardless of where future research may be focused, we must reiterate that an absolute standard of proficiency must be maintained from observation to observation, as in the proficiency grading system.

CONCLUSIONS

The decision aid adapted from Wald's binomial probability ratio test and sequential sampling technique provides a useful method for augmenting the current proficiency as-

sessment method. Similar to the current method, information is examined one task trial at a time until a decision regarding student proficiency levels can be determined. This feature permits concurrent use of the decision aid with the current proficiency assessment method. Mathematical computations are not complex and, when performed using established parameter values, do not require repeated calculations. The PGS does not place difficult additional requirements on the flight instructor which would interfere with training and safety responsibilities.

The decision aid can effectively supplement instructor assessments of student proficiency, especially for students who exhibit inconsistent task performance. The increased accuracy of predicting the flight evaluation performance of students during training and practice most likely results from the established, constant parameters used in the decision aid. The decision aid is most useful for integrating student task performance information over time.

Flight instructor proficiency assessments of student task performance are often premature if the student exhibits inconsistent and substandard performance. Although specific reasons are unclear, it is plausible that a high proportion of substandard performances on difficult flight tasks represent considerable risk to personnel and equipment. Considering these risks, the instructor may be more willing to declare proficiency when, in fact, proficiency has not been attained. Future research should focus on replication of this finding and extend the investigation of possible reasons for these premature assessments.

This decision aid has potential for training instructors to integrate a series of judgments to reach a decision that proficiency has been achieved or that further training is necessary. Variability in proficiency assessments attributable to intra- or interinstructor differences and conditions under which a proficiency assessment are made can be studied using this decision aid as a research tool.

ACKNOWLEDGMENTS

This article is based on research conducted by the Training Analysis and Evaluation Group, Orlando, Florida, during an in-situ evaluation of an operational flight simulator for the SH-3 aircraft. Special thanks are given to the Training Department, HS-1, Naval Air Station, Jacksonville, Florida, for their cooperation and assistance in this study. The authors appreciate the many helpful comments of the two anonymous reviewers and Ruth Ireland of the Navy Personnel Research and Development Center in the preparation of this manuscript.

The opinions expressed in this article are those of the authors, are not official, and do not necessarily reflect the views of the Navy Department.

REFERENCES

Becker, G. (1958). Sequential decision making: Wald's model and estimates of parameters. *Journal of Experimental Psychology, 5,* 628–636.

Browning, R. F., Ryan, L. E., and Scott, P. G. (1978). *Utilization of device 2F87F OFT to achieve flight hour reductions in P-3 fleet replacement pilot training* (TAEG Report 54). Orlando, FL: Training Analysis and Evaluation Group.

Browning, R. F., Ryan, L. E., Scott, P. G., and Smode, A. F. (1977). *Training effectiveness evaluation of device 2F87F, P-3C operational flight trainer* (TAEG Report 42). Orlando, FL: Training Analysis and Evaluation Group.

Caro, P. W. (1988). Flight training and simulation. In E. L. Wiener and D. C. Nagel (Eds.), *Human factors in aviation* (pp. 229–261). San Diego, CA: Academic.

Caro, P. W., Shelnutt, J. B., and Spears, W. D. (1981). *Aircrew training device utilization* (AFHRL-TR-80-35). Brooks Air Force Base, TX: Air Force Human Resources Laboratory.

Dawes, R. M. (1979). The robust beauty of improper linear models in decision making. *American Psychologist, 34,* 571–582.

Dawes, R. M., and Corrigan, B. (1974). Linear models in decision making. *Psychological Bulletin, 81,* 95–106.

Dreyfus, H. L., and Dreyfus, S. E. (1986). *Mind over machine.* New York: Free Press.

Ferguson, R. (1970). A model for computer-assisted criterion-referenced measurement. *Education, 91,* 25–31.

Fitts, P. M., and Posner, M. J. (1968). *Human performance.* Belmont, CA: Brooks/Cole.

Flexman, R. R., and Stark, E. A. (1987). Training simulators. In G. Salvendy (Ed.), *Handbook of human factors* (pp. 1012–1038). New York: Wiley Interscience.

Goldstein, I. L. (1987). The relationship of training goals and training systems. In G. Salvendy (Ed.), *Handbook of human factors* (pp. 963–975). New York: Wiley Interscience.

Helmreich, M. S., Wilhelm, R. A., Gregorich, S. E., and

Chidester, T. R. (1990). Preliminary results from the evaluation of cockpit resource management training: Performance ratings of flight crews. *Aviation, Space, and Environmental Medicine, 61,* 576–579.

Hoel, P. G. (1971). *Introduction to mathematical statistics.* New York: Wiley.

Hogarth, R. M. (1987). *Judgment and choice.* New York: Wiley.

Kahneman, D., Slovic, P., and Tversky, A. (1982). *Judgment under uncertainty: Heuristics and biases.* New York: Cambridge University Press.

Kahneman, D., and Tversky, A. (1973). On the psychology of prediction. *Psychological Review, 80,* 237–251.

Kalisch, S. J. (1980). *Computerized instructional adaptive testing model: Formulation and validation* (AFHRL-TR-79-33). Brooks Air Force Base, TX: Air Force Human Resources Laboratory.

Klein, G. A. (1989). Do decision biases explain too much? *Human Factors Society Bulletin, 32*(5), 1–3.

Lintern, G., Roscoe, S. N., Koonce, J. M., and Segal, L. D. (1990). Transfer of landing skills in beginning flight training. *Human Factors, 32,* 319–327.

Meehl, P. E. (1954). *Clinical versus statistical prediction: A theoretical analysis and a review of the evidence.* Minneapolis: University of Minnesota Press.

Meister, D. (1985). *Behavioral analysis and measurement methods.* New York: Wiley.

Padgett, M. Y., and Ilgen, D. R. (1989). The impact of ratee performance characteristics on rater cognitive processes and alternative measures of rater accuracy. *Organizational Behavior and Human Decision Processes, 44,* 232–260.

Roscoe, S. N., Jensen, R. S., and Gawron, V. J. (1980). Introduction to training systems. In S. N. Roscoe (Ed.), *Aviation psychology* (pp. 173–181). Ames: Iowa State University Press.

Sawyer, J. (1966). Measurement and prediction, clinical and statistical. *Psychological Bulletin, 66,* 178–200.

Sidman, M. (1960). *Tactics of scientific research.* New York: Basic Books.

Slovic, P. (1976). Toward an understanding and improving decision. In E. I. Salkovitz (Ed.), *Science, technology, and the modern Navy, thirtieth anniversary 1946–1976* (ONR Report 37, pp. 220–240). Arlington, VA: Office of Naval Research.

Slovic, P. (1988). Testimony. In U.S. Congress, *Iran Air Flight 655 compensation: Hearings before the Defense Policy Panel of the Committee on Armed Services, U.S. House of Representatives, Washington, DC, October 6, 1988* (pp. 193–217). Washington, DC: U.S. Government Printing Office.

Smith, P. C. (1976). Behaviors, results, and organizational effectiveness: The problem of criteria. In M. D. Dunnette (Ed.), *Handbook of industrial and organizational psychology* (pp. 745–775). New York: Wiley.

Tversky, A., and Kahneman, D. (1974). Judgment under uncertainty: Heuristics and biases. *Science, 185,* 1124–1131.

Wald, A. (1947). *Sequential analysis.* New York: Wiley. (Reprinted by Dover Publications, 1973)

Wisudha, A. D. (1985). Design of decision-aiding systems. In G. Wright (Ed.), *Behavioral decision making* (pp. 235–256). New York: Plenum.

Symposium on Distributed Simulation for Military Training of Teams/Groups: **A Case Study of Distributed Training and Training Performance**

Daniel J. Dwyer and **Randall L. Oser**, Naval Air Warfare Center Training Systems Division, Orlando, FL, and **Jennifer E. Fowlkes**, Enzian Technology, Inc., Orlando, FL

This paper describes the first actual application of a distributed training network to the military mission called Close Air Support (CAS). It represents a "case study" and is based upon a set of data collected on military personnel during a one-week series of exercises in a distributed training environment. We describe the objectives of the measurement process, discuss the development and use of the measurement tools, provide several observations based upon the data collected, and offer several preliminary conclusions related to measuring training performance in distributed environments.

INTRODUCTION

Nowhere is the coordination, synchronization, and interaction among different units who must work together more evident than in the military. Combat missions such as those encountered in Operation Desert Storm have highlighted the critical need for coordination among multiple services during military operations. During mission execution, the integration, timing, and synchronization of assets within a single service are formidable; between services they are an order of magnitude more difficult to achieve.

Likewise, teams consisting of several different units which involve emergency and disaster relief, hostage negotiation, firefighting, and medical response (to name a few) face similar types of challenges. The time to discover and resolve the between unit differences is not during a crisis situation, but rather in the training environment. Fortunately, technologies such as Distributed Interactive Simulation (DIS) provide access to tools that promise to bridge the gap between operational needs and simulation capabilities.

DIS technology allows multiple simulators to be networked in a manner that permits geographically-dispersed participants to "con-gregate" in a virtual environment in order to conduct interactive training. One effort that capitalizes on DIS technology to perform research-based studies for multi-service military training is the Multi-Service Distributed Training Testbed (MDT2). MDT2 was the first application of DIS technology for training military personnel in multi-service close air support (CAS). CAS involves air strikes against hostile targets which are in close proximity to friendly forces. This effort produced the first team performance data collected in this type of environment. This paper describes the performance measurement techniques used and reports the results of the data collected in this distributed training environment.

METHOD

MDT2 Configuration

The MDT2 configuration linked simulators at three sites: Armstrong Lab (Mesa, AZ), Naval Air Warfare Center (Patuxent River, MD), and Ft. Knox (KY). The network included two F-16 aircraft simulators, a Laser Designator simulator, an OV-10 aircraft simulator, and several tank simulators. The network was established via the Defense Simulation Internet

Reprinted from Proceedings of the Human Factors and Ergonomics Society 39th Annual Meeting (pp. 1316–1320).

and T-1 lines. This setup allowed key participants at distant sites and across the services to perform CAS in real-time.

Scenarios and Participants

Two scenarios, one offensive and one defensive, were developed to support the February 1995 case study. Each scenario provided opportunities for participants to demonstrate their ability to perform critical tasks associated with 25 training objectives identified during front end analyses. The training objectives and tasks focused on the interservice interactions involved in multi-service CAS. Scenario duration was approximately 1.5–2.0 hours per day for each of five days. Within each scenario, 3–5 CAS missions were typically conducted. Each CAS mission was divided into three phases – Planning, Contact Point (CP), and Attack. The measurement tools were designed to accommodate these three phases. The offensive scenario was run on Monday, Wednesday, and Friday; the defensive scenario was run on Tuesday and Thursday. Participants received feedback at the completion of each scenario during an After Action Review (AAR) conducted via video teleconference. A team of eighteen individuals from the USAF, USMC, and USA served as the single unit of measure.

O/Cs

A team of seven observer/controllers (O/Cs), located at the three sites and composed of CAS subject matter experts (SMEs), was responsible for collecting the measurement data. Because of their different viewing locations and limited access to communication channels, various O/Cs were assigned to complete different measurement tools for different phases of the CAS missions.

Performance Measurement Tools

The goals for measurement were (1) to provide a source of diagnostic performance feedback to support AARs, and (2) to examine performance trends across training days. Two measurement tools, each of which supported both goals, were developed. The techniques provided different perspectives of the team processes that were exhibited during training. Each technique is described below.

TARGETs. The TARGETs (Targeted Acceptable Responses to Generated Events or Tasks) instrument is a behaviorally-focused checklist for collecting structured observations of team behaviors (Fowlkes, Lane, Salas, Franz, & Oser, 1994). The checklist was developed around the 25 training objectives and their associated tasks. A key characteristic of the TARGETs method is that an appropriate set of behaviors for each task is identified *a priori* and arranged in a checklist format. As an exercise unfolds, assessors score each of the checklist items as either present or absent. Because of the nature of the TARGETs checklist, team performance can be analyzed in a number of ways. One method involves calculating the proportion of behaviors performed correctly relative to the total set of behaviors that should have been performed. This provides a global indication of performance. A second method involves dividing groups of behaviors into functionally-related clusters, for example those pertinent to a specific training objective. These clusters can then be examined separately in order to provide an indication of how teams performed within the functional areas.

TOM. The purpose of the TOM (Teamwork Observation Measure) was to identify strengths and weaknesses and to obtain ratings on the teamwork dimensions of Communication, Coordination, Adaptability, and Situational Awareness. Each of the four teamwork dimensions were defined and then decomposed into key factors. For example, Communication was divided into factors such as format, terminology, clarity, and acknowledgements. For each key factor within a given dimension, the O/Cs provided comments which highlighted critical teaching points for feedback. In order to track performance trends, the O/Cs also provided ratings of how well the key participants interacted with each other on each of the four teamwork dimensions. The ratings included "1" (Needs Work), "2" (Satisfactory), "3" (Very Good), and "4" (Outstanding).

RESULTS AND DISCUSSION

The results of the data collection are presented separately for each of the three CAS mission phases: Planning, CP, and Attack. For each phase, we address data from both the TARGETs and TOM measurement tools.

The Planning Phase

The left side of Figure 1 presents performance trends from the TARGETs data. Three O/Cs at Ft. Knox provided these data. The data for each day represent average performance scores provided by all O/Cs across all CAS missions run that day. As can be seen in Figure 1, overall performance generally improved across the five days. This trend clearly indicates that task performance improved as a function of MDT2 practice. In an attempt to further examine specific clusters of behaviors, the TARGETs data were broken out into the functional areas of Target Selection, Airspace Coordination Areas (ACA), Control of Aircraft, and Synchronization. All components except Control of Aircraft appeared to asymptote at high levels by day 3; performance related to the Control of Aircraft peaked on day 4 before declining slightly on day 5.

The TOM data similarly revealed improved performance across the five exercise days (see the right side of Figure 1). These data were collected by two O/Cs at Ft. Knox. Although the TARGETs data revealed consistent upward trends in all areas assessed, the TOM data depicted a drop in performance on day 2 for Communication. Specific examples of the Communication decline were addressed during the AAR. Other than this decline, performance on each of the teamwork dimensions steadily increased up to day 4, where performance appeared to level off. Interestingly, not until day 3 was performance consistently rated above the satisfactory ("2") level.

The CP Phase

Five O/Cs (two at Ft. Knox, two at Armstrong Lab, and one at Patuxent River) provided TARGETs data for the CP phase. These data are presented on the left side of Figure 2. In general, overall performance, once again, improved over the five days. The data were further analyzed to determine the extent to which team performance was seen to vary as a function of O/C perspective (i.e., the site from which the O/C observed training). As can be seen in this plot, the assessments were fairly consistent, however two of the plots show a day 4 decline.

An examination of the TOM data (see the right side of Figure 2) also depicts a drop in performance on day 4. These data suggest that the decline may have been attributable to shortfalls in Coordination and Situational Awareness. Only one O/C provided TOM data during this phase. While these data are not definitive, they served as a useful discussion area during the AAR.

Also noteworthy is the observation that despite the relatively high performance indicated from TARGETs, the TOM data suggested that there was considerable room for improvement in all areas measured. In fact, the TOM data never reached a rating of "3" (very good), and not until day 3 did the ratings surpass level "2" (satisfactory). Apparently, the tasks that should have been performed were performed, however Communication, Coordination, and Situational Awareness could have been enhanced.

The Attack Phase

Five O/Cs (the same ones as in the CP Phase) completed the TARGETs instrument for the Attack Phase. These data are plotted on the left side of Figure 3. In general, overall performance improved over the five days of training. (Note that there are missing data during this phase, perhaps due to the additional workload required of the O/Cs). Similar to the data for the CP Phase, the Attack Phase data were examined to determine the extent to which team performance varied depending on O/C location. One O/C noted a performance drop on day 4. Again, by examining the TOM data (see the right side of Figure 3), one sees a corresponding drop in Coordination behaviors on day 4. This information was useful in supporting feedback to exercise participants. Only one O/C provided TOM data during this phase.

CONCLUSIONS

Overall, several positive observations emerged from this case study. First, the measurement tools were able to support both of our research goals: to diagnose performance

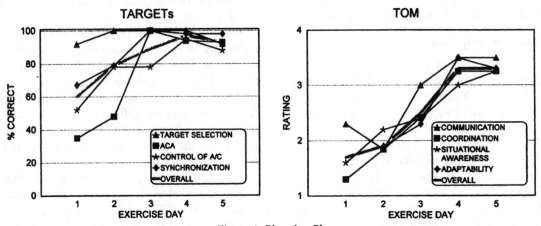

Figure 1. Planning Phase.

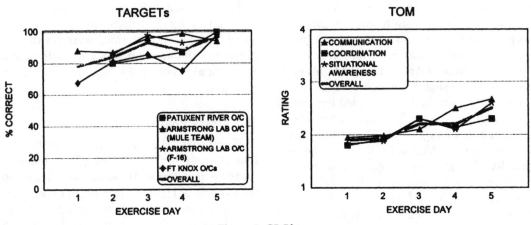

Figure 2. CP Phase.

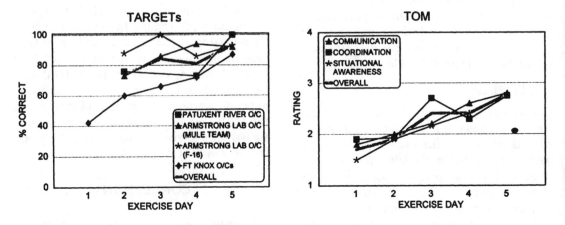

Figure 3. Attack Phase.

deficiencies to support the AAR, and to track trends in performance. In many instances, the measurement tools identified areas which served as the basis for AAR focus. Also, the improvement over the course of the training week suggests that MDT2 was effective in increasing multi-service CAS performance. Furthermore, the trends from the TARGETs checklist data and from the TOM data were generally parallel to one another, and the pattern in the data provided by geographically-separated O/Cs also tended to mimic each other. These patterns are indicative of sensitive and reliable instruments. This is particularly encouraging when one considers the use of O/Cs who were often separated from one another, who were from different military services, and who had to assess performance in high workload, dynamic, and noisy environments.

Second, much of the team performance data that has been collected in the past has been based on SMEs who review archival recordings, *post hoc*. However, in this effort, there was a need to rapidly provide feedback to participants based on the TARGETs and TOM instruments. As a result, the instruments had to be completed on-line and in real-time – a process which has not been used frequently in the past. In this application, the O/Cs demonstrated that they were able to collect the data in this manner, representing a significant accomplishment which is necessary before tools of these types can be used in applied settings.

Finally, the potential utility of using multiple measurement techniques in this type of setting is also suggested by our observations. An inspection of both the TARGETs checklist data and the TOM rating data allowed us to triangulate on problem areas and provided a diagnostic method for pinpointing specific performance shortfalls. This was especially useful for the AAR.

Data collection in field settings is often problematic. The problem in drawing valid conclusions and generalizing to other domains from field research is compounded when case studies are used. These types of problems were certainly present in this case study, and the data are subject to many threats to validity (Cook and Campbell, 1979). However, no other study has systematically collected performance data of this type in a DIS-based training environment. This represents a first step in this area.

Many of our findings are encouraging, however they must yet be confirmed in more controlled settings. The types of techniques and tools we have demonstrated have been well-received by the users, however they must continue to evolve through refinement and improvements. Additional research in these areas is needed before these processes can be transitioned to more applied settings.

REFERENCES

Cook, T.D. & Campbell, D.T. (1979). *Quasi-experimentation: Design & analysis for field settings.* Boston: Houghton Mifflin Co.

Fowlkes, J.E., Lane, N.E., Salas, E., Franz, T., & Oser, R.(1994). Improving the measurement of team performance: The TARGETs methodology. *Military Psychology, 6,* 47-61.

SUBJECT INDEX